EARTHQUAKE AND VOLCANO DEFORMATION

EARTHQUAKE AND VOLCANO DEFORMATION

PAUL SEGALL

Princeton University Press
Princeton and Oxford

Library of Congress Cataloging-in-Publication Data

Segall, Paul, 1954–
Earthquake and volcano deformation / Paul Segall.
    p. cm.
Includes bibliographical references and index.
ISBN 978-0-691-13302-7 (hardcover : alk. paper)
1. Rock deformation—Mathematical models.   2. Strains and stress—Mathematical models.
3. Volcanism.   4. Earthquakes.   5. Deformations (Mechanics)   I. Title.
  QE 604.S44 2010
   551.8–dc22   2009013202

British Library Cataloging-in-Publication Data is available

To Joan, Kim, and Matt

Contents

Preface

The past decades have witnessed a tremendous improvement in our ability to precisely measure deformation of the earth's crust. Prior to the advent of space-based geodetic systems, crustal strains were measured by triangulation, leveling, and following the advent of lasers, Electronic Distance Meters (EDM). Very Long Baseline Interferometry (VLBI) was the first system to determine relative motion of sites at intercontinental distance scales and thus measure contemporary rates of plate motion. VLBI determines relative positions by measuring time delays from extragalactic sources recorded at widely separated radio telescopes. The large antennas required for VLBI restricted application to a few major research groups. Beginning in the late 1980s, the satellite-based Global Positioning System (GPS) allowed for precise relative three-dimensional positioning over a broad range of distance scales with relatively compact receivers and antennas. Early GPS measurements were conducted in campaign mode, in which researchers reoccupied fixed benchmarks to determine changes in position over time. As the price of GPS receivers dropped, due to widespread commercial application, it became cost effective to deploy GPS receivers in permanent networks with automated data processing, such that position changes were determined daily with an accuracy of millimeters. At the time of this writing, continuously recording GPS networks for the study of crustal deformation with more than 1,000 stations have been developed in the United States and Japan. Smaller networks are operational in a number of other countries. With the maturing of GPS processing strategies, it has become possible to determine precise relative positions over time periods much shorter than the nominal 24-hour solutions. With rapid data sampling, it is now possible to measure dynamic ground motions during earthquakes, in effect bridging crustal deformation monitoring with seismology. The future promises increased accuracy and flexibility as additional GPS frequencies as well as new Global Navigation Satellite Systems (GNSS) become available.

The concurrent development of Interferometric Synthetic Aperture Radar (InSAR) now allows researchers to map earth surface deformations over wide areas with very high spatial resolution. InSAR employs imaging radars in which both the amplitude and phase of the reflected electromagnetic signals are compared from two (or more) orbital passes. If, hypothetically, the spacecraft returned to precisely the same position in space and the ground surface had not deformed, the measured phase would not change between passes. On the other hand, if the ground deforms between radar acquisitions, the phase changes in proportion to the displacement in the radar line-of-sight direction. InSAR and GPS are fully complementary in that GPS provides precise three-dimensional displacements with frequent temporal sampling. Spatial coverage with GPS is limited, however, by the number of receivers that can be deployed. InSAR provides scalar range-change measurements that are dense in space but temporally limited by the orbital repeat period of the spacecraft. The precision of InSAR is comparable to that of GPS; both systems employ related parts of the electromagnetic spectrum and are thus similarly influenced by atmospheric path delays. InSAR requires that the scattering characteristics of the ground surface not change significantly between radar acquisitions. If this is not the case, the coherent sum of contributions from individual scatterers within an image pixel has phase that varies randomly from image to image, a phenomenon known as *temporal decorrelation*. Temporal decorrelation is a function of surface characteristics, particularly vegetation cover, the time between radar passes, and the radar wavelength. (Longer wavelengths scatter off larger, and hence more stable, objects.) Early studies maintained coherence over many years by focusing on arid regions with sparse vegetation cover. The number of InSAR-capable spacecraft is currently increasing; one system presently employs an L-band radar (roughly 24-cm wavelength, as with GPS) that maintains good coherence over heavily vegetated regions.

While the unique capabilities of GPS and InSAR have largely been responsible for the rapid increase in the number of researchers studying earthquake and volcano deformation, as well as contemporary plate motions and plate boundary deformation, other measurement systems provide invaluable data. Borehole and long-baseline strainmeters and tiltmeters provide local measurements that, at least within some frequency band, are considerably more precise than either GPS or InSAR. Instrumental strain and tilt measurements usually lack the long-term stability of GPS but are very powerful for studying transient deformation phenomena. Sea-floor geodetic systems, including acoustic ranging systems and combined acoustic-GPS-based systems, are beginning to open the offshore environment to deformation monitoring. Gravimetric measurements also provide significant constraints on earthquake and particularly volcanic processes. Technological advances in this field have also been significant. Absolute gravimeters that measure the local acceleration of gravity experienced by a falling test mass in an evacuated chamber have been deployed in a number of settings and provide long-term stability not previously achieved with relative gravimeters. Superconducting instruments provide the capability for continuous gravity monitoring. The recently launched Gravity Recovery and Climate Experiment (GRACE) satellite mission measures global changes in gravity with unprecedented precision. GRACE data have been used to detect changes in gravity associated with a magnitude 9 earthquake; future missions will undoubtedly push this capability to smaller events. In short, rapid technological advances have changed the study of active crustal deformation from data-impoverished to, while perhaps not data-rich, at least decidedly middle class.

Measurements of crustal deformations, including relative displacements, strains, tilts, and rotations, are generally not an end in themselves. Rather, we strive to better understand the mechanics of tectonic and volcanic systems by comparing predictions of physically based models with observations. In many instances, we seek to estimate unknown parameters, or fields, at depth within the earth—for example, the distribution of slip on a fault or the amount of opening on a sill or dike—by comparison to measurements made at or near the earth's surface. This is the domain of geophysical inverse theory. Research into the mechanics of earthquakes and volcanoes requires familiarity with data collection methods, physically based models, and appropriate inverse methodologies. Nearly 10 years ago, I set out to organize disparate course notes into a textbook that would cover these three principal areas. I soon realized that this was neither practical nor necessary. There are a number of excellent books on GPS, and similarly informative texts for students wanting to learn inverse methods. There were, however, no books on physical models used to interpret the terabytes of data collected globally for the study of earthquake and volcanic processes. This text is my attempt to fill that gap. The scope is limited to what is usually described as forward models—given a description of the earth and internal sources of deformation, how does the earth's surface deform in space and time? I do not address, other than tangentially, data collection methods or inversion strategies. Nor is this work intended to be a comprehensive review of recent research results. Rather, the intent is to present a summary of the fundamentals of the discipline, at least from my perspective. The emphasis is on analytical and quasi-analytical methods. Analytical solutions, whenever available, provide far greater insight into the underlying physics, and of course serve as benchmarks for numerical methods. Many interesting and important problems require numerical solution, and powerful methods, including finite element, finite volume, discrete element, and spectral element methods, are available in these cases. With minor exceptions, numerical methods are, however, beyond our present scope.

The text is directed toward graduate students, advanced undergraduates, and researchers in geophysics and related disciplines. The reader is assumed to be familiar with ordinary and partial differential equations as well as linear algebra. Integral transform methods (Fourier and Laplace) are used extensively; familiarity with complex variable methods including contour integration is also desirable. A few fundamental concepts and results are reviewed in the appendixes. It is assumed that the reader has been exposed to basic concepts of continuum mechanics, including stress and infinitesimal strain tensors and elasticity at the level of an introductory undergraduate course.

The book begins with a review of the principles of continuum mechanics, with a slant toward geophysical applications. The goal here is to be neither exhaustive nor formal, but rather to summarize concepts and results that are used throughout the remainder of the text. Chapter 2 develops the solution for a long strike-slip fault with uniform slip in an elastic half-space and compares model predictions to data from the San Andreas and North Anatolian faults. Chapter 3 extends the analysis to general elastic dislocation theory. Here, we derive several forms of Volterra's formula that allow one to compute displacements due to arbitrary dislocation sources. Results are derived for normal and reverse faults in two dimensions and for point and rectangular dislocations in three-dimensional half-spaces. Chapter 4 introduces crack models of faulting, in which the stress acting on the fault surface is specified rather than displacement discontinuity, and presents a number of methods for analyzing the associated mixed boundary value problems. Chapter 5 presents methods for computing the surface deformations resulting from dislocations in elastically heterogeneous earth models, including image, quasi-static propagator matrix, and first-order perturbation methods. Chapter 6 begins a discussion of time-dependent deformation with an introduction to viscoelasticity. The problem of an infinitely long strike-slip fault in an elastic layer overlying a viscoelastic half-space is developed. This is then extended to include normal and reverse faulting, making extensive use of propagator matrix solutions derived in chapter 6. Chapter 7 explores models of volcano deformation, beginning with the solution for a pressurized spherical magma chamber in an elastic half-space. The solution for a pressurized ellipsoidal magma chamber, which can be specialized to model volcanic pipes and sills, is developed. Previously introduced dislocation and crack solutions are specialized to model dikes and sills. This chapter ends with an analysis of viscoelastic relaxation of a shell surrounding a spherical magma chamber. Chapter 8 explores the effect of irregular surface topography on deformation, as well as corrections to account for earth curvature. We do not delve into spherical earth models, in part due to mathematical complexity, but also because, except in rare cases, such models are not required. Chapter 9 is focused on gravitational effects, both on the influence of gravitational coupling on the deformation field and on how the gravitational field changes during deformation. The latter is of particular interest in volcanic regions, where simultaneous measurements of gravity and deformation provide stronger constraints on the nature of intruding fluids than can be achieved by either measurement alone. Chapter 10 explores coupled deformation and pore-fluid effects through the theory of linear poroelasticity. These effects are central in areas where the subsurface pore pressure is artificially perturbed due to fluid extraction or injection; however, because deformation and pore-fluid flow are fully coupled, poroelastic effects are expected to be more generally present in tectonic and volcanic environments. Chapter 11 reviews modern concepts of fault friction, including an introduction to rate- and state-dependent friction, with application to pre- and postseismic slip. The book closes by pulling together threads from a number of previous chapters to discuss interseismic deformation earthquake cycle models, including both viscoelastic and friction effects.

The material can be taught, as it is at Stanford, as a two-quarter sequence. The first quarter covers elastic solutions, as discussed in chapters 1 through 5 and most of chapter 7, and provides students, including those interested in volcano deformation, with the basics of the discipline. The second quarter covers more advanced topics, including viscoelasticity, elastogravitational coupling, and friction. This material is covered in chapter 6 and chapters 8 through 12.

Instructors and students can visit the book's website at http://press.princeton.edu/titles/9093.html to find supplemental materials relating to the text.

Acknowledgments

I am very much indebted to the many colleagues and students who have contributed to my understanding of the subjects discussed here and have materially aided the writing of this book. Any list will be necessarily incomplete; my sincere apologies to those whose contributions I have inadvertently omitted.

The field of crustal deformation builds heavily on foundational results from the fields of solid mechanics and seismology. I have leaned heavily on *Quantitative Seismology: Theory and Methods*, by Keitti Aki and Paul Richards; *Theoretical Global Seismology*, by Tony Dahlen and Jeroen Tromp; *Introduction to the Mechanics of Continuous Media*, by Lawrence Malvern; and *Foundations of Solid Mechanics*, by Y. C. Fung.

This book began taking form during a sabbatical at the Institute de Physique du Globe, Paris, in the spring of 2000; I would like to thank my host François Cornet and the then-director Claude Jaupart for making my visit so enjoyable. The discussion of image sources in chapter 5 is based on lecture notes of David Barnett. The section on propagator matrix methods in chapter 5 benefited greatly from discussions with Steve Ward. Kaj Johnson worked through important material on propagator matrix methods in plane strain and made significant contributions to the viscoelastic models discussed in chapters 6 and 12, including running calculations and checking my results. Shin-ichi Miyazaki helped with the perturbation calculations for earth's sphericity in chapter 8, and Fred Pollitz computed reference displacements in spherical earth models. The material on gravitational effects in chapter 9 benefited greatly from discussions with Steve Ward, Fred Pollitz, Rongjiang Wang, Volker Klemann, Jeroen Tromp, and John Wahr. Discussions with John Rudnicki and Evelyn Roeloffs improved my understanding of the theory of linear poroelasticity. Andy Ruina introduced me to the intricacies of fault friction; Jim Rice, Jim Dieterich, and Allan Rubin contributed to a deeper understanding of the subject. Arvid Johnson and Dave Pollard introduced me to the fundamentals. Benjamin Segall was always available to consult on mathematical complexities.

The manuscript benefited tremendously from the careful reviews by Jim Rice, Barry Parsons, John Rudnicki, and Wayne Thatcher. Jim Rice, in particular, significantly clarified the development in a number of important areas. Individual chapters were reviewed by John Rudnicki, Jim Savage, Fred Pollitz, Yuri Fialko, Charles Williams, Rongjiang Wang, Volker Klemann, Allan Rubin, Nick Beeler, Kaj Johnson, Tappan Mukerji, Evelyn Roeloffs, Yukitoshi Fukahata, Jeff Freymueller, Roland Burgmann, Jerry Mitrovica, and Shin-ichi Miyazaki. Despite their and my best efforts, errors undoubtedly remain; for these I alone am responsible.

I would also like to thank the following who either spotted errors, helped in coding sample calculations, or contributed in other important ways: Howard Zebker, Steve Kirby, Yoshiyuki Tanaka, John Townend, Sang-ho Yun, Chris Weinberger, Shuo Ma, Stuart Schmitt, Danica Dralus, Peter Lovely, Kyle Anderson, Mark Matthews, Maurizio Battaglia, Tim Dixon, Shelley Kenner, Peter Cervelli, Andy Hooper, and Yosuke Aoki.

While I received no funding especially for this project, my research has been generously supported over the years by the National Science Foundation, the U.S. Geological Survey's National Earthquake Hazards Reduction Program, NASA, and the Department of Energy.

Finally, I would like to thank the staff at Princeton University Press for their help in bringing this project to fruition. Senior editor Ingrid Gnerlich and production editor Natalie Baan shepherded the project. Jennifer Harris did a wonderful job of editing, and Dimitri Karetnikov turned rough figures into objects worthy of publication.

Origins

The birth of geodetic studies of active deformation is generally associated with the publication of Harry Fielding Reid's elastic rebound hypothesis following the great 1906 San Francisco earthquake (Reid 1910, 1911). While elastic rebound has remained a cornerstone of fault mechanics for a century now, the history is, not surprisingly longer and considerably more interesting.

The first geodetic observations of ground movements associated with an earthquake were reported by the Dutch surveyor J. J. A. Müller, following the 1892 Tapanuli earthquake in Sumatra. The earthquake occurred during a triangulation survey, and Müller deduced that the triangulation stations had been displaced during the event, noting that "it seems very likely indeed that this is the first time that dislocation phenomena of the earth's surface are disclosed by geodetic means" (English translation by Bonafede et al. [1992]). These authors note that Müller was unlikely to have used the word *dislocation* in the modern geophysical sense; nevertheless, his observation is striking.) Reid himself (1913), analyzed the Sumatra data, interpreting the motion as resulting from right-lateral slip on a northwest-trending fault, although ground rupture was not observed in this area. We now understand Reid's hypothesized fault to be part of the great Sumatran fault. One might conclude that Reid conducted the first geodetic fault inversion, albeit not in the quantitative sense that such inversions are done today. Nevertheless, he obtained essentially the correct result, and with data from only three triangulation stations.

In the years immediately prior to the Tapanuli earthquake, fault offsets had been observed in a number of tectonic environments. Alexander McKay observed 2.4-meter lateral offsets following the 1888 Glennwye earthquake on the south island of New Zealand. Shortly after that, Bunjiro Koto reported large vertical and lateral offsets following the 1891 Nobi earthquake in central Japan. Koto clearly recognized that fault slip was the cause of the earthquake and not, as was commonly assumed at the time, an effect of strong ground shaking. He wrote, "This vertical movement and horizontal shifting seem to have been the sole cause of the late catastrophe" (English translation 1990).

Shortly thereafter, Oldham (1981) reported on the great Assam earthquake of 1897 in the foothills of the Indian Himalayas, including repeated triangulation and leveling surveys. He appears to have struggled to reconcile the triangulation measurements with his belief that the Himalayas resulted from north–south compression. Oldham was aware of fault-bounded mountain ranges associated with crustal extension and wrote, "It is conceivable, though with difficulty, that in such a region tensional strains might be set up of sufficient greatness to give rise to a severe earthquake by their sudden relief." He appears not to have recognized that strain relief leads to strain changes opposite in sign to the tectonic strain. However, his reference to strain relief as a cause of earthquakes is striking.

It was the great American geologist Grove Karl Gilbert, however, who in 1884 was the first to articulate that earthquakes were caused by sudden release of strain and that these strains accumulate slowly over long periods of time. Gilbert studied normal faults in the Basin and Range province of the American West, and in particular examined fault scarps formed by the 1872 Owens Valley earthquake, at the western margin of the Great Basin. Gilbert underestimated the amount of strike-slip motion along the 1872 rupture, focusing his attention on the vertical offsets. Nevertheless, he gave a particularly clear description of the underlying physics:

> The upthrust produces a local strain in the crust ... and this strain increases until it is sufficient to overcome the starting friction on the fractured surface. Suddenly, and almost instantaneously, there is an amount of motion sufficient to relieve the strain, and this is followed by a long period of quiet, during which the strain is gradually reimposed. The motion at the instant of yielding is so swift and so abruptly terminated as to constitute a shock, and this shock vibrates through the crust with diminishing force in all directions. (Gilbert 1884, p. 50)

What is so fascinating about Gilbert is that he chose to publish his work first in the *Salt Lake Tribune*. Gilbert recognized that the scarps along the Wasatch fault were analogous to the scarps he had studied in Owens Valley and that this indicated a significant seismic risk for Salt Lake City. He anticipated the concept of seismic gaps, writing that "any locality on the fault line of a large mountain range, which has been exempt from earthquake for a long time, is by so much nearer to the date of recurrence."

Gilbert's insights were so profound that he presaged by nearly a century what has become the canonical spring-block model so often used to discuss earthquake slip instabilities:

> Attach a rope to a heavy box and drag it slowly, by means of a windlass, across a floor. As the crank is turned, the tension in the rope gradually increases until it suffices to overcome the starting friction, as it is called. Once started, the box moves easily, because sliding friction is less than starting friction. The rope shortens or sags until its tension is only sufficient for the sliding friction, and it would continue in that state but that the box, having acquired momentum, is carried a little too far. This slacks the rope still more, and the box stops, to be started only when the tension again equals the starting friction. In this way the box receives an uneven, jerky motion.

> Something of this sort happens with the mountain. (Gilbert 1884, p. 50)

Gilbert had the concept of elastic rebound, but Reid had the data. In particular, he was able to make use of triangulation surveys in the San Francisco Bay Area, first conducted during the gold rush from 1851 to 1865 and then extended during the period 1874 to 1892. Following the 1906 San Francisco earthquake, it was recognized that geodetic stations had displaced significantly, to the point that the "triangulation would no longer be of value as a means of control for accurate surveys.... It was, therefore, decided to repair the old triangulation damaged by the earthquake, by doing new triangulation" (Hayford and Baldwin 1908). The surveys were completed in July 1907, and the resulting displacements were published by Hayford and Baldwin in 1908, two years after the earthquake.

Reid (1910) noted that the displacement of stations far from the fault, including the Farallon Islands, had displaced significantly and reasoned that these motions were only in part due to the earthquake itself: "We must therefore conclude that the strains were set up by a slow relative displacement of the land on opposite sides of the fault and practically parallel with it." Reid could not conceive of the possibility of far-field relative displacements, which we now understand arise from relative plate motion, and assumed that the interseismic displacements gradually decayed to zero beyond the extent of the triangulation network.

Reid further noted that the total relative displacement of distant points on opposite sides of the fault since the surveys of 1851 to 1865 represented only half the average slip during the earthquake. He concluded that "it seems not improbable, therefore, that the strain was accumulating for 100 years, although there is not satisfactory reason to suppose that it accumulated at a uniform rate." Reid thus laid down a framework for long-term earthquake forecasting, by comparing the coseismic strain release with the measured rate of strain accumulation between major earthquakes.

Reid computed the stresses released in the earthquake from the measured strains and laboratory values of the elastic modulus of granite. He noted that the stresses were less than that required to fracture fresh rock, concluding that "former ruptures of the fault plane were by no means entirely healed, but that this plane was somewhat less strong than the surrounding rock and yielded to a smaller force than would have been necessary to break the solid rock." He went on to estimate the mechanical work done during the earthquake, noting that "this energy was stored up in the rock as potential energy of elastic strain immediately before the rupture; when the rupture occurred, it was transformed into the kinetic energy of the moving mass, into heat and into energy of vibrations."

Reid also conducted experiments with a layer of jelly sheared between two pieces of wood. He found that initially fault-perpendicular lines were displaced in the correct sense but that they did not warp into the profiles observed in the triangulation measurements. He concluded that the deforming forces were more likely to be applied from below, rather than from the sides. In a footnote, he refers to a suggestion by G. K. Gilbert to make the cut representing the fault only partway through the jelly. While these experiments failed, they might be considered the first hint of dislocation theory applied to faulting!

In 1911, Reid summarized his elastic rebound model in five statements:

1. The fracture of the rock, which causes a tectonic earthquake, is the result of elastic strains, greater than the strength of the rock can withstand, produced by the relative displacements of neighboring portions of the earth's crust.

2. These relative displacements are not produced suddenly at the time of the fracture, but attain their maximum amounts gradually during a more or less long period of time.

3. The only mass movements that occur at the time of the earthquake are the sudden elastic rebounds of the sides of the fracture towards position of no elastic strain; and these movements extend to distances of only a few miles from the fracture.

4. The earthquake vibrations originate in the surface of fracture; the surface from which they start has at first a very small area, which may quickly become very large, but at a rate not greater than the velocity of compressional elastic waves in the rock.

5. The energy liberated at the time of an earthquake was, immediately before the rupture, in the form of energy of elastic strain of the rock. (Reid 1911, p. 436)

Reid's idea that geodesy could lead to long-term earthquake forecasting led Arthur L. Day, director of the Geophysical Laboratory of the Carnegie Institution of Washington, to request the director of the U.S. Coast and Geodetic Survey to resurvey the California triangulation network. This work commenced in 1922 under the direction of William Bowie, chief of the Division of Geodesy of the Coast and Geodetic Survey. By 1948, Charles A. Whitten reported that the surveys of 1947 revealed "positive evidence of a slow continuous movement of the area west of the San Andreas Fault, relative to the area east of the fault" (Whitten 1948). He estimated the rate of relative motion at 5 cm/yr and concluded that 3 meters of relative motion had accumulated since 1880.

In the meantime, geodetic methods were beginning to be applied to the study of volcanoes (see review in Poland et al. 2006). Uplift and subsidence in volcanic areas had long been known from visual observations, especially where coastlines presented readily accessible level surfaces. Undoubtedly the most famous example is the ruins of a Roman market (Macellum), also known as the Temple of Serapis or Serapeum, located in Pozzuoli (just west of Naples, Italy), in the center of the Campi Flegri caldera. Perforations and marine shells on standing marble columns have long been recognized as evidence for many meters of subsidence and subsequent uplift. (The history of observations and interpretations is reviewed by Dvorak and Mastrolorenzo [1991].) Most spectacularly, several hundred meters of coastline recession was documented prior to the 1538 eruption of Monte Nuovo, within the caldera. Measurements of water depth on the floor of Serapis, beginning in the early 1800s, augmented by modern leveling measurements beginning in 1905, revealed subsidence at a rate of 14 mm/yr during this time period (Dvorak and Mastrolorenzo 1991).

In Japan, Fusakichi Omori may have been the first to use geodetic measurements to document volcano deformation. In 1913, he reported on vertical deformation at Usu volcano on the island of Hokkaido from repeated leveling surveys. He also documented volcanic tilt, by measuring differential lake-level changes in nearby Toya Lake. Subsequently, Omori (1918) used tide gauge measurements and repeated leveling surveys to document broad-scale subsidence accompanying the 1914 eruption of Sakurajima volcano, extending roughly 20 kilometers from the volcano.

In Hawaii, Thomas A. Jaggar, founder and director of the Hawaiian Volcano Observatory, noted that the horizontal pendulums in a Bosch-Omori seismograph could be used to measure ground tilt. He found that tilts on Kilauea volcano showed significant variations depending on the level of lava in the Halema'uma'u crater, including a dramatic crater-ward tilt accompanying the draining of the lava lake in February 1924 and the subsequent explosive eruption in May of that year. Jaggar and Finch (1929) noted that Hawaiian volcanic activity was cyclic, with "an inflation of the edifice with new magma, and ending with a yielding of the edifice, [and] deflation of the magma." This is possibly the first report of inflation–deflation cycles on an active volcano. Combining tilt recordings with leveling surveys, Jaggar and Finch (1929) concluded that Kilauea volcano rose by 0.6 meter between 1913 and 1920, with uplift diminishing to zero at a radial distance of 10 kilometers, and then subsided roughly 4 meters between 1920 and 1926, associated with the 1924 lava lake drainage and explosive eruption. Corroborating evidence for substantial deflation came from repeated triangulation surveys (Wilson 1935) that showed large-scale, inward-directed displacements between 1922 and 1926.

Modeling of volcano deformation remained largely qualitative until the publication of what came to be called the "Mogi model" by Kiyoo Mogi in 1958. Mogi made use of the mathematical solution of Yamakawa (1955) for a pressurized spherical cavity in an elastic half-space. The same solution, also known as a point center of dilatation in an elastic half-space, was found earlier by E. M. Anderson (1936) but was applied to the stress state around magma chambers rather than surface deformation. The first publication of this important result, however, was even earlier, by Katsutada Sezawa in 1931. What is remarkable about Sezawa's result is that he analyzed a more complex problem, including the effect of a spherical plastic shell with constant yield stress, surrounding the magma chamber. His result for the surface displacements reduces to the classical Yamakawa-Mogi result in the limit that the thickness of the plastic shell goes to zero. While he presents no data in his paper, Sezawa compared model predictions with leveling data from Komagatake volcano and concluded that the source depth was within the upper 3 kilometers of the crust. Mogi (1958) was able to make use of far more extensive data sets. He modeled the subsidence accompanying the 1914 Sakurajima eruption, as well as subsequent uplift during the periods 1915 to 1919 and 1919 to 1932, and showed that the data were reasonably well explained by a pressure source at a depth of 10 km. Mogi also modeled the 1924 eruption of Kilauea, suggesting that the leveling data indicated sources at depths of 3.5 and 25 km. Last, he concluded that Wilson's triangulation measurements were consistent with the source model inferred from the leveling data.

The next major theoretical advance came with the application of elastic dislocation theory to the study of earthquakes. Dislocation theory dates to the Italian mathematicians Vito Volterra and Carlo Somigliana in the early part of the twentieth century; however, it was J. A. Steketee (1958) who first suggested that elastic dislocations could be used to model deformation due to faulting and was the first to use Volterra's formula for this application. Steketee was also the first to derive the displacements due to a point-source representation of a fault—in this case, a vertical strike-slip fault. It was left to Michael Chinnery (1961) to derive the displacements resulting from a finite rectangular dislocation, again for a vertical strike-slip fault. Chinnery also found the displacements for the limiting case of an infinitely long fault and compared model predictions with geodetic data from the Tango and North Idu earthquakes and Japan as well as the 1906 San Francisco earthquake. Shortly thereafter, Maruyama (1964) derived both point-source and finite dislocation solutions for strike-slip and dip-slip faulting on both vertical and horizontal faults. Savage and Hastie (1966) derived the vertical displacements due to pure dip slip on a dipping fault and compared predictions from a number of earthquakes. In particular, they were able to demonstrate thrust motion accompanying the great 1964 Alaskan earthquake. Mansinha and Smylie (1971) derived results for general slip on a finite dipping fault. Around this time, Weertman (1965) introduced models of surface deformation resulting from changes in frictional stress on faults in two dimensions, and Ben-Meneham and Singh (1968) developed methods for computing deformation in layered half-spaces.

The early 1960s also brought the plate tectonic revolution. Reid saw the slow northwestward drift of the crust northwest of the San Andreas fault in the triangulation measurements but could only speculate on the mechanism. Plate tectonics provided the engine for strain accumulation, the arm to crank Gilbert's windlass. The history of plate tectonics, well beyond the scope of this discussion, is forever tied to the names Harry H. Hess and Robert S. Dietz for the concept of sea-floor spreading, to F. Vine and D. Matthews for deciphering the record of sea-floor spreading in the marine magnetic anomaly record, and to J. Tuzo Wilson for elucidating transform faults. Lynn Sykes used earthquake focal mechanisms to confirm the sense of slip on transform faults predicted by Tuzo Wilson. Dan McKenzie, Robert L. Parker, and W. Jason Morgan derived the kinematics of relative plate motions, based on sea-floor magnetic anomalies, transform fault azimuths, and earthquake slip vectors.

In 1969, Walter Elsasser considered the diffusion of strain in an elastic plate coupled to a viscous asthenosphere resulting from sudden displacement at a plate boundary, initiating the study of time-dependent deformation that continues to this day. Transient postseismic deformation was reported as early as 1931, following the 1927 Tango earthquake in Japan (Tsuboi 1931), and was well documented following the 1946 Nankaido earthquake (e.g., Okada and Nagata 1953). Fitch and Scholz (1971) modeled deformation measurements before, during, and after the 1946 Nankaido earthquake and discussed the earthquake deformation cycle in subduction zones. Accelerated postseismic deformation was also recognized by Wayne Thatcher (1975) based on triangulation surveys in the decades following the 1906 earthquake.

In the early 1960s, laser distance measuring devices began to be used to measure strain along the San Andreas fault in California, and by the mid 1960s, this technology was being deployed at Kilauea volcano. In 1970, James C. Savage and William H. Prescott of the U.S. Geological Survey initiated a program to use laser distance measuring devices with aircraft-flown profiles of temperature and humidity to correct for atmospheric refraction, allowing for crustal strain measurements an order of magnitude more precise than those determined by triangulation. In 1973, Savage and Burford introduced the buried screw dislocation model of interseismic deformation for long strike-slip faults. The combination of accurate measurements and physically based models for interpreting the data ushered in the modern era of crustal deformation studies.

References

Anderson, E. M. 1936. The dynamics of the formation of cone-sheets, ring-dykes, and caldron subsidence. *Proceedings of the Royal Society of Edinburgh* **56**, 128–157.

Ben-Meneham, A., and S. J. Singh. 1968. Multipolar elastic fields in a layered half-space. *Bulletin of the Seismological Society of America* **58**, 1519–1572.

Bonafede, M., J. Strehlau, and A. R. Ritsema. 1992. Geophysical and structural aspects of fault mechanics—a brief historical review. *Terra Nova* **4**, 458–463.

Chinnery M. A. 1961. The deformation of the ground around surface faults. *Bulletin of the Seismological Society of America* **51**, 355–372.

Dvorak, J. J., and G. Mastrolorenzo. 1991. The mechanisms of recent vertical crustal movements in Campi Flegrei caldera, southern Italy. *Geological Society of America*, Special Paper 263.

Elsasser, W. M. 1969. Convection and stress propagation in the upper mantle. In S. K. Runcorn (Ed.), *The Application of Modern Physics to the Earth and Planetary Interiors*. New York: John Wiley, pp. 223–246.

Fitch, T. J., and C. H. Scholz. 1971. Mechanism of underthrusting in southwest Japan: a model of convergent plate interactions. *Journal of Geophysical Research* **76**, 7260–7292.

Gilbert, G. K. 1884. A theory of the earthquakes of the Great Basin with a practical application. *The American Journal of Science* **27**, 49–53, reprinted from the *Salt Lake Tribune* of 20 Sept. 1883.

Hayford, J. F., and A. L. Baldwin. 1908. Geodetic measurements of earth movements. In A. C. Lawson (Ed.), *The California Earthquake of April 18, 1906, Report of the State Earthquake Investigation Commission*, vol. I., Washington, DC: Carnegie Institute of Washington, pp. 114–145.

Jaggar, T. A., and R. H. Finch. 1929. Tilt records for thirteen years at the Hawaiian Volcano Observatory. *Bulletin of the Seismological Society of America* **19**, 38–51.

Koto, B. 1990. On the cause of the great earthquake in central Japan. *Journal of the College of Science, Imperial University of Japan* **5**, 296–353, 1893; excerpted in *Terra Nova* **2**, 301–305.

Mansinha, L., and D. E. Smylie. 1971. The displacement fields of inclined faults. *Bulletin of the Seismological Society of America* **61**, 1433–1440.

Maruyama, T. 1964. Static elastic dislocations in an infinite and semi-infinite medium. *Bulletin of the Earthquake Research Institute, University of Tokyo* **42**, 289–368.

Mogi, K. 1958. Relations between the eruptions of various volcanoes and the deformations of the ground surfaces around them. *Bulletin of the Earthquake Research Institute, University of Tokyo* **36**, 99–134.

Okada, A., and T. Nagata. 1953. Land deformation of the neighborhood of Muroto Point after the Nankaido great earthquake in 1946. *Bulletin of the Earthquake Research Institute, Tokyo University* **31**, 169–177.

Oldham, R. D. 1981. Report on the Great Earthquake of 12th June 1897. *Memoirs of the Geological Survey of India* **XXIX**, reprint, pp. 361–371.

Omori, F. 1918. The Sakura-jima eruptions and earthquakes. *Bulletin of the Imperial Earthquake Investigation Committee* **8**, 152–179.

Poland, M., M. Hamburger, and A. Newman. 2006. The changing shapes of active volcanoes: history, evolution, and future challenges for volcano geodesy. *Journal of Volcanology and Geothermal Research* **150**, 1–13.

Reid, H. F. 1910. The mechanics of the earthquake. In A. C. Lawson (Ed.), *The California earthquake of April 18, 1906, Report of the State Earthquake Investigation Commission*, vol. II., Washington, DC: Carnegie Institute of Washington, pp. 3–55.

———. 1911. The elastic rebound theory of earthquakes. *Bulletin of the Department of Geology, University of California Publications* **6**(19), 413–444.

———. 1913. Sudden earth-movements in Sumatra in 1892. *Bulletin of the Seismological Society of America* **3**, 72–79.

Savage, J. C., and R. O. Burford. 1973. Geodetic determination of relative plate motion in central California. *Journal of Geophysical Research* **78**, 832–845.

Savage, J. C., and L. M. Hastie. 1971. Surface deformation associated with dip-slip faulting. *Journal of Geophysical Research* **71**, 4897–4904.

Sezawa, K. 1931. The plastico-elastic deformation of a semi-infinite solid body due to an internal force. *Bulletin of the Earthquake Research Institute, University of Tokyo* **9**, 398–406.

Steketee, J. A. 1958. Some geophysical applications of the elasticity theory of dislocations. *The Canadian Journal of Physics* **36**, 1168–1198.

Thatcher, W. 1975. Strain accumulation and release mechanism of the 1906 San Francisco earthquake. *Journal of Geophysical Research* **80**, 4862–4872.

Tsuboi, C. 1931. Investigation on the deformation of the earth's crust in the Tango district connected with the Tango earthquake of 1927. Part III. *Bulletin of the Earthquake Research Institute, University of Tokyo* **9**, 423–434.

Weertman, J. 1965. Relationship between displacements on a free surface and the stress on a fault. *Bulletin of the Seismological Society of America* **55**, 945–953.

Whitten, C. A. 1948. Horizontal earth movement, vicinity of San Francisco, California. *Transactions, American Geophysical Union* **29**, 318–323.

Wilson, R. M. 1935. *Ground Surface Movements at Kilauea Volcano*. Hawaii: University of Hawaii Research Publication 10.

Yamakawa, N. 1955. On the strain produced in a semi-infinite elastic solid by an interior source of stress. *Zisin (Journal of the Seismological Society of Japan)* **8**, 84–98.

1

Deformation, Stress, and Conservation Laws

In this chapter, we will develop a mathematical description of deformation. Our focus is on relating deformation to quantities that can be measured in the field, such as the change in distance between two points, the change in orientation of a line, or the change in volume of a borehole strain sensor. We will also review the Cauchy stress tensor and the conservation laws that generalize conservation of mass and momentum to continuous media. Last, we will consider constitutive equations that relate the stresses acting on a material element to the resultant strains and/or rates of strain. This necessarily abbreviated review of continuum mechanics borrows from a number of excellent textbooks on the subject to which the reader is referred for more detail (references are given at the end of the chapter).

If all the points that make up a body (say, a tectonic plate) move together without a change in the shape of the body, we will refer to this as *rigid body motion*. On the other hand, if the shape of the body changes, we will refer to this as *deformation*. To differentiate between deformation and rigid body motion, we will consider the relative motion of neighboring points. To make this mathematically precise, we will consider the displacement **u** of a point at **x** relative to an arbitrary origin $\mathbf{x}_0$ (figure 1.1). Note that vector **u** appears in boldface, whereas the components of the vector u_i do not.

Taking the first two terms in a Taylor's series expansion about $\mathbf{x}_0$,

$$u_i(\mathbf{x}) = u_i(\mathbf{x}_0) + \left. \frac{\partial u_i}{\partial x_j} \right|_{x_0} dx_j + \cdots \qquad i = 1, 2, 3, \tag{1.1}$$

in equation (1.1), and in what follows, summation on repeated indices is implied. The first term, $u_i(\mathbf{x}_0)$, represents a rigid body translation since all points in the neighborhood of $\mathbf{x}_0$ share the same displacement. The second term gives the relative displacement in terms of the gradient of the displacements $\partial u_i / \partial x_j$, or $\nabla \mathbf{u}$ in vector notation. The partial derivatives $\partial u_i / \partial x_j$ make up the *displacement gradient tensor*, a second rank tensor with nine independent components:

$$\begin{bmatrix} \partial u_1/\partial x_1 & \partial u_1/\partial x_2 & \partial u_1/\partial x_3 \\ \partial u_2/\partial x_1 & \partial u_2/\partial x_2 & \partial u_2/\partial x_3 \\ \partial u_3/\partial x_1 & \partial u_3/\partial x_2 & \partial u_3/\partial x_3 \end{bmatrix}. \tag{1.2}$$

The displacement gradient tensor can be separated into symmetric and antisymmetric parts as follows:

$$u_i(\mathbf{x}) = u_i(\mathbf{x}_0) + \frac{1}{2} \left(\frac{\partial u_i}{\partial x_j} + \frac{\partial u_j}{\partial x_i} \right) dx_j + \frac{1}{2} \left(\frac{\partial u_i}{\partial x_j} - \frac{\partial u_j}{\partial x_i} \right) dx_j + \cdots. \tag{1.3}$$

If the magnitudes of the displacement gradients are small, $|\partial u_i / \partial x_j| \ll 1$, the symmetric and antisymmetric parts of the displacement gradient tensor can be associated with the small strain and rotation tensors, as in equations (1.4) and (1.5). Fortunately, the assumption of small strain and rotation is nearly always satisfied in the study of active crustal deformation, with the important exceptions of deformation within the cores of active fault zones, where finite strain measures are required.

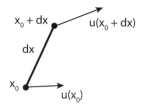

Figure 1.1. Generalized displacement of a line segment **dx**. Point $\mathbf{x}_0$ displaces by amount $\mathbf{u}(\mathbf{x}_0)$, whereas the other end point, $\mathbf{x}_0 + \mathbf{dx}$, displaces by $\mathbf{u}(\mathbf{x}_0 + \mathbf{dx})$.

Define the symmetric part of the displacement gradient tensor, with six independent components, to be the *infinitesimal strain tensor*, ϵ_{ij}:

$$\epsilon_{ij} \equiv \frac{1}{2}\left(\frac{\partial u_i}{\partial x_j} + \frac{\partial u_j}{\partial x_i}\right). \tag{1.4}$$

The antisymmetric part of the displacement gradient tensor is defined to be the *rotation*, ω_{ij}, which includes the remaining three independent components:

$$\omega_{ij} \equiv \frac{1}{2}\left(\frac{\partial u_i}{\partial x_j} - \frac{\partial u_j}{\partial x_i}\right). \tag{1.5}$$

By substituting equations (1.4) and (1.5) into (1.3) and neglecting terms of order $(dx)^2$, we see that the displacement $\mathbf{u}(\mathbf{x})$ is composed of three components—rigid body translation, strain, and rigid body rotation:

$$\mathbf{u}(\mathbf{x}) = \mathbf{u}(\mathbf{x}_0) + \epsilon \cdot \mathbf{dx} + \omega \cdot \mathbf{dx},$$
$$u_i(\mathbf{x}) = u_i(\mathbf{x}_0) + \epsilon_{ij}dx_j + \omega_{ij}dx_j. \tag{1.6}$$

We must still show that ϵ_{ij} is sensibly identified with strain and ω_{ij} with rigid body rotation. This will be the subject of the next two sections.

1.1 Strain

To see why ϵ_{ij} is associated with strain, consider a line segment with end points $\mathbf{x}$ and $\mathbf{x} + \mathbf{dx}$ (figure 1.2). Following deformation, the coordinates of the end points are ξ and $\xi + \mathbf{d\xi}$. The length of the line segment changes from $dl_0 \equiv |\mathbf{dx}|$ to $dl \equiv |\mathbf{d\xi}|$.

The strain can be defined in terms of the change in the squared length of the segment—that is, $dl^2 - dl_0^2$. The squared length of the segment prior to deformation is

$$dl_0^2 = dx_i dx_i = \delta_{ij}dx_i dx_j, \tag{1.7}$$

where the Kronecker delta $\delta_{ij} = 1$ if $i = j$ and $\delta_{ij} = 0$ otherwise. Similarly, the squared length of the segment following deformation is

$$dl^2 = d\xi_m d\xi_m = \delta_{mn}d\xi_m d\xi_n. \tag{1.8}$$

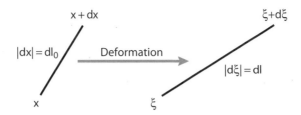

Figure 1.2. Strained line segment. Line segment with endpoints $\mathbf{x}$ and $\mathbf{x} + d\mathbf{x}$ in the undeformed state transforms to a line with endpoints $\boldsymbol{\xi}$ and $\boldsymbol{\xi} + d\boldsymbol{\xi}$ in the deformed state.

We can consider the final coordinates to be a function of the initial coordinates $\xi_i = \xi_i(\mathbf{x})$ or, conversely, the initial coordinates to be a function of the final coordinates $x_i = x_i(\boldsymbol{\xi})$. Adopting the former approach, for the moment, and writing the total differential $d\xi_m$ as

$$d\xi_m = \frac{\partial \xi_m}{\partial x_j} dx_j, \tag{1.9}$$

we find

$$dl^2 = \delta_{mn} \frac{\partial \xi_m}{\partial x_i} \frac{\partial \xi_n}{\partial x_j} dx_i dx_j. \tag{1.10}$$

Thus, the change in the squared length of the line segment is

$$dl^2 - dl_0^2 = \left(\delta_{mn} \frac{\partial \xi_m}{\partial x_i} \frac{\partial \xi_n}{\partial x_j} - \delta_{ij} \right) dx_i dx_j \equiv 2E_{ij} dx_i dx_j. \tag{1.11}$$

E_{ij}, as defined here, is known as the *Green's strain tensor*. It is a so-called *Lagrangian* formulation with derivatives of final position with respect to initial position.

Employing the inverse relationship $x_i = x_i(\boldsymbol{\xi})$, which as described here is an *Eulerian* formulation, yields the so-called *Almansi strain tensor*, e_{ij}:

$$dl^2 - dl_0^2 = \left(\delta_{ij} - \delta_{mn} \frac{\partial x_m}{\partial \xi_i} \frac{\partial x_n}{\partial \xi_j} \right) d\xi_i d\xi_j \equiv 2e_{ij} d\xi_i d\xi_j. \tag{1.12}$$

Note that both equations (1.11) and (1.12) are valid, whether or not $|\partial u_i / \partial x_j| \ll 1$. We do not yet have the strains described in terms of displacements. To do so, note that the displacement is the difference between the initial position and the final position $u_i = \xi_i - x_i$. Differentiating,

$$\frac{\partial \xi_i}{\partial x_j} = \frac{\partial u_i}{\partial x_j} + \delta_{ij}. \tag{1.13}$$

Substituting this into equation (1.11) gives

$$2E_{ij} = \delta_{mn} \left(\frac{\partial u_m}{\partial x_i} + \delta_{mi} \right) \left(\frac{\partial u_n}{\partial x_j} + \delta_{nj} \right) - \delta_{ij}, \tag{1.14}$$

$$= \delta_{mn} \left(\delta_{mi} \frac{\partial u_n}{\partial x_j} + \delta_{nj} \frac{\partial u_m}{\partial x_i} + \frac{\partial u_m}{\partial x_i} \frac{\partial u_n}{\partial x_j} + \delta_{mi} \delta_{nj} \right) - \delta_{ij}. \tag{1.15}$$

Thus, the Green strain is given by

$$E_{ij} = \frac{1}{2} \left(\frac{\partial u_i}{\partial x_j} + \frac{\partial u_j}{\partial x_i} + \frac{\partial u_n}{\partial x_i} \frac{\partial u_n}{\partial x_j} \right). \tag{1.16}$$

The corresponding Almansi (Eulerian) strain is given by

$$e_{ij} = \frac{1}{2} \left(\frac{\partial u_i}{\partial \xi_j} + \frac{\partial u_j}{\partial \xi_i} - \frac{\partial u_n}{\partial \xi_i} \frac{\partial u_n}{\partial \xi_j} \right). \tag{1.17}$$

Notice that the third terms in both equations (1.16) and (1.17) are products of gradients in displacement, and are thus second order. These terms are safely neglected as long as the displacement gradients are small compared to unity. Furthermore, in this limit, we do not need to distinguish between derivatives taken with respect to initial or final coordinates. The infinitesimal strain is thus given by equation (1.4).

More generally, the Lagrangian formulation tracks a material element, or a collection of particles. It is an appropriate choice for elasticity problems where the reference state is naturally defined as the unstrained configuration. The Eulerian formulation refers to the properties at a given point in space and is naturally used in fluid mechanics where one could imagine monitoring the velocity, density, temperature, and so on at fixed points in space as the flowing fluid moves by. In contrast, the Lagrangian formulation would track these properties as one moved with the flow. For example, fixed weather stations that measure wind velocity, barometric pressure, and temperature yield an Eulerian description of the atmosphere. Neutrally buoyant weather balloons moving with the wind yield a corresponding Lagrangian description.

Geodetic observing stations are fixed to the solid earth, suggesting that a Lagrangian formulation is sensible. Furthermore, constitutive equations, which refer to a parcel of material, are Lagrangian. However, the familiar Cauchy stress, to be defined shortly, is inherently Eulerian. (Dahlen and Tromp 1998 give a more complete discussion of these issues.) These distinctions turn out not to be important as long as the initial stresses are of the same order as, or smaller than, the incremental stresses (and any rotations are not much larger than the strains). For the most part in this book, we will restrict our attention to infinitesimal strains and will not find it necessary to distinguish between Lagrangian and Eulerian formulations. In the earth, however, the initial stresses, at least their isotropic components, are likely to be large, and a more thorough accounting of the effects of this prestress is required. This arises in chapter 9, where we will consider the role of gravitation in the equilibrium equations.

An important measure of strain is the change in the length of an arbitrarily oriented line segment. In practice, linear extension is measured by a variety of extensometers, ranging from strain gauges in laboratory rock mechanics experiments to field extensometers and laser strainmeters. To the degree that the strain is spatially uniform, geodimeters, or Electronic Distance Meters (EDM) measure linear extension. Consider the change in length of a reference line $\mathbf{dx}$, with initial length l_0 and final length l. In the small strain limit, we have

$$l^2 - l_0^2 = (l + l_0)(l - l_0) = 2\epsilon_{ij} dx_i dx_j,$$

$$2l_0(l - l_0) \simeq 2\epsilon_{ij} dx_i dx_j,$$

$$\frac{\Delta l}{l_0} \simeq \epsilon_{ij} \widehat{dx_i} \widehat{dx_j}, \tag{1.18}$$

where $\widehat{dx_i}$ are the components of a unit vector in the direction of the baseline $\mathbf{dx}$. The change in length per unit length of the baseline $\mathbf{dx}$ is given by the scalar product of the strain tensor

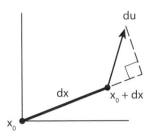

Figure 1.3. Extension and rotation of a line segment. The extension is the dot product of the relative displacement **du** with the unit vector in the direction of the line segment $\widehat{\mathbf{dx}} = \mathbf{dx}/|\mathbf{dx}|$. The rotation of the line segment is the cross product of the relative displacement with the same unit vector.

and the unit vectors $\widehat{\mathbf{dx}}$. This can also be written as $\widehat{\mathbf{dx}} \cdot \epsilon \cdot \widehat{\mathbf{dx}}$. This result has a simple geometric interpretation. The change in length of the baseline is the projection of the relative displacement vector **du** onto the baseline direction $\Delta l = du_i \widehat{dx}_i$ (figure 1.3). In an irrotational field, the relative displacement between nearby points is given by $du_i = \epsilon_{ij} dx_j$ (equation [1.6]). Thus, $\Delta l = \epsilon_{ij} \widehat{dx}_i dx_j$. Dividing both sides by l_0 yields the last equation in equation (1.18).

Geodimeters, which measure distance with great accuracy, can be used to determine the change in distance Δl between two benchmarks separated by up to several tens of kilometers. If the baseline lies in a horizontal plane, $\mathbf{dx}^T = (dx_1, dx_2, 0)$, and the strain is spatially uniform between the two end points, the extensional strain is given by

$$\frac{\Delta l}{l_0} = \epsilon_{11} \cos^2 \alpha + 2\epsilon_{12} \cos \alpha \sin \alpha + \epsilon_{22} \sin^2 \alpha, \tag{1.19}$$

where α is the angle between the x_1 axis and the baseline. Equation (1.19) shows how the elongation varies with orientation. In the small strain limit, we can relate the rate of line length change to the *strain rate* by differentiating both sides of equation (1.19) with respect to time; to first-order changes in orientation, $\partial \alpha / \partial t$ can be neglected. The exact relationship corresponding to equation (1.18) valid for finite strain is $(1/l)dl/dt = \widehat{\mathbf{d\xi}} \cdot D \cdot \widehat{\mathbf{d\xi}}$, (e.g., Malvern 1969), where the *rate of deformation tensor* is defined as

$$D_{ij} \equiv \frac{1}{2} \left(\frac{\partial v_i}{\partial \xi_j} + \frac{\partial v_j}{\partial \xi_i} \right), \tag{1.20}$$

where **v** is the instantaneous particle velocity with components v_i. The rate of deformation tensor, in which derivatives are taken with respect to the current configuration is not generally equivalent to the time derivative of a strain tensor in which derivatives are taken with respect to the initial configuration. The distinction is negligible, however, if one takes the initial configuration to be that at the time of the first geodetic measurement, not an unrealizable, undeformed state, so that generally, $|\partial u_i / \partial x_j| \ll 1$.

If the rate of elongation is measured on a number of baselines with different orientations α_i, it is possible to determine the components of the average strain-rate tensor by least squares solution of the following equations:

$$\begin{bmatrix} (1/l_i)dl_i/dt \\ \vdots \end{bmatrix} = \begin{bmatrix} \cos^2 \alpha_i & 2 \cos \alpha_i \sin \alpha_i & \sin^2 \alpha_i \\ \vdots & \vdots & \vdots \end{bmatrix} \begin{bmatrix} \dot{\epsilon}_{11} \\ \dot{\epsilon}_{12} \\ \dot{\epsilon}_{22} \end{bmatrix}, \tag{1.21}$$

where the superimposed dot indicates differentiation with respect to time.

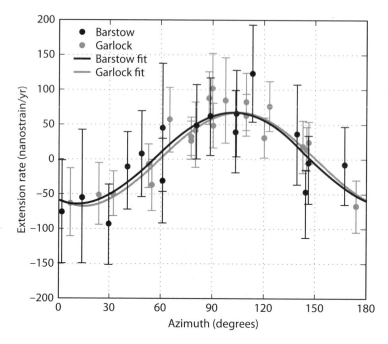

Figure 1.4. Linear extension rate as a function of azimuth for precise distance measurements from the Eastern California Shear Zone. Data from two networks are shown. The Garlock network is located at the eastern end of the Garlock fault. The estimated strain-rate fields for the two networks (shown by the continuous curves) are very similar. From Savage et al. (2001).

In figure 1.4, linear extension rates vary as a function of azimuth in a manner consistent with a uniform strain-rate field, within rather large errors given the magnitude of the signal. These data were collected in the Southern California Shear Zone using precise distance measuring devices. Data from the U.S. Geological Survey (USGS) Geodolite program, which measured distance repeatedly using an Electronic Distance Meter (EDM), were analyzed by Johnson et al. (1994) to determine the average strain rate in southern California. Figure 1.5 shows the principal strain-rate directions and magnitudes for different polygons assuming that the strain rate is spatially uniform within each polygon.

Borehole dilatometers measure volume change, or dilatational strain, in the earth's crust (figure 1.6). The volume change of an element is related to the extension in the three coordinate directions. Let V_0 be the initial volume and V be the final volume, and define the *dilatation* as the change in volume over the initial volume $\Delta \equiv (V - V_0)/V_0$. If x_i represent the sides of a small cubical element in the undeformed state, then $V_0 = x_1 x_2 x_3$. Similarly, if ξ_i represent the sides in the deformed state, then $V = \xi_1 \xi_2 \xi_3$. The final dimensions are related to the undeformed dimensions by $\xi_1 = x_1(1 + \epsilon_{11})$, $\xi_2 = x_2(1 + \epsilon_{22})$, and $\xi_3 = x_3(1 + \epsilon_{33})$, so that

$$\Delta = \frac{x_1 x_2 x_3 (1 + \epsilon_{11})(1 + \epsilon_{22})(1 + \epsilon_{33}) - x_1 x_2 x_3}{x_1 x_2 x_3}. \tag{1.22}$$

To first order in strain, this yields $\Delta = \epsilon_{11} + \epsilon_{22} + \epsilon_{33} = \epsilon_{ii} = \text{trace}(\epsilon)$, which can be shown to be independent of the coordinate system. In other words, the dilatation is an *invariant* of the strain tensor. Note further that $\epsilon_{ii} = \partial u_i / \partial x_i$, so volumetric strain is equivalent to the divergence of the displacement field $\Delta = \nabla \cdot \mathbf{u}$.

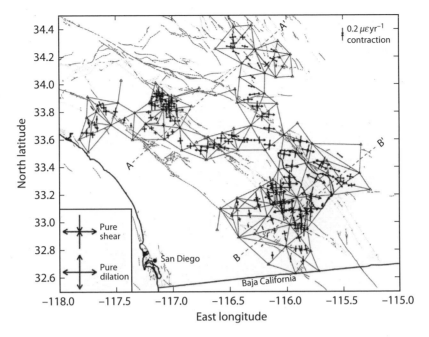

Figure 1.5. Strain-rate distribution in southern California determined from the rate of linear extension measured by the U.S. Geological Survey Geodolite project. Maximum and minimum extension rates are average for the different quadrilaterals. Scale bar is located in the upper-right corner. From Johnson et al. (1994).

We have seen that the three normal strains ϵ_{ij}, $i = j$ represent stretching in the three coordinate directions. The off-diagonal components ϵ_{ij}, $i \neq j$, or shear strains, represent a change in shape. It can be shown that the shear strain is equal to the change in an initially right angle. In section 1.2, you will see how horizontal angle measurements determined by geodetic triangulation can be used to measure crustal shear strain.

1.1.1 Strains in Curvilinear Coordinates

Due to the symmetry of a particular problem, it is often convenient to express the strains in cylindrical polar or spherical coordinates. We present the results without derivation; details are given in, for example, Malvern (1969). In cylindrical polar coordinates, the strains are

$$\epsilon_{rr} = \frac{\partial u_r}{\partial r},$$

$$\epsilon_{\theta\theta} = \frac{1}{r}\frac{\partial u_\theta}{\partial \theta} + \frac{u_r}{r},$$

$$\epsilon_{zz} = \frac{\partial u_z}{\partial z},$$

$$\epsilon_{r\theta} = \frac{1}{2}\left(\frac{1}{r}\frac{\partial u_r}{\partial \theta} + \frac{\partial u_\theta}{\partial r} - \frac{u_\theta}{r}\right),$$

$$\epsilon_{rz} = \frac{1}{2}\left(\frac{\partial u_z}{\partial r} + \frac{\partial u_r}{\partial z}\right),$$

$$\epsilon_{z\theta} = \frac{1}{2}\left(\frac{\partial u_\theta}{\partial z} + \frac{1}{r}\frac{\partial u_z}{\partial \theta}\right). \tag{1.23}$$

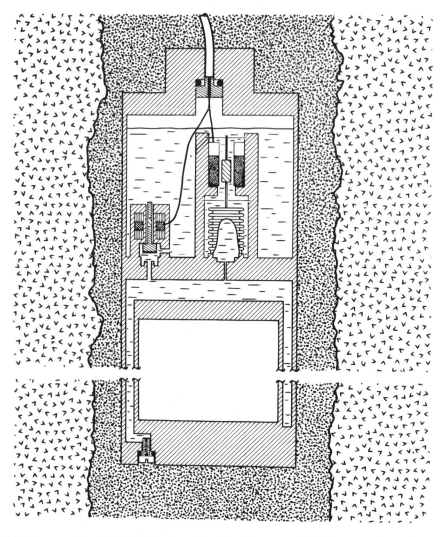

Figure 1.6. Cross section of a Sacks-Evertson borehole dilatometer. The lower sensing volume is filled with a relatively incompressible fluid. The upper reservoir is partially filled with a highly compressible, inert gas and is connected to the sensing volume by a thin tube and bellows. The bellows is attached to a displacement transducer that records the flow of fluid into or out of the sensing volume. As the instrument is compressed, fluid flows out of the sensing volume. The output of the displacement transducer is calibrated to strain by comparing observed solid earth tides to theoretically predicted tides. After Agnew (1985).

In spherical coordinates, the strains are given by

$$\epsilon_{rr} = \frac{\partial u_r}{\partial r},$$

$$\epsilon_{\theta\theta} = \frac{1}{r}\frac{\partial u_\theta}{\partial \theta} + \frac{u_r}{r},$$

$$\epsilon_{\phi\phi} = \frac{1}{r\sin\theta}\frac{\partial u_\phi}{\partial \phi} + \frac{u_r}{r} + \frac{\cot\theta}{r}u_\theta,$$

$$\epsilon_{r\theta} = \frac{1}{2} \left(\frac{1}{r} \frac{\partial u_r}{\partial \theta} + \frac{\partial u_\theta}{\partial r} - \frac{u_\theta}{r} \right),$$

$$\epsilon_{r\phi} = \frac{1}{2} \left(\frac{1}{r \sin \theta} \frac{\partial u_r}{\partial \phi} + \frac{\partial u_\phi}{\partial r} - \frac{u_\phi}{r} \right),$$

$$\epsilon_{\theta\phi} = \frac{1}{2} \left(\frac{1}{r \sin \theta} \frac{\partial u_\theta}{\partial \phi} + \frac{1}{r} \frac{\partial u_\phi}{\partial \theta} - \frac{\cot \phi}{r} u_\phi \right). \tag{1.24}$$

1.2 Rotation

There are only three independent components in the rotation tensor as defined by equation (1.5), so the infinitesimal rotation may be represented as a vector $\mathbf{\Omega}$:

$$\Omega_k \equiv -\frac{1}{2} e_{kij} \omega_{ij}, \tag{1.25}$$

where e_{ijk} is the permutation tensor. e_{ijk} vanishes if any index is repeated, is equal to $+1$ for e_{123} or for any cyclic permutation of the indices, and is equal to -1 for e_{321} or for any cyclic permutation of the indices. Because it is a rank-three tensor, the permutation tensor is easily distinguished from the Almansi strain. From the definition of ω_{ij} (1.5), we can write the rotation $\mathbf{\Omega}$ as

$$\Omega_k = \frac{1}{2} \left(-\frac{1}{2} e_{kij} \frac{\partial u_i}{\partial x_j} + \frac{1}{2} e_{kij} \frac{\partial u_j}{\partial x_i} \right). \tag{1.26}$$

Now $-e_{kij} = e_{kji}$, so (noting that i and j are dummy indices)

$$\Omega_k = \frac{1}{2} \left(\frac{1}{2} e_{kji} \frac{\partial u_i}{\partial x_j} + \frac{1}{2} e_{kij} \frac{\partial u_j}{\partial x_i} \right), \tag{1.27}$$

$$\Omega_k = \frac{1}{2} \left(e_{kij} \frac{\partial u_j}{\partial x_i} \right), \tag{1.28}$$

$$\mathbf{\Omega} = \frac{1}{2} (\nabla \times \mathbf{u}). \tag{1.29}$$

That is, the rotation vector is half the curl of the displacement field.

Equation (1.25) can be inverted to write ω_{ij} in terms of Ω_k. To do so, we make use of the so-called e-δ identity:

$$e_{knm} e_{kij} = \delta_{mi} \delta_{nj} - \delta_{mj} \delta_{ni}. \tag{1.30}$$

Multiplying both sides of equation (1.25) by e_{mnk}:

$$e_{mnk} \Omega_k = -\frac{1}{2} e_{mnk} e_{kij} \omega_{ij},$$

$$= -\frac{1}{2} \left(\delta_{mi} \delta_{nj} \omega_{ij} - \delta_{mj} \delta_{ni} \omega_{ij} \right),$$

$$= -\frac{1}{2} (\omega_{mn} - \omega_{nm}) = -\omega_{mn}. \tag{1.31}$$

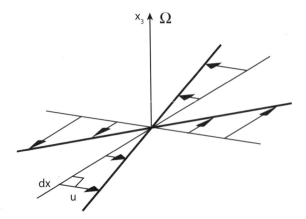

Figure 1.7. Rotational deformation.

We can now see clearly the meaning of the term $\omega_{ij}dx_j$ in the Taylor's series expansion (1.6) for the displacements

$$du_i = \omega_{ij}dx_j = -e_{ijk}\Omega_k dx_j = e_{ikj}\Omega_k dx_j, \tag{1.32}$$

which in vector notation is $\mathbf{du} = \boldsymbol{\Omega} \times \mathbf{dx}$. This is illustrated in figure 1.7.

Consider now how an arbitrarily oriented baseline rotates during deformation. This could represent the change in orientation of a tiltmeter or the change in orientation of a geodetic baseline determined by repeated triangulation surveys. We will denote the rotation of a baseline as $\boldsymbol{\Theta}$ and recognize that the rotation of a line segment is different from the rotational component of the deformation $\boldsymbol{\Omega}$. Define the rotation of a line segment $\boldsymbol{\Theta}$ as

$$\boldsymbol{\Theta} \equiv \frac{\mathbf{dx} \times \mathbf{d\xi}}{dl_o^2} \tag{1.33}$$

(Agnew 1985). Note that this definition is reasonable since in the limit that the deformation gradients are small, the magnitude of $\boldsymbol{\Theta}$ is

$$|\boldsymbol{\Theta}| = \frac{|dl||dl_o|}{dl_o^2} \sin \Theta \sim \Theta, \tag{1.34}$$

where Θ is the angle between $\mathbf{dx}$ and $\mathbf{d\xi}$. In indicial notation, equation (1.33) is written as

$$\Theta_i = \frac{e_{ijk}dx_j d\xi_k}{dx_n dx_n}. \tag{1.35}$$

As before, we write the final position as the initial position plus the displacement $\xi_i = x_i + u_i$, so

$$d\xi_k = \frac{\partial \xi_k}{\partial x_m}dx_m = \left(\frac{\partial u_k}{\partial x_m} + \delta_{km}\right)dx_m. \tag{1.36}$$

Substituting equation (1.36) into (1.35) yields

$$\mathrm{d}x_n\mathrm{d}x_n\Theta_i = e_{ijk}\mathrm{d}x_j\left(\frac{\partial u_k}{\partial x_m} + \delta_{km}\right)\mathrm{d}x_m,$$

$$= e_{ijk}\left(\frac{\partial u_k}{\partial x_m}\right)\mathrm{d}x_j\mathrm{d}x_m + e_{ijk}\mathrm{d}x_j\mathrm{d}x_k,$$

$$= e_{ijk}(\epsilon_{km} + \omega_{km})\mathrm{d}x_j\mathrm{d}x_m. \tag{1.37}$$

Note that the term $e_{ijk}\mathrm{d}x_j\mathrm{d}x_k$ is the cross product of $\mathbf{dx}$ with itself and is therefore zero. We have also expanded the displacement gradients as the sum of strain and rotation, so finally

$$\Theta_i = e_{ijk}(\epsilon_{km} + \omega_{km})\widehat{\mathrm{d}x_j}\widehat{\mathrm{d}x_m}, \tag{1.38}$$

where as before the $\widehat{\mathrm{d}x_j}$ are unit vectors.

Consider the rotational term in equation (1.38):

$$e_{ijk}\omega_{km}\widehat{\mathrm{d}x_j}\widehat{\mathrm{d}x_m} = -e_{ijk}e_{kmn}\Omega_n\widehat{\mathrm{d}x_j}\widehat{\mathrm{d}x_m},$$

$$= -(\delta_{im}\delta_{jn} - \delta_{in}\delta_{jm})\Omega_n\widehat{\mathrm{d}x_j}\widehat{\mathrm{d}x_m},$$

$$= \Omega_i - \Omega_j\widehat{\mathrm{d}x_j}\widehat{\mathrm{d}x_i}, \tag{1.39}$$

where in the first equation we have made use of equation (1.31), and in the last that $\widehat{\mathrm{d}x_m}\widehat{\mathrm{d}x_m} = 1$.

Thus, the general expression for the rotation of a line segment (1.38) can be written as

$$\Theta_i = e_{ijk}\epsilon_{km}\widehat{\mathrm{d}x_j}\widehat{\mathrm{d}x_m} + \Omega_i - \Omega_j\widehat{\mathrm{d}x_j}\widehat{\mathrm{d}x_i}, \tag{1.40}$$

or in vector form as

$$\boxed{\boldsymbol{\Theta} = \widehat{\mathbf{dx}} \times (\epsilon \cdot \widehat{\mathbf{dx}}) + \boldsymbol{\Omega} - (\boldsymbol{\Omega} \cdot \widehat{\mathbf{dx}})\widehat{\mathbf{dx}}.} \tag{1.41}$$

The strain component of the rotation has a simple interpretation. Recall from figure 1.3 that the elongation is equal to the dot product of the relative displacement and a unit vector parallel to the baseline. The rotation is the cross product of the unit vector and the relative displacement $\mathrm{d}u_i = \epsilon_{ij}\widehat{\mathrm{d}x_j}$.

To understand the dependence of $\boldsymbol{\Theta}$ on $\boldsymbol{\Omega}$ geometrically, consider two end-member cases when the strain vanishes. In the first case, the rotation vector is normal to the baseline. In this case, $\boldsymbol{\Omega} \cdot \widehat{\mathbf{dx}} = 0$, so $\boldsymbol{\Theta} = \boldsymbol{\Omega}$; the rotation of the line segment is simply $\boldsymbol{\Omega}$ (figure 1.8). (Notice from the right-hand rule for cross products that the sign of the rotation due to strain is consistent with the sign of $\boldsymbol{\Omega}$.) If, on the other hand, the rotation vector is parallel to the line segment, then $(\boldsymbol{\Omega} \cdot \widehat{\mathbf{dx}})\widehat{\mathbf{dx}} = |\boldsymbol{\Omega}|\widehat{\mathbf{dx}} = \boldsymbol{\Omega}$ so that $\boldsymbol{\Theta} = 0$. When the rotation vector is parallel to the baseline, the baseline does not rotate at all (figure 1.8). For intermediate cases, $0 < \boldsymbol{\Theta} < \boldsymbol{\Omega}$.

Consider the vertical-axis rotation of a horizontal line segment $\mathbf{dx}^{\mathrm{T}} = (\mathrm{d}x_1, \mathrm{d}x_2, 0)$. In this case, equation (1.40) reduces to

$$\Theta_3 = \epsilon_{12}(\cos^2\alpha - \sin^2\alpha) + (\epsilon_{22} - \epsilon_{11})\cos\alpha\sin\alpha + \Omega_3, \tag{1.42}$$

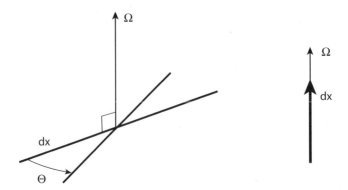

Figure 1.8. Rotation of a line segment. In the left figure, the rotation vector is perpendicular to the baseline, and $\Theta = \Omega$. In the right figure, the rotation vector is parallel to the baseline, and $\Theta = 0$.

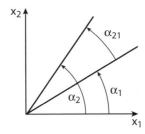

Figure 1.9. Angle α_{21} between two direction measurements α_1 and α_2.

where α is the angle between the x_1 axis and the baseline. Note that the rotation Θ_3 is equivalent to the change in orientation dα. With geodetic triangulation, surveyors were able to determine the angle between three observation points with a *theodolite*. Triangulation measurements were widely made during the late nineteenth and early twentieth centuries before the advent of laser distance measuring devices. Surveying at different times, it was possible to determine the change in the angle. Denoting the angle formed by two baselines oriented at α_1 and α_2 as $\alpha_{21} = \alpha_2 - \alpha_1$ (figure 1.9), then

$$\delta\alpha_{21} = \delta\alpha_2 - \delta\alpha_1 = \epsilon_{12} \left(\cos 2\alpha_2 - \cos 2\alpha_1 \right) + \left(\frac{\epsilon_{22} - \epsilon_{11}}{2} \right) \left(\sin 2\alpha_2 - \sin 2\alpha_1 \right). \tag{1.43}$$

Note that the rotation Ω_3 is common to both $\delta\alpha_1$ and $\delta\alpha_2$ and thus drops out of the difference. The change in the angle thus depends only on the strain field. Defining engineering shear strains in terms of the tensor strains as $\gamma_1 = (\epsilon_{11} - \epsilon_{22})$ and $\gamma_2 = 2\epsilon_{12}$:

$$\delta\alpha_{21} = \frac{1}{2}\gamma_2 \left(\cos 2\alpha_2 - \cos 2\alpha_1 \right) - \frac{1}{2}\gamma_1 \left(\sin 2\alpha_2 - \sin 2\alpha_1 \right). \tag{1.44}$$

Notice from figure 1.10 that γ_1 is a north–south compression and an east–west extension, equivalent to right-lateral shear across planes trending 45 degrees west of north. γ_2 is right-lateral shear across east–west planes.

Determination of shear strain using the change in angles measured during a geodetic survey was first proposed by Frank (1966). Assuming that the strain is spatially uniform over some

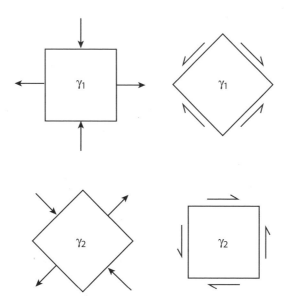

Figure 1.10. Definition of engineering shear strains.

area, it is possible to collect a series of angle changes in a vector of observations related to the two shear strains:

$$\begin{bmatrix} \delta\alpha_{ij} \\ \vdots \end{bmatrix} = \begin{bmatrix} -\frac{1}{2}\left(\sin 2\alpha_i - \sin 2\alpha_j\right) & \frac{1}{2}\left(\cos 2\alpha_i - \cos 2\alpha_j\right) \\ \vdots & \vdots \end{bmatrix} \begin{bmatrix} \gamma_1 \\ \gamma_2 \end{bmatrix}. \qquad (1.45)$$

Given a system of equations such as this, the two shear strains are easily estimated by least squares. As expected, the angle changes are insensitive to areal strain. A significant advantage of *trilateration* using geodimeters, which measure distance, over triangulation, which measures only angles, is that with trilateration it became possible to determine areal strain as well as shear strain.

Savage and Burford (1970) used Frank's method to determine the strain in small quadrilaterals that cross the San Andreas fault near the town of Cholame, California. The triangulation network is shown in figure 1.11. The quads are designated by a letter (A to X) from southwest to northeast across the fault. The shear strain estimates with error bars are shown in figure 1.12. The data show a positive γ_1 centered over the fault, as expected for a locked right-lateral fault. The γ_2 distribution shows negative values on the southwest side of the fault and positive values on the northeast side of the fault. Savage and Burford (1970) showed that these strains were caused by the 1934 Parkfield earthquake, which ruptured the San Andreas just northwest of the triangulation network. The data are discussed in more detail by Segall and Du (1993).

1.3 Stress

While this text focuses on deformation parameters that can be measured by modern geophysical methods (Global Positioning System [GPS], strainmeters, Interferometric Synthetic Aperture Radar [InSAR], and so on), it is impossible to develop physically consistent models of earthquake and volcano processes that do not involve the forces and stresses acting within the earth. It will be assumed that the reader has been introduced to the concept of stress in

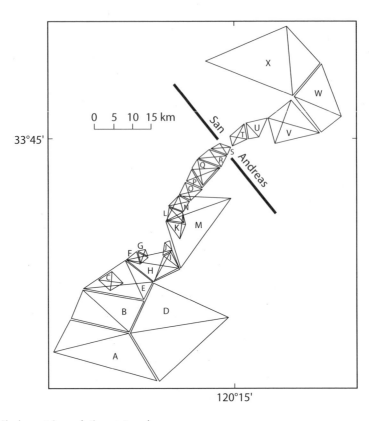

Figure 1.11. Cholame triangulation network.

the mechanics of continuous deformation. We will distinguish between body forces that act at all points in the earth, such as gravity, from surface forces. Surface forces are those that act either on actual surfaces within the earth, such as a fault or an igneous dike, or forces that one part of the earth exerts on an adjoining part. The *traction vector* **T** is defined as the limit of the surface force per unit area dF acting on a surface element dA, with unit normal **n** as the size of the area element tends to zero (figure 1.13). The traction thus depends not only on the forces acting within the body, but also on the orientation of the surface elements. The traction components acting on three mutually orthogonal surfaces (nine components in all) populate a second-rank *stress tensor*, σ_{ij}:

$$\sigma = \begin{bmatrix} \sigma_{11} & \sigma_{12} & \sigma_{13} \\ \sigma_{21} & \sigma_{22} & \sigma_{23} \\ \sigma_{31} & \sigma_{32} & \sigma_{33} \end{bmatrix}.$$

We will adopt here the convention that the first subscript refers to the direction of the force, whereas the second subscript refers to the direction of the surface normal (figure 1.14). Also, we will adopt the sign convention that a positive stress is one in which the traction vector acts in a positive direction when the outward-pointing normal to the surface points in a positive direction, and the traction vector points in a negative direction when the unit normal points in a negative direction. Thus, unless otherwise specified, tensile stresses are positive.

Note that this definition of stress makes no mention of the undeformed coordinate system, so this stress tensor, known as the *Cauchy stress*, is a Eulerian description. Lagrangian stress

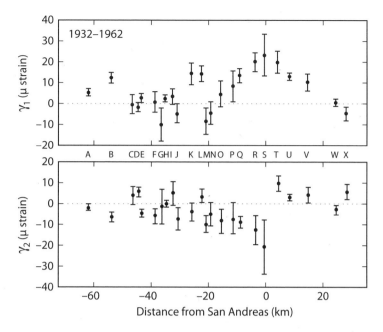

Figure 1.12. Shear strain changes in the Cholame network between 1932 and 1962. After Savage and Burford (1970). The positive values of γ_1 centered on the fault, at quadrilateral S, result from the 1934 Parkfield earthquake northwest of the triangulation network. Slip in this earthquake also caused fault-parallel compression on the east side of the network (positive γ_2) and fault-parallel extension on the west side (negative γ_2).

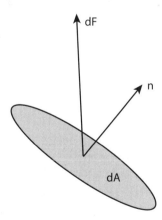

Figure 1.13. Definition of the traction vector. Force dF acts on surface dA with unit normal n. The traction vector is defined as the limit as dA tends to zero of the ratio dF/dA.

tensors, known as *Piola-Kirchoff stresses*, can be defined where the surface element and unit normal are measured relative to the undeformed state (e.g., Malvern 1969). It is shown in all continuum mechanics texts, and thus will not be repeated here, that in the absence of distributed couples acting on surface elements or within the body of the material, the balance of moments requires that the Cauchy stress tensor be symmetric, $\sigma_{ij} = \sigma_{ji}$. This is not necessarily true for other stress tensors—in particular, the first Piola-Kirchoff stress is not symmetric.

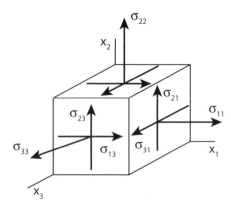

Figure 1.14. Stresses acting on the faces of a cubical element. All of the stress components shown are positive—that is, the traction vector and outward-pointing surface normal both point in positive coordinate directions.

It is also shown in standard texts, by considering a tetrahedron with three surfaces normal to the coordinate axes and one normal to the unit vector $\mathbf{n}$, that equilibrium requires that

$$T_i = \sigma_{ij} n_j \qquad \mathbf{T} = \sigma \cdot \mathbf{n}. \tag{1.46}$$

This is known as *Cauchy's formula*; it states that the traction vector $\mathbf{T}$ acting across a surface is found by the dot product of the stress and the unit normal.

The mean normal stress is equal to minus the pressure $\sigma_{kk}/3 = -p$, the negative sign arising because of our sign convention. The mean stress is a *stress invariant*—that is, $\sigma_{kk}/3$ is independent of the coordinate system. It is often useful to subtract the mean stress to isolate the so-called *deviatoric stress*:

$$\tau_{ij} = \sigma_{ij} - \frac{\sigma_{kk}}{3} \delta_{ij}. \tag{1.47}$$

There are, in fact, three stress invariants. The second invariant is given by $\mathrm{II} = \frac{1}{2}(\sigma_{ij}\sigma_{ij} - \sigma_{ii}\sigma_{jj})$. The second invariant of the deviatoric stress, often written as $J_2 = \frac{1}{2}(\tau_{ij}\tau_{ij})$ measures the squared maximum shear stress. It can be shown that

$$J_2 = \frac{1}{6}\left[(\sigma_{11} - \sigma_{22})^2 + (\sigma_{22} - \sigma_{33})^2 + (\sigma_{11} - \sigma_{33})^2\right] + \sigma_{12}^2 + \sigma_{23}^2 + \sigma_{13}^2. \tag{1.48}$$

The third invariant is given by the determinant of the stress tensor, although it is not widely used.

1.4 Coordinate Transformations

It is often necessary to transform stress or strain tensors from one coordinate system to another—for example, between a local North, East, Up system and a coordinate system that is oriented parallel and perpendicular to a major fault. To begin, we will consider the familiar rotation of a vector to a new coordinate system. For example, a rotation about the x_3 axis from

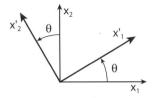

Figure 1.15. Coordinate rotation about the x_3 axis.

the original (x_1, x_2) system to the new (x'_1, x'_2) system (figure 1.15) can be written in matrix form as

$$\begin{bmatrix} v'_1 \\ v'_2 \\ v'_3 \end{bmatrix} = \begin{bmatrix} \cos\theta & \sin\theta & 0 \\ -\sin\theta & \cos\theta & 0 \\ 0 & 0 & 1 \end{bmatrix} \begin{bmatrix} v_1 \\ v_2 \\ v_3 \end{bmatrix}, \tag{1.49}$$

or compactly as $[\mathbf{v}]' = A[\mathbf{v}]$, where $[\mathbf{v}]'$ and $[\mathbf{v}]$ represent the vector components with respect to the primed and unprimed coordinates, respectively. In indicial notation,

$$v'_i = a_{ij} v_j. \tag{1.50}$$

The elements a_{ij} are referred to as *direction cosines*. It is easy to verify that the inverse transformation is given by $[\mathbf{v}] = A^T [\mathbf{v}]'$:

$$v_i = a_{ji} v'_j. \tag{1.51}$$

The matrix rotating the components of the vector from the primed to the unprimed coordinates is the inverse of the matrix in equation (1.49). Clearly, $a_{ij} = a_{ji}^T$, but also comparing equations (1.50) and (1.51), $a_{ij} = a_{ji}^{-1}$. Therefore, $a_{ji}^T = a_{ji}^{-1}$, so that the transformation is *orthogonal*.

To determine the a_{ij}, let $\mathbf{e}_i$ represent basis vectors in the original coordinate system and $\mathbf{e}'_i$ represent basis vectors in the new coordinates. One can express a vector in either system:

$$\mathbf{x} = x_j \mathbf{e}_j = x'_j \mathbf{e}'_j. \tag{1.52}$$

Taking the dot product of both sides with respect to $\mathbf{e}_i$, we have

$$x_j (\mathbf{e}_j \cdot \mathbf{e}_i) = x'_j (\mathbf{e}'_j \cdot \mathbf{e}_i),$$
$$x_j \delta_{ij} = x'_j (\mathbf{e}'_j \cdot \mathbf{e}_i),$$
$$x_i = x'_j (\mathbf{e}'_j \cdot \mathbf{e}_i), \tag{1.53}$$

so that $a_{ji} = (\mathbf{e}'_j \cdot \mathbf{e}_i)$. A similar derivation gives $a_{ij} = (\mathbf{e}_j \cdot \mathbf{e}'_i)$. For example, for a rotation about the x_3 axis, we have

$$a_{11} = \mathbf{e}_1 \cdot \mathbf{e}'_1 = \cos(\theta),$$
$$a_{12} = \mathbf{e}_2 \cdot \mathbf{e}'_1 = \sin(\theta),$$
$$a_{21} = \mathbf{e}_1 \cdot \mathbf{e}'_2 = -\sin(\theta),$$
$$a_{22} = \mathbf{e}_2 \cdot \mathbf{e}'_2 = \cos(\theta). \tag{1.54}$$

We can now show that second-rank tensors transform by

$$T'_{ij} = a_{ik}a_{jl}T_{kl},$$ (1.55)

$$T_{ij} = a_{ki}a_{lj}T'_{kl}.$$ (1.56)

To do so, consider the second-rank tensor T_{kl} that relates two vectors u and v:

$$v_k = T_{kl}u_l.$$ (1.57)

For example, u could be pressure gradient, v could be fluid flux, and T would be the permeability tensor. The components of the two vectors transform according to

$$v'_i = a_{ik}v_k,$$ (1.58)

and

$$u_l = a_{jl}u'_j.$$ (1.59)

Substituting equations (1.58) and (1.59), into (1.57), we have

$$v'_i = a_{ik}v_k = a_{ik}T_{kl}u_l,$$
$$v'_i = a_{ik}T_{kl}a_{jl}u'_j,$$
$$v'_i = a_{ik}a_{jl}T_{kl}u'_j.$$ (1.60)

This can be written as $v'_i = T'_{ij}u'_j$, if we define T in the rotated coordinate system as

$$\boxed{T'_{ij} \equiv a_{ik}a_{jl}T_{kl},}$$ (1.61)

or in matrix notation as

$$[T]' = A[T]A^T,$$ (1.62)

where $[T]'$ refers to the matrix of the tensor with components taken relative to the primed coordinates, whereas $[T]$ are the components with respect to the unprimed coordinates. Note that all second-rank tensors, including stress, transform according to the same rules. The derivation can be extended to show that higher rank tensors transform according to

$$\Gamma'_{ijkl\,...} = a_{i\alpha}a_{j\beta}a_{k\gamma}\,...\,\Gamma_{\alpha\beta\gamma\,...}\,.$$ (1.63)

1.5 Principal Strains and Stresses

The strain tensor can be put into diagonal form by rotating into proper coordinates in which the shear strains vanish:

$$\begin{bmatrix} \epsilon_1 & 0 & 0 \\ 0 & \epsilon_2 & 0 \\ 0 & 0 & \epsilon_3 \end{bmatrix}.$$ (1.64)

The strain tensor itself has six independent components. In order to determine the strain, we must specify not only the three principal values, $\epsilon^{(k)}$, $k = 1, 2, 3$, but also the three principal strain directions, so that it still takes six parameters to describe the state of strain. The principal directions are mutually orthogonal (see the following). Thus, while it requires two angles to specify the first axis, the second axis must lie in the perpendicular plane, requiring only one additional angle. The third axis is uniquely determined by orthogonality and the right-hand sign convention.

The principal strains are also the extrema in the linear extension $\Delta l / l_0$. For simplicity, we show this for the case of irrotational deformation, $\omega = 0$; however, see problem 5. In the small strain limit, the linear extension is proportional to the dot product of the relative displacement vector and a unit vector parallel to the baseline:

$$\frac{\Delta l}{l_0} = \frac{\mathrm{d}u_i \mathrm{d}x_i}{\mathrm{d}x_k \mathrm{d}x_k} = \frac{\mathbf{du} \cdot \mathbf{dx}}{\mathbf{dx} \cdot \mathbf{dx}} \tag{1.65}$$

(equation [1.18]). The maximum and minimum extensions occur when $\mathbf{du}$ and $\mathbf{dx}$ are parallel—that is, $\mathbf{du} = \epsilon \mathbf{dx}$, where ϵ is a constant, or

$$\mathrm{d}u_i = \epsilon \mathrm{d}x_i = \epsilon \delta_{ij} \mathrm{d}x_j. \tag{1.66}$$

From equation (1.6), the relative displacement of adjacent points in the absence of rotation is $\mathrm{d}u_i = \epsilon_{ij} \mathrm{d}x_j$. Equating with (1.66) yields

$$\epsilon_{ij} \mathrm{d}x_j = \epsilon \delta_{ij} \mathrm{d}x_j, \tag{1.67}$$

or rearranging,

$$\left(\epsilon_{ij} - \epsilon \delta_{ij} \right) \mathrm{d}x_j = 0. \tag{1.68}$$

Thus, the principal strains ϵ in equation (1.68) are the eigenvalues of the matrix ϵ_{ij}, and the $\mathrm{d}x_j$ are the eigenvectors, or principal strain directions. Similarly, the principal stresses and the principal stress directions are the eigenvalues and eigenvectors of the matrix σ_{ij}.

To determine the principal strains and strain directions, we must find the eigenvalues and the associated eigenvectors. Note that the equations (1.68) have nontrivial solutions if and only if

$$\det \left(\epsilon_{ij} - \epsilon \delta_{ij} \right) = 0. \tag{1.69}$$

In general, this leads to a cubic equation in ϵ. The strain tensor is symmetric, which means that all three eigenvalues are real (e.g., Strang 1976). They may, however, be either positive, negative, or zero; some eigenvalues may be repeated. For the distinct (nonrepeated) eigenvalues, the corresponding eigenvectors are mutually orthogonal. With repeated eigenvalues, there exists sufficient arbitrariness in choosing corresponding eigenvectors that they can be chosen to be orthogonal.

For simplicity in presentation, we will examine the two-dimensional case, which reduces to

$$\det \begin{pmatrix} \epsilon_{11} - \epsilon & \epsilon_{12} \\ \epsilon_{21} & \epsilon_{22} - \epsilon \end{pmatrix} = 0, \tag{1.70}$$

which has the following two roots given by solution of the preceding quadratic:

$$\epsilon = \left(\frac{\epsilon_{11} + \epsilon_{22}}{2} \right) \pm \sqrt{\left(\frac{\epsilon_{11} - \epsilon_{22}}{2} \right)^2 + \epsilon_{12}^2}. \tag{1.71}$$

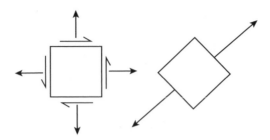

Figure 1.16. Left: Two-dimensional strain state given by equation (1.72), where the arrows denote relative displacement of the two sides of the reference square. Right: Same strain state in the principal coordinate system.

The principal strain directions are found by substituting the principal values back into equation (1.68) and solving for the $dx^{(k)}$. As an example, consider the simple two-dimensional strain tensor illustrated in figure 1.16:

$$\epsilon_{ij} = \begin{bmatrix} 1 & 1 \\ 1 & 1 \end{bmatrix}. \tag{1.72}$$

From equation (1.71), we find that $\epsilon^{(1)} = 2$ and $\epsilon^{(2)} = 0$. Now to find the principal strain direction corresponding to the maximum principal strain, we substitute $\epsilon^{(1)} = 2$ into equation (1.68), yielding

$$\begin{pmatrix} 1-2 & 1 \\ 1 & 1-2 \end{pmatrix} \begin{pmatrix} dx_1 \\ dx_2 \end{pmatrix} = 0. \tag{1.73}$$

These are two redundant equations that relate dx_1 and dx_2:

$$-dx_1 + dx_2 = 0, \tag{1.74}$$

$$dx_1 - dx_2 = 0. \tag{1.75}$$

The eigenvectors are usually normalized so that

$$\mathbf{dx}^{(1)} = \begin{pmatrix} \frac{1}{\sqrt{2}} \\ \frac{1}{\sqrt{2}} \end{pmatrix} \text{ or } \begin{pmatrix} -\frac{1}{\sqrt{2}} \\ -\frac{1}{\sqrt{2}} \end{pmatrix}. \tag{1.76}$$

For this particular example, the principal direction corresponding to the maximum extension bisects the x_1 and x_2 axes in the first and third quadrants (figure 1.16). We know that the second principal direction has to be orthogonal to the first, but we could find it also by the same procedure. Substituting $\epsilon^{(2)} = 0$ into equation (1.68) yields

$$\begin{pmatrix} 1-0 & 1 \\ 1 & 1-0 \end{pmatrix} \begin{pmatrix} dx_1 \\ dx_2 \end{pmatrix} = 0, \tag{1.77}$$

or $dx_1 = -dx_2$. Again, we normalize the eigenvector so that

$$\mathbf{dx}^{(2)} = \begin{pmatrix} \frac{1}{\sqrt{2}} \\ -\frac{1}{\sqrt{2}} \end{pmatrix} \text{ or } \begin{pmatrix} -\frac{1}{\sqrt{2}} \\ \frac{1}{\sqrt{2}} \end{pmatrix}, \tag{1.78}$$

showing that the second eigenvector, corresponding to the minimum extension, is also 45 degrees from the coordinate axes, but in the second and fourth quadrants.

1.6 Compatibility Equations

Recall the kinematic equations relating strain to displacement (1.4). Given the strains, these represent six equations in only three unknowns, the three displacement components. This implies that there must be relationships between the strains. These relations, known as *compatibility equations*, guarantee that the displacement field be single valued. To be more concrete, this means that the displacement field does not admit gaps opening or material interpenetrating.

In two dimensions, we have

$$\epsilon_{xx} = \frac{\partial u_x}{\partial x}, \tag{1.79}$$

$$\epsilon_{yy} = \frac{\partial u_y}{\partial y}, \tag{1.80}$$

$$\epsilon_{xy} = \frac{1}{2}\left(\frac{\partial u_x}{\partial y} + \frac{\partial u_y}{\partial x}\right). \tag{1.81}$$

Notice that

$$2\frac{\partial^2 \epsilon_{xy}}{\partial x \partial y} = \frac{\partial^3 u_x}{\partial y^2 \partial x} + \frac{\partial^3 u_y}{\partial x^2 \partial y} = \frac{\partial^2 \epsilon_{xx}}{\partial y^2} + \frac{\partial^2 \epsilon_{yy}}{\partial x^2}. \tag{1.82}$$

An example of a strain field that is not compatible is as follows. Assume that the only nonzero strain is $\epsilon_{xx} = Ay^2$, where A is a constant. Note that such a strain could not occur without a shear strain, nor in fact without gradients in shear strain. This follows from equation (1.82).

In three dimensions, the compatibility equations can be written compactly as:

$$\boxed{\epsilon_{ij,kl} + \epsilon_{kl,ij} - \epsilon_{ik,jl} - \epsilon_{jl,ik} = 0,} \tag{1.83}$$

where we have used the notation $_{,i} \equiv \partial/\partial x_i$. The symmetry of the strain tensor can be used to reduce the 81 equations (1.83) to 6 independent equations. Furthermore, there are three relations among the *incompatibility elements*, so there are only three truly independent compatibility relations. The interested reader is referred to Malvern (1969).

When do you have to worry about compatibility? If you solve the governing equations in terms of stress, then you must guarantee that the implied strain fields are compatible. If, on the other hand, the governing equations are written in terms of displacements, then compatibility is automatically satisfied—assuming that the displacements are continuous (true except perhaps along a fault or crack), one can always derive the strains by differentiation.

1.7 Conservation Laws

The behavior of a deforming body is governed by conservation laws that generalize the concepts of conservation of mass and momentum. Before detailing these, we will examine some important relations that will be used in subsequent derivations.

The *divergence theorem* relates the integral of the outward component of a vector $\mathbf{f}$ over a closed surface S to the volume integral of the divergence of the vector over the volume V

enclosed by S:

$$\int_S f_i n_i \, dS = \int_V \frac{\partial f_i}{\partial x_i} \, dV, \tag{1.84}$$

$$\int_S \mathbf{f} \cdot \mathbf{n} \, dS = \int_V \nabla \cdot \mathbf{f} \, dV. \tag{1.85}$$

The time rate of change of a quantity, f, is different in a fixed Eulerian frame from a moving Lagrangian frame. Any physical quantity must be the same, regardless of the system. This equivalence can be expressed as

$$f^L(\mathbf{x}, t) = f^E[\boldsymbol{\xi}(\mathbf{x}, t), t], \tag{1.86}$$

where the superscripts L and E indicate Lagrangian and Eulerian descriptions, respectively. Here f can be a scalar, vector, or tensor. Recall also that $\mathbf{x}$ refers to initial coordinates, whereas $\boldsymbol{\xi}$ refers to current coordinates. Differentiating with respect to time and using the chain rule yields

$$\frac{df^L}{dt} = \frac{\partial f^E}{\partial t} + \frac{\partial f^E}{\partial \xi_i} \frac{\partial \xi_i}{\partial t}. \tag{1.87}$$

The first term on the right-hand side refers to the time rate of change of f at a fixed point in space, whereas the second term refers to the convective rate of change as particles move to a place having different properties. The Lagrangian derivative is often written, where using an Eulerian description, as Df/Dt, and is also known as the *material time derivative*:

$$\frac{Df}{Dt} = \frac{\partial f^E}{\partial t} + \mathbf{v} \cdot \nabla f^E. \tag{1.88}$$

A linearized, time-integrated form of equation (1.88) valid for small displacements:

$$\delta f^L = \delta f^E + \mathbf{u} \cdot \nabla f^E, \tag{1.89}$$

relates the Lagrangian description of a perturbation in f to the Eulerian description (Dahlen and Tromp 1998, equation 3.16). The spatial derivative in equation (1.89) is with respect to the current coordinates, although this distinction can be dropped in a linearized analysis.

We will also state here, without proof, the *Reynolds transport theorem*, which gives the material time derivative of a volume integral—that is, following a set of points in volume V—as the region moves and distorts. The theorem states (e.g., Malvern 1969) that for a property $\mathcal{A}$,

$$\frac{D}{Dt} \int_V \rho \mathcal{A} dV = \int_V \rho \frac{D\mathcal{A}}{Dt} dV, \tag{1.90}$$

where ρ is the Eulerian mass density.

We are now ready to state the important conservation laws of continuum mechanics. Mass must be conserved during deformation. The time rate of change of mass in a current element with volume V, $\int_V (\partial \rho / \partial t) dV$, must be balanced by the flow of material out of the element

$\int_S \rho \mathbf{v} \cdot \mathbf{n} dS$, where S is the surface bounding V. Making use of the divergence theorem (1.85):

$$\int_V \frac{\partial \rho}{\partial t} dV + \int_S \rho \mathbf{v} \cdot \mathbf{n} dS = 0,$$

$$\int_V \left[\frac{\partial \rho}{\partial t} + \nabla \cdot \rho \mathbf{v} \right] dV = 0. \tag{1.91}$$

This must hold for every choice of volume element, which yields the *continuity equation*:

$$\boxed{\frac{\partial \rho}{\partial t} + \frac{\partial \rho v_i}{\partial x_i} = 0.} \tag{1.92}$$

Note that for an incompressible material with uniform density, the continuity equation reduces to $\nabla \cdot \mathbf{v} = 0$.

Conservation of linear momentum for the current volume element V surrounded by surface S requires that the rate of change of momentum be balanced by the sum of surface forces and body forces:

$$\frac{D}{Dt} \int_V \rho \mathbf{v} dV = \int_S \mathbf{T} dS + \int_V \rho \mathbf{f} dV. \tag{1.93}$$

Here, $\mathbf{T}$ is the traction acting on the surface S, and $\mathbf{f}$ are body forces per unit mass, such as gravity or electromagnetic forces. The traction is related to the stress via $\mathbf{T} = \sigma \cdot n$ (equation [1.46]). Application of the divergence theorem (1.85) yields

$$\frac{D}{Dt} \int_V \rho \mathbf{v} dV = \int_V [\nabla \cdot \sigma + \rho \mathbf{f}] dV. \tag{1.94}$$

Application of the Reynolds transport theorem (1.90) yields

$$\int_V \left[\nabla \cdot \sigma + \rho \mathbf{f} - \rho \frac{D\mathbf{v}}{Dt} \right] dV = 0. \tag{1.95}$$

Again, this must hold for all volume elements so that

$$\boxed{\nabla \cdot \sigma + \rho \mathbf{f} = \rho \frac{D\mathbf{v}}{Dt}.} \tag{1.96}$$

For small deformations, $D\mathbf{v}/Dt = \ddot{\mathbf{u}}$. Furthermore, if accelerations can be neglected, the equations of motion (1.96) reduce to the *quasi-static* equilibrium equations:

$$\boxed{\begin{aligned} &\frac{\partial \sigma_{ij}}{\partial x_j} + \rho f_i = 0, \\ &\nabla \cdot \sigma + \rho \mathbf{f} = 0. \end{aligned}} \tag{1.97}$$

Note that this is an Eulerian description. In chapter 9, we will write Lagrangian equations of motion for perturbations in stress and displacement relative to a gravitationally equilibrated reference state. In this formulation, an advective stress term appears, which from equation (1.89) is of the form $\mathbf{u} \cdot \nabla \sigma^0$, where σ^0 is the stress in the reference state.

1.7.1 Equilibrium Equations in Curvilinear Coordinates

In cylindrical polar coordinates,

$$\frac{1}{r}\frac{\partial}{\partial r}(r\sigma_{rr}) + \frac{1}{r}\frac{\partial \sigma_{r\theta}}{\partial \theta} + \frac{\partial \sigma_{rz}}{\partial z} - \frac{\sigma_{\theta\theta}}{r} = 0,$$

$$\frac{1}{r^2}\frac{\partial}{\partial r}(r^2\sigma_{r\theta}) + \frac{1}{r}\frac{\partial \sigma_{\theta\theta}}{\partial \theta} + \frac{\partial \sigma_{\theta z}}{\partial z} = 0,$$

$$\frac{1}{r}\frac{\partial}{\partial r}(r\sigma_{zr}) + \frac{1}{r}\frac{\partial \sigma_{\theta z}}{\partial \theta} + \frac{\partial \sigma_{zz}}{\partial z} = 0. \tag{1.98}$$

In spherical coordinates,

$$\frac{\partial \sigma_{rr}}{\partial r} + \frac{1}{r}\frac{\partial \sigma_{r\theta}}{\partial \theta} + \frac{1}{r\sin\theta}\frac{\partial \sigma_{r\phi}}{\partial \phi} + \frac{1}{r}(2\sigma_{rr} - \sigma_{\theta\theta} - \sigma_{\phi\phi} + \sigma_{r\theta}\cot\phi) = 0,$$

$$\frac{\partial \sigma_{r\theta}}{\partial r} + \frac{1}{r}\frac{\partial \sigma_{\theta\theta}}{\partial \theta} + \frac{1}{r\sin\theta}\frac{\partial \sigma_{\theta\phi}}{\partial \phi} + \frac{1}{r}[3\sigma_{r\theta} + (\sigma_{\theta\theta} - \sigma_{\phi\phi})\cot\theta] = 0,$$

$$\frac{\partial \sigma_{r\phi}}{\partial r} + \frac{1}{r}\frac{\partial \sigma_{\theta\phi}}{\partial \theta} + \frac{1}{r\sin\theta}\frac{\partial \sigma_{\phi\phi}}{\partial \phi} + \frac{1}{r}(3\sigma_{r\phi} + 2\sigma_{\theta\phi}\cot\theta) = 0. \tag{1.99}$$

All equations are shown without body forces (Malvern 1969).

1.8 Constitutive Laws

It is worth summarizing the variables and governing equations discussed to this point. There are three displacement components, u_i; six strains, ϵ_{ij}; six stresses, σ_{ij}; and density, ρ, for a total of 16 variables (all of which are generally functions of position and time). In terms of the governing equations, we have the six kinematic equations that relate strain and displacement (1.4), the three equilibrium equations (1.97), and the single continuity equation (1.92), for a total of 10. At this point, we have six more unknowns than equations, which makes sense because we have not yet defined the material behavior. The so-called *constitutive equations* provide the remaining six equations relating stress and strain (possibly including time derivatives) so that we have a closed system of equations. It should be noted that this discussion is limited to a purely mechanical theory. A more complete, thermomechanical description, including temperature, entropy, and conservation of energy, is discussed briefly in chapter 10.

For a linear elastic material, the strain is proportional to stress:

$$\boxed{\sigma_{ij} = C_{ijkl}\epsilon_{kl},} \tag{1.100}$$

where C_{ijkl} is the symmetric fourth-rank stiffness tensor. Equation (1.100) is referred to as generalized *Hooke's law*. The symmetry of the strain tensor makes it irrelevant to consider cases other than for which the C_{ijkl} are symmetric in k and l. The stress is also symmetric, so the C_{ijkl} must be symmetric in i and j. Last, the existence of a strain energy function requires $C_{ijkl} = C_{klij}$, as follows. The strain energy per unit volume in the undeformed state is defined as

$$dW = \sigma_{ij} d\epsilon_{ij}, \tag{1.101}$$

so that $\sigma_{ij} = \partial W / \partial \epsilon_{ij}$. From the mixed partial derivatives $\partial^2 W / \partial \epsilon_{ij} \partial \epsilon_{kl}$, it follows that $\partial \sigma_{ij} / \partial \epsilon_{kl} = \partial \sigma_{kl} / \partial \epsilon_{ij}$. Making use of equation (1.100) leads directly to $C_{ijkl} = C_{klij}$.

It can be shown that for an isotropic medium—that is, one in which the mechanical properties do not vary with direction—the most general fourth-rank tensor exhibiting the preceding symmetries is

$$C_{ijkl} = \lambda \delta_{ij} \delta_{kl} + \mu(\delta_{ik} \delta_{jl} + \delta_{il} \delta_{jk}) \tag{1.102}$$

(e.g., Malvern 1969). Substituting equation (1.102) into (1.100) yields the isotropic form of Hooke's law:

$$\boxed{\sigma_{ij} = 2\mu \epsilon_{ij} + \lambda \epsilon_{kk} \delta_{ij},} \tag{1.103}$$

where μ and λ are the Lame constants, μ being the shear modulus that relates shear stress to shear strain.

Other elastic constants, which are convenient for particular applications, can be defined. For example, the bulk modulus, K, relates mean normal stress $\sigma_{kk}/3$ to volumetric strain ϵ_{kk}:

$$\sigma_{kk} = 3K \epsilon_{kk}. \tag{1.104}$$

Alternatively, equation (1.103) can be written in terms of Young's modulus E and Poisson's ratio v as

$$E \epsilon_{ij} = (1 + v)\sigma_{ij} - v\sigma_{kk} \delta_{ij}. \tag{1.105}$$

Young's modulus relates stress and strain for the special case of uniaxial stress. Applying stress in one direction generally leads to strain in an orthogonal direction. Poisson's ratio measures the ratio of strain in the orthogonal direction to that in the direction the stress is applied. Notice that while there are five widely used elastic constants μ, λ, K, v, and E, only two are independent. Sometimes the symbol G is used for shear modulus, but we reserve this for the universal gravitational constant. There are a host of relationships between the five elastic constants—some prove particularly useful:

$$K = \frac{2\mu(1 + v)}{3(1 - 2v)} = \lambda + \frac{2}{3}\mu,$$

$$\lambda = \frac{2\mu v}{(1 - 2v)},$$

$$E = 2\mu(1 + v),$$

$$\frac{\mu}{\lambda + \mu} = 1 - 2v,$$

$$\frac{\lambda}{\lambda + 2\mu} = \frac{v}{1 - v}. \tag{1.106}$$

The existence of a positive definite strain energy requires that both μ and K are nonnegative. This implies, from the first of equations (1.106), that ν is bounded by $-1 \le \nu \le 0.5$. For $\nu = 0.5$, $1/K = 0$, and the material is incompressible.

It is possible to write the compatibility equations in terms of stresses using Hooke's law. Inverting equations (1.103) to write strain in terms of stress and substituting into equation (1.83) yields

$$\sigma_{ij,kl} + \sigma_{kl,ij} - \sigma_{ik,jl} - \sigma_{jl,ik} = \frac{\nu}{1+\nu} \left(\sigma_{nn,kl}\delta_{ij} + \sigma_{nn,ij}\delta_{kl} - \sigma_{nn,jl}\delta_{ik} - \sigma_{nn,ik}\delta_{jl} \right) . \tag{1.107}$$

As before, only six of these equations are independent. Setting $k = l$ and summing,

$$\sigma_{ij,kk} + \sigma_{kk,ij} - \sigma_{ik,jk} - \sigma_{jk,ik} = \frac{\nu}{1+\nu} \left(\sigma_{nn,kk}\delta_{ij} + \sigma_{nn,ij} \right) . \tag{1.108}$$

This result can be further simplified using the equilibrium equations (1.97), yielding the *Beltrami-Michell compatibility equations*:

$$\sigma_{ij,kk} + \frac{1}{1+\nu}\sigma_{kk,ij} - \frac{\nu}{1+\nu}\sigma_{kk,mn}\delta_{ij} = -(\rho f_i)_{,j} - (\rho f_j)_{,i}. \tag{1.109}$$

A useful relationship can be obtained by further contracting equation (1.109), setting $i = j$ and summing, which gives

$$\nabla^2 \sigma_{kk} = - \left(\frac{1+\nu}{1-\nu} \right) \nabla \cdot (\rho \mathbf{f}). \tag{1.110}$$

There will be occasion to discuss fluidlike behavior of rocks at high temperature within the earth. The constitutive equations for a linear Newtonian fluid are

$$\sigma_{ij} = 2\eta D'_{ij} - p\delta_{ij}, \tag{1.111}$$

where η is the viscosity, p is the fluid pressure, and D'_{ij} are the components of the deviatoric rate of deformation, defined by equation (1.20) and

$$D'_{ij} = D_{ij} - \frac{D_{kk}}{3}\delta_{ij}. \tag{1.112}$$

Equation (1.111) is appropriate for fluids that may undergo arbitrarily large strains. In chapter 6, we will consider viscoelastic materials and implicitly make the approximation that displacement gradients remain small so that the rate of deformation tensor can be approximated by the strain rate, $D_{ij} \approx \dot{\epsilon}_{ij}$. Constitutive equations for viscoelastic and poroelastic materials will be introduced in chapters 6 and 10.

We will now return to the question of the number of equations and the number of unknowns for an elastic medium. If it is desirable to explicitly consider the displacements—for example, if the boundary conditions are given in terms of displacement—then the kinematic equations (1.4) can be used to eliminate the strains from Hooke's law, leaving 10 unknowns (three displacement, six stresses, and density) and 10 equations (three equilibrium equations, six constitutive equations, and continuity). If, on the other hand, the boundary conditions are given in terms of tractions, it may not be necessary to consider the displacements explicitly. In this case, there are 13 variables (six stresses, six strains, and density). The governing equations are: equilibrium (three), the constitutive equations (six), continuity (one), and compatibility (three), leaving 13 equations and 13 unknowns.

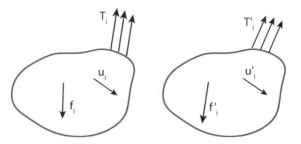

Figure 1.17. Two sets of body forces and surface tractions used to illustrate the reciprocal theorem. Left: Unprimed system. Right: Primed system.

In general, solving 13 coupled partial differential equations can be a formidable task. Fortunately, we can begin—in the next chapter—with a far simpler two-dimensional problem that does a reasonably good job of describing deformation adjacent to long strike-slip faults.

1.9 Reciprocal Theorem

The *Betti-Rayleigh reciprocal theorem* is an important tool from which many useful results can be derived. As one example, in chapter 3, the reciprocal theorem is employed to derive *Volterra's formula*, which gives the displacements due to slip on a general planar discontinuity in an elastic medium. We will present a derivation of the reciprocal theorem here, following, for example, Sokolnikoff (1983). Imagine two sets of self-equilibrating body forces and surface tractions, f_i, T_i and f_i', T_i', acting on the same body (figure 1.17). The unprimed forces and tractions generate displacements, u_i, whereas the primed generate u_i', both unique to within rigid body motions. We will consider the work done by forces and tractions in the unprimed state acting through the displacements in the primed state:

$$\int_A T_i u_i' \mathrm{d}A + \int_V f_i u_i' \mathrm{d}V = \int_A T_i u_i' \mathrm{d}A - \int_V \sigma_{ij,j} u_i' \mathrm{d}V, \tag{1.113}$$

where we have rewritten the volume integral making use of the quasi-static equilibrium equations (1.97). Use of the divergence theorem on the first integral on the right-hand side leads to

$$\int_A T_i u_i' \mathrm{d}A + \int_V f_i u_i' \mathrm{d}V = \int_V (\sigma_{ij} u_i')_{,j} \mathrm{d}V - \int_V \sigma_{ij,j} u_i' \mathrm{d}V,$$

$$= \int_V \sigma_{ij} u_{i,j}' \mathrm{d}V,$$

$$= \int_V C_{ijkl} u_{k,l} u_{i,j}' \mathrm{d}V, \tag{1.114}$$

the last step resulting from application of Hooke's law (1.100). You have seen that the C_{ijkl} have the symmetry $C_{ijkl} = C_{klij}$, implying that the left-hand side of equation (1.114) is symmetric with respect to primed and unprimed coordinates. In other words, we can interchange the primed and unprimed states:

$$\int_A T_i u_i' \mathrm{d}A + \int_V f_i u_i' \mathrm{d}V = \int_A T_i' u_i \mathrm{d}A + \int_V f_i' u_i \mathrm{d}V. \tag{1.115}$$

Equation (1.115) is one form of the reciprocal theorem. It states that the work done by the primed body forces and tractions acting through the unprimed displacements is equal to the work done by the unprimed body forces and tractions acting through the primed displacements.

1.10 Problems

1. (a) Derive expression (1.12) for the Almansi strain.
   (b) Show that in terms of the displacement u_i, the Almansi strain tensor is given by equation (1.17).

2. Consider a line segment $\widehat{\mathbf{dx}}$ that lies in the (x_1, x_2) plane. Show that the elongation is given by

$$\frac{\Delta l}{l} = \epsilon_{11}\cos^2\theta + 2\epsilon_{12}\cos\theta\sin\theta + \epsilon_{22}\sin^2\theta,$$

where θ is the orientation of the baseline with respect to the x_1 axis. Given the strain

$$\begin{bmatrix} \epsilon_{11} & \epsilon_{12} \\ \epsilon_{21} & \epsilon_{22} \end{bmatrix} = \begin{bmatrix} 0.1 & 0.1 \\ 0.1 & 0.1 \end{bmatrix},$$

graph $\Delta l/l$ between 0 and π. What is the meaning of the maximum and minimum?

3. Prove that a pure rotation involves no extension in any direction.
   Hint: Replace ω_{ij} for ϵ_{ij} in equation (1.18), and note the antisymmetry of ω_{ij}.

4. Consider the rotation of a line segment due to strain. Derive an expression for Θ_3 in terms of the strain components. For simplicity, take $\mathbf{dx}$ to lie in the (x_1, x_2) plane:

$$\mathbf{dx}^\mathrm{T} = (dx_1, dx_2, 0).$$

Draw diagrams showing that the sign of the rotation you compute for each strain component makes sense physically.

5. Show formally that the principal strains are the extrema in the linear extension. Recall from equation (1.18) that for small deformations,

$$\frac{\Delta l}{l_0} = \frac{\epsilon_{ij}x_i x_j}{x_k x_k}.$$

To find the extrema, set the gradient of $\Delta l/l_0$ with respect to x_n to zero. First show that the gradient is given by

$$\frac{\partial(\Delta l/l_0)}{\partial x_n} = \frac{2\epsilon_{in}x_i}{x_k x_k} - \left(\frac{\epsilon_{ij}x_i x_j}{x_k x_k}\right)\frac{2x_n}{x_m x_m}. \tag{1.116}$$

Notice that the quantity in parentheses, $\epsilon_{ij}x_i x_j/x_k x_k$, is a scalar. It is the value of $\Delta l/l_0$ at the extremum, which we denote ϵ. Show that the condition that the gradient (1.116) vanish leads to

$$(\epsilon_{in} - \epsilon\delta_{in})x_i = 0,$$

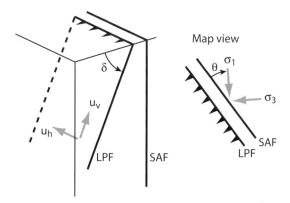

Figure 1.18. Loma Prieta fault geometry. SAF = San Andreas fault; LPF = Loma Prieta fault.

thus demonstrating that the maximum and minimum elongations are the eigenvalues of the strain.

6. In two dimensions, show that the principal strain directions θ are given by

$$\tan 2\theta = \frac{2\epsilon_{12}}{\epsilon_{11} - \epsilon_{22}},$$

where θ is measured counterclockwise from the x_1 axis.
Hint: if you use an eigenvector approach, the following identity may be useful:

$$\tan 2\theta = \frac{2 \tan \theta}{1 - \tan^2 \theta}.$$

However, other methods also lead to the same result.

7. This problem explores Hooke's law for two-dimensional deformation. For *plane strain* deformation, all strains in the 3-direction are assumed to be zero, $\epsilon_{13} = \epsilon_{23} = \epsilon_{33} = 0$.

   (a) Show that for this case, $\sigma_{33} = \nu(\sigma_{11} + \sigma_{22})$.

   (b) Show that for plane strain, Hooke's law (1.103) for an isotropic medium reduces to

$$2\mu\epsilon_{\alpha\beta} = \sigma_{\alpha\beta} - \nu(\sigma_{11} + \sigma_{22})\delta_{\alpha\beta} \qquad \alpha, \beta = 1, 2. \tag{1.117}$$

   (c) Show that for plane stress conditions $\sigma_{13} = \sigma_{23} = \sigma_{33} = 0$, Hooke's law reduces to the same form as in (b), but with ν replaced by $\nu/(1 + \nu)$.

8. The M 6.9 1989 Loma Prieta, California, earthquake occurred with oblique slip on a fault that strikes parallel to the San Andreas fault (SAF) and dips to the southwest approximately 70 degrees (figure 1.18). We believe that the SAF last slipped in a purely right-lateral event, in 1906, on a vertical fault plane. The problem is to find whether a uniform stress state could drive pure, right-lateral, strike slip on a vertical fault *and* the observed oblique slip on the parallel Loma Prieta fault (LPF).

   The important points to note are that the SAF is vertical; the LPF dips to the southwest ($\delta \approx 70$ degrees); slip in 1906 was pure, right-lateral, strike slip; and slip in 1989 was $u_h = \beta u_v$, where u_h is horizontal displacement, u_v is vertical displacement, and $\beta \approx 1.4$.

Assume that the fault slips in the direction of the shear stress acting on the fault plane, so that $\tau_h = \beta \tau_v$, where τ_h and τ_v are, respectively, the horizontal and vertical components of the resolved shear stress on the LPF. You can assume that the intermediate principal stress, σ_2, is vertical.

Hint: Solve for τ_h and τ_v in terms of the principal stresses. The constraint that $\tau_h = \beta \tau_v$ will allow you to find a family of solutions (stress states) parameterized by

$$\phi \equiv \frac{\sigma_2 - \sigma_3}{\sigma_1 - \sigma_3}.$$

Note that

$$\phi \to 1 \quad \text{as} \quad \sigma_2 \to \sigma_1,$$
$$\phi \to 0 \quad \text{as} \quad \sigma_2 \to \sigma_3,$$

and thus $0 < \phi < 1$. Express the family of admissible stress states in terms of ϕ as a function of θ.

9. Use the strain-displacement equations (1.4) and the isotropic form of Hooke's law (1.103) to show that the quasi-static equilibrium equations (1.97) can be written as

$$\mu \nabla^2 u_i + (\lambda + \mu) \frac{\partial u_{k,k}}{\partial x_i} + \rho f_i = 0. \tag{1.118}$$

These equations are known as the *Navier form* of the equilibrium equations in elasticity theory.

10. Consider a solid elastic body with volume V and surface S with unit normal n_i. Tractions T_i are applied to the S; in general, they are nonuniform and may be zero in places. We want to find the volume change

$$\Delta V = \int_S u_i n_i \, \mathrm{d}s,$$

where u_i is the displacement. Use the reciprocal theorem to show that the volume change can be computed by

$$\Delta V = \frac{1}{3K} \int_S T_i x_i \, \mathrm{d}s,$$

where K is the bulk modulus.

1.11 References

Agnew, D. C. 1985. Strainmeters and tiltmeters. *Reviews of Geophysics* **24**(3), 579–624.

Dahlen, F. A., and J. Tromp. 1998. *Theoretical global seismology.* Princeton, NJ: Princeton University Press.

Frank, F. C. 1966. Deduction of earth strains from survey data. *Bulletin of the Seismological Society of America* **56**, 35–42.

Johnson, H. O., D. C. Agnew, and F. K. Wyatt. 1994. Present-day crustal deformation in Southern California. *Journal of Geophysical Research B* **99**(12), 23,951–23,974.

Malvern, L. E. 1969. *Introduction to the mechanics of continuous media.* Englewood Cliffs, NJ: Prentice Hall.

Savage, J. C., and R. O. Burford. 1970. Accumulation of tectonic strain in California. *Bulletin of the Seismological Society of America* **60**, 1877–1896.

Savage, J. C., W. Gan, and J. L. Svarc. 2001. Strain accumulation and rotation in the Eastern California Shear Zone. *Journal of Geophysical Research* **106**, 21,995–22,007.

Segall, P., and Y. Du. 1993. How similar were the 1934 and 1966 Parkfield earthquakes? *Journal of Geophysical Research* **98**, 4527–4538.

Sokolnikoff, I. S. 1983. *Mathematical theory of elasticity*. New York: McGraw-Hill; reprinted Malabar, FL: Robert E. Krieger Publishing, pp. 390–391.

Strang, G. 1976. *Linear algebra and its applications*. New York: Academic Press, pp. 206–220.

2

Dislocation Models of Strike-Slip Faults

In this chapter, we will begin by considering a simple two-dimensional model of a very long strike-slip fault (figure 2.1). We consider a homogeneous, linear elastic half-space. The fault is taken to be infinitely long in the x_3 direction, and the deformation is antiplane strain. This means that the only nonzero displacement is parallel to the fault in the x_3 direction, and it varies only in the plane perpendicular to the fault. That is, u_3 varies only with x_1 and x_2, $\mathbf{u}(\mathbf{x}) = u_3(x_1, x_2)\hat{\mathbf{e}}_3$. As a starting point, we will consider the slip to be uniform with depth along the fault. Later, we will consider the more realistic situation in which the slip varies with depth.

2.1 Full-Space Solution

We will begin by ignoring the earth's surface and considering a fault in a full-space. The solution for a half-space is then rather easily constructed. Notice from figure 2.1 that the displacement field is discontinuous across the fault surface; the x_2, x_3 plane. A surface of imposed displacement discontinuity is known as a *dislocation*. In the antiplane geometry, the displacement traces a helical motion around the dislocation (figure 2.2), which is therefore known as a *screw dislocation*. The line parallel to the x_3 direction where the displacement jumps discontinuously from zero to some value s is the *dislocation line*. The half-plane of the fault, $x_1 = 0$, $x_2 > 0$, is the *dislocation surface*.

The boundary condition at the dislocation is that the displacement discontinuity across the dislocation surface is equal to the fault slip, s. Adopting a radial coordinate system centered on the dislocation line (the x_3 axis),

$$u_3(\theta = 2\pi) - u_3(\theta = 0) = s. \tag{2.1}$$

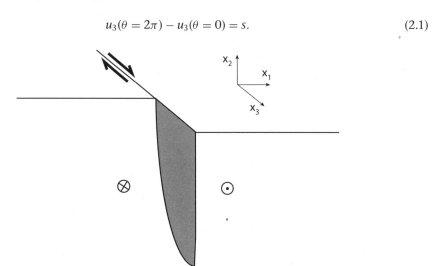

Figure 2.1. Infinitely long strike-slip fault. The fault lies in the plane normal to x_1, with slip in the x_3 direction. The earth's surface is the plane $x_2 = 0$. In this illustration, slip tapers with depth. This chapter begins with the case in which slip is uniform with depth.

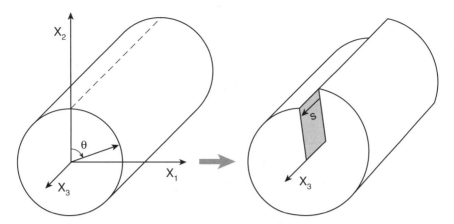

Figure 2.2. Screw dislocation, with dislocation line along x_3. The slip vector s shows the displacement of the back surface with respect to the front; the indicated slip is left-lateral.

In antiplane strain, there are only two nonzero strains, which from equations (1.4) are

$$\epsilon_{13} = \frac{1}{2}\frac{\partial u_3}{\partial x_1},$$

$$\epsilon_{23} = \frac{1}{2}\frac{\partial u_3}{\partial x_2}. \tag{2.2}$$

Hooke's law for a homogeneous, isotropic elastic material (1.103) is

$$\sigma_{ij} = 2\mu\epsilon_{ij} + \lambda\epsilon_{kk}\delta_{ij}. \tag{2.3}$$

In the antiplane case, there are only two nonzero stresses, corresponding to the shear strains (2.2):

$$\sigma_{13} = \mu\frac{\partial u_3}{\partial x_1},$$

$$\sigma_{23} = \mu\frac{\partial u_3}{\partial x_2}. \tag{2.4}$$

The stresses must satisfy the equilibrium equations (1.97). Neglecting inertial terms—that is, equivalent to assuming that the displacements develop slowly compared to elastic wave speeds—and body forces f_i, including gravity, we have

$$\frac{\partial \sigma_{ij}}{\partial x_j} = 0. \tag{2.5}$$

While in general there are three equilibrium equations, in the antiplane case, two are solved automatically, and there is only one nontrivial equation:

$$\frac{\partial \sigma_{31}}{\partial x_1} + \frac{\partial \sigma_{32}}{\partial x_2} = 0. \tag{2.6}$$

Substituting the stresses (2.4) into the equilibrium equation (2.6) yields

$$\frac{\partial^2 u_3}{\partial x_1^2} + \frac{\partial^2 u_3}{\partial x_2^2} = \boxed{\nabla^2 u_3(x_1, x_2) = 0.} \tag{2.7}$$

The displacement field satisfies Laplace's equation. This is a general result for antiplane elasticity problems. Note that since the solution is written in terms of the displacement, it is not necessary to introduce the compatibility equations.

From figure 2.2, we notice that the displacement increases smoothly with θ except at the dislocation surface. Thus, one might guess a solution of the form

$$u_3 = \pm \frac{s\theta}{2\pi}. \tag{2.8}$$

Note from equation (2.8) that u_3 increments by $\pm s$, as θ varies from 0 to 2π. In the following, we will adopt the $-s$ solution, making the sense of slip opposite to that shown in figure 2.2. Changing the sign of s simply reverses the sense of slip. For equation (2.8) to be a valid solution, it must match the boundary conditions and satisfy the equilibrium equations. Equation (2.8) was in fact constructed to satisfy the boundary condition (2.1). To see whether it also satisfies the equilibrium equations, write Laplace's equation in polar coordinates:

$$\nabla^2 u_3 = \frac{\partial^2 u_3}{\partial r^2} + \frac{1}{r} \frac{\partial u_3}{\partial r} + \frac{1}{r^2} \frac{\partial^2 u_3}{\partial \theta^2}. \tag{2.9}$$

In this form, it is clear that $\nabla^2 u_3 = 0$. Thus, equation (2.8) satisfies the governing equation and the slip boundary condition on the fault surface. If we had not recalled the Laplacian in polar coordinates, we could also check that equation (2.8) satisfies equation (2.7) by converting the displacements to Cartesian coordinates:

$$u_3 = -\frac{s\theta}{2\pi} = -\frac{s}{2\pi} \tan^{-1} \left(\frac{x_1}{x_2} \right). \tag{2.10}$$

The relevant partial derivatives are

$$\frac{\partial u_3}{\partial x_1} = -\frac{s}{2\pi} \frac{x_2}{x_1^2 + x_2^2},$$

$$\frac{\partial u_3}{\partial x_2} = \frac{s}{2\pi} \frac{x_1}{x_1^2 + x_2^2}. \tag{2.11}$$

A further differentiation of equation (2.11) verifies that the equilibrium equations are indeed satisfied.

Solution (2.8) satisfies equilibrium and the boundary conditions on the dislocation surface. What about the stress far from the dislocation—that is, as $r \to \infty$? The stresses can be found from equations (2.4) and (2.11):

$$\sigma_{23} = \mu \frac{\partial u_3}{\partial x_2} = \frac{s\mu}{2\pi} \frac{x_1}{x_1^2 + x_2^2} = \frac{s\mu}{2\pi} \frac{\sin \theta}{r}, \tag{2.12}$$

$$\sigma_{13} = \mu \frac{\partial u_3}{\partial x_1} = \frac{-s\mu}{2\pi} \frac{x_2}{x_1^2 + x_2^2} = \frac{-s\mu}{2\pi} \frac{\cos \theta}{r}. \tag{2.13}$$

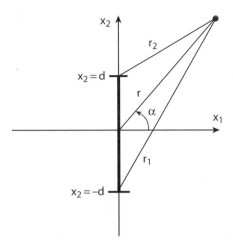

Figure 2.3. Dislocation dipole. A pair of oppositely signed dislocations are located at $x_2 = -d$ and $x_2 = +d$, causing uniform slip in the enclosed interval.

Notice that the stresses decay as $1/r$ from the *dislocation line*, not the dislocation surface. It is the dislocation line that is the source of strain, not the dislocation surface.

We will next consider slip extending over a finite interval in the x_2 direction. This solution is easily obtained summing the solutions for two oppositely signed dislocations. The governing differential equation (2.7) is linear. Thus, if two displacement fields $u^{(1)}$ and $u^{(2)}$ independently solve equation (2.7)—that is, $\nabla^2 u^{(1)} = 0$ and $\nabla^2 u^{(2)} = 0$—their sum also satisfies the equation $\nabla^2(u^{(1)} + u^{(2)}) = 0$. This is known as the *principle of superposition*. In particular, add a screw dislocation at depth $x_2 = -d$ and an oppositely signed screw dislocation at depth $x_2 = +d$ (figure 2.3). A dislocation at $x_2 = d$ is obtained by replacing x_2 with $x_2 - d$ in equation (2.10). This leads to

$$u_3 = \frac{-s}{2\pi} \left[\tan^{-1}\left(\frac{x_1}{x_2 + d} \right) - \tan^{-1}\left(\frac{x_1}{x_2 - d} \right) \right]. \tag{2.14}$$

The stress due to slip over the finite interval is thus

$$\sigma_{13} = \frac{-s\mu}{2\pi} \left(\frac{x_2 + d}{r_1^2} - \frac{x_2 - d}{r_2^2} \right), \tag{2.15}$$

where the distance from the lower dislocation is $r_1^2 = x_1^2 + (x_2 + d)^2$, and the distance from the upper dislocation is $r_2^2 = x_1^2 + (x_2 - d)^2$. On the plane of the fault, $x_1 = 0$, this becomes

$$\sigma_{13}(x_1 = 0) = \frac{s\mu d}{\pi(x_2^2 - d^2)} = \frac{s\mu d}{\pi(x_2 - d)(x_2 + d)} \tag{2.16}$$

(see figure 2.4). Note that equation (2.16) gives the change in stress due to slip on the fault. The total stress is the sum of any preexisting stresses and the changes due to fault slip. As expected, the stress change is negative inside the slipping zone and positive outside this region. We interpret the negative stress as a decrease relative to some initial state and the positive stresses as a stress increase. Physically, it makes sense for fault slip to relax the stress. Because the slip changes discontinuously from s to 0, the strains, and thus stresses, are infinite at the dislocation line. Note from equations (2.12) and (2.13) that the dislocation stresses have a $1/r$

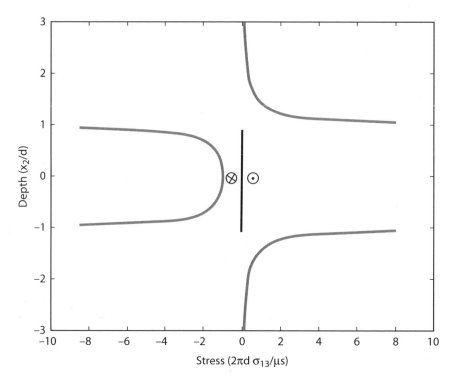

Figure 2.4. Stress change due to a pair of dislocations at $x_2 = -1$ and $x_2 = +1$.

singularity. For a finite fault represented by a pair of dislocations, the stress is singular both inside and outside the fault, as seen in figure 2.4 and equation (2.16), where $x_2 - d$ and $x_2 + d$ measure the distance from each end of the fault.

For a pair of dislocations (a dipole), representing slip over a finite interval in x_2, the far-field stress decays more rapidly with distance than it does for a single dislocation. To see this, first note that from equation (2.13), the stress resulting from a dislocation at $x_2 = -d$ is

$$\sigma_{13} = -\frac{s\mu}{2\pi} \left[\frac{x_2 + d}{x_1^2 + (x_2 + d)^2} \right]. \tag{2.17}$$

For slip due to a pair of oppositely signed dislocations, one at d and a second at slightly greater depth $d + \Delta d$, the stress is thus

$$\sigma_{13} = -\frac{s\mu}{2\pi} \left[\frac{x_2 + d + \Delta d}{x_1^2 + (x_2 + d + \Delta d)^2} - \frac{x_2 + d}{x_1^2 + (x_2 + d)^2} \right]. \tag{2.18}$$

Noticing that in the limit as $\Delta d \to 0$, this is the formal definition of a differential, we can write the stress as

$$\sigma_{13} = -\frac{s\mu \Delta d}{2\pi} \frac{\partial}{\partial d} \left[\frac{x_2 + d}{x_1^2 + (x_2 + d)^2} \right],$$

$$= -\frac{M_0}{2\pi} \frac{\cos(2\alpha)}{r^2}, \tag{2.19}$$

where M_0 is the seismic moment (see chapter 3) per unit length of fault; $M_0 = \mu s \Delta d$, the product of the shear modulus, the slip, and the fault area per unit length in the x_3

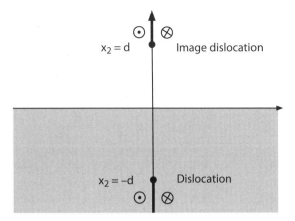

Figure 2.5. Dislocation at $x_2 = -d$, and image dislocation at $x_2 = d$.

direction; and α is the angle between the x_1 axis and the vector pointing from the center of the dislocation dipole to the observation point, as in figure 2.3. In summary, the dislocation stresses decay proportional to $1/r$ in two dimensions, and the stresses due to a dislocation dipole, in two dimensions, as $1/r^2$, as long as $r \gg \Delta d$.

2.2 Half-Space Solution

To a reasonable approximation, no stress is transmitted between the solid earth and the atmosphere. We will thus model the earth's surface as one across which the shear and normal tractions vanish. Furthermore, if all of the relevant length scales in the problem are small compared with the earth's radius, we are able to ignore earth curvature and model the free surface as planar. Here, the free surface is taken to be $x_2 = 0$, so the governing equations must be satisfied for all $x_2 < 0$. For antiplane strain problems, the half-space solution is easily constructed using the *method of images*.

What are the stresses acting on the surface $x_2 = 0$ due to a dislocation in a full-space at $x_2 = -d$? The only nonzero stress acting on the surface is σ_{23}, and from equation (2.12),

$$\sigma_{23} = \frac{s\mu}{2\pi} \frac{x_1}{x_1^2 + (x_2 + d)^2}, \tag{2.20}$$

$$\sigma_{23}(x_2 = 0) = \frac{s\mu}{2\pi} \frac{x_1}{x_1^2 + d^2}. \tag{2.21}$$

Clearly, a single dislocation itself does not satisfy the free-surface boundary condition. The remedy is to add a (fictitious) image dislocation at $x_2 = d$ with opposite sign, equidistant from the putative free surface (figure 2.5). We must ensure that the image dislocation surface is restricted to the region $x_2 > 0$ so that it does not generate discontinuities within the earth ($x_2 < 0$). Note that in figure 2.5, the actual dislocation points along the negative x_2 axis. In section 2.5, you will see that changing the direction of the dislocation surface 180 degrees reverses the sense of slip, so that by comparing to figure 2.2, it is in fact the $-s$ solution that is shown in figure 2.5.

The σ_{23} stress due to the dislocation and its image is

$$\sigma_{23} = \frac{s\mu}{2\pi}\left[\frac{x_1}{x_1^2 + (x_2 + d)^2} - \frac{x_1}{x_1^2 + (x_2 - d)^2}\right],\tag{2.22}$$

so the stress on the surface $x_2 = 0$ is

$$\sigma_{23}(x_2 = 0) = \frac{s\mu}{2\pi}\left(\frac{x_1}{x_1^2 + d^2} - \frac{x_1}{x_1^2 + d^2}\right) = 0.\tag{2.23}$$

By summing the contribution from the dislocation and its image, the free-surface boundary condition is now satisfied. The displacements due to the dislocation and its image are given by

$$u_3 = \frac{-s}{2\pi}\left[\tan^{-1}\left(\frac{x_1}{x_2 + d}\right) - \tan^{-1}\left(\frac{x_1}{x_2 - d}\right)\right].\tag{2.24}$$

We can now construct a solution for slip over a finite interval in a half-space by summing the contributions from two dislocations and their respective images. In particular, assume that slip is uniform from depth d_2 to depth d_1, where $|d_1| > |d_2|$. Using the principle of superposition,

$$u_3 = \frac{-s}{2\pi}\left[\underbrace{\tan^{-1}\left(\frac{x_1}{x_2 + d_1}\right)}_{deep\ dislocation} - \underbrace{\tan^{-1}\left(\frac{x_1}{x_2 - d_1}\right)}_{deep\ image} - \underbrace{\tan^{-1}\left(\frac{x_1}{x_2 + d_2}\right)}_{shallow\ dislocation} + \underbrace{\tan^{-1}\left(\frac{x_1}{x_2 - d_2}\right)}_{shallow\ image}\right].$$

$$\tag{2.25}$$

Displacements on the free surface are of particular interest, since this is where data is collected:

$$u_3(x_2 = 0) = \frac{-s}{\pi}\left[\tan^{-1}\left(\frac{x_1}{d_1}\right) - \tan^{-1}\left(\frac{x_1}{d_2}\right)\right].\tag{2.26}$$

In the next sections, we will consider two limiting cases of this result.

2.2.1 Coseismic Faulting

Let the shallow dislocation come to the surface—that is, $d_2 \to 0$. This describes uniform slip from the surface to depth d_1 (figure 2.6). First note that

$$\lim_{d_2 \to 0}\tan^{-1}\left(\frac{x_1}{d_2}\right) = \frac{\pi}{2}\,\text{sgn}(x_1),\tag{2.27}$$

so that

$$u_3(x_2 = 0) = \frac{-s}{\pi}\left[\tan^{-1}\left(\frac{x_1}{d_1}\right) - \frac{\pi}{2}\,\text{sgn}(x_1)\right].\tag{2.28}$$

Now notice that

$$\tan^{-1}\left(\frac{x_1}{d_1}\right) + \tan^{-1}\left(\frac{d_1}{x_1}\right) = \frac{\pi}{2}\,\text{sgn}(x_1),\tag{2.29}$$

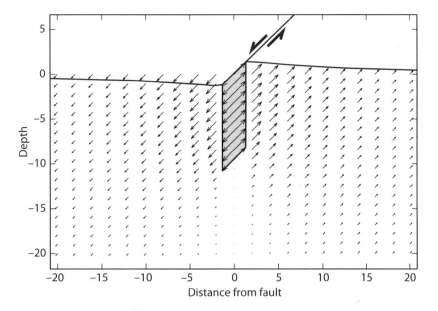

Figure 2.6. Dislocation slips from the earth's surface to depth $d_1 = 10$ km. Displacements are directed in and out of the page but are plotted to simulate a three-dimensional perspective.

so that

$$u_3(x_2 = 0) = \frac{s}{\pi} \tan^{-1}\left(\frac{d_1}{x_1}\right), \tag{2.30}$$

which holds for all values of x_1 (figure 2.7). We can now compute the shear strain, except at the fault trace ($x_1 = 0$), where the displacements are discontinuous and therefore nondifferentiable:

$$\epsilon_{13} = \frac{1}{2}\frac{\partial u_3}{\partial x_1} = \frac{-s}{2\pi d_1}\left[\frac{1}{1 + \left(\frac{x_1}{d_1}\right)^2}\right]. \tag{2.31}$$

Note that the *coseismic strain*, or strain change in large strike-slip earthquakes (equation [2.31]), is everywhere negative. That is because earthquakes release shear strain, as expected. The maximum strain, which occurs at the fault trace, is $-s/2\pi d_1$. Note that the displacement at the fault trace is discontinuous, so the strain there must be defined at $x_1 = \pm\delta$ in the limit as δ tends to zero. The strain change decays with distance from the fault such that the magnitude of the strain is half the maximum when $x_1 = d_1$ (figure 2.7).

2.2.2 Interseismic Deformation

Now let the deep dislocation go to infinity, $d_1 \to \infty$. In this case, the fault is locked from the surface to depth d_2 but slips by a constant amount below that depth (figure 2.8). This was proposed by Savage and Burford (1970) as a first-order model of interseismic

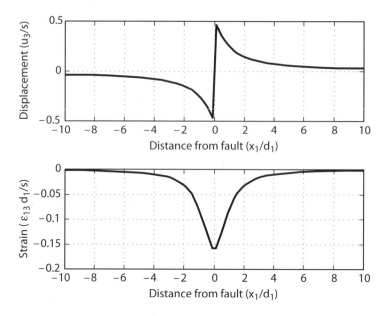

Figure 2.7. Coseismic displacement and strain.

deformation—that is, deformation between large plate rupturing events—on the San Andreas fault. In this limit, the displacement is

$$u_3(x_2 = 0) = \frac{s}{\pi} \tan^{-1}\left(\frac{x_1}{d_2}\right). \tag{2.32}$$

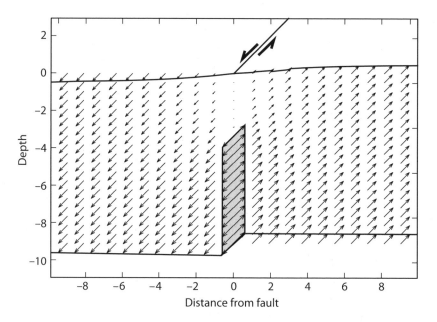

Figure 2.8. Fault is locked from the earth's surface to depth $d_2 = 3$ km. Displacements are directed in and out of the page but are plotted to simulate a three-dimensional perspective.

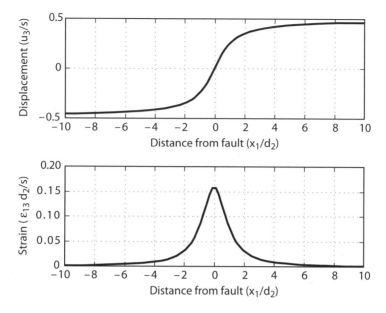

Figure 2.9. Interseismic displacement and strain.

Differentiating equation (2.32) with respect to time yields the surface velocity v_3. If the geometry is time-invariant—that is, if the depth d_2 does not change with time,

$$v_3(x_2 = 0) = \frac{\dot{s}}{\pi} \tan^{-1}\left(\frac{x_1}{d_2}\right), \tag{2.33}$$

where $\dot{s} = ds/dt$ is the fault slip rate. Note from equation (2.33) and figure 2.9 that the velocity far from the fault is equal to half the fault slip rate.

Differentiating equation (2.33) with respect to the spatial coordinate perpendicular to the fault yields the shear strain rate at the free surface:

$$\dot{\epsilon}_{13} = \frac{1}{2}\frac{\partial v_3}{\partial x_1} = \frac{\dot{s}}{2\pi d_2}\left[\frac{1}{1 + \left(\dfrac{x_1}{d_2}\right)^2}\right]. \tag{2.34}$$

Notice that the shear strain rate is everywhere positive; thus, shear strain accumulates on the fault between earthquakes. In fact, the *interseismic strain* (2.34) is exactly equal in magnitude and opposite in sign to the coseismic strain (2.31) if the cumulative deep slip integrated over the time between earthquakes is equal to the coseismic slip (figure 2.10). Thus, after a complete earthquake cycle, there is no strain in the intervening blocks, simply rigid block translation of the two sides of the fault.

It is important to note that, as mentioned earlier, the deformation is caused by the dislocation line and not by the dislocation surface. The implication of this is that the deformation field at the surface is independent of the dip of the fault, if the slip on the fault is spatially

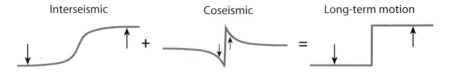

Figure 2.10. Interseismic displacement plus coseismic displacement equals the long-term fault motion.

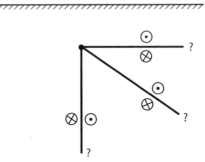

Figure 2.11. The surface displacements and strain are independent of the dip of the dislocation surface for a single screw dislocation.

uniform and extends to infinite depth, as long as the fault does not breach the surface. In particular, the deformation at the surface is the same whether the fault plane is vertical (as we have modeled it) or horizontal. A horizontal dislocation might represent a zone of decoupling. Our results show, unfortunately, that we cannot distinguish between these models on the basis of surface displacement or strain for infinitely long strike-slip faults (figure 2.11).

There is an alternative perspective on interseismic deformation that is commonly used in both elastic and viscoelastic models. One can consider that over the long term, the fault moves in essentially rigid block motion, with no strain outside a narrow fault zone. Between large earthquakes, the upper part of the fault is locked. Kinematically, this can be accomplished by starting with rigid block motion and superimposing *back slip* at a rate equal to the long-term slip rate on the seismogenic part of the fault. From equation (2.30), the velocity is thus

$$v_3(x_2 = 0) = \frac{\dot{s}}{2} \, \mathrm{sgn}(x_1) - \frac{\dot{s}}{\pi} \tan^{-1} \left(\frac{d}{x_1} \right),$$

$$= \frac{\dot{s}}{\pi} \tan^{-1} \left(\frac{x_1}{d} \right), \tag{2.35}$$

using the identity in equation (2.29). The back-slip model yields the same result for the interseismic velocity derived previously (2.33) for a steadily creeping dislocation extending infinitely below locking depth d.

2.2.3 Postseismic Slip

Last, consider buried slip with $d_1 > d_2$. Thatcher (1975) modeled postseismic deformation along the San Andreas fault assuming slip over a confined interval beneath the coseismic rupture surface. During an earthquake, stress decreases along the slipping fault and is shed to the lower part of the fault. If the fault in the lower crust creeps in response to the imposed stress, there may be transient buried slip. This model is a very simple representation of that process. More realistic models, which account for stress-driven frictional creep or distributed

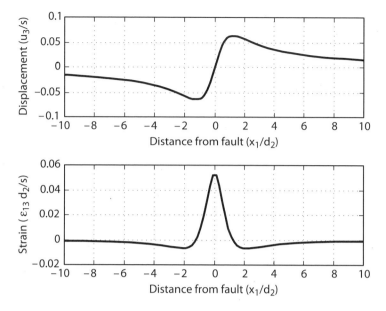

Figure 2.12. Postseismic displacement and strain.

viscoelastic deformation, will be considered in later chapters. This present model is useful primarily because of its simplicity.

For confined slip between depths d_1 and d_2, the displacements at the free surface are

$$u_3(x_2 = 0) = \frac{-s}{\pi} \left[\tan^{-1}\left(\frac{x_1}{d_1}\right) - \tan^{-1}\left(\frac{x_1}{d_2}\right) \right], \tag{2.36}$$

while the surface strains are

$$\epsilon_{13} = \frac{s}{2\pi} \left(\frac{d_2}{d_2^2 + x_1^2} - \frac{d_1}{d_1^2 + x_1^2} \right). \tag{2.37}$$

Note that the strain, for buried slip, undergoes a sign change (figure 2.12), being positive near the fault and negative away from the fault. This feature turns out to be characteristic of deformation occurring over a finite depth interval. It is common to more sophisticated and realistic models of postseismic deformation.

2.3 Distributed Slip

We can use the results in the previous section to derive a general expression for deformation due to an arbitrary distribution of fault slip with depth. Consider the case where slip is confined to the interval from d to $d + \Delta d$:

$$u_3(x_2 = 0) = \frac{-s}{\pi} \left[\tan^{-1}\left(\frac{x_1}{d + \Delta d}\right) - \tan^{-1}\left(\frac{x_1}{d}\right) \right]. \tag{2.38}$$

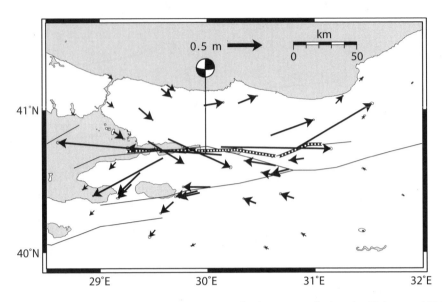

Figure 2.13. Horizontal components of the coseismic displacements during the 17 August 1999 Izmit, Turkey, earthquake. Dashed line indicates the rupture zone of the earthquake. Data from Reilinger et al. (2000).

Noticing that in the limit as $\Delta d \to 0$, this is the formal definition of a differential, we can write the displacement due to slip over Δd as

$$u_3(x_1, x_2 = 0) = \frac{-s\Delta d}{\pi} \frac{\partial}{\partial d} \tan^{-1}\left(\frac{x_1}{d}\right), \tag{2.39}$$

$$= \frac{s\Delta d}{\pi} \frac{x_1}{x_1^2 + d^2}. \tag{2.40}$$

Thus, the displacement due to an arbitrary distribution of slip with depth, $s(\xi)$, is

$$u_3(x_1, x_2 = 0) = \frac{1}{\pi} \int_0^\infty s(\xi) \frac{x_1}{x_1^2 + \xi^2} d\xi. \tag{2.41}$$

The forward problem, predicting the surface displacements at x_1 given the slip distribution $s(\xi)$, is solved simply by integrating. Commonly, we have measured displacements at the earth's surface and want to determine as much as possible about the distribution of slip at depth in the earth. This inverse problem, estimating $s(\xi)$ from measured values of $u_3(x_1, x_2 = 0)$, was addressed by Matthews and Segall (1993) among others.

2.4 Application to the San Andreas and Other Strike-Slip Faults

The simple models explored in this chapter can be compared to actual displacements recorded in large strike-slip earthquakes. Displacements during the 17 August 1999, Izmit, Turkey, earthquake, a magnitude 7.5 event, were precisely measured using Global Positioning System (GPS) receivers. The horizontal components of the displacements during the earthquake are shown in figure 2.13. The earthquake rupture was approximately 150 km long, hardly infinite. However, if we focus on measurements made near the central part of the east–west trending rupture, end effects can be minimized.

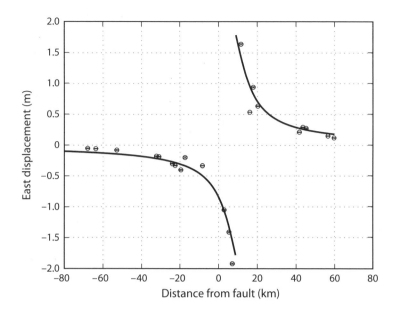

Figure 2.14. Coseismic displacements during the 17 August 1999 Izmit, Turkey, earthquake. Data from Reilinger et al. (2000). The fault parallel (east–west) component of the displacement is plotted as a function of (north–south) distance perpendicular to the fault. Only stations near the middle of the rupture are plotted to minimize effects of the fault ends. Also shown is the prediction of the simple infinitely long screw dislocation equation (2.30), with slip of 3.6 m and a fault depth of 8 km.

Figure 2.14 shows the fault-parallel (east–west) displacements as a function of north–south distance perpendicular to the fault. Also shown is the prediction from a simple two-dimensional screw dislocation extending from the earth's surface to a depth of 8 km with 3.6 m of slip. It is quite remarkable that this exceedingly simple model fits the data so well. A three-dimensional model is required to fit the observations more accurately. Inversions of the GPS measurements, which allow the slip to vary in magnitude over the fault plane, yield a maximum slip of 5.7 m and an average slip of 3.6 m over the central 75 km of the rupture (Reilinger et al. 2000).

We can also use the results of this chapter to estimate the interseismic slip rate on the San Andreas fault and the depth to which the fault is locked between earthquakes—the so-called *locking depth*. A summary of strain-rate data shown in Figure 2.15, from Thatcher (1990), shows a broad zone of shearing centered on the San Andreas fault. We can interpret the data using the simple model of a single-screw dislocation slipping at a constant rate given in equation (2.34). This equation shows that the shear strain drops to half its peak value at a distance d from the fault. From the figure, we estimate that half-width to be approximately 30 km. Given this, the interseismic slip rate $\dot{s}$ is given by

$$\dot{s} = \pi d(2\dot{\epsilon}_{13})_{max}. \tag{2.42}$$

The data show a maximum engineering shear strain rate, $(2\dot{\epsilon}_{13})_{max}$, of approximately 5×10^{-7} 1/yr. This leads to a slip rate of roughly 4.7 cm/yr.

These estimates are higher than our best current estimates. The main reason is that, in most places in California, there is not a single fault but multiple parallel faults. This causes the strain to be spread out over a broader area than would be predicted from a single dislocation

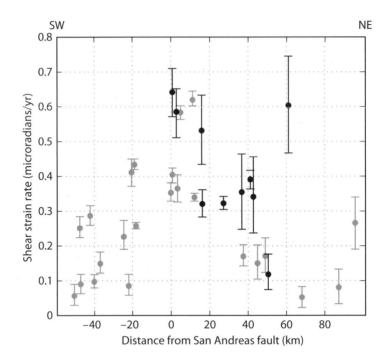

Figure 2.15. Shear strain rate as a function of distance from the San Andreas fault. In this figure, the strain rate is given in engineering strain, which is twice the tensor strain used in this text. (Light symbols, southern California, dark symbols, northern California.) After Thatcher (1990).

and biases to large values the locking depth. Overestimating d also causes the slip rate to be overestimated.

The slip rate $\dot{s}$ is estimated to be closer to 3.5 cm/yr, from both geologic and geodetic studies in the central creeping section of the San Andreas. With this slip rate and the observed maximum shear strain rate of 5×10^{-7} 1/yr, we estimate d to be approximately 20 km.

One can do a better job by considering measurements from a more limited geographic region. The Southern California Earthquake Center (SCEC) has produced a crustal velocity field using GPS, Very Long Baseline Interferometry (VLBI), and Electronic Distance Meter (EDM) measurements. The GPS data span the time interval 1986 to 1997 and consist of a mixture of episodic campaign measurements and continuously recording stations. VLBI measurements at 10 southern California sites were collected between 1980 and 1994, whereas the EDM surveys were conducted predominantly from 1970 to 1992.

In order to isolate a relatively simple part of the San Andreas fault system, we focus on data from the Carrizo Plain section, north of the Big Bend in the San Andreas (figure 2.16). In this region, the San Andreas fault trace is relatively straight and simple. Also the area is not complicated by additional faults, except perhaps in the Santa Maria basin. The velocity vectors, which are displayed in a North America fixed reference frame, are very nearly parallel to the San Andreas and increase in magnitude from east to west.

In order to model these velocities, we need to account for the fact that the data and model are in different reference frames. The data are in a North America fixed frame; sites on stable North America are expected to have zero motion. The model given by equation (2.33), however, predicts half the motion on one side of the fault, and half on the other side. By symmetry, the fault trace has zero velocity. One must account for this difference in comparing the model predictions to measured velocities.

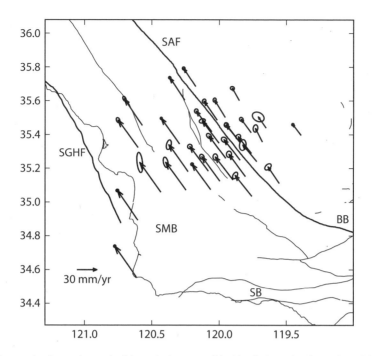

Figure 2.16. Interseismic station velocities relative to stable North America from the 1980s and 1990s as measured by a variety of techniques, analyzed and compiled by the Southern California Earthquake Center. SAF = San Andreas fault; SGHF = San Gregorio Hosgri fault; BB = Big Bend of San Andreas; SMB = Santa Maria basin; SB = Santa Barbara.

Figure 2.17 shows a profile of the fault-parallel component of velocity perpendicular to the trend of the San Andreas fault. The San Andreas is located at $x = 0$. Also shown is the theoretical prediction from equation (2.33) for a fault-locking depth of 20 km and a slip rate of 39.5 mm/yr. While there are significant misfits at points farthest from the fault, the comparison between the observations and the prediction of the simple model is satisfactory. Discrepancies could result from larger than expected observational errors or defects in the model, such as additional deformation sources, nonuniform elastic properties, or complexities in fault geometry, to name a few. We will revisit these data in chapter 12 in the context of viscoelastic earthquake cycle models. With these models, the surface velocity field can be fit with slip rates consistent with paleoseismic observations and locking depths consistent with the depths of major strike-slip earthquakes in California.

2.5 Displacement at Depth

The displacements within the earth can be computed from equation (2.25). Care must be taken when evaluating the $\tan^{-1}$ functions because they are discontinuous. Mathematically, the discontinuities are referred to as *branch cuts*. We must ensure that the branch cuts occur on the fault surface and not elsewhere.

First, recall that figure 2.2 defined the dislocation to extend along the positive x_2 axis, with the angle θ measured positive clockwise from x_2. Looking from above the fault, the $+s\theta/2\pi$ solution is left-lateral, with the negative x_1 side moving in the positive x_3 direction. Add a right-lateral rigid body offset along the fault with slip that exactly negates the slip due to the dislocation for $x_2 > 0$ (figure 2.18). The sum is a right-lateral dislocation that extends along the

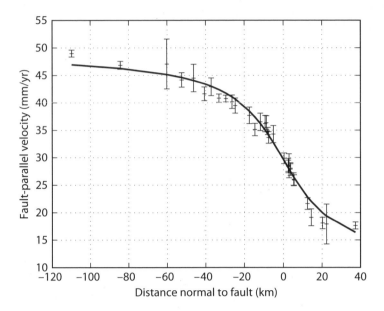

Figure 2.17. Velocity profile across the San Andreas fault. Fault-parallel component of velocity for the stations shown in figure 2.16. Error bars are one standard deviation. The San Andreas fault is approximately located at $x = 0$. Also shown is the predicted velocity of an infinitely long buried screw dislocation from equation (2.33) for a fault-locking depth of 20 km and a slip rate of 39.5 mm/yr.

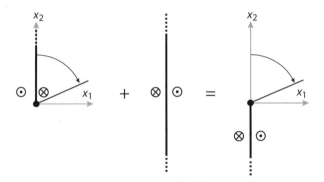

Figure 2.18. Definition of branch cuts for a strike-slip fault. Figure on the left shows the dislocation extending along the positive x_2 axis. Addition of rigid body slip of opposite sign cancels the slip in the domain $x_2 > 0$ and generates an oppositely signed dislocation for $x_2 < 0$.

negative x_2 axis. Thus, flipping the dislocation plane 180 degrees changes the sense of slip on the fault. For example, for a buried dislocation extending in the negative x_2 direction, its image dislocation extending in the positive x_2 direction, as in figure 2.5, has the *same* sense of slip, since changing the orientation of the plane 180 degrees is equivalent to changing the sense of slip.

For the dislocation extending along the negative x_2 direction, the branch cut is located at $\theta = \pi$, so we must define the argument of the function θ to be $-\pi < \theta < \pi$. Figure 2.19 shows the branch cuts for a buried dislocation between depths d_1 and d_2. Numerical calculations that compute the displacements in the body should use an arctangent function that takes two arguments and produces the correct sign to represent an angle in each of the four quadrants.

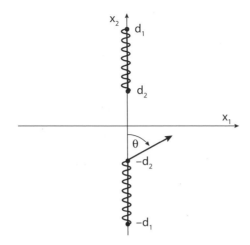

Figure 2.19. Definition of branch cuts (wavy lines) for a strike-slip fault in a half-plane.

Figures 2.6 and 2.8 show the displacement field for uniform-slip dislocations. Notice that the displacements decay quite rapidly below the dislocation and with distance away from the dislocation.

2.6 Summary and Perspective

It is worth reviewing the approach and assumptions made in this chapter. We assumed that the earth could be modeled as a homogeneous, isotropic, elastic half-space. Our simple fault model involved specifying the displacement on the fault surface. For two-dimensional antiplane geometry, as might approximate a very long strike-slip fault, we found that the displacements must satisfy Laplace's equation. Solutions in a full-space are easily constructed, and half-space solutions are then found using the method of images. The displacements at the free surface are found to follow the familiar $\tan^{-1}$ distribution, with argument either x/d or d/x, depending on whether the fault is locked or slipping at the earth's surface.

The limitations to these simple models are manifold. The main benefits of the results developed in this chapter are their simplicity and the fact that to first order, they fit some geodetic observations quite well. However, faults are never infinitely long, and we may be very interested in effects near the end of an earthquake rupture. Three-dimensional dislocations will be the subject of chapter 3. Furthermore, we have so far ignored the earth's curvature and topography; this will be considered in chapter 8. The earth is not homogeneous and isotropic, nor perfectly elastic, for that matter. Methods for determining elastic fields due to dislocations in heterogeneous earth models are discussed in chapter 5. The effects of viscoelastic relaxation are described in chapter 6. Last, we should note that by specifying the fault slip, we are unable to say anything about *how* that slip comes about. A more complete description of faulting involves specifying a constitutive law for the fault surface, or fault zone, and then computing how the slip develops as a function of time for some loading geometry (as in chapter 11). As a first step in that direction, we will consider crack models in which the stress change (stress drop) on the fault surface is specified in chapter 4.

In the next chapter, we will begin by deriving dislocation solutions for two-dimensional dip-slip faults and faults in three dimensions.

2.7 Problems

1. Show that the stress due to a pair of screw dislocations is proportional to $\cos(2\theta)/r^2$, as in equation (2.19).

2. Use the principle of superposition to derive expressions for the surface velocity and strain rate for two parallel faults that are locked from the earth's surface to some depth d. The first fault is located at $x_1 = \zeta_1$ and has slip rate $\dot{s}_1$, and the second fault is located at $x_1 = \zeta_2$, with slip-rate $\dot{s}_2$. Generalize this to an arbitrary number of faults. This simple model can be compared to geodetic measurements collected across the San Andreas fault system in northern California.

3. Show that the surface displacements for a slip distribution that varies linearly with depth as $s(\xi) = s_0(1 - \xi/D)$ are given by

$$u_3(x_1, x_2 = 0) = \frac{s_0}{\pi} \left[\tan^{-1}\left(\frac{D}{x_1}\right) - \frac{x_1}{2D} \log\left(\frac{D^2 + x_1^2}{x_1^2}\right) \right]. \tag{2.43}$$

4. Prove that for a single screw dislocation buried in an elastic half-space that does not cut the free surface, the displacement field at the free surface is independent of the dip of the dislocation plane (see figure 2.11).

5. Write a script to plot the interseismic displacements within the earth due to an infinitely long, buried nonvertical strike-slip fault. Include the contributions from the image sources so that the plane $x_2 = 0$ is traction free.

2.8 References

Matthews, M. V., and P. Segall. 1993. Statistical inversion of crustal deformation data and estimation of the depth distribution of slip in the 1906 earthquake. *Journal of Geophysical Research* **98**, 12,153–12,163.

Reilinger, R. E., S. Ergintav, R. Bürgmann, S. McClusky, O. Lenk, A. Barka, O. Gurkan, L. Hearn, K. L. Feigl, R. Cakmak, B. Aktug, H. Ozener, and M. N. Töksoz. 2000. Coseismic and postseismic fault slip for the 17 August 1999, M = 7.5, Izmit, Turkey, earthquake. *Science* **289**, 1519–1524.

Savage, J. C., and R. O. Burford. 1970. Accumulation of tectonic strain in California. *Bulletin of the Seismological Society of America* **60** (6), 1877–1896.

Thatcher, W. 1975. Strain accumulation and release mechanism of the 1906 San Francisco earthquake. *Journal of Geophysical Research* **80**, 4862–4872.

———. 1990. Present-day crustal movements and the mechanics of cyclic deformation. In R. E. Wallace, Ed., *The San Andreas Fault system*, U.S. Geological Survey Professional Paper 1515. Washington, DC: U.S. Government Printing Office, pp. 189–205.

3

Dip-Slip Faults and Dislocations in Three Dimensions

In chapter 2, we developed methods for analyzing very long strike-slip faults. This chapter begins by deriving expressions for infinitely long thrust and normal faults, as shown in figure 3.1, and eventually presents methods for computing displacements for general three-dimensional dislocations.

The fault shown in figure 3.1 is understood to extend indefinitely in and out of the page. In this case, the displacements are restricted to the x_1, x_2 plane—that is, $u_3 = 0$. Furthermore, the remaining displacement components do not vary in the x_3 direction: $\partial u_1/\partial x_3 = \partial u_2/\partial x_3 = 0$. The resulting state of deformation is one of *plane strain*.

Plane-strain problems are significantly more complex than antiplane problems. To see why, consider the governing equations. First, we have the kinematic equations relating displacement and strain (1.4). For plane strain, these are $\epsilon_{11} = u_{1,1}$, $\epsilon_{22} = u_{2,2}$, and $\epsilon_{12} = (u_{1,2} + u_{2,1})/2$, where, as usual, the comma indicates differentiation with respect to spatial coordinates. All the other strain components are zero. Similarly, there are three independent stress components, σ_{11}, σ_{12}, and σ_{22}. The stress acting normal to the x_3 plane, σ_{33}, is determined by the constraint that $\epsilon_{33} = 0$.

For plane strain, there are two independent equilibrium equations (1.97) as opposed to only one for antiplane strain. There are thus eight unknowns (two displacements, three stresses, and three strains) to be solved for using the three kinematic equations, two equilibrium equations, and three nontrivial constitutive equations. Using Hooke's law (1.103) and the kinematic equations (1.4) relating strain to displacement (as in problem 9 in chapter 1) results in the two Navier equilibrium equations for the two independent displacement components:

$$\mu \nabla^2 u_1 + (\lambda + \mu)\frac{\partial u_{k,k}}{\partial x_1} + f_1 = 0,$$

$$\mu \nabla^2 u_2 + (\lambda + \mu)\frac{\partial u_{k,k}}{\partial x_2} + f_2 = 0. \qquad k = 1, 2. \qquad (3.1)$$

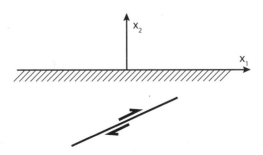

Figure 3.1. Infinitely long dip-slip fault.

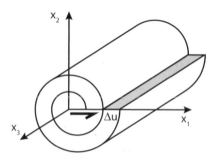

Figure 3.2. Geometry for edge dislocation, generated by slip of an amount Δu parallel to the dislocation surface.

Equations (3.1) are more difficult to solve than Laplace's equation for the antiplane strain case. Notice that they represent two *coupled* differential equations in two unknowns u_1 and u_2. Solution to plane-strain problems in a full-space can be found, writing equations in terms of stress, using either Airy's stress functions (e.g., Timoshenko and Goodier 1970) or complex potential methods (Muskhelishvili 1975; Sokolnikoff 1983; Barber 1992).

The solution in the half-space is again somewhat more complex. In plane strain, there are two stress components that exert tractions on the surface $x_2 = 0$: a normal stress, σ_{22}, and a shear stress, σ_{12}. It is straightforward to show that an image source does not cancel both components of the traction vector acting on the free surface. One can introduce an image to remove one component of the traction, but it has the effect of doubling the other. It is possible to use higher order images to cancel the second traction component or to use Fourier transform or other methods to find a traction distribution that exactly cancels the remaining stress acting on the putative free surface.

Rather than follow either of these approaches, we will derive a general method, *Volterra's formula*, that allows us to directly compute the displacements from an arbitrary dislocation surface. The reason for adopting this approach is that it readily generalizes to three dimensions, whereas the other approaches may not. In particular, it is possible to directly integrate to obtain the displacements as long as the *elastostatic Green's functions* for the medium are known. Fortunately, the Green's functions are well known for two- and three-dimensional, homogeneous, isotropic half-spaces, and our efforts are therefore considerably reduced over other methods.

In terms of dislocations, dip-slip faults in two dimensions are described by *edge dislocations*. Consider figure 3.2, in which the dislocation line is again parallel to the x_3 axis. The edge dislocation is generated by cutting the half-plane ($x_2 = 0$, $x_1 > 0$) and displacing it parallel to the dislocation surface, but *perpendicular* to the dislocation line. (Recall that for the screw dislocation, the displacement was parallel to the dislocation line.)

An edge dislocation can also be created by cutting the material and displacing the cut *perpendicular* to the dislocation surface. In figure 3.3, the medium has been cut along the positive x_2 axis, and a semi-infinite slab of material with thickness Δu has been inserted and welded into place. In this way, a pair of edge dislocations can be employed as a simple two-dimensional model of a volcanic dike or sill (see chapter 7). While not obvious from figures 3.2 and 3.3, the stresses generated by the two different edge dislocations are in fact the same (see problem 5).

3.1 Volterra's Formula

Our derivation, following Steketee (1958) and Savage (1980), is based on the reciprocal theorem (1.115). Take the unprimed system to be a dislocation in a body with displacements $u^{(F)}$

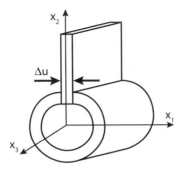

Figure 3.3. Edge dislocation generated by cutting along the positive x_2 axis and inserting a half-plane of material with uniform thickness Δu.

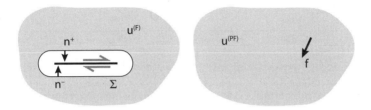

Figure 3.4. Definition sketch for Volterra's formula.

and tractions $\sigma_{ij}^{(F)} n_j$ acting on the fault surface Σ (figure 3.4). Let the primed system be an isolated point force, $f^{(PF)}$, associated with displacements $u^{(PF)}$ and tractions $\sigma_{ij}^{(PF)} n_j$ acting on Σ. We represent the point force of unit magnitude as a *Dirac delta function* (figure 3.5) at $\mathbf{x}$ acting in the k direction as $\delta_{ik}\delta(\mathbf{x} - \boldsymbol{\xi})$. Application of the reciprocal theorem (1.115) leads to

$$\int_V f_i^{(PF)} u_i^{(F)} \mathrm{d}V + \int_\Sigma u_i^{(F)} \sigma_{ij}^{(PF)} n_j \mathrm{d}A = \int_\Sigma u_i^{(PF)} \sigma_{ij}^{(F)} n_j \mathrm{d}A. \tag{3.2}$$

Consider the first integral. Integration over the delta function yields the displacement at x:

$$\int_V u_i^{(F)}(\boldsymbol{\xi}) \delta_{ik}\delta(\mathbf{x} - \boldsymbol{\xi}) \mathrm{d}V(\boldsymbol{\xi}) = u_k^{(F)}(\mathbf{x}). \tag{3.3}$$

Next consider the second integral on the left-hand side, which we break into two halves, one above the dislocation with unit normal n^+ and one below the dislocation with unit normal n^-. Note that the normals always point out of the body. The fault displacements $u_i^{(F)}$ are discontinuous across the boundary, whereas the stresses due to the point force are continuous. Thus, taking $\mathbf{n} = \mathbf{n}^- = -\mathbf{n}^+$,

$$\int_\Sigma u_i^{(F)} \sigma_{ij}^{(PF)} n_j \mathrm{d}A(\boldsymbol{\xi}) = \int_\Sigma u_i^{-(F)} \sigma_{ij}^{(PF)} n_j \mathrm{d}A(\boldsymbol{\xi}) - \int_\Sigma u_i^{+(F)} \sigma_{ij}^{(PF)} n_j \mathrm{d}A(\boldsymbol{\xi}),$$

$$= -\int_\Sigma (u_i^{+(F)} - u_i^{-(F)}) \sigma_{ij}^{(PF)} n_j \mathrm{d}A(\boldsymbol{\xi}),$$

$$= -\int_\Sigma s_i \sigma_{ij}^{(PF)} n_j \mathrm{d}A(\boldsymbol{\xi}), \tag{3.4}$$

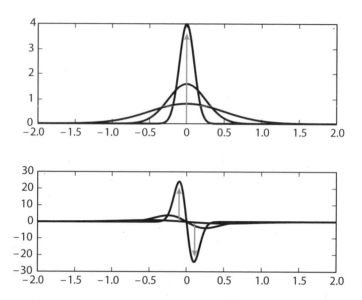

Figure 3.5. Dirac delta function (top) and its spatial derivative (bottom) as the limit of a Gaussian distribution, with standard deviation σ in the limit $\sigma \to 0$. Curves are drawn for $\sigma = 0.5, 0.25$, and 0.1. Gaussian distributions are normalized to unit area; the delta function, which can be defined as a limit of such distributions, thus has unit area but vanishes for all $x \neq 0$. The derivative of the delta function, found from the same limiting process, is equal to a pair of oppositely signed impulses.

where the displacement discontinuity vector, s_i, is defined as the difference in displacement between the upper and lower dislocation faces:

$$s_i \equiv u_i^{+(F)} - u_i^{-(F)}. \tag{3.5}$$

s_i corresponds to a fault slip vector when there is no opening or closing component to the displacement discontinuity.

The corresponding integral on the right-hand side vanishes because both the displacement field due to the point force $u^{(PF)}$ and the traction due to the fault $\sigma_{ij}^{(F)} n_j$ must be continuous across the fault surface. Thus, the integrals over the positive and negative sides of Σ cancel. Substituting equations (3.3) and (3.4) into (3.2) gives Volterra's formula:

$$\boxed{u_k(\mathbf{x}) = \int_\Sigma s_i(\boldsymbol{\xi}) \sigma_{ij}^k(\boldsymbol{\xi}, \mathbf{x}) n_j \mathrm{d}\Sigma(\boldsymbol{\xi}).} \tag{3.6}$$

We will refer to equation (3.6) as the *traction form* of Volterra's formula. The notation is that $\sigma_{ij}^k(\boldsymbol{\xi}, \mathbf{x})$ denotes the stress at $\boldsymbol{\xi}$ due to a point force in the k direction at $\mathbf{x}$. The result is not changed if we consider work done at remote boundaries far from the dislocation, as in the limit the relevant integrals all vanish. Similarly, inclusion of a free surface does not alter the result; since the tractions vanish on the free surface, no work is done there.

3.1.1 Body Force Equivalents and Moment Tensors

Volterra's formula allows us to compute the displacements due to an arbitrary dislocation, as long as the elastic stress due to a concentrated point force in the medium of interest is known.

Volterra's formula (3.6) can also be written in a more familiar form if we express the stress $\sigma_{ij}^k(\boldsymbol{\xi}, \mathbf{x})$ in terms of the elastostatic Green's tensors.

The Green's tensor $g_i^k(\mathbf{x}, \boldsymbol{\xi})$ is the displacement at $\mathbf{x}$ in the i direction, due to a concentrated unit force at $\boldsymbol{\xi}$ acting in the k direction. Given the Green's tensor, it is possible to compute the displacement due to an arbitrary force distributed by integrating. For example, given a distribution of forces $f_j(\boldsymbol{\xi})$, the displacement is

$$u_i(\mathbf{x}) = \int_V f_j(\boldsymbol{\xi}) g_i^j(\mathbf{x}, \boldsymbol{\xi}) dv_\xi. \tag{3.7}$$

Suppose that the distribution $f_j(\boldsymbol{\xi})$ is a localized unit point force at $\boldsymbol{\xi}_1$ acting in the kth direction. Then, $f_j(\boldsymbol{\xi}) = \delta_{jk}\delta(\boldsymbol{\xi} - \boldsymbol{\xi}_1)$. Substituting into equation (3.7), we find that $u_i(\mathbf{x}) = g_i^k(\mathbf{x}, \boldsymbol{\xi}_1)$, which recovers the definition of the Green's function.

The stress associated with the displacement, g_i^k is from Hooke's law $\sigma_{ij} = C_{ijpq}\epsilon_{pq}$. So Volterra's formula (3.6) can be written as

$$u_k(\mathbf{x}) = \int_\Sigma C_{ijpq} n_j s_i(\boldsymbol{\xi}) \frac{\partial g_p^k(\boldsymbol{\xi}, \mathbf{x})}{\partial \xi_q} d\Sigma(\boldsymbol{\xi}). \tag{3.8}$$

Equation (3.8) can be cast in a more convenient form, by making use of the fact that Green's tensors exhibit source–receiver *reciprocity*. Consider two states: The first state involves a concentrated force f_j at $\mathbf{x} = \boldsymbol{\xi}_1$ with associated displacements $g_i^j(\mathbf{x}, \boldsymbol{\xi}_1)$. The second state involves a concentrated force f_k at $\mathbf{x} = \boldsymbol{\xi}_2$ with associated displacements $g_i^k(\mathbf{x}, \boldsymbol{\xi}_2)$. Application of the reciprocal theorem leads to

$$\int_V \delta_{ij}\delta(\mathbf{x} - \boldsymbol{\xi}_1) g_i^k(\mathbf{x}, \boldsymbol{\xi}_2) dV = \int_V \delta_{ik}\delta(\mathbf{x} - \boldsymbol{\xi}_2) g_i^j(\mathbf{x}, \boldsymbol{\xi}_1) dV,$$

$$g_j^k(\boldsymbol{\xi}_1, \boldsymbol{\xi}_2) = g_k^j(\boldsymbol{\xi}_2, \boldsymbol{\xi}_1). \tag{3.9}$$

In other words, the displacement at $\boldsymbol{\xi}_1$ in the j direction due to a force at $\boldsymbol{\xi}_2$ in the k direction is exactly equal to the displacement at $\boldsymbol{\xi}_2$ in the k direction due to a force at $\boldsymbol{\xi}_1$ in the j direction. We are now in a position to make use of reciprocity to write equation (3.8) as

$$u_k(\mathbf{x}) = \int_\Sigma C_{ijpq} s_i(\boldsymbol{\xi}) n_j \frac{\partial g_k^p(\mathbf{x}, \boldsymbol{\xi})}{\partial \xi_q} d\Sigma(\boldsymbol{\xi}), \tag{3.10}$$

where the Green's tensors now give the displacement in the k direction at observation point $\mathbf{x}$, due to a point force acting in the p direction at $\boldsymbol{\xi}$. The term $C_{ijpq} s_i n_j$ is a second-rank tensor:

$$m_{pq} = C_{ijpq} s_i n_j, \tag{3.11}$$

which is known as the *areal moment tensor density* (e.g., Aki and Richards 1980). With this definition, Volterra's formula can be written compactly as

$$u_k(\mathbf{x}) = \int_\Sigma m_{pq}(\boldsymbol{\xi}) \frac{\partial g_k^p(\mathbf{x}, \boldsymbol{\xi})}{\partial \xi_q} d\Sigma(\boldsymbol{\xi}). \tag{3.12}$$

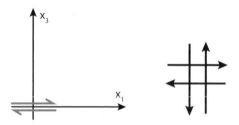

Figure 3.6. Slip in the x_1 direction on a fault plane perpendicular to the x_3 axis on the left corresponds to the double-couple representation on the right.

Equations (3.10) and (3.12) are the principal results of this section. Equation (3.12) shows that the displacements can be written as a convolution of the moment tensor density with the derivatives of the Green's tensors. The *moment tensor* gives a measure of the size and average mechanism of an earthquake:

$$M_{pq} = \int_{\Sigma} m_{pq} d\Sigma. \tag{3.13}$$

For an isotropic medium, equation (1.102) can be used for the elastic stiffness tensor so that the moment density (3.11) is written as

$$m_{ij} = \mu(s_i n_j + s_j n_i) + \lambda s_k n_k \delta_{ij}. \tag{3.14}$$

Substituting equation (3.14) into (3.12) yields another form of Volterra's equation appropriate for an isotropic medium:

$$u_k(\mathbf{x}) = \int_{\Sigma} s_i(\boldsymbol{\xi}) \left[\mu \left(\frac{\partial g_k^i}{\partial \xi_j} + \frac{\partial g_k^j}{\partial \xi_i} \right) + \lambda \delta_{ij} \frac{\partial g_k^m}{\partial \xi_m} \right] n_j d\Sigma(\boldsymbol{\xi}). \tag{3.15}$$

For a purely shearing source—that is, one in which there is no component of displacement discontinuity normal to the dislocation surface—$s_i n_i = 0$, and the third term in the integral (3.15) vanishes. Note however, that for a dike or sill with opening normal to the plane, we need to include this term.

Equation (3.15) shows that the displacements are due to *double-couple* sources. For example, take $i = 1$ and $j = 3$—that is, a fault plane perpendicular to the x_3 axis, with slip in the x_1 direction. According to equation (3.15), the displacement in the k direction is related to $\partial g_k^1/\partial \xi_3 + \partial g_k^3/\partial \xi_1$, which is the displacement due to a pair of force couples acting as shown in figure 3.6. To understand why this is so, consider a point force acting in the positive direction at $\xi + \Delta\xi$. Add a force acting in the negative direction at ξ, and take the limit as $\Delta\xi \to 0$. The resulting expression is the formal definition of a derivative. Thus, differentiating a point-source solution with respect to the *source* coordinates leads to a force couple. (Note the derivative of the Green's function with respect to the *observation* coordinates is a strain, not the displacement due to a force couple.) Adding force couples in two orthogonal directions gives the well-known double-couple representation of an earthquake source. The two force couples balance so that there is no net moment about the source. We thus refer to equation (3.15) as the *double-couple form* of Volterra's formula.

It is possible to determine a distribution of body forces that produces displacements equivalent to a dislocation (e.g., Aki and Richards 1980). Return to equation (3.12), but write the derivative of the Green's functions as

$$\frac{\partial g_k^i(\mathbf{x}, \boldsymbol{\xi})}{\partial \xi_j} = -\int_V g_k^i(\mathbf{x}, \boldsymbol{\eta}) \frac{\partial}{\partial \eta_j} \delta(\boldsymbol{\eta} - \boldsymbol{\xi}) \mathrm{d}V(\boldsymbol{\eta}), \tag{3.16}$$

which can be seen by integration by parts,

$$-\int_V g_k^i(\mathbf{x}, \boldsymbol{\eta}) \frac{\partial}{\partial \eta_j} \delta(\boldsymbol{\eta} - \boldsymbol{\xi}) \mathrm{d}V(\boldsymbol{\eta}) = \int_V \delta(\boldsymbol{\eta} - \boldsymbol{\xi}) \frac{\partial}{\partial \eta_j} g_k^i(\mathbf{x}, \boldsymbol{\eta}) \mathrm{d}V(\boldsymbol{\eta}),$$

$$= \frac{\partial g_k^i(\mathbf{x}, \boldsymbol{\xi})}{\partial \xi_j}. \tag{3.17}$$

The derivative of the Dirac delta function can be understood as the derivative of a Gaussian function in the limit that the width of the distribution (the standard deviation) goes to zero (figure 3.5). Substituting this expression into equation (3.12) and exchanging order of integration yields

$$u_k(\mathbf{x}) = \int_V \left[-\int_\Sigma m_{ij}(\boldsymbol{\xi}) \frac{\partial}{\partial \eta_j} \delta(\boldsymbol{\eta} - \boldsymbol{\xi}) \mathrm{d}\Sigma(\boldsymbol{\xi}) \right] g_k^i(\mathbf{x}, \boldsymbol{\eta}) \mathrm{d}V(\boldsymbol{\eta}). \tag{3.18}$$

This shows that the displacements anywhere in the medium can be computed as an appropriate distribution of body forces convolved with the Green's functions. The equivalent body forces are

$$f_i(\boldsymbol{\eta}) = -\int_\Sigma m_{ij}(\boldsymbol{\xi}) \frac{\partial}{\partial \eta_j} \delta(\boldsymbol{\eta} - \boldsymbol{\xi}) \mathrm{d}\Sigma(\boldsymbol{\xi}). \tag{3.19}$$

The delta function in equation (3.19) can be expanded as the product of delta functions in the three coordinate directions; assuming without loss of generality that the fault plane is normal to the x_3 direction, $\delta(\boldsymbol{\eta} - \boldsymbol{\xi}) = \delta(\eta_1 - \xi_1)\delta(\eta_2 - \xi_2)\delta(\eta_3)$. Given this, and expanding the sum over j in equation (3.19), we have

$$f_i(\boldsymbol{\eta}) = -\delta(\eta_3) \frac{\partial}{\partial \eta_1} \int \int m_{i1}(\xi_1, \xi_2) \delta(\eta_1 - \xi_1)\delta(\eta_2 - \xi_2) \mathrm{d}\xi_1 \mathrm{d}\xi_2$$

$$-\delta(\eta_3) \frac{\partial}{\partial \eta_2} \int \int m_{i2}(\xi_1, \xi_2) \delta(\eta_1 - \xi_1)\delta(\eta_2 - \xi_2) \mathrm{d}\xi_1 \mathrm{d}\xi_2$$

$$-\frac{\partial}{\partial \eta_3} \delta(\eta_3) \int \int m_{i3}(\xi_1, \xi_2) \delta(\eta_1 - \xi_1)\delta(\eta_2 - \xi_2) \mathrm{d}\xi_1 \mathrm{d}\xi_2,$$

$$= -\delta(\eta_3) \frac{\partial}{\partial \eta_1} m_{i1}(\eta_1, \eta_2) - \delta(\eta_3) \frac{\partial}{\partial \eta_2} m_{i2}(\eta_1, \eta_2) - m_{i3}(\eta_1, \eta_2) \frac{\partial \delta(\eta_3)}{\partial \eta_3}. \tag{3.20}$$

Note that $f_i(\boldsymbol{\eta})$ is a spatial distribution of body force densities, with units force per unit volume.

If we consider the special case in figure 3.6 in which slip acts in the x_1 direction and the fault is normal to the x_3 direction, then from equation (3.14), the only nonzero components of the moment tensor density are $m_{13} = m_{31}$. The net moment about the x_2 axis has two contributions, one due to forces acting in the 1-direction, the other due to forces acting in the 3-direction. It can be shown (problem 1) that these two contributions are of equal magnitude but opposite sign, such that the net moment vanishes. The magnitude of the individual terms is defined as the *scalar seismic moment*, M_0, which for a homogeneous, isotropic medium is

$$M_0 = \mu \int_\Sigma s(\xi_1, \xi_2) d\Sigma(\xi) = \mu \bar{s} A \qquad (3.21)$$

(problem 1), where $\bar{s}$ is the average slip, and A is the fault area.

For a point source at the origin, the areal moment tensor density is $m_{ij}(\eta_1, \eta_2) = M_{ij}\delta(\eta_1)\delta(\eta_2)$, where M is the moment tensor defined in equation (3.13). In this case, equation (3.20) becomes

$$f_i(\eta) = -M_{ij} \frac{\partial}{\partial \eta_j} \delta(\eta). \qquad (3.22)$$

One can then write the equilibrium equations using the body force equivalent of a point source as

$$\boxed{\nabla \cdot \sigma - M \cdot \nabla \delta(\mathbf{x} - \mathbf{x}_s) = 0,} \qquad (3.23)$$

where $\mathbf{x}_s$ is the source location. To model finite sources, one simply integrates the point source result over the fault surface.

To understand the physical interpretation of equation (3.23), note that the spatial derivative of the Dirac delta function is a pair of oppositely signed impulses (figure 3.5). Consider again the geometry shown in figure 3.6. The body force equivalent of a point moment tensor at $\mathbf{x}_s = 0$ is $-M \cdot \nabla \delta(\mathbf{x})$. For the geometry in figure 3.6, the only nonzero components of the moment tensor are $M_{13} = M_{31} = \mu s A$, where A is the source area. The equivalent force in the 1-direction is thus $-M_{13}\partial\delta(\mathbf{x})/\partial x_3 = -\mu s A \delta(x_1)\delta(x_2)\partial\delta(x_3)/\partial x_3$. The product $\delta(x_1)\delta(x_2)$ localizes the source at the origin $x_1 = 0$, $x_2 = 0$. The term $\partial\delta(x_3)/\partial x_3$ represents a force couple with forces acting in the 1-direction and lever arm in the 3-direction; the negative sign causes the force on the $x_3 > 0$ side of the fault to act in the positive x_1 direction (compare to figure 3.5). The equivalent force in the 3-direction is $-M_{31}\partial\delta(\mathbf{x})/\partial x_1 = -\mu s A \delta(x_2)\delta(x_3)\partial\delta(x_1)/\partial x_1$, which is a force couple, with forces in the 3-direction and lever arm in the 1-direction.

While the moment tensor description is widely used in seismology, the moment tensor is not well defined when there is a contrast in elastic properties across the fault, as might well exist if the fault juxtaposes two different rock types. Should one use the elastic moduli on one side or the other? Or an average? There is no unique choice. Notice that this ambiguity does not exist for the traction form of Volterra's formula (3.6), because the tractions must be continuous across the fault surface even if the elastic properties are not. Some workers advocate replacing the moment tensor with a *potency*, which is independent of the elastic properties at the fault. For example, equation (3.6) could be written as

$$u_k(\mathbf{x}) = \int_\Sigma p_{ij}(\xi)\sigma_{ij}^k(\xi, \mathbf{x})d\Sigma(\xi), \qquad (3.24)$$

where due to the symmetry in the stress, the *potency density tensor* is

$$p_{ij} = \frac{1}{2}\left(s_i n_j + s_j n_i\right) \qquad (3.25)$$

(e.g., Ampuero and Dahlen 2005).

3.2 Screw Dislocations

As a first application of Volterra's formula, we compute the displacements due to a screw dislocation, which were previously derived, more or less by inspection, in chapter 2. Because the problem is two-dimensional, the appropriate Green's function is equivalent to a *line source*, rather than a point source. Adopting the coordinate system in figure 2.1, a vertical fault lies in the plane $\xi_1 = 0$, and extends infinitely in the 3-direction. The unit normal is $n_j = \delta_{1j}$, and the slip vector is $s_i = s\delta_{i3}$. Thus, Volterra's formula (3.6) becomes

$$u_3(x_1, x_2) = \int_{-\infty}^{0} s_3(\xi_2) \left[\int_{-\infty}^{\infty} \sigma_{13}^3(\xi_1 = 0, \xi_2; x_1, x_2, x_3 - \xi_3)d\xi_3 \right] d\xi_2, \qquad (3.26)$$

since the slip is independent of ξ_3. The stress depends only on the difference between the source and field coordinates, $x_3 - \xi_3$. Thus, integrating with respect to ξ_3 is equivalent to integrating over $-x_3$. As the source is located at $\mathbf{x}$, integrating the point force solution along x_3 gives the displacements due to a line source extending infinitely in the 3-direction. Thus, equation (3.26) can be written as

$$u_3(x_1, x_2) = \int_{-\infty}^{0} s_3(\xi_2)\tilde{\sigma}_{13}^3(\xi_1 = 0, \xi_2; x_1, x_2)d\xi_2, \qquad (3.27)$$

where $\tilde{\sigma}_{13}^3$ is the shear stress acting on the plane $\xi_1 = 0$ due to a *line force* in the 3-direction at x_1, x_2.

Rather than determine the stress due to the line force by integrating the three-dimensional Green's function, we derive the antiplane result here independently. As in chapter 2, we begin with the solution in a full-space and construct the half-space solution using the method of images. Recalling the equilibrium equations for antiplane strain (2.7), and noting that for the full-space, symmetry requires the displacements be independent of θ, the governing equation is (see equation [2.9])

$$\mu \left(\frac{\partial^2 \tilde{g}_3^3}{\partial r^2} + \frac{1}{r} \frac{\partial \tilde{g}_3^3}{\partial r} \right) = -\delta(r), \qquad (3.28)$$

where $\delta(r)$ is the Dirac delta function representing a concentrated line force at the origin acting in the x_3 direction, and $\tilde{g}_3^3$ denotes the displacement in the x_3 direction due to a concentrated line force acting in the x_3 direction. The free-surface boundary condition requires that the shear stress $\tilde{\sigma}_{23}$ vanish on the surface $\xi_2 = 0$.

The homogeneous equation has the following solution:

$$\tilde{g}_3^3 = A \ln r, \quad r > 0. \qquad (3.29)$$

From equation (1.23), the strain is $\epsilon_{3r} = (1/2)\partial u_3/\partial r$, and from Hooke's law, the corresponding stress is

$$\tilde{\sigma}_{3r} = \frac{\mu A}{r}, \qquad (3.30)$$

where $\tilde{\sigma}_{3r}$ results from forces in the x_3 direction acting on surfaces with unit normal in the r direction. The constant A is determined by noting that the integral of the stress on a cylinder about the origin must balance the concentrated line force, which has unit magnitude

$$1 = \int_0^{2\pi} \tilde{\sigma}_{3r} r \, d\theta = 2\pi \mu A. \tag{3.31}$$

Thus, $A = 1/2\pi\mu$, and the Green's function is

$$\tilde{g}_3^3 = \frac{1}{2\pi\mu} \ln r = \frac{1}{2\pi\mu} \ln \sqrt{(\xi_1 - x_1)^2 + (\xi_2 + x_2)^2}. \tag{3.32}$$

Here, the point source has been offset from the origin to the point $(x_1, -x_2)$. The horizontal shear stress is

$$\tilde{\sigma}_{23}^3 = \mu \frac{\partial \tilde{g}_3^3}{\partial x_2} = \frac{1}{2\pi} \frac{\xi_2 + x_2}{(\xi_1 - x_1)^2 + (\xi_2 + x_2)^2}, \tag{3.33}$$

which does not match the free-surface boundary condition $\tilde{\sigma}_{23}(\xi_2 = 0) = 0$. However, adding an image source of the same sign at (x_1, x_2) does cause the shear stress to vanish.

Equation (3.27) involves the shear stress $\tilde{\sigma}_{13}^3$ acting on the fault surface, which we take to be the plane $\xi_1 = 0$. Including the concentrated line force and its image,

$$\tilde{\sigma}_{13}^3 = -\frac{1}{2\pi} \left[\frac{x_1}{x_1^2 + (\xi_2 + x_2)^2} + \frac{x_1}{x_1^2 + (\xi_2 - x_2)^2} \right]. \tag{3.34}$$

If we measure the displacement only on the free surface, then $x_2 = 0$, and equation (3.27) becomes

$$u_3(x_1, x_2 = 0) = -\frac{1}{\pi} \int_{-\infty}^0 s_3(\xi_2) \frac{x_1}{x_1^2 + \xi_2^2} \, d\xi_2, \tag{3.35}$$

which is consistent with equation (2.41).

3.3 Two-Dimensional Edge Dislocations

In order to use Volterra's formula to compute the displacements due to an infinitely long edge dislocation, we require the plane strain Green's functions. These correspond to the displacements from a concentrated line force acting in the plane. Rather than derive the result as for the antiplane strain case earlier, we make use of Melan's (1932; note correction in 1940) results. (See also Dundurs [1962], who gives results for concentrated line forces in joined elastic half-planes; the half-plane is a special case in which the shear modulus of the region not containing the localized line force vanishes.)

The Green's functions could be derived starting from the well-known solution for a concentrated force in a full-plane (Love 1944, article 148). An image source of the same sign located symmetrically about the plane $x_2 = 0$ cancels one component of the traction acting on $x_2 = 0$. The problem then is to remove the remaining traction component. Melan's approach is to use the result for a concentrated force acting on the boundary of a half-plane. An appropriative distribution of such forces is constructed that exactly cancels the tractions from the full-space line force and its image. An alternative approach is to use Fourier transform methods to remove the resulting tractions. Such an approach is used in section 10.6 to derive the solution for a

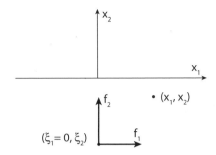

Figure 3.7. Line forces at $(0, \xi_2)$, extending infinitely in the 3-direction, give rise to displacements measured at (x_1, x_2).

line center of dilatation, a simple model of an expanding magma chamber, in a poroelastic half-plane.

For a line force in the x_1 direction, the displacements at (x_1, x_2) resulting from a line force acting at $(0, \xi_2)$ (figure 3.7) are

$$g_1^1 = \frac{-1}{2\pi\mu\,(1-\nu)}\left[\frac{3-4\nu}{4}\ln r_1 + \frac{8\nu^2 - 12\nu + 5}{4}\ln r_2 + \frac{(x_2 - \xi_2)^2}{4r_1^2}\right.$$
$$\left. + \frac{(3-4\nu)\,(x_2+\xi_2)^2 + 2\xi_2\,(x_2+\xi_2) - 2\xi_2^2}{4r_2^2} - \frac{\xi_2 x_2\,(x_2+\xi_2)^2}{r_2^4}\right], \qquad (3.36)$$

$$g_2^1 = \frac{1}{2\pi\mu\,(1-\nu)}\left[(1-2\nu)\,(1-\nu)\,\theta_2 + \frac{(x_2-\xi_2)\,x_1}{4r_1^2} + (3-4\nu)\,\frac{(x_2-\xi_2)\,x_1}{4r_2^2} - \frac{\xi_2 x_2 x_1\,(x_2+\xi_2)}{r_2^4}\right]. \qquad (3.37)$$

For a line force in the x_2 direction,

$$g_1^2 = \frac{1}{2\pi\mu\,(1-\nu)}\left[-(1-2\nu)\,(1-\nu)\,\theta_2 + \frac{(x_2-\xi_2)\,x_1}{4r_1^2} + (3-4\nu)\,\frac{(x_2-\xi_2)\,x_1}{4r_2^2} + \frac{\xi_2 x_2 x_1\,(x_2+\xi_2)}{r_2^4}\right], \qquad (3.38)$$

$$g_2^2 = \frac{1}{2\pi\mu\,(1-\nu)}\left[-\frac{3-4\nu}{4}\ln r_1 - \frac{8\nu^2 - 12\nu + 5}{4}\ln r_2 - \frac{x_1^2}{4r_1^2} + \frac{2\xi_2 x_2 - (3-4\nu)\,x_1^2}{4r_2^2} - \frac{\xi_2 x_2 x_1^2}{r_2^4}\right], \qquad (3.39)$$

where

$$r_1^2 = x_1^2 + (x_2 - \xi_2)^2,$$
$$r_2^2 = x_1^2 + (x_2 + \xi_2)^2, \qquad (3.40)$$

and

$$\theta_2 = \tan^{-1}\left(\frac{x_1 - \xi_1}{x_2 + \xi_2}\right). \qquad (3.41)$$

Note that if the line force is located at $x_1 = \xi_1$ rather than at $x_1 = 0$, we simply replace x_1 with $x_1 - \xi_1$.

For a line force in the x_1 direction, the corresponding stresses are

$$
\sigma_{11}^1 = \frac{-1}{2\pi(1-v)}\left\{\frac{x_1^3}{r_1^4} + \frac{x_1\left(x_1^2 - 4\xi_2 x_2 - 2\xi_2^2\right)}{r_2^4} + \frac{8\xi_2 x_2 x_1\left(x_2 + \xi_2\right)^2}{r_2^6}\right.
$$
$$
\left. + \frac{1-2v}{2}\left[\frac{x_1}{r_1^2} + \frac{3x_1}{r_2^2} - \frac{4x_2 x_1\left(x_2 + \xi_2\right)}{r_2^4}\right]\right\}, \tag{3.42}
$$

$$
\sigma_{12}^1 = \frac{-1}{2\pi(1-v)}\left\{\frac{\left(x_2 - \xi_2\right)x_1^2}{r_1^4} + \frac{\left(x_2 + \xi_2\right)\left(2\xi_2 x_2 + x_1^2\right)}{r_2^4} - \frac{8\xi_2 x_2 x_1^2\left(x_2 + \xi_2\right)}{r_2^6}\right.
$$
$$
\left. + \frac{1-2v}{2}\left[\frac{x_2 - \xi_2}{r_1^2} + \frac{3x_2 + \xi_2}{r_2^2} - \frac{4x_2\left(x_2 + \xi_2\right)^2}{r_2^4}\right]\right\}, \tag{3.43}
$$

$$
\sigma_{22}^1 = \frac{-1}{2\pi(1-v)}\left\{\frac{\left(x_2 - \xi_2\right)^2 x_1}{r_1^4} - \frac{x_1\left(\xi_2^2 - x_2^2 + 6\xi_2 x_2\right)}{r_2^4} + \frac{8\xi_2 x_2 x_1^3}{r_2^6}\right.
$$
$$
\left. - \frac{1-2v}{2}\left[\frac{x_1}{r_1^2} - \frac{x_1}{r_2^2} - \frac{4x_2 x_1\left(x_2 + \xi_2\right)}{r_2^4}\right]\right\}. \tag{3.44}
$$

For a line force in the x_2 direction,

$$
\sigma_{11}^2 = \frac{-1}{2\pi(1-v)}\left\{\frac{\left(x_2 - \xi_2\right)x_1^2}{r_1^4} + \frac{\left(x_2 + \xi_2\right)\left(x_1^2 + 2\xi_2^2\right) - 2\xi_2 x_1^2}{r_2^4} + \frac{8\xi_2 x_2\left(x_2 + \xi_2\right)x_1^2}{r_2^6}\right.
$$
$$
\left. + \frac{1-2v}{2}\left[-\frac{x_2 - \xi_2}{r_1^2} + \frac{3\xi_2 + x_2}{r_2^2} + \frac{4x_2 x_1^2}{r_2^4}\right]\right\}, \tag{3.45}
$$

$$
\sigma_{12}^2 = \frac{-x_1}{2\pi(1-v)}\left\{\frac{\left(x_2 - \xi_2\right)^2}{r_1^4} + \frac{x_2^2 - 2\xi_2 x_2 - \xi_2^2}{r_2^4} + \frac{8\xi_2 x_2\left(x_2 + \xi_2\right)^2}{r_2^6}\right.
$$
$$
\left. + \frac{1-2v}{2}\left[\frac{1}{r_1^2} - \frac{1}{r_2^2} + \frac{4x_2\left(x_2 + \xi_2\right)}{r_2^4}\right]\right\}, \tag{3.46}
$$

$$
\sigma_{22}^2 = \frac{-1}{2\pi(1-v)}\left\{\frac{\left(x_2 - \xi_2\right)^3}{r_1^4} + \frac{\left(x_2 + \xi_2\right)\left[\left(x_2 + \xi_2\right)^2 + 2\xi_2 x_2\right]}{r_2^4} - \frac{8\xi_2 x_2\left(x_2 + \xi_2\right)x_1^2}{r_2^6}\right.
$$
$$
\left. + \frac{1-2v}{2}\left[\frac{x_2 - \xi_2}{r_1^2} + \frac{3x_2 + \xi_2}{r_2^2} - \frac{4x_2 x_1^2}{r_2^4}\right]\right\}. \tag{3.47}
$$

We are now in a position to evaluate the displacements for dip-slip faults. For simplicity, observation points are restricted to the free surface where field measurements are collected. Volterra's formula (3.6) requires the stress acting on the fault due to a line force acting at the observation point. In the Melan solution, the line force is applied at ξ, and the stress is resolved at $\mathbf{x}$. To be consistent with our notation for Volterra's formula, we swap $\mathbf{x}$ and ξ in equation (3.6). For a vertical fault $n_j = \delta_{1j}$ and $s_i = s\delta_{2i}$, so with the current notation, equation (3.6) becomes

$$
u_k(\xi_1, \xi_2 = 0) = \int s(x_2)\tilde{\sigma}_{12}^k(x_1, x_2; \xi_1, \xi_2 = 0)dx_2 \quad k = 1, 2. \tag{3.48}
$$

As in section 3.2, we have integrated the point force solution along x_3 to produce a line source. The tilde notation is dropped in the following; however, it should be understood that displacement and stress Green's functions are associated with line sources rather than point sources. Given that observations are restricted to the free surface, we simplify Melan's solution for a line source acting on the free surface $\xi_2 = 0$:

$$\sigma_{12}^1(\xi_1, \xi_2 = 0) = -\frac{2}{\pi} \frac{(x_1 - \xi_1)^2 x_2}{[(x_1 - \xi_1)^2 + x_2^2]^2} \tag{3.49}$$

$$\sigma_{12}^2(\xi_1, \xi_2 = 0) = -\frac{2}{\pi} \frac{(x_1 - \xi_1) x_2^2}{[(x_1 - \xi_1)^2 + x_2^2]^2}. \tag{3.50}$$

For a vertical fault located at the origin of the coordinate system, take $x_1 = 0$, without loss of generality, and $x_2 = z$, where z is the dummy variable on the fault surface. Assuming uniform slip below depth d, substituting equation (3.50) into (3.48) and integrating leads to

$$\begin{aligned}
u_1(\xi_1, \xi_2 = 0) &= \frac{-2s\xi_1^2}{\pi} \int_{-\infty}^{-d} \frac{z}{(\xi_1^2 + z^2)^2} dz, \\
&= \frac{2s\xi_1^2}{\pi} \left[\frac{1}{2(\xi_1^2 + z^2)} \right]_{z=-\infty}^{z=-d}, \\
&= \frac{s}{\pi} \frac{\xi_1^2}{\xi_1^2 + d^2}.
\end{aligned} \tag{3.51}$$

For the vertical component of displacement due to slip on a vertical fault,

$$\begin{aligned}
u_2(\xi_1, \xi_2 = 0) &= \frac{2s\xi_1}{\pi} \int_{-\infty}^{-d} \frac{z^2}{(\xi_1^2 + z^2)^2} dz, \\
&= \frac{2s\xi_1}{\pi} \left[\frac{-z}{2(\xi_1^2 + z^2)} + \frac{1}{2\xi_1} \tan^{-1}\left(\frac{z}{\xi_1}\right) \right]_{z=-\infty}^{z=-d}, \\
&= \frac{s}{\pi} \left[\frac{\xi_1 d}{\xi_1^2 + d^2} + \tan^{-1}\left(\frac{\xi_1}{d}\right) \right].
\end{aligned} \tag{3.52}$$

Notice that the horizontal displacement u_1 goes to s/π for $|\xi_1| \gg d$. We can remove this rigid body motion, without affecting the stresses or strains. After doing so and introducing a distance parameter scaled by the fault depth $\zeta = \xi_1/d$, we have

$$u_1 = -\frac{s}{\pi} \frac{1}{1 + \zeta^2},$$

$$u_2 = \frac{s}{\pi} \left[\frac{\zeta}{1 + \zeta^2} + \tan^{-1}(\zeta) \right]. \tag{3.53}$$

3.3.1 Dipping Fault

We now consider the more general case of a fault dipping at an angle δ with respect to the horizontal. The most direct way to derive the displacements at the free surface is to compute the displacements for a horizontal dislocation (see problem 2). The displacements due to a dipping dislocation are found by a vector sum of the horizontal and vertical components of the slip vector (see problem 3). Vector summation of the two components is valid because, as has been emphasized earlier, the dislocation line is the source of deformation, not the dislocation plane. This approach is not valid for the displacements within the earth, however, where we we must ensure that the branch cut coincides with the dislocation surface.

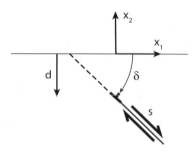

Figure 3.8. Geometry for a dipping edge dislocation in a half-space. The dislocation line, corresponding to the top of the fault, is at depth d, and the dislocation dips at angle δ from the free surface.

Here we will examine a more general, if somewhat more involved, approach using the double-couple form of Volterra's formula that is valid everywhere, not only at the free surface. From figure 3.8, the fault normal is $\mathbf{n} = [\sin \delta, \cos \delta, 0]^T$, and the fault slip is $\mathbf{s} = s[\cos \delta, -\sin \delta, 0]^T$.

Assuming constant slip over a confined interval, equation (3.15) becomes

$$
u_k(\mathbf{x}) = \mu s \int \left[2 \sin \delta \cos \delta \left(\frac{\partial g_k^1}{\partial \xi_1} - \frac{\partial g_k^2}{\partial \xi_2} \right) + (\cos^2 \delta - \sin^2 \delta) \left(\frac{\partial g_k^1}{\partial \xi_2} + \frac{\partial g_k^2}{\partial \xi_1} \right) \right] d\Sigma,
$$

$$
= \mu s \int \left[\sin 2\delta \left(\frac{\partial g_k^1}{\partial \xi_1} - \frac{\partial g_k^2}{\partial \xi_2} \right) + \cos 2\delta \left(\frac{\partial g_k^1}{\partial \xi_2} + \frac{\partial g_k^2}{\partial \xi_1} \right) \right] d\Sigma. \tag{3.54}
$$

The derivatives of the plane strain Green's tensors for displacements in the x_1 direction, evaluated at the free surface ($x_2 = 0$), are

$$
\frac{\partial g_1^1}{\partial \xi_1} = \frac{-1}{\pi \mu} \left\{ \frac{(x_1 - \xi_1)[\nu \xi_2^2 - (1 - \nu)(x_1 - \xi_1)^2]}{[(x_1 - \xi_1)^2 + \xi_2^2]^2} \right\},
$$

$$
\frac{\partial g_1^1}{\partial \xi_2} = \frac{-1}{\pi \mu} \left\{ \frac{\xi_2[(2 - \nu)(x_1 - \xi_1)^2 + (1 - \nu)\xi_2^2]}{[(x_1 - \xi_1)^2 + \xi_2^2]^2} \right\},
$$

$$
\frac{\partial g_1^2}{\partial \xi_1} = \frac{1}{\pi \mu} \left\{ \frac{\xi_2[(1 - \nu)\xi_2^2 - \nu(x_1 - \xi_1)^2]}{[(x_1 - \xi_1)^2 + \xi_2^2]^2} \right\},
$$

$$
\frac{\partial g_1^2}{\partial \xi_2} = \frac{1}{\pi \mu} \left\{ \frac{(x_1 - \xi_1)[(1 - \nu)\xi_2^2 - \nu(x_1 - \xi_1)^2]}{[(x_1 - \xi_1)^2 + \xi_2^2]^2} \right\}, \tag{3.55}
$$

whereas displacements in the x_2 direction are

$$
\frac{\partial g_2^1}{\partial \xi_1} = \frac{1}{\pi \mu} \left\{ \frac{\xi_2[\nu \xi_2^2 - (1 - \nu)(x_1 - \xi_1)^2]}{[(x_1 - \xi_1)^2 + \xi_2^2]^2} \right\},
$$

$$
\frac{\partial g_2^1}{\partial \xi_2} = \frac{1}{\pi \mu} \left\{ \frac{(x_1 - \xi_1)[\nu \xi_2^2 - (1 - \nu)(x_1 - \xi_1)^2]}{[(x_1 - \xi_1)^2 + \xi_2^2]^2} \right\},
$$

$$
\frac{\partial g_2^2}{\partial \xi_1} = \frac{1}{\pi \mu} \left\{ \frac{(x_1 - \xi_1)[(1 - \nu)(x_1 - \xi_1)^2 + (2 - \nu)\xi_2^2]}{[(x_1 - \xi_1)^2 + \xi_2^2]^2} \right\},
$$

$$
\frac{\partial g_2^2}{\partial \xi_2} = \frac{-1}{\pi \mu} \left\{ \frac{\xi_2[(1 - \nu)\xi_2^2 - \nu(x_1 - \xi_1)^2]}{[(x_1 - \xi_1)^2 + \xi_2^2]^2} \right\}. \tag{3.56}
$$

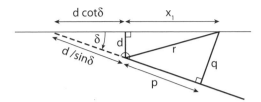

Figure 3.9. Coordinate system in which the vector from the fault to the observation point r is decomposed into perpendicular component q and parallel component p.

Combining with preceding equations (3.54) yields

$$u_1 = \frac{s}{\pi} \int \sin 2\delta \left[\frac{(x_1 - \xi_1)^3 - (x_1 - \xi_1)\xi_2^2}{r^4} \right] - 2 \cos 2\delta \left[\frac{(x_1 - \xi_1)^2 \xi_2}{r^4} \right] d\Sigma,$$

$$u_2 = \frac{s}{\pi} \int \sin 2\delta \left[\frac{\xi_2^3 - (x_1 - \xi_1)^2 \xi_2}{r^4} \right] + 2 \cos 2\delta \left[\frac{(x_1 - \xi_1)\xi_2^2}{r^4} \right] d\Sigma, \tag{3.57}$$

where

$$r^2 = (x_1 - \xi_1)^2 + \xi_2^2. \tag{3.58}$$

To perform the integration along the fault surface, we introduce a new coordinate η that runs along the fault surface. The coordinates of points on the fault are specified by

$$\xi_1 = \eta \cos \delta,$$

$$\xi_2 = -d - \eta \sin \delta, \tag{3.59}$$

where d is the depth to the top of the fault. In this coordinate system, the distance from the fault to the observation point is

$$r^2 = (x_1 - \eta \cos \delta)^2 + (d + \eta \sin \delta)^2. \tag{3.60}$$

The fact that all of the terms in equation (3.57) have r^4 in the denominator makes integration difficult. The problem can be simplified greatly by adopting a coordinate system that is parallel and perpendicular to the fault, following Sato and Matsu'ura (1974). Define q as the coordinate perpendicular to the fault and p as the coordinate parallel to the fault (figure 3.9).

With reference to this figure, $\sin \delta$ is $q/(x + d \cot \delta)$, so that $q = x \sin \delta + d \cos \delta$, where for notational simplicity, we take $x \equiv x_1$. From figure 3.9, $r^2 = x^2 + d^2 = p^2 + q^2$. The fault parallel coordinate p is found by

$$p^2 = r^2 - q^2 = r^2 - (x \sin \delta + d \cos \delta)^2,$$

$$= x^2 \cos^2 \delta + d^2 \sin^2 \delta - 2xd \sin \delta \cos \delta,$$

$$= (x \cos \delta - d \sin \delta)^2. \tag{3.61}$$

In summary,

$$q = x \sin \delta + d \cos \delta,$$

$$p = x \cos \delta - d \sin \delta, \tag{3.62}$$

and the inverse

$$x = p \cos \delta + q \sin \delta,$$
$$d = -p \sin \delta + q \cos \delta. \tag{3.63}$$

Introducing a new coordinate $\eta' = p - \eta$, the distance from a point on the fault to an observation point, from equation (3.60), is

$$
\begin{aligned}
r^2 &= [x - (p - \eta') \cos \delta]^2 + [d + (p - \eta') \sin \delta]^2, \\
&= x^2 + d^2 + (p - \eta')^2 - 2p(p - \eta'), \\
&= p^2 + q^2 + (p - \eta')^2 - 2p(p - \eta'), \\
&= q^2 + \eta'^2,
\end{aligned} \tag{3.64}
$$

which greatly simplifies the integration. If W is the downdip extent of slip, the limits of integration change from $\int_0^W d\eta$ to $\int_p^{p-W} d\eta'$. Expressing the integrals (3.57) in terms of η', we have

$$u_1 = -\frac{2s}{\pi} \int_p^{p-W} \frac{q\eta'(\eta' \cos \delta + q \sin \delta)}{(q^2 + \eta'^2)^2} d\eta',$$

$$u_2 = -\frac{2s}{\pi} \int_p^{p-W} \frac{q\eta'(q \cos \delta - \eta' \sin \delta)}{(q^2 + \eta'^2)^2} d\eta'. \tag{3.65}$$

These integrals are of the same form as those in equations (3.51) and (3.52), so equations (3.65) integrate to

$$u_1 = -\frac{s}{\pi} \left[\cos \delta \tan^{-1} \left(\frac{\eta'}{q} \right) - \frac{q\eta' \cos \delta + q^2 \sin \delta}{q^2 + \eta'^2} \right]_p^{p-W},$$

$$u_2 = \frac{s}{\pi} \left[\sin \delta \tan^{-1} \left(\frac{\eta'}{q} \right) + \frac{q^2 \cos \delta - q\eta' \sin \delta}{q^2 + \eta'^2} \right]_p^{p-W}. \tag{3.66}$$

For a single dislocation, let $W \to \infty$. Noting that $pq \cos \delta + q^2 \sin \delta = x(d \cos \delta + x \sin \delta)$ and $q^2 \cos \delta - qp \sin \delta = d(d \cos \delta + x \sin \delta)$, equation (3.66) becomes

$$u_1 = -\frac{s}{\pi} \left\{ \cos \delta \left[-\frac{\pi}{2} - \tan^{-1} \left(\frac{p}{q} \right) \right] + \frac{x(d \cos \delta + x \sin \delta)}{x^2 + d^2} \right\},$$

$$u_2 = \frac{s}{\pi} \left\{ \sin \delta \left[-\frac{\pi}{2} - \tan^{-1} \left(\frac{p}{q} \right) \right] - \frac{d(d \cos \delta + x \sin \delta)}{x^2 + d^2} \right\}. \tag{3.67}$$

Note from figure 3.9 that the following angles sum to π:

$$[\pi/2 - \tan^{-1}(p/q)] + \tan^{-1}(x/d) + (\pi/2 - \delta) = \pi. \tag{3.68}$$

where the three terms identify the three angles indicated in figure 3.9. Thus,

$$u_1 = \frac{s}{\pi} \left\{ \cos \delta [\tan^{-1}(x/d) - \delta + \pi/2] - \frac{x(d \cos \delta + x \sin \delta)}{x^2 + d^2} \right\},$$

$$u_2 = -\frac{s}{\pi} \left\{ \sin \delta [\tan^{-1}(x/d) - \delta + \pi/2] + \frac{d(d \cos \delta + x \sin \delta)}{x^2 + d^2} \right\}. \tag{3.69}$$

Once again, we can remove rigid body terms. In this case, subtract $(s/\pi)[(\pi/2 - \delta) \cos \delta - \sin \delta]$ from the horizontal displacement and $(s/\pi)(\delta - \pi/2) \sin \delta$ from the vertical. This yields

$$u_1(x_1, x_2 = 0) = \frac{s}{\pi} \left[\cos \delta \tan^{-1}(\zeta) + \frac{\sin \delta - \zeta \cos \delta}{1 + \zeta^2} \right],$$

$$u_2(x_1, x_2 = 0) = -\frac{s}{\pi} \left[\sin \delta \tan^{-1}(\zeta) + \frac{\cos \delta + \zeta \sin \delta}{1 + \zeta^2} \right], \tag{3.70}$$

where the parameter ζ is the distance from the dislocation scaled by the depth,

$$\zeta \equiv \frac{x_1 - \xi_1}{d}. \tag{3.71}$$

Note that as written here, $s > 0$ corresponds to normal faulting for $0 \le \delta \le \pi/2$ and reverse faulting for $\pi/2 \le \delta \le \pi$, while the opposite sense follows for $s < 0$.

Last, the surface parallel normal strain is

$$\epsilon_{11}(x_1, x_2 = 0) = \frac{\partial}{\partial x_1} u_1(x_1, x_2 = 0) = \frac{2s}{\pi d} \left[\frac{\zeta^2 \cos \delta - \zeta \sin \delta}{(1 + \zeta^2)^2} \right]. \tag{3.72}$$

Equations (3.70) apply to the edge dislocation geometry shown in figure 3.8. In order to model uniform slip over a confined interval, we superimpose two dislocations with opposite sign at either end of the fault surface. For a dislocation that breaks the surface, take a positive dislocation at $\xi_1 = 0$, $d = 0$, a negative dislocation at $\xi_1 = x_d \equiv d/\tan \delta = L \cos \delta$, and depth d, where L is the downdip length of the fault. Taking the limit of equations (3.70) as $d \to 0$, the only nonzero terms are the arctangents, which go to $\pi/2 \, \mathrm{sgn}(\zeta)$ in the limit. Here $\mathrm{sgn}(z)$ gives the sign of z. Thus, the displacements for a surface breaking dislocation are, taking $\xi_1 = x_d$ so that $\zeta = (x_1 - x_d)/d$,

$$u_1(x_1, x_2 = 0) = -\frac{s}{\pi} \left\{ \cos \delta \left[\tan^{-1}(\zeta) - (\pi/2) \, \mathrm{sgn}(x) \right] + \frac{\sin \delta - \zeta \cos \delta}{1 + \zeta^2} \right\},$$

$$u_2(x_1, x_2 = 0) = \frac{s}{\pi} \left\{ \sin \delta \left[\tan^{-1}(\zeta) - (\pi/2) \, \mathrm{sgn}(x) \right] + \frac{\cos \delta + \zeta \sin \delta}{1 + \zeta^2} \right\}. \tag{3.73}$$

3.4 Coseismic Deformation Associated with Dipping Faults

The displacements and horizontal strain due to a 20-degree dipping thrust fault are shown in figure 3.10. The geometry is characteristic of large subduction zone earthquakes. Note that

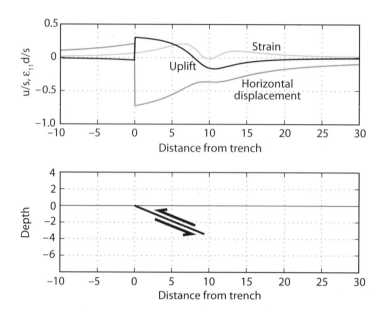

Figure 3.10. Coseismic deformation for a thrust fault. The top panel shows the horizontal and vertical displacements and the horizontal strain. The bottom panel shows the fault geometry.

the horizontal displacement is negative, meaning directed toward the fault, on the hanging-wall side of the fault and positive on the foot-wall side. There is uplift at the fault on the hanging wall, but subsidence above the downdip end of the fault. Generally, we anticipate tilting of the land away from the trench, except landward of the downdip projection of the fault, which tilts toward the trench. The horizontal strain is extensional, due to elastic rebound of stored compression, except for a small region above the downdip end of the fault. For most subduction zone earthquakes, much of the displacement field is underwater and therefore difficult to measure. There are many excellent comparisons between observations and model predictions for stations on the hanging wall of subduction earthquakes.

Data exist on both sides of the fault for continental thrust events. For example, the 20 September 1999 Chi-Chi, Taiwan, earthquake ($M_W = 7.6$) generated very large displacements that were measured with high accuracy using GPS. The earthquake occurred along the north–south trending Chelungpu fault (Kao and Chen 2000) and resulted in 2,440 fatalities, over 11,000 injuries, and the collapse of at least 50,000 structures. Coseismic displacements were determined by Yu et al. (2001) at 128 stations of the Taiwan GPS Network. The horizontal coseismic displacements increase from 1 m at the southern end of the fault to 9 m at the northern end (figure 3.11).

The vertical offsets (not shown) were largest near the fault trace, where the hanging wall was uplifted as much as 4.4 meters at the northern end of the rupture (plate 1), and decrease rapidly toward the east. The stations on the foot-wall side show smaller amounts of subsidence ranging from 0.02 to 0.32 meter. Net displacements at the ground surface range up to nearly 12 meters, at the northwest corner of the rupture. Along the north–south trending fault trace, the ground is warped into a monoclinal flexure indicative of shallow thrusting. Aftershocks illuminate a fault that dips 20 to 30 degrees to the east (Kao and Chen 2000).

Because the strike of the main rupture was nearly north–south, we compare the east and vertical components of the displacement field to predictions of a plane strain model (figure 3.12). In actuality, the Chi-Chi earthquake involved a significant amount of oblique

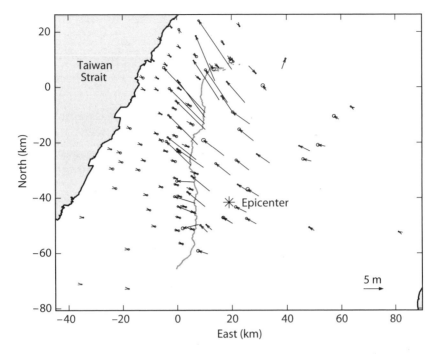

Figure 3.11. GPS-derived coseismic displacements during the 1999 Chi-Chi, Taiwan, earthquake. Horizontal displacement vectors with 95% confidence ellipses at tips. Data from Yu et al. (2001).

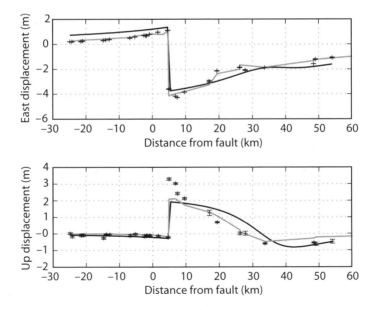

Figure 3.12. Displacements associated with the 1999 Chi-Chi, Taiwan, earthquake determined from GPS surveys. Black curves show prediction from a two-dimensional edge dislocation model with 5.6 meters of slip. Gray curves show prediction from a three-dimensional model with spatially variable fault slip, after Johnson et al. (2001). GPS data from Yu et al. (2001).

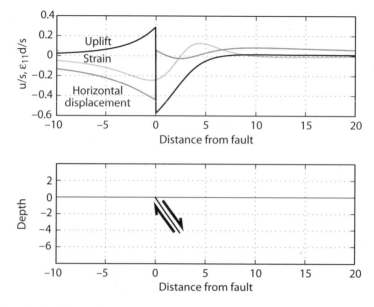

Figure 3.13. Coseismic deformation for a normal fault. The lower panel shows the fault geometry.

motion, so the plane strain assumption is not fully valid. Furthermore, the main rupture is only 75 km long, with east–west trending ruptures at the northern end of the fault further complicating the fault geometry. Last, we know from analysis of both GPS and seismic data that the slip was strongly concentrated near the northern end of the fault rupture, as can be clearly seen in the displacements in figure 3.11. Nevertheless, the simple two-dimensional model captures the essential features of the observed displacements. The observed horizontal displacements decay with distance more rapidly on the foot-wall side than predicted by the simple two-dimensional model, due to the three dimensionality of the fault. Furthermore, the uplift close to the fault is underpredicted, whereas the uplift between 10 and 20 km east of the fault trace is overpredicted. A three-dimensional model with distributed slip on the fault, however, fits the observations quite well except for the large uplift close to the fault trace (figure 3.12).

The deformation due to normal faulting earthquakes is also distinctive. The displacements and horizontal strain for a 60-degree dipping normal fault are shown in figure 3.13. The vertical displacements are generally straightforward—down on the hanging wall and up on the foot wall, as expected. There is, however, a slight uplift far from the fault on the hanging-wall side of the fault. The horizontal motions are generally directed away from the fault. Elastic rebound gives rise to an increase in compressive strain near the fault, with a smaller amount of extension above the downdip end of the fault. Changing the dip of the fault changes the ratio of the uplift to subsidence, as illustrated in figure 3.14, demonstrating that vertical displacement data are useful for determining fault dip.

As an example of the displacements associated with a normal fault, we consider the M 7.0 28 October 1983 Borah Peak, Idaho, earthquake. The Borah Peak earthquake ruptured the roughly 30-km-long Lost River fault. The vertical displacements determined by repeated leveling surveys are shown in figure 3.15, after Stein and Barrientos (1985). Notice that relative to benchmarks far from the fault, the range was uplifted by up to 0.2 m, whereas the basin subsided by over 1.0 m. The calculated displacements from a three-dimensional dislocation with uniform slip, as developed in the next section, are shown with a solid line. A simple

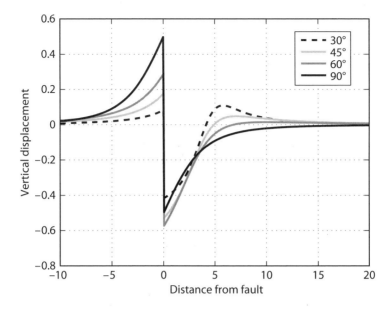

Figure 3.14. Effect of dip on the vertical component of displacement for a uniform-slip normal fault. Displacement normalized by slip.

plane strain calculation captures the essence of the coseismic deformation. Comparison to figure 3.14 suggests a fault dip of close to 45 degrees, consistent with the modeling of Stein and Barrientos (1985). The other panels in figure 3.15 show a cross section of the fault geometry with aftershocks and mainshock focal mechanism and a geologic cross section. Notice that the geologic structure mimics the displacements in an individual earthquake, suggesting that the basin and range structures have grown due to repeated faulting events.

3.5 Displacements and Stresses Due to Edge Dislocation at Depth

Volterra's formula together with the Melan Green's functions can be integrated to give the displacements due to dislocations at depth within an elastic half-space. The dislocation is located at $\xi_1 = 0$, ξ_2 (figure 3.16). For a dislocation not at $\xi_1 = 0$, simply replace x_1 with $x_1 - \xi_1$ in the following expressions.

The displacements are

$$u_1 = \frac{s_1}{\pi(1-\nu)} \left\{ \frac{(1-\nu)}{2}(\theta_2 - \theta_1) + \frac{x_1(x_2 - \xi_2)}{4r_1^2} - \frac{x_1[x_2 + (3 - 4\nu)\xi_2]}{4r_2^2} + \frac{\xi_2 x_2 x_1(x_2 + \xi_2)}{r_2^4} \right\}$$

$$+ \frac{s_2}{\pi(1-\nu)} \left\{ \frac{(1-2\nu)}{4}\log(r_2/r_1) - \frac{(x_2 - \xi_2)^2}{4r_1^2} + \frac{[x_2^2 + \xi_2^2 - 4(1-\nu)\xi_2(x_2 + \xi_2)]}{4r_2^2} + \frac{x_2\xi_2(x_2 + \xi_2)^2}{r_2^4} \right\},$$

$$u_2 = \frac{s_1}{\pi(1-\nu)} \left\{ \frac{(1-2\nu)}{4}\log(r_2/r_1) + \frac{(x_2 - \xi_2)^2}{4r_1^2} - \frac{[(x_2 + \xi_2)^2 - 2\xi_2^2 - 2(1 - 2\nu)\xi_2(x_2 + \xi_2)]}{4r_2^2} \right.$$

$$\left. + \frac{x_2\xi_2(x_2 + \xi_2)^2}{r_2^4} \right\} + \frac{s_2}{\pi(1-\nu)} \left\{ \frac{(1-\nu)}{2}(\theta_1 - \theta_2) + \frac{x_1(x_2 - \xi_2)}{4r_1^2} - \frac{x_1[x_2 + (3 - 4\nu)\xi_2]}{4r_2^2} \right.$$

$$\left. - \frac{\xi_2 x_2 x_1(x_2 + \xi_2)}{r_2^4} \right\}, \tag{3.74}$$

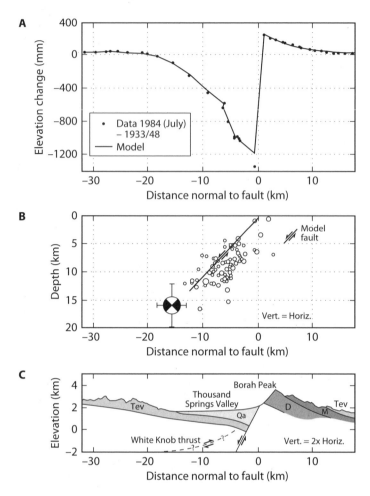

Figure 3.15. The 1983 Borah Peak, Idaho, earthquake. A: Vertical displacements from repeated leveling projected onto a profile normal to the fault. The leveling route is not perpendicular to the fault, resulting in kinks in the profile. B: Cross section showing model dislocation, aftershocks, and mainshock focal mechanism. C: Geologic cross section. After Stein and Barrientos (1985).

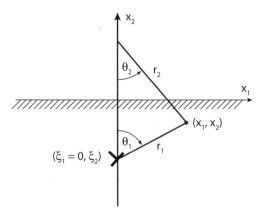

Figure 3.16. Geometry of edge dislocation at depth in an elastic half-space. r_1 and r_2 measure the distance from the dislocation and the image dislocation, respectively.

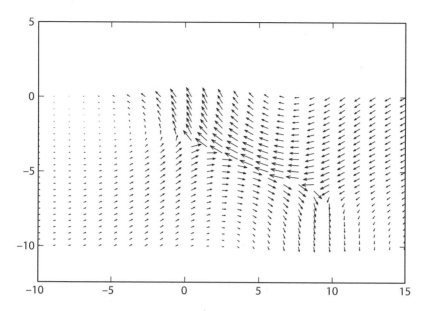

Figure 3.17. Displacements within the half-space from a 20-degree dipping thrust fault.

where

$$r_1^2 = x_1^2 + (x_2 - \xi_2)^2,$$
$$r_2^2 = x_1^2 + (x_2 + \xi_2)^2, \tag{3.75}$$

measure the squared distance from the observation point to the dislocation and the image dislocation, respectively. θ_1 refers to the angle about the dislocation, and θ_2 refers to the angle about the image dislocation:

$$\theta_1 = \tan^{-1}\left(\frac{x_1 - \xi_1}{x_2 - \xi_2}\right),$$

$$\theta_2 = \tan^{-1}\left(\frac{x_1 - \xi_1}{x_2 + \xi_2}\right). \tag{3.76}$$

As was the case for the screw dislocation model of strike-slip faulting, one needs to be careful about the position of the branch cuts. Branch cuts arise from the terms proportional to θ, which can be defined for any interval const $\leq \theta \leq$ const $+ 2\pi$. The choice of the constant does not affect the strains or stresses; however, for the displacements, one needs to ensure that the branch cuts coincide with the fault surface.

The displacements for a 20-degree dipping thrust fault are shown in figure 3.17. The discontinuity in displacement is evident and coincides with the position of the branch cut in θ_1. We clearly see the uplift over the upper edge of the thrust as well as the subsidence over the lower edge of the dislocation. Displacements are concentrated in the hanging wall, a pattern that becomes more marked for shallow faults as the dip decreases.

The displacements due to a 60-degree dipping normal fault are shown in figure 3.18. Again, the position of the fault surface is clearly seen in the discontinuity in the displacement field. As expected, the hanging wall is generally displaced downward, while the foot wall moves up. Notice that the horizontal displacements are not uniformly away from the fault. Above the

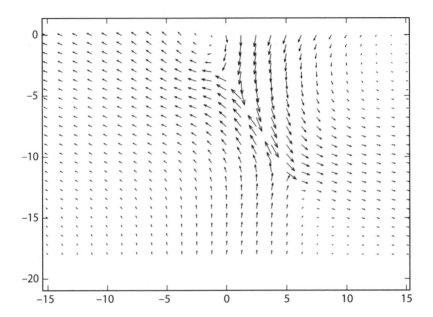

Figure 3.18. Displacements within the half-space from a 60-degree dipping normal fault.

upper edge of the dislocation, the horizontal component of motion actually moves toward the updip projection of the fault.

The stresses at (x_1, x_2) due to a long edge dislocation at $(0, \xi_2)$ can be computed by differentiating the displacements to obtain the strains following equation (1.4) and then substituting into Hooke's law (1.103). This yields

$$
\begin{aligned}
\sigma_{11} = &\frac{\mu s_2}{2\pi\,(1-\nu)} \left\{ \frac{x_1 \left[(x_2 - \xi_2)^2 - x_1^2 \right]}{r_1^4} - \frac{x_1 \left[(x_2 + \xi_2)^2 - x_1^2 \right]}{r_2^4} \right. \\
&+ \left. \frac{4\xi_2 x_1}{r_2^6} \left[(2\xi_2 - x_2)\,(x_2 + \xi_2)^2 + (3x_2 + 2\xi_2)\,x_1^2 \right] \right\} \\
&- \frac{\mu s_1}{2\pi\,(1-\nu)} \left\{ \frac{(x_2 - \xi_2) \left[(x_2 - \xi_2)^2 + 3x_1^2 \right]}{r_1^4} - \frac{(x_2 + \xi_2) \left[(x_2 + \xi_2)^2 + 3x_1^2 \right]}{r_2^4} \right. \\
&+ \left. \frac{2\xi_2}{r_2^6} \left[6x_2\,(x_2 + \xi_2)\,x_1^2 - (x_2 - \xi_2)\,(x_2 + \xi_2)^3 - x_1^4 \right] \right\}, \quad (3.77)
\end{aligned}
$$

$$
\begin{aligned}
\sigma_{22} = &\frac{-\mu s_2}{2\pi\,(1-\nu)} \left\{ \frac{x_1 \left[3\,(x_2 - \xi_2)^2 + x_1^2 \right]}{r_1^4} d - \frac{x_1 \left[3\,(x_2 + \xi_2)^2 + x_1^2 \right]}{r_2^4} - \frac{4\xi_2 x_2 x_1}{r_2^6} \left[3\,(x_2 + \xi_2)^2 - x_1^2 \right] \right\} \\
&- \frac{\mu s_1}{2\pi\,(1-\nu)} \left\{ \frac{(x_2 - \xi_2) \left[(x_2 - \xi_2)^2 - x_1^2 \right]}{r_1^4} - \frac{(x_2 + \xi_2) \left[(x_2 + \xi_2)^2 - x_1^2 \right]}{r_2^4} \right. \\
&- \left. \frac{2\xi_2}{r_2^6} \left[6x_2\,(x_2 + \xi_2)\,x_1^2 - (3x_2 + \xi_2)\,(x_2 + \xi_2)^3 + x_1^4 \right] \right\}, \quad (3.78)
\end{aligned}
$$

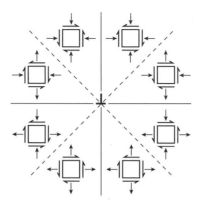

Figure 3.19. Signs of the stresses due to an edge dislocation. After Hirthe and Lothe (1992).

$$
\sigma_{12} = \frac{\mu s_2}{2\pi\,(1-\nu)}\left\{\frac{(x_2-\xi_2)\left[(x_2-\xi_2)^2-x_1^2\right]}{r_1^4} - \frac{(x_2+\xi_2)\left[(x_2+\xi_2)^2-x_1^2\right]}{r_2^4}\right.
$$

$$
\left.+\frac{2\xi_2}{r_2^6}\left[6x_2\,(x_2+\xi_2)\,x_1^2 - x_1^4 + (\xi_2-x_2)\,(x_2+\xi_2)^3\right]\right\}
$$

$$
-\frac{\mu s_1}{2\pi\,(1-\nu)}\left\{\frac{x_1\left[(x_2-\xi_2)^2-x_1^2\right]}{r_1^4} - \frac{x_1\left[(x_2+\xi_2)^2-x_1^2\right]}{r_2^4} + \frac{4\xi_2 x_2 x_1}{r_2^6}\left[3\,(x_2+\xi_2)^2-x_1^2\right]\right\}.
$$

$$(3.79)$$

The stresses due to a single dislocation are all of order $1/r$, as was the case for the screw dislocation. In the limit that the distance from the dislocation line is small compared to the depth, the r_2 terms vanish and we are left with the stresses due to a dislocation in a (two-dimensional) full-space. Plate 2 illustrates the stress distributions for an edge dislocation far from the free surface, while figure 3.19 shows the signs of the stresses. In plate 2, the dislocation-parallel normal stress, σ_{11}, is compressive above the dislocation line, reflecting the convergence of material, while the stress below the dislocation is tensile. The shear stress σ_{12} decreases behind the dislocation line (fault tip) and increases in front of the dislocation line, as expected, but shows a more complex distribution above and below the dislocation line. The fault-perpendicular normal stress σ_{22} is compressive (tensile) above (below) the dislocation line due to the Poisson effect, changing sign at the plane of the dislocation.

3.6 Dislocations in Three Dimensions

We are now in a position to determine the elastic fields due to dislocations in three dimensions. The Green's functions for the elastic half-space were first derived by Mindlin in 1936. The derivation is beyond the present scope. We will investigate a derivation for the Green's function in a full-space, which was originally found by Kelvin in 1848.

3.6.1 Full-Space Green's Functions

A derivation of the elastostatic Green's function in a three-dimensional full-space is presented here. $g_i^j(\mathbf{x}, \boldsymbol{\xi})$ is the displacement at $\mathbf{x}$ in the i direction, due to a point force, with magnitude F, acting at $\boldsymbol{\xi}$ in the j direction. Specifically, the Green's tensor satisfies the equilibrium equations:

$$\mu \nabla^2 g_i^j + (\lambda + \mu) \frac{\partial^2 g_k^j}{\partial x_i \partial x_k} + F \delta_{ij} \delta(\mathbf{x} - \boldsymbol{\xi}) = 0. \tag{3.80}$$

On dimensional grounds, the Green's tensors must scale with force over shear modulus and inversely with distance R between the source and the receiver:

$$R = |\mathbf{r}| = |\mathbf{x} - \boldsymbol{\xi}| = \sqrt{(x_1 - \xi_1)^2 + (x_2 - \xi_2)^2 + (x_3 - \xi_3)^2}. \tag{3.81}$$

One such solution is proportional to $F \delta_{ij}/\mu R$. The only other solution that is a second-rank tensor and scales with $1/R$ is proportional to $F \mathbf{rr}^T/\mu R^3$. Thus, the general solution must be of the form

$$g_i^j(\mathbf{x}, \boldsymbol{\xi}) = \alpha \frac{F \delta_{ij}}{\mu R} + \beta \frac{F r_i r_j}{\mu R^3}, \tag{3.82}$$

where α and β are dimensionless constants that can depend on ratios of elastic moduli (i.e., Poisson's ratio). The constants are determined by substituting into the equilibrium equations (3.80). A fair amount of calculation is saved by noting that

$$\frac{\partial}{\partial x_j} \frac{\partial R}{\partial x_i} = \frac{\delta_{ij}}{R} - \frac{r_i r_j}{R^3}. \tag{3.83}$$

The Green's tensor can thus be written as

$$g_i^j(\mathbf{x}, \boldsymbol{\xi}) = \frac{(\alpha + \beta)}{\mu} \frac{F \delta_{ij}}{R} - \frac{\beta}{\mu} \frac{\partial}{\partial x_j} \frac{\partial R}{\partial x_i}. \tag{3.84}$$

Substituting into equation (3.80) yields

$$(\alpha + \beta) \delta_{ij} \nabla^2 \left(\frac{1}{R}\right) - \beta \frac{\partial^2}{\partial x_i \partial x_j} \nabla^2 R + \frac{1}{(1 - 2v)} \left[(\alpha + \beta) \frac{\partial^2}{\partial x_i \partial x_j} \left(\frac{1}{R}\right) - \beta \frac{\partial^2}{\partial x_i \partial x_j} \nabla^2 R \right] + \delta_{ij} \delta(\mathbf{x} - \boldsymbol{\xi}) = 0. \tag{3.85}$$

It is straightforward to show that $\nabla^2 R = 2/R$. Using this result, we show that equation (3.85) is satisfied for $i \neq j$ only if

$$\alpha = (3 - 4v)\beta. \tag{3.86}$$

For $i = j$, equation (3.85) reduces to

$$4\beta(1 - v)\nabla^2 \frac{1}{R} + \delta(\mathbf{x} - \boldsymbol{\xi}) = 0. \tag{3.87}$$

Equation (3.87) is satisfied for all $R \neq 0$. To determine the constant β, integrate over a unit volume surrounding the point force:

$$4\beta(1 - v) \int_V \nabla^2 \left(\frac{1}{R}\right) dV + 1 = 0. \tag{3.88}$$

Using the divergence theorem (1.84) to change the volume integral into a surface integral, we have

$$4\beta(1-v)\int_S \nabla\left(\frac{1}{R}\right)\cdot dS + 1 = 0,$$

$$4\beta(1-v)\left(-\frac{1}{R^2}\right)4\pi R^2 = -1,$$

$$\beta = \frac{1}{16\pi\mu(1-v)}. \tag{3.89}$$

The Green's function for the full-space is thus given by

$$g_i^j(\mathbf{x},\xi) = \frac{1}{16\pi\mu(1-v)}\left[(3-4v)\frac{\delta_{ij}}{R} + \frac{r_i r_j}{R^3}\right], \tag{3.90}$$

or alternatively,

$$g_i^j(\mathbf{x},\xi) = \frac{1}{4\pi\mu}\left[\frac{\delta_{ij}}{R} - \frac{1}{4(1-v)}\frac{\partial}{\partial x_j}\frac{\partial R}{\partial x_i}\right]. \tag{3.91}$$

3.6.2 Half-Space Green's Functions

The Green's functions for the half-space, found by Mindlin (1936), are given, but not derived, here. The notation is as defined previously, with the subscript referring to the component of the displacement and the superscript the direction of the force. Our coordinate system is one in which the positive x_3 axis points upward. Thus, $x_3 < 0$ for points within the earth. The source is located at $\xi_1 = \xi_2 = 0$, ξ_3. The distance from the source r_1 and the distance from the image source r_2 are given by

$$r_1 = \sqrt{x_1^2 + x_2^2 + (x_3-\xi_3)^2},$$

$$r_2 = \sqrt{x_1^2 + x_2^2 + (x_3+\xi_3)^2}. \tag{3.92}$$

With this geometry, the displacements due to a vertical force are given by

$$g_1^3 = \frac{x_1}{16\pi\mu(1-v)}\left[\frac{(x_3-\xi_3)}{r_1^3} + \frac{(3-4v)(x_3-\xi_3)}{r_2^3} + \frac{4(1-v)(1-2v)}{r_2(r_2-x_3-\xi_3)} + \frac{6x_3\xi_3(x_3+\xi_3)}{r_2^5}\right],$$

$$g_2^3 = \frac{x_2}{16\pi\mu(1-v)}\left[\frac{(x_3-\xi_3)}{r_1^3} + \frac{(3-4v)(x_3-\xi_3)}{r_2^3} + \frac{4(1-v)(1-2v)}{r_2(r_2-x_3-\xi_3)} + \frac{6x_3\xi_3(x_3+\xi_3)}{r_2^5}\right],$$

$$g_3^3 = \frac{1}{16\pi\mu(1-v)}\left[\frac{3-4v}{r_1} + \frac{5-12v+8v^2}{r_2} + \frac{(x_3-\xi_3)^2}{r_1^3} + \frac{(3-4v)(x_3+\xi_3)^2 - 2x_3\xi_3}{r_2^3}\right.$$
$$\left. + \frac{6x_3\xi_3(x_3+\xi_3)^2}{r_2^5}\right]. \tag{3.93}$$

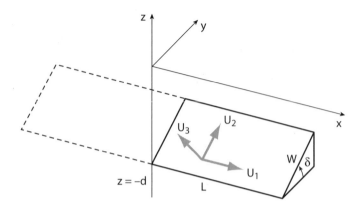

Figure 3.20. Definition of geometric parameters for a finite dislocation, after Okada (1985). Note that Okada (1985) denotes the components of the slip as U_k, whereas here we refer to the fault slip as s_k and the displacement at the earth's surface as u_k.

The displacements due to a horizontal force acting in the x_1 direction are given by

$$g_1^1 = \frac{1}{16\pi\mu(1-\nu)} \left\{ \frac{3-4\nu}{r_1} + \frac{1}{r_2} + \frac{x_1^2}{r_1^3} + \frac{(3-4\nu)x_1^2}{r_2^3} + \frac{2(r_2^2 - 3x_1^2)x_3\xi_3}{r_2^5} \right.$$
$$\left. + \frac{4(1-\nu)(1-2\nu)[r_2^2 - x_1^2 - r_2(x_3 + \xi_3)]}{r_2(r_2 - x_3 - \xi_3)^2} \right\},$$

$$g_2^1 = \frac{x_1 x_2}{16\pi\mu(1-\nu)} \left[\frac{1}{r_1^3} + \frac{(3-4\nu)}{r_2^3} - \frac{6x_3\xi_3}{r_2^5} - \frac{4(1-\nu)(1-2\nu)}{r_2(r_2 - x_3 - \xi_3)^2} \right],$$

$$g_3^1 = \frac{x_1}{16\pi\mu(1-\nu)} \left[\frac{(x_3 - \xi_3)}{r_1^3} + \frac{(3-4\nu)(x_3 - \xi_3)}{r_2^3} - \frac{4(1-\nu)(1-2\nu)}{r_2(r_2 - x_3 - \xi_3)} - \frac{6x_3\xi_3(x_3 + \xi_3)}{r_2^5} \right].$$
$$(3.94)$$

Due to the symmetry in the problem, the displacements due to a horizontal force acting in the x_2 direction are given by an appropriate change of the x_1 and x_2 axes. For a source not located at the origin, replace x_1 with $x_1 - \xi_1$ and x_2 with $x_2 - \xi_2$.

3.6.3 Point-Source Dislocations

Given the appropriate Green's functions, we are finally in a position to use Volterra's formula (3.15) to derive the displacements due to an arbitrary dislocation in a three-dimensional half-space. The point-source solutions are given here, where we adopt the geometry defined by Okada (1985), as shown in figure 3.20.

Given the fault geometry defined in figure 3.20, Volterra's formula (3.15) reduces to the following expressions.

For strike slip ($s_2 = s_3 = 0$),

$$u_i(\mathbf{x}) = \mu s_1 d\Sigma \left[-\left(\frac{\partial g_i^1}{\partial \xi_2} + \frac{\partial g_i^2}{\partial \xi_1} \right) \sin\delta + \left(\frac{\partial g_i^1}{\partial \xi_3} + \frac{\partial g_i^3}{\partial \xi_1} \right) \cos\delta \right]. \qquad (3.95)$$

For dip slip ($s_1 = s_3 = 0$),

$$u_i(\mathbf{x}) = \mu s_2 d\Sigma \left[\left(\frac{\partial g_i^2}{\partial \xi_3} + \frac{\partial g_i^3}{\partial \xi_2} \right) \cos 2\delta + \left(\frac{\partial g_i^3}{\partial \xi_3} - \frac{\partial g_i^2}{\partial \xi_2} \right) \sin 2\delta \right]. \qquad (3.96)$$

For dilatant sources ($s_1 = s_2 = 0$),

$$u_i(\mathbf{x}) = s_3 d\Sigma \left[\lambda \frac{\partial g_i^n}{\partial \xi_n} + 2\mu \left(\frac{\partial g_i^2}{\partial \xi_2} \sin^2 \delta + \frac{\partial g_i^3}{\partial \xi_3} \cos^2 \delta \right) - \mu \left(\frac{\partial g_i^2}{\partial \xi_3} + \frac{\partial g_i^3}{\partial \xi_2} \right) \sin 2\delta \right], \quad (3.97)$$

where s_i is slip, $d\Sigma$ is the fault area increment, such that the product $s d\Sigma$ is the source strength, or *potency*, and δ is the fault dip. Introducing the Mindlin Green's functions (3.94) and (3.93) into the preceding expressions and evaluating for a source located at ($\xi_1 = \xi_2 = 0$, $\xi_3 = -d$) and for an observation point at the earth's surface (x_1, x_2, $x_3 = 0$) yields the following.

For strike slip,

$$u_1 = -\frac{s_1 d\Sigma}{2\pi} \left(\frac{3x_1^2 q}{r^5} + I_1^0 \sin \delta \right),$$

$$u_2 = -\frac{s_1 d\Sigma}{2\pi} \left(\frac{3x_1 x_2 q}{r^5} + I_2^0 \sin \delta \right),$$

$$u_3 = -\frac{s_1 d\Sigma}{2\pi} \left(\frac{3x_1 dq}{r^5} + I_4^0 \sin \delta \right). \quad (3.98)$$

For dip slip,

$$u_1 = -\frac{s_2 d\Sigma}{2\pi} \left(\frac{3x_1 pq}{r^5} - I_3^0 \sin \delta \cos \delta \right),$$

$$u_2 = -\frac{s_2 d\Sigma}{2\pi} \left(\frac{3x_2 pq}{r^5} - I_1^0 \sin \delta \cos \delta \right),$$

$$u_3 = -\frac{s_2 d\Sigma}{2\pi} \left(\frac{3dpq}{r^5} - I_5^0 \sin \delta \cos \delta \right). \quad (3.99)$$

For an opening dislocation,

$$u_1 = \frac{s_3 d\Sigma}{2\pi} \left(\frac{3x_1 q^2}{r^5} - I_3^0 \sin^2 \delta \right),$$

$$u_2 = \frac{s_3 d\Sigma}{2\pi} \left(\frac{3x_2 q^2}{r^5} - I_1^0 \sin^2 \delta \right),$$

$$u_3 = \frac{s_3 d\Sigma}{2\pi} \left(\frac{3dq^2}{r^5} - I_5^0 \sin^2 \delta \right), \quad (3.100)$$

where

$$I_1^0 = (1 - 2\nu)x_2 \left[\frac{1}{r(r+d)^2} - x_1^2 \frac{3r+d}{r^3(r+d)^3} \right],$$

$$I_2^0 = (1 - 2\nu)x_1 \left[\frac{1}{r(r+d)^2} - x_2^2 \frac{3r+d}{r^3(r+d)^3} \right],$$

$$I_3^0 = (1 - 2\nu) \left[\frac{x_1}{r^3} \right] - I_2^0,$$

$$I_4^0 = -(1 - 2\nu) \left[x_1 x_2 \frac{2r+d}{r^3(r+d)^2} \right],$$

$$I_5^0 = (1 - 2\nu) \left[\frac{1}{r(r+d)} - x_1^2 \frac{2r+d}{r^3(r+d)^2} \right], \quad (3.101)$$

and

$$p = x_2 \cos \delta + d \sin \delta,$$

$$q = x_2 \sin \delta - d \cos \delta,$$

$$r^2 = x_1^2 + x_2^2 + d^2 = x_1^2 + p^2 + q^2. \tag{3.102}$$

3.6.4 Finite Rectangular Dislocations

According to Volterra's formula (3.15), the solution for a finite source can be obtained by integrating the point-source solution over the fault surface. For general three-dimensional fault geometries, this is usually difficult to do analytically. For the particular case of a rectangular dislocation surface of length L and downdip width W (figure 3.20) and spatially uniform slip or opening, the integration can be accomplished in closed form. (See Okada 1985 for a summary of early work on this subject.) While the integration is in principle straightforward, the calculations are tedious. Okada (1985), while far from the first to complete these integrations, did a very careful job and is now, properly, widely cited. Following Okada (1985), point sources can be distributed on the fault by substituting $x \to x - \xi'$, $y \to y - \eta' \cos \delta$, $d \to d - \eta' \sin \delta$ as in equation (3.59) when deriving the results for a two-dimensional dislocation. The elastic fields due to the rectangular fault are then constructed by integrating ξ' from 0 to L, and η' from 0 to W. Okada (1985), following Sato and Matsu'ura (1974), changes variables to $\xi = x - \xi'$ and $\eta = p - \eta'$, where $p = y \cos \delta + d \sin \delta$. (See discussion around equation [3.64].) The integration is then over the domain

$$\int_x^{x-L} d\xi \int_p^{p-W} d\eta. \tag{3.103}$$

The displacements are given here following the notation of Chinnery (1961), where

$$f(\xi, \eta)\| = f(x, p) - f(x, p - W) - f(x - L, p) + f(x - L, p - W). \tag{3.104}$$

For strike slip,

$$u_1 = -\frac{s_1}{2\pi} \left[\frac{\xi q}{R(R + \eta)} + \tan^{-1} \frac{\xi \eta}{qR} + I_1 \sin \delta \right] \Big\|,$$

$$u_2 = -\frac{s_1}{2\pi} \left[\frac{\tilde{y} q}{R(R + \eta)} + \frac{q \cos \delta}{R + \eta} + I_2 \sin \delta \right] \Big\|,$$

$$u_3 = -\frac{s_1}{2\pi} \left[\frac{\tilde{d} q}{R(R + \eta)} + \frac{q \sin \delta}{R + \eta} + I_4 \sin \delta \right] \Big\|. \tag{3.105}$$

For dip slip,

$$u_1 = -\frac{s_2}{2\pi} \left[\frac{q}{R} - I_3 \sin \delta \cos \delta \right] \Big\|,$$

$$u_2 = -\frac{s_2}{2\pi} \left[\frac{\tilde{y} q}{R(R + \xi)} + \cos \delta \tan^{-1} \frac{\xi \eta}{qR} - I_1 \sin \delta \cos \delta \right] \Big\|,$$

$$u_3 = -\frac{s_2}{2\pi} \left[\frac{\tilde{d} q}{R(R + \xi)} + \sin \delta \tan^{-1} \frac{\xi \eta}{qR} - I_5 \sin \delta \cos \delta \right] \Big\|. \tag{3.106}$$

For opening, dikelike, dislocation,

$$u_1 = \frac{s_3}{2\pi} \left[\frac{q^2}{R(R+\eta)} - I_3 \sin^2 \delta \right] \Big|\Big|,$$

$$u_2 = \frac{s_3}{2\pi} \left\{ \frac{-\tilde{d}q}{R(R+\xi)} - \sin \delta \left[\frac{\xi q}{R(R+\eta)} - \tan^{-1} \frac{\xi\eta}{qR} \right] - I_1 \sin^2 \delta \right\} \Big|\Big|,$$

$$u_3 = \frac{s_3}{2\pi} \left\{ \frac{\tilde{y}q}{R(R+\xi)} + \cos \delta \left[\frac{\xi q}{R(R+\eta)} - \tan^{-1} \frac{\xi\eta}{qR} \right] - I_5 \sin^2 \delta \right\} \Big|\Big|, \qquad (3.107)$$

where

$$I_1 = \frac{\mu}{\lambda+\mu} \left[\frac{-\xi}{\cos \delta (R+\tilde{d})} \right] - \frac{\sin \delta}{\cos \delta} I_5,$$

$$I_2 = \frac{\mu}{\lambda+\mu} \left[-\ln(R+\eta) \right] - I_3,$$

$$I_3 = \frac{\mu}{\lambda+\mu} \left[\frac{\tilde{y}}{\cos \delta (R+\tilde{d})} - \ln(R+\eta) \right] + \frac{\sin \delta}{\cos \delta} I_4,$$

$$I_4 = \frac{\mu}{\lambda+\mu} \frac{1}{\cos \delta} \left[\ln(R+\tilde{d}) - \sin \delta \ln(R+\eta) \right],$$

$$I_5 = \frac{\mu}{\lambda+\mu} \frac{2}{\cos \delta} \tan^{-1} \left[\frac{\eta(X+q\cos \delta) + X(R+X)\sin \delta}{\xi(R+X)\cos \delta} \right], \qquad (3.108)$$

and p, q, r are as defined in equation (3.102), and

$$\tilde{y} = \eta \cos \delta + q \sin \delta,$$

$$\tilde{d} = \eta \sin \delta - q \cos \delta,$$

$$R^2 = \xi^2 + \eta^2 + q^2 = \xi^2 + \tilde{y}^2 + \tilde{d}^2,$$

$$X^2 = \xi^2 + q^2. \qquad (3.109)$$

Several of the terms in equation (3.108) have $\cos \delta$ in the denominator. For faults with vertical dip ($\cos \delta = 0$), these terms should be replaced by

$$I_1 = -\frac{\mu}{2(\lambda+\mu)} \left[\frac{\xi q}{(R+\tilde{d})^2} \right],$$

$$I_3 = \frac{\mu}{2(\lambda+\mu)} \left[\frac{\eta}{R+\tilde{d}} + \frac{\tilde{y}q}{(R+\tilde{d})^2} - \ln(R+\eta) \right],$$

$$I_4 = -\frac{\mu}{\lambda+\mu} \frac{q}{R+\tilde{d}},$$

$$I_5 = -\frac{\mu}{\lambda+\mu} \frac{\xi \sin \delta}{R+\tilde{d}}. \qquad (3.110)$$

The strains and tilts can be computed by differentiating the displacements with respect to the field coordinates, **x**; expressions are given in Okada (1985). The displacements in equations

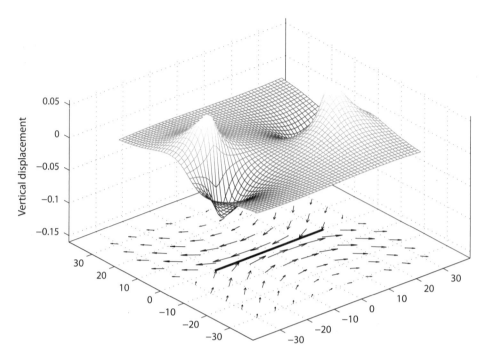

Figure 3.21. Displacements due to a vertical strike-slip fault, from depth 2 to 20.

(3.105), (3.106), and (3.107) are restricted to the free surface. The procedure for computing the displacements within the earth, which can be used to calculate the strains and stresses, is the same, although care must be taken to ensure that the branch cuts coincide with the dislocation surface; expressions are given by Okada (1992).

3.6.5 Examples

This section illustrates the predicted surface deformation for several typical fault geometries. To start, we will examine the displacement fields for simple uniform-slip dislocations, meaning that the slip vector is everywhere the same on the fault surface, as in equations (3.105), (3.106), and (3.107). For a vertical strike-slip fault, the deformation pattern is illustrated in figure 3.21. Notice that the displacements are not everywhere parallel to the fault. The displacements are directed away from the fault in the two compressional quadrants and toward the fault in the two extensional quadrants. For the same reason, there is uplift in the two compressional quadrants and subsidence in the two extensional quadrants. The horizontal displacements decay with distance—rapidly for shallow faulting, more slowly if the slip extends to greater depth. Similarly, the position of the maximum uplift and subsidence depends on the depth distribution of slip.

We can compare predictions from dislocation models to geodetic observations of actual strike-slip earthquakes. One particularly well studied earthquake is the 1999 Hector Mine, California, earthquake, M 7.1. Plate 3A shows the horizontal displacements determined from both GPS and satellite radar measurements (InSAR). The radar observations were processed to determine the range change in the line-of-sight (LOS) direction between the satellite and the ground using interferometric methods on both ascending and descending passes. So-called *azimuth offset fields* were also determined, which yield the horizontal component of motion in the direction of the spacecraft motion. While the offset measurements are considerably

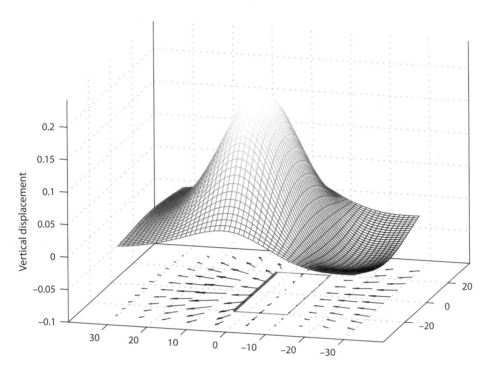

Figure 3.22. Displacements due to a blind thrust fault. The rectangle marks the surface projection of the dislocation, with the heavy line indicating the updip edge Width is 15, depth to bottom 22.

noisier than the interferometric phase determinations, they do provide useful independent data. Fialko et al. (2001) combined the four data types to determine the three-dimensional displacement field. Notice that the orientation and magnitude of the displacements agree quite well with the completely independent GPS measurements.

The observations compare quite well with the predictions of a simple rectangular dislocation; compare to figure 3.21. For a more precise comparison, we must account for the nonplanar fault geometry in this earthquake and the nonuniform distribution of slip, as in plate 3B.

Figure 3.22 shows the vertical and horizontal displacement for a deep blind thrust fault geometry. Notice the uplift of the hanging wall above the updip end of the fault and the lesser subsidence above the downdip end of the fault. The pattern of horizontal displacements may not be completely intuitive, in that the displacements are smaller immediately above the fault than they are to either side. In addition, the horizontal displacements on the footwall side of the updip edge act in the same direction as the hanging-wall motion. This is visible in the two-dimensional solution (figure 3.17). In two dimensions, the horizontal displacements change sign where the updip projection of the fault plane intersects the free surface.

A nice example of the deformation resulting from a buried thrust fault is the M_w 6.7 1994 Northridge, California, earthquake. The coseismic displacement field was well recorded by a combination of GPS measurements, leveling, and InSAR. Figure 3.23 illustrates the displacement field from GPS and leveling. The south-dipping blind fault caused uplift of as much as 417 ± 5 mm and horizontal displacements of as much as 216 ± 3 mm (Hudnut et al. 1996). Notice that the horizontal displacements north of the epicenter are directed

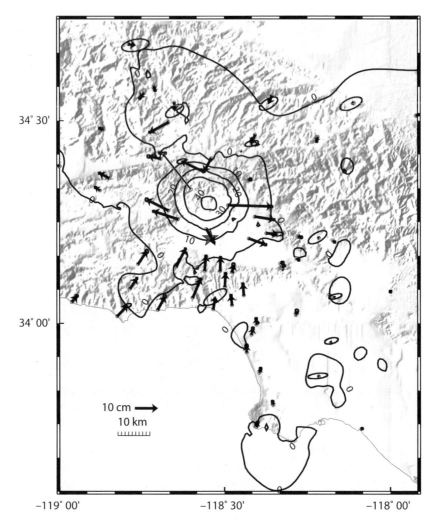

Figure 3.23. Horizontal and vertical displacements due to the 1994 Northridge, California, earthquake, M_w 6.7. The contours show the vertical displacement from a combination of GPS and spirit leveling. Horizontal displacement vectors are from GPS, and the star represents the epicenter. The earthquake ruptured updip to the north, so the epicenter is located above the downdip end of the fault. No correction has been made for minor subsidence due to groundwater withdrawal. Figure courtesy of Ken Hudnut.

east–west, nearly perpendicular to the slip vector, which is essentially pure reverse faulting. These observations can be compared with the prediction of a simple uniform-slip dislocation dipping 40 degrees to the south–southwest, as in figure 3.24. The simple model captures most of the important features of the deformation field, indicating the validity of elastic dislocations as a description of coseismic deformation.

3.6.6 Distributed Slip
According to Volterra's formula (3.15), the displacements can be computed by integrating the slip distribution against the derivatives of the Green's tensors for any distribution of

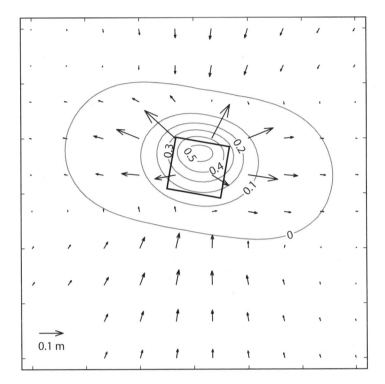

Figure 3.24. Displacements predicted from a simple uniform-slip dislocation model with geometry comparable to the 1994 Northridge earthquake. The contours indicate vertical displacement in meters. The rectangle marks the surface projection of the dislocation, and the heavy line is the updip edge. Compare with observations in figure 3.23.

slip; however, some form of numerical approximation is often necessary. One commonly used approach is to approximate the fault surface with a series of small rectangular elements (plate 4). The displacements are then given by a finite sum of terms of the form (3.105), (3.106), and (3.107). Rectangles, however, are awkward for approximating curved surfaces. Curved surfaces are better represented with triangular elements; solutions for triangular dislocations are given by Jeyakumaran et al. (1992) and employed by Maerten et al. (2005).

It is worth pointing out that for many purposes, one does not need to tessalate the fault surface with finite dimensioned elements. Again referring to Volterra's formula (3.15), one can simply approximate the integral as a finite sum of point sources. This approach is generally adequate for computing elastic fields at the free surface, although some care must be taken near the fault trace if it breaks the free surface. Special care also needs to be taken when evaluating the stresses near the fault, since the Green's tensors, and thus their derivatives, are singular. With finite sources, it is possible to compute the stress in the center of the dislocation element, thus avoiding singularities at the dislocation edges.

Gaussian quadrature provides a useful numerical procedure for integrating point sources. An example of numerically integrating point-source solutions is shown in figure 3.25. Gaussian quadrature was used to compute the solution for a rectangular dislocation with equal amounts of strike-slip and dip-slip motion. The numerical result agrees very well with the analytical solution even with only five integration points in each coordinate direction. In this particular case, the maximum differences are less than 0.01%.

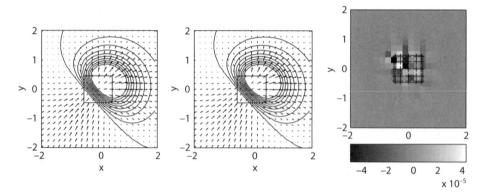

Figure 3.25. An example of numerical integration using Gaussian quadrature. The left panel shows the horizontal and vertical (contoured) displacements from the numerical integration. The maximum horizontal displacement is roughly 7.7% of the slip amplitude. The maximum vertical displacement is about 14% of the slip amplitude, and the contour interval is 0.012 in units of slip. The middle panel shows the same for the analytical results. The right panel shows the normalized difference in the magnitude of the displacement, (numerical–analytical)/analytical. The solid rectangle is the surface projection of the dislocation surface. The Gaussian points are shown as crosses. Calculation courtesy of J. Townend.

3.7 Strain Energy Change Due to Faulting

Faulting changes the elastic strain energy stored in the earth. The stress work per unit area of the dislocation is

$$dW = -\int_0^s \mathbf{T} \cdot \mathbf{ds'},$$ (3.111)

where $\mathbf{T}$ is the traction, $\mathbf{s}$ is fault slip, and the negative sign arises because the tractions and displacements act in opposite directions. Imagine that the tractions are relaxed slowly from an initial value $\mathbf{T}^0$ to the final value $\mathbf{T}^f < \mathbf{T}^0$ such that the displacements are always proportional to the change in traction:

$$\mathbf{T}(\mathbf{s'}) = \mathbf{T}^0 + (\mathbf{T}^f - \mathbf{T}^0)\frac{\mathbf{s'}}{\mathbf{s}},$$ (3.112)

where $\mathbf{s}$ is the final slip. Thus, equation (3.111) can be written as

$$dW = -\left[T_k^0 s_k + \frac{1}{2}(T_k^f - T_k^0)s_k\right] = -\frac{s_k}{2}(T_k^f + T_k^0).$$ (3.113)

The total work done is found by integrating over the dislocation surface:

$$W = \int_\Sigma dW = -\frac{1}{2}\int_\Sigma s_k(\sigma_{kj}^0 + \sigma_{kj}^f)n_j d\Sigma.$$ (3.114)

If no work is done at other boundaries, equation (3.114) represents the total change in strain energy. Notice that the change in elastic strain energy depends on the average stress acting through the fault displacements and is thus dependent on the total stress, not simply the stress change accompanying fault slip. It should be noted here that for *uniform-slip dislocations* of the type discussed mainly in this chapter, the integral in equation (3.114) is formally divergent. This is discussed briefly in section 4.3.

3.8 Summary and Perspective

Elastic dislocations are the backbone of crustal deformation modeling. They provide useful models of coseismic deformation accompanying earthquakes, as well as deformation resulting from planar intrusions, including dikes and sills, as discussed in chapter 7.

The central results in this chapter are the various forms of Volterra's formula, which allow one to compute the elastic fields due to dislocations with either spatially uniform or variable slip. Application of Volterra's formula requires the elastostatic Green's tensors, which are well known for a uniform, isotropic half-space. Analytical results can be obtained by integrating slip over rectangular surfaces, and numerical codes that compute displacements, strains, and tilts from uniform-slip rectangular dislocations are widely used in analyzing crustal deformation data. One should not, however, be limited to uniform-slip representations. The elastic fields associated with nonuniform slip, or opening, can be approximated either by integrating point dislocation solutions or by summing the contributions of small rectangular or triangular dislocation elements, as discussed in section 3.6.6.

While the results in this chapter have been restricted to elastic half-spaces with uniform properties, Volterra's formula applies more generally for heterogeneous earth models. The appropriate Green's tensor is, of course, required. Chapter 5 presents several methods for computing the elastic fields due to dislocations in heterogeneous earth models. In chapter 8, we will explore approximate corrections for the effects of earth curvature and topography.

A fundamental feature of all of the dislocation solutions presented to this point is that the slip, or opening, has been specified as a displacement discontinuity boundary condition on the dislocation surface. There has been no discussion of why the particular slip or opening should develop. In the next chapter, we will explore crack models, in which the slip or opening is determined by the shear stress drop on the surface in the case of faults, or by the magma pressure in the case of planar dikes and sills.

3.9 Problems

1. Show that for slip in the ξ_1 direction on a fault with unit normal in the ξ_3 direction, the only nonzero components of the moment tensor density in an isotropic medium are $m_{13} = m_{31}$.

    Given the body force equivalent distribution to the fault slip, one can compute the net moment as $\int \boldsymbol{\xi} \times \mathbf{f} \, dV$. The net moment about the ξ_2 axis is thus (e.g., Aki and Richards 1980),

$$\int_V \epsilon_{213}\xi_1 f_3 \, dV + \int_V \epsilon_{231}\xi_3 f_1 \, dV.$$

From the body force equivalents defined in equation (3.20), show that this yields

$$\int_V \xi_1 \delta(\xi_3) \frac{\partial}{\partial \xi_1} m_{31}(\xi_1, \xi_2) d\xi_1 d\xi_2 d\xi_3 - \int_V \xi_3 m_{13}(\xi_1, \xi_2) \frac{\partial}{\partial \xi_3} \delta(\xi_3) d\xi_1 d\xi_2 d\xi_3.$$

Carrying out the integration with respect to ξ_3, show that this leads to

$$-\int_\Sigma m_{31}(\xi_1, \xi_2) d\xi_1 d\xi_2 + \int_\Sigma m_{13}(\xi_1, \xi_2) d\xi_1 d\xi_2,$$

which vanishes due to the symmetry in m. Show that for an isotropic medium, each of these integrals has magnitude

$$M_0 = \mu \int_\Sigma s_1(\xi_1, \xi_2)d\xi_1 d\xi_2.$$

2. Use Volterra's formula to derive the free-surface displacements due to a horizontal dislocation with uniform slip s at depth d. Show that the horizontal and vertical displacements are given by

$$u_1(\xi_1, \xi_2 = 0) = \frac{s_1}{\pi}\left[\frac{-\xi_1 d}{\xi_1^2 + d^2} + \tan^{-1}\left(\frac{\xi_1}{d}\right)\right],$$

$$u_2(\xi_1, \xi_2 = 0) = -\frac{s_1}{\pi}\frac{d^2}{\xi_1^2 + d^2}.$$

3. For a dipping fault with uniform slip, sum the contributions from the horizontal and vertical components of fault slip from problem 2 and equations (3.51) and (3.52) to show that the free-surface displacements are given by

$$u_1 = \frac{1}{\pi}\left[s_1 \tan^{-1}(\zeta) - \frac{s_2 + s_1\zeta}{1 + \zeta^2}\right],$$

$$u_2 = \frac{1}{\pi}\left[s_2 \tan^{-1}(\zeta) + \frac{s_2\zeta - s_1}{1 + \zeta^2}\right], \tag{3.115}$$

where

$$\zeta \equiv \frac{x_1 - \xi_1}{d}.$$

Use this result to derive the general expression (3.70) for displacement due to a fault dipping at an angle δ from the horizontal. Note that in the derivation for slip on a vertical fault, the positive side displaced upward, equations (3.51) and (3.52), so the sign of the vertical component of slip in equation (3.115) is opposite to that in equation (3.70); see figure 3.8.

4. Show that the displacements for a dip-slip fault with uniform slip from the earth's surface to depth d, as in equation (3.73), can be written in the following form:

$$u_1(x_1, x_2 = 0) = \frac{s \cos \delta}{\pi}\left[\tan^{-1}\left(\frac{x_1 - x_d}{d}\right) - (\pi/2)\mathrm{sgn}(x_1) - \frac{(x_1 - x_p)d}{d^2 + (x_1 - x_d)^2}\right],$$

$$u_2(x_1, x_2 = 0) = \frac{s \sin \delta}{\pi}\left[(\pi/2)\mathrm{sgn}(x_1) - \tan^{-1}\left(\frac{x_1 - x_d}{d}\right) - \frac{x_1 d}{d^2 + (x_1 - x_d)^2}\right]$$

(Cohen 1996), where $x_d = d/\tan \delta$, and $x_p = d/(\cos \delta \sin \delta)$ is the distance from the fault trace to the perpendicular projection of the bottom of the fault to the free surface.

5. Prove that the two representations of an edge dislocation shown in figure 3.26 are equivalent by showing that the strains, and therefore the stresses, in both cases are identical. Use the stress form of Volterra's formula and Green's functions appropriate for a full-space (that is, let $r_2 \to \infty$ in the Green's functions given in section 3.3). For the representation on the left-side of figure 3.26, take the opening to be uniform along the x_2 axis from 0 to ∞. For the representation on the right-side of figure 3.26, take the slip to be uniform along the x_1 axis from $-\infty$ to 0. To compute the strains, it is appropriate to take the spatial

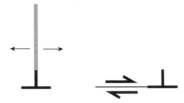

Figure 3.26. A single edge dislocation can be interpreted as due to the insertion of a slab of material of uniform thickness Δu extending along the x_2 axis from 0 to ∞ (left) or the slip of Δu extending along the x_1 axis from $-\infty$ to 0 (right).

derivatives inside the integral in Volterra's formula; however, be sure to take the derivatives with respect to the *observation* coordinates, not the coordinates defining the position of the dislocation.

6. In chapter 2, it was argued that the deformation due to a single buried screw dislocation is independent of the dip of the dislocation surface. Prove this using Volterra's formula by solving for the displacement due to slip on a dipping interface, and show that the result does not depend on dip.

7. Solve for surface displacements due to a two-dimensional dike. Model the dike as infinitely long along the strike, and assume plane strain deformation. Take a vertical edge dislocation with horizontal displacement discontinuity to model the dike. Use Volterra's formula to determine the horizontal and vertical displacements at the earth's surface. Plot the displacements for a dike that extends from the surface to depth D.

8. Solve for surface displacements due to a two-dimensional sill. Model the sill as infinitely long along the strike, and assume plane strain deformation. Take a horizontal edge dislocation with vertical displacement discontinuity to model the sill. Use Volterra's formula to determine the horizontal and vertical displacements at the earth's surface. Plot the displacements for a sill of length L located at depth D.

9. In polar coordinates centered on the dislocation, the stresses due to an edge dislocation in a two-dimensional infinite elastic medium, subject to plane strain conditions (see equation [1.117]), can be written as

$$\sigma_{rr} = \sigma_{\theta\theta} = D\frac{\sin\theta}{r},$$

$$\sigma_{r\theta} = -D\frac{\cos\theta}{r},$$

$$D = \frac{\mu s}{2\pi(1-\nu)}.$$

Derive the displacements in polar coordinates from the strains. You will find an equation for the shear strain of the form

$$\int f(\theta)\,d\theta + \frac{df(\theta)}{d\theta} + r\frac{dg(r)}{dr} - g(r) = \frac{-s}{\pi}\cos\theta,$$

where $f(\theta)$ and $g(r)$ are functions of θ and r only. Note that this can be separated into two equations, one a function of θ only and one a function of r only. This then leads to the solution

$$u_r = \frac{s}{2\pi} \left[\frac{(1-2v)}{2(1-v)} \sin\theta \ln(r) - \theta \cos\theta \right],$$

$$u_\theta = \frac{s}{2\pi} \left[\frac{(1-2v)}{2(1-v)} \cos\theta \ln(r) + \theta \sin\theta + \frac{1}{2(1-v)} \cos\theta \right], \tag{3.116}$$

where the terms corresponding to rigid body translations and rotations have been excluded.

3.10 References

Aki, K., and P. G. Richards. 1980. *Quantitative seismology: theory and methods*. New York: W. H. Freeman.

Ampuero, J.-P., and F. A. Dahlen. 2005. Ambiguity of the moment tensor. *Bulletin of the Seismological Society of America* **95**, 390–400.

Barber, J. R. 1992. *Elasticity*. Boston: Kluwer.

Chinnery, M. A. 1961. The deformation of the ground around surface faults. *Bulletin of the Seismological Society of America* **51**, 355–372.

Cohen, S. 1996. Convenient formulas for determining dip-slip fault parameters from geophysical observables. *Bulletin of the Seismological Society of America* **86**, 1642–1644.

Dundurs, J. 1962. Force in smoothly joined elastic half-planes. *Journal of Engineering, Mechanics Division, Proceedings of the American Society of Civil Engineers* **88**(EM5), 25–40.

Fialko, Y., M. Simons, and D. C. Agnew. 2001. The complete (3-D) surface displacement field in the epicentral area of the 1999 M_W 7.1 Hector Mine earthquake, California, from space geodetic observations. *Geophysical Research Letters* **28**(16), 3063–3066.

Hirth, J. P., and J. Lothe. 1992. *Theory of dislocations*. Malabar, FL: Krieger Publishing Co., pp. 73–78.

Hudnut, K. W., Z. Shen, M. Murray, S. McClusky, R. King, T. Herring, B. Hager, Y. Feng, P. Fang, A. Donnellan, and Y. Bock. 1996. Co-seismic displacements of the 1994 Northridge, California, earthquake. *Bulletin of the Seismological Society of America* **86**, S19–S36.

Jeyakumaran, M., J. W. Rudnicki, and L. M. Keer. 1992. Modeling of slip zones with triangular dislocation elements. *Bulletin of the Seismological Society of America* **82**(5), 2153–2169.

Johnson, K. M., Y. J. Hsu, P. Segall, and S. B. Yu. 2001. Fault geometry and slip distribution of the 1999 Chi-Chi, Taiwan, earthquake imaged from inversion of GPS data. *Geophysical Research Letters* **28**, 2285–2288.

Jónsson, S., H. Zebker, P. Segall, and F. Amelung. 2002. Fault slip distribution of the Mw 7.2 Hector Mine earthquake estimated from satellite radar and GPS measurements. *Bulletin of the Seismological Society of America* **92**, 1377–1389.

Kao, H., and W. P. Chen. 2000. The Chi-Chi earthquake sequence: active, out-of-sequence thrust faulting in Taiwan. *Science* **288**, 2346–2349.

Kelvin, Sir W. Thompson. "Cambridge and Dublin Mathematical Journal, 1848," reprinted in *Mathematical and Physical Papers* **1**, 97.

Love, A. E. H. 1944. *A treatise on the mathematical theory of elasticity*. New York: Dover.

Maerten, F., P. Resor, D. D. Pollard, and L. Maerten. 2005. Inverting for slip on three-dimensional fault surfaces using angular dislocations. *Bulletin of the Seismological Society of America* **95**, 1654–1665.

Melan, E. 1932. Der Spannungszustand der durch eine Einzelkraft im Innern beanspruchten Halb-scheibe, *Z. Angew. Math. Mech.* **12**, 343–346; correction in *Z. Angew. Math. Mech.* **20**, 368, 1940.

Mindlin, R. D. 1936. Force at a point in the interior of a semi-infinite solid. *Physics* **7**, 195–202.

Muskhelishvili, N. I., 1975. *Some basic problems of the mathematical theory of elasticity*, 4th ed. Published in Moscow, 1954, English ed., Leyden: Noordhoff International Publishing.

Okada, Y. 1985. Surface deformation due to shear and tensile faults in a half-space. *Bulletin of the Seismological Society of America* **75**, 1135–1154.

———. 1992. Internal deformation due to shear and tensile faults in a half-space. *Bulletin of the Seismological Society of America* **82**, 1018–1040.

Sato, R., and M. Matsu'ura. 1974. Strains and tilts on the surface of a semi-infinite medium. *Journal of Physics of the Earth* **22**, 213–221.

Savage, J. C. 1980. Dislocations in seismology. In F.R.N. Nabarro (Ed.), *Dislocations in solids*. New York: North-Holland Publishing Company, pp. 251–339.

Sokolnikoff, I. S. 1983. *Mathematical theory of elasticity*. New York: McGraw-Hill, 1956; reprinted Malabar, FL: Robert E. Krieger Publishing Co.

Stein, R. S., and S. E. Barrientos. 1985. Planar high-angle faulting in the basin and range: geodetic analysis of the 1983 Borah Peak, Idaho, earthquake. *Journal of Geophysical Research* **90**, 11, 355–11,366.

Steketee, J. A. 1958. Some geophysical applications of the elasticity theory of dislocations. *Canadian Journal of Physics* **36**, 1168.

Timoshenko, S. P., and J. N. Goodier. 1970. *Theory of elasticity*, New York: McGraw-Hill.

Yu, S. B., L. C. Kuo, Y. J. Hsu, H. H. Su, C. C. Liu, C. S. Huo, J. F. Lee, T. C. Lai, C. L. Liu, T. F. Tseng, C. S. Tsai, and T. C. Shin. 2001. Preseismic deformation and coseismic displacements associated with the 1999 Chi-Chi, Taiwan, earthquake. *Bulletin of the Seismological Society of America* **91**, 995–1012.

4

Crack Models of Faults

With the dislocation models discussed in the previous two chapters, the displacement discontinuity, or slip, was prescribed as a boundary condition. As discussed previously, such models cannot hope to explain *why* the fault slip occurs. We have also seen that constant slip on the fault surface leads to nonphysical stress singularities at the fault tip.

In the earth, shear stress accumulates on a fault until the stress exceeds the fault's strength and the fault slips. During the earthquake, the shear stress acting on the fault decreases by some amount $\Delta\tau = \tau_i - \tau_f$, where τ_i is the initial shear stress acting on the fault, and τ_f is the final stress. If τ_f is measured after all seismic waves have propagated away from the fault, then $\Delta\tau$ is referred to as the *static stress drop*. In this chapter, we will investigate models in which the stress drop on the fault is specified, rather than the amount of slip (figure 4.1). To begin, we will consider the case where the stress drop is uniform inside the slipping zone or *crack*. Later, we will relax this assumption. For a magma-filled dike, it may be sensible to consider a uniform pressure acting on the dike walls. Whether the resistive stress on a fault is spatially uniform, however, cannot be answered without a more thorough description of the frictional behavior of the fault.

Using the superposition principle, we can construct the solution for a crack subject to remote stress τ^∞ with residual shear stress acting on the fault $\tau^0 < \tau^\infty$ (figure 4.2A), by summing the solution for an uncracked body subject to uniform remote stress τ^∞ (figure 4.2C) and a crack loaded by internal tractions $-(\tau^\infty - \tau^0)$ and no remote stress (figure 4.2B). In what follows, we specify $\Delta\tau$ on the crack faces; it is understood that addition of a uniform tectonic stress also solves the problem specified in figure 4.2A.

The shear stress drop is specified only inside the crack; the stress change outside the crack must be solved for. The appropriate boundary condition outside the crack is that there be no displacement discontinuity. Thus, even the simplest crack problem involves a *mixed boundary condition* in which the traction is specified on part of the plane and the displacement discontinuity is specified on the remainder (figure 4.1). For an infinitely long vertical strike-slip fault parallel to the x_3 direction, as in figure 4.1, the mathematical problem is

$$\nabla^2 u_3 = 0,$$
$$\sigma_{13}(x_1 = 0) = -\Delta\tau \qquad |x_2| < a,$$
$$[u] = 0 \qquad |x_2| > a, \tag{4.1}$$

where the fault extends to depth a. Here, $[u]$ indicates displacement discontinuity $[u] = u^+ - u^-$.

4.1 Boundary Integral Method

We will now see that it is possible to construct a solution to the antiplane strain crack problem (4.1) using known solutions for screw dislocations. Each dislocation satisfies Laplace's equation, so a superposition of many dislocations also satisfies Laplace's equation. Recall from

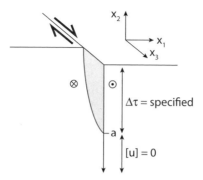

Figure 4.1. Crack model of a strike-slip fault. Within the slipping zone, the stress drop $\Delta\tau$ is specified; the slip must be solved for. Outside the slipping zone, the slip is known, and is identically zero; however, the stress change must be solved for.

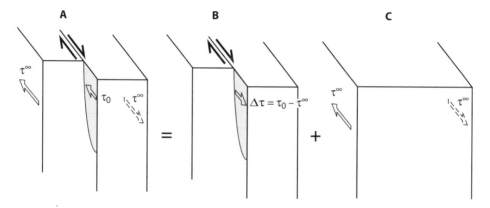

Figure 4.2. A: Fault loaded by remote stress τ^∞, with residual fault surface traction τ^0, can be constructed from the solutions for B, a crack loaded by internal tractions $-(\tau^\infty - \tau^0)$, and C, a uniformly loaded earth subject to far-field stress τ^∞.

equation (2.13) that the stress on the fault plane due to a single dislocation is

$$\sigma_{13}(x_1 = 0, x_2) = \frac{-\mu}{2\pi} \frac{s}{x_2}. \tag{4.2}$$

We can imagine summing many dislocations to mimic an arbitrary (discrete) slip distribution. Summing N dislocations with variable strength δs_i and location on the fault ξ_i (figure 4.3), the stress is

$$\sigma_{13}(x_1 = 0, x_2) = \frac{-\mu}{2\pi} \sum_{i=1}^{N} \frac{\delta s_i}{x_2 - \xi_i}. \tag{4.3}$$

Now, pass to the limit of a continuous distribution of infinitesimal dislocations, which have strength

$$\delta s = \frac{\partial s}{\partial \xi} \, d\xi = B(\xi) \, d\xi, \tag{4.4}$$

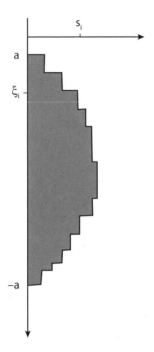

Figure 4.3. A discrete distribution of dislocations with strength s_i located at position ξ_i on the fault. Notice that there are an equal number of positively signed dislocations and negatively signed dislocations, so the crack closes at both ends. In this figure, the distribution is not symmetric about the $x_2 = 0$ axis.

where $B(\xi) = \partial s / \partial \xi$ is known as the *Burger's vector distribution*. The terminology comes from the material science application of dislocation theory, where the slip vector in a crystal lattice is known as the *Burger's vector*. The limit of equation (4.3) is thus

$$\sigma_{13}(z) = \frac{-\mu}{2\pi} \int_{-a}^{a} \frac{B(\xi)}{z - \xi} \, d\xi, \tag{4.5}$$

where we have let $x_2 = z$ be the depth on the fault. Notice that the displacement boundary condition outside the crack $[u] = 0$ for $|z| > a$ is automatically satisfied; there are no dislocations in that region, so the displacements there are continuous.

According to the boundary condition on the crack face (4.1):

$$-\Delta \tau = -(\tau^\infty - \tau^0) = \frac{-\mu}{2\pi} \int_{-a}^{a} \frac{B(\xi)}{z - \xi} \, d\xi \qquad |z| \le a. \tag{4.6}$$

For specified $\Delta \tau$, equation (4.6) represents an integral equation for $B(\xi)$. The boundary condition is specified as the value of the definite integral, while the unknown function $B(\xi)$ is in the integrand. The solution to this integral equation was given by Bilby and Eshelby (1968; see also Rice 1968; Barber 1992). We will cite the result here, postponing a derivation to section 4.1.1:

$$B(z) = \frac{-2}{\mu \pi} \frac{1}{\sqrt{a^2 - z^2}} \int_{-a}^{a} \frac{\Delta \tau(\xi) \sqrt{a^2 - \xi^2}}{z - \xi} \, d\xi + \frac{C}{\sqrt{a^2 - z^2}}, \qquad (4.7)$$

where C is a constant proportional to the net dislocation

$$C = \frac{1}{\pi} \int_{-a}^{a} B(\xi) \, d\xi \qquad (4.8)$$

(see problem 4). For cracks that close on both ends, the net dislocation vanishes $\int_{-a}^{a} B(\xi) \, d\xi = 0$ —that is, there are as many positive dislocations as negative dislocations.

To this point, we have said nothing about the traction-free boundary condition on the earth's surface. In chapter 2, we found that for antiplane strain problems, adding an oppositely signed dislocation an equal distance above the earth's surface was sufficient to render it traction free. Thus, if we now restrict attention to dislocation distributions that are symmetric about the plane $z = 0$, this guarantees that the shear stress there vanishes. For every dislocation of one sign, there is an oppositely signed dislocation symmetric about the free surface.

Determination of $B(z)$ requires integration of the expression in equation (4.7). Fortunately, these integrals are easily performed by expanding the stress in a particular set of polynomial functions known as *Chebyshev polynomials*. Chebyshev polynomials come in two varieties, the first are designated $T_n(x)$, while the second are designated $U_n(x)$. They are both orthogonal functions—that is, they satisfy

$$\int_{a}^{b} w(x) f_n(x) f_m(x) \, dx = h \delta_{mn}, \qquad (4.9)$$

where $w(x)$ are weight functions. Specifically, for the $T_n(x)$, $w(x) = (1 - x^2)^{-1/2}$ and $h = \pi/2$, unless $n = 0$ where $h = \pi$. For the $U_n(x)$, $w(x) = (1 - x^2)^{1/2}$ and $h = \pi/2$. The Chebyshev polynomials have the following recurrence: relationships

$$F_{n+1} = 2x F_n - F_{n-1}, \qquad (4.10)$$

where F_n refers to Chebyshev polynomials of both first and second kinds. The first few polynomials are given by

$$\begin{aligned} T_0 &= 1, & U_0 &= 1, \\ T_1 &= x, & U_1 &= 2x, \\ T_2 &= 2x^2 - 1, & U_2 &= 4x^2 - 1. \end{aligned} \qquad (4.11)$$

An extremely useful property of Chebyshev polynomials, for our purposes, is that they satisfy the following integral relationships (Abramowitz and Stegun 1972; Mason and Handscomb 2003; section 9.5.1):

$$\int_{-1}^{1} \frac{\sqrt{1 - t^2} U_{n-1}(t)}{z - t} \, dt = \pi T_n(z); \quad |z| \le 1, \qquad (4.12)$$

$$\int_{-1}^{1} \frac{T_n(t)}{\sqrt{1-t^2}(z-t)}\, dt = \begin{cases} -\pi U_{n-1}(z); & |z| \leq 1, n \geq 1 \\ 0; & n = 0 \end{cases}. \tag{4.13}$$

Notice that the first of these integrals is precisely the form of the integral in equation (4.7). Thus, if we expand the stresses in the U_{n-1}, the integral is done trivially. In particular, normalizing distance scales by a, let the stress be given by

$$\Delta\tau(z/a) = \sum_{n=1}^{\infty} c_n U_{n-1}(z/a). \tag{4.14}$$

Assuming that the net dislocation is zero ($C = 0$)—that is, we have one image for each dislocation—equation (4.7) becomes

$$B(z) = \frac{-2}{\mu} \sum_{n=1}^{\infty} \frac{c_n T_n(z)}{\sqrt{a^2 - z^2}}. \tag{4.15}$$

The simplest case is one in which the stress change is uniform within the interior of the crack. From equation (4.14), this corresponds to $c_1 = \Delta\tau$ and $c_n = 0$ for all $n > 1$. Since $T_1(z) = z$, equation (4.15) in this case reduces immediately to

$$B(z) = \frac{-2\Delta\tau}{\mu} \frac{z}{\sqrt{a^2 - z^2}}. \tag{4.16}$$

Equation (4.16) gives the appropriate dislocation density such that the uniform stress drop boundary condition is satisfied everywhere inside the crack. Given $B(\xi)$, the slip is computed from equation (4.4):

$$s(z) = \int_{-a}^{z} B(\xi)\, d\xi,$$

$$= \frac{-2\Delta\tau}{\mu} \int_{-a}^{z} \frac{\xi}{\sqrt{a^2 - \xi^2}}\, d\xi,$$

$$= \frac{2\Delta\tau}{\mu} \sqrt{a^2 - \xi^2} \Big|_{-a}^{z},$$

$$s(z) = \frac{2\Delta\tau}{\mu} \sqrt{a^2 - z^2} \qquad |z| < a. \tag{4.17}$$

Thus, the slip distribution for a uniform stress drop is elliptical. Note that the maximum slip is at the earth's surface and is proportional to the depth of the fault a and the strain drop $\Delta\tau/\mu$:

$$s(z = 0) = \frac{2\Delta\tau a}{\mu}. \tag{4.18}$$

For example, for a strain drop of 10^{-4} ($\Delta\tau = 3$ MPa, $\mu = 3 \times 10^4$ MPa) and $a = 10$ km, the maximum slip at the earth's surface is 2 meters.

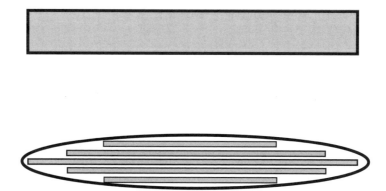

Figure 4.4. Representation of an opening mode crack such as a dike or sill with a summation of dislocations. Top: A sill modeled as a single pair of dislocations, equivalent to a slab of material with constant thickness inserted between the dislocations. Bottom: Representation of a sill subject to a uniform pressure boundary condition by a series of dislocation pairs with the appropriate distribution to satisfy the pressure boundary condition.

Note that the same procedure works for two-dimensional crack problems under plane strain loading by simply replacing the screw dislocations with edge dislocations (figure 4.4), with one minor change and one important exception. The stress due to a single edge dislocation is such that we replace μ in equation (4.6), and all subsequent equations, with $\mu/(1 - v)$. The important difficulty arises in matching the free-surface boundary condition. Because the image dislocation cancels only one component of the surface traction in plane strain, forcing the dislocation distribution to be antisymmetric about the $x_2 = 0$ plane does not satisfy the free surface-boundary conditions. If, however, one sums elemental solutions for dislocations in a half-space (as opposed to a full space), then the free-surface conditions are automatically satisfied. The solution to the resulting integral equation will, however, be more complex. As you will see, it is straightforward to develop numerical procedures based on this approach for solving general crack problems in both two and three dimensions.

4.1.1 Inversion of the Integral Equation

Here, we will consider a derivation of equation (4.7) that makes extensive use of the integral relations (4.12) and (4.13). First, normalize the length scales in equation (4.6) by a:

$$\Delta\tau = \frac{\mu}{2\pi} \int_{-1}^{1} \frac{B(\xi)}{z - \xi} \, d\xi. \tag{4.19}$$

Next, expand the Burgers vector distribution in the following:

$$B(\xi) = \frac{1}{\sqrt{1 - \xi^2}} \sum_{n=0}^{\infty} c_n T_n(\xi) = \frac{c_0}{\sqrt{1 - \xi^2}} + \frac{1}{\sqrt{1 - \xi^2}} \sum_{n=1}^{\infty} c_n T_n(\xi). \tag{4.20}$$

This choice can be justified as follows. We require the displacement to be discontinuous across the crack. The polynomials are of course continuous; however, the term $(1 - \xi^2)^{1/2}$ introduces a branch cut across the crack (see appendix C). Choosing $(1 - \xi^2)^{-1/2}$ causes the dislocation density to be singular at the crack tips. Indeed, it can be shown that $(1 - \xi^2)^{-1/2}$ behavior near the crack tip generates uniform stress within the crack in the vicinity of the tip and the appropriate stress singularity outside the crack (see problem 2). Substituting into equation (4.19) and making use of (4.13) leads to

$$\Delta\tau(z) = \frac{\mu}{2\pi} \sum_{n=1}^{\infty} c_n \int_{-1}^{1} \frac{T_n(\xi)}{\sqrt{1-\xi^2}(z-\xi)} d\xi,$$

$$= -\frac{\mu}{2} \sum_{n=1}^{\infty} c_n U_{n-1}(z). \tag{4.21}$$

Notice that the c_0 term drops out. Now multiply both sides of equation (4.21) by $(1-\xi^2)^{1/2}/(z-\xi)$ and integrate between -1 and $+1$:

$$\int_{-1}^{1} \frac{\Delta\tau(\xi)\sqrt{1-\xi^2}}{z-\xi} d\xi = -\frac{\mu}{2} \sum_{n=1}^{\infty} c_n \int_{-1}^{1} \frac{U_{n-1}(\xi)\sqrt{1-\xi^2}}{z-\xi} d\xi,$$

$$= -\frac{\mu\pi}{2} \sum_{n=1}^{\infty} c_n T_n(z),$$

$$= -\frac{\mu\pi}{2} \left[\sqrt{1-z^2} B(z) - c_0, \right], \tag{4.22}$$

where the first step follows by equation (4.12) and the second by (4.20). Reverting to dimensional form leads directly to equation (4.7).

4.2 Displacement on the Earth's Surface

To compute the displacement at the earth's surface, recall from equation (2.41) that

$$u(x) = \frac{1}{\pi} \int_{-a}^{0} s(\xi) \frac{x}{x^2 + \xi^2} d\xi. \tag{4.23}$$

For a uniform stress drop, we input the slip distribution (4.17) into (4.23), which leads to

$$u(x) = \frac{2\Delta\tau x}{\mu\pi} \int_{-a}^{0} \frac{\sqrt{a^2 - \xi^2}}{x^2 + \xi^2} d\xi,$$

$$= \frac{\Delta\tau x}{\mu\pi} \int_{-a}^{a} \frac{\sqrt{a^2 - \xi^2}}{x^2 + \xi^2} d\xi. \tag{4.24}$$

This integral is solved using contour integration methods in appendix C, equation (C.8), yielding

$$u(x) = \frac{\Delta\tau}{\mu} \left[\text{sgn}(x)\sqrt{a^2 + x^2} - x \right]. \tag{4.25}$$

The displacements from the constant stress drop crack model are compared with those of the uniform-slip dislocation in section 4.5.

4.3 A Brief Introduction to Fracture Mechanics

The spatially uniform stress drop crack considered earlier involves infinite stresses at the crack tip. To see this, note that the gradient of the crack surface displacement, the Burger's vector distribution, is singular. From equation (4.16), note that $|B(z)| \to \infty$ as $|z| \to a$. Thus, the strain, and therefore the stress, is infinite at the crack tip. While it is certainly true that stresses remain finite in the earth, uniform stress drop crack models have played a key role in the development of engineering fracture mechanics, and many of these concepts are used to describe faults as well as dikes and sills within the earth.

Tensile fracture of engineering materials occurs at stresses much lower than predicted by the theoretical strength of molecular bonds. As a rule, calculations based on bond strengths suggest fracture stresses of order $E/10$, where E is the Young's modulus of the material. Observed fracture strengths are factors of ten to a thousand times less than these values. In 1920, Griffith suggested that minute flaws or cracks concentrate stress, so that the theoretical strength is reached only locally at the crack tip. He demonstrated this in the laboratory by drawing pristine thin glass fibers whose strength approached the theoretical tensile strength. These experiments eventually led to the development of strong fiber-reinforced composites.

Griffith suggested that if the energetics favor crack growth, then the material fails at a low macroscopic (but high local) stress. He considered the energy of a cracked body of the form

$$U_{total} = U_0 + U_{strain} - W_L + U_{surf}, \tag{4.26}$$

where U_0 is the energy of the uncracked, unstressed body; U_{strain} is the elastic strain energy; W_L is the potential energy of the loading system associated with applied boundary forces; and U_{surf} is the surface energy associated with the crack faces. This expression ignores kinetic, thermal, chemical, and other forms of energy. For a crack of length $2a$, the surface energy is $4\gamma a$, where γ is the surface free energy. Together, the elastic strain energy and the work done at the boundaries represent the mechanical potential energy of the body.

For a two-dimensional opening crack subject to uniform remote stress σ, the change in strain energy can be computed as follows. Imagine making a cut perpendicular to the applied stress and applying tractions to the crack faces that are equal and opposite to the remote stress. In this configuration, the crack does not open. If we *slowly* (so there is no kinetic energy involved) relax the crack–surface tractions, the crack will open. When the tractions are relaxed to zero, the displacements on the crack will be

$$u(x) = \frac{(1-\nu)\sigma}{\mu}\sqrt{a^2 - x^2} \qquad |x| < a. \tag{4.27}$$

This result can be derived by the same procedure as used previously in the antiplane case, by noting that the stresses acting on the crack plane due to an edge dislocation (equation [3.78], considering only the full-space, r_1, terms) are related to those due to a screw dislocation (4.2) simply by replacing μ with $\mu/(1-\nu)$. When comparing equation (4.27) to the antiplane case, note that equation (4.17) gives the total slip, whereas (4.27) gives the displacement of one crack face.

As the crack opens, work is done on the loading mechanism that supplies the crack surface tractions, thereby decreasing the strain energy of the elastic body. The work done per unit area of crack is

$$dW = -\int_0^u \sigma(u')\,du'. \tag{4.28}$$

The negative sign arises because the tractions and displacements act in opposite directions. We assume that the crack is inhibited from propagating as the stresses relax, so the crack displacements are always proportional to the change in traction. Thus, equation (4.28) can be written as

$$dW = -\int_0^u \sigma \cdot \left(1 - \frac{u'}{u}\right)\,du' = -\frac{\sigma}{2}u. \tag{4.29}$$

The total work done is found by integrating over both crack faces:

$$
\begin{aligned}
W = 2\int_{-a}^a dW &= -\frac{(1-\nu)\sigma^2}{\mu}\int_{-a}^a \sqrt{a^2 - x^2}\,dx \\
&= -\frac{(1-\nu)\sigma^2}{\mu}\left[\frac{x}{2}\sqrt{a^2 - x^2} + \frac{a^2}{2}\sin^{-1}\left(\frac{x}{a}\right)\right]_{-a}^a, \\
&= -\frac{\pi(1-\nu)\sigma^2 a^2}{2\mu}.
\end{aligned} \tag{4.30}
$$

If the remote boundaries are sufficiently far away, or as in the earth, traction free, then there is no work done on these surfaces, and equation (4.30) gives the change in strain energy of the elastic body. If, as in a laboratory test, loads are applied to the boundaries and these are not prevented from displacing, then work is done on these boundaries. It turns out that in this case, the strain energy actually increases on introduction of the crack. This may seem paradoxical, but it arises because work is done on the body by the remotely applied stress. The total change in mechanical potential energy $U_{strain} - W_L$, however, does decrease, and is in fact given by equation (4.30).

Introducing the crack decreases the net mechanical potential energy (equation [4.30]). In a naturally formed crack, this energy is absorbed in creating new crack surface, increasing the surface energy. The change in total energy on a small increment of crack growth is

$$\delta U_{total} = \frac{\partial U_{mech}}{\partial a}\delta a + \frac{\partial U_{surf}}{\partial a}\delta a, \tag{4.31}$$

ignoring terms related to kinetic energy, thermal energy, and so forth. The surface energy increases linearly with crack length, whereas the mechanical energy decreases quadratically with length (equation [4.30]), leading to a critical crack length and stress (figure 4.5). For a given stress, cracks longer than the critical length were predicted by Griffith to grow unstably. Interestingly, cracks less than the critical length are predicted to close unstably (that is, to heal), although adsorption of chemical constituents on the crack faces may inhibit crack healing (Rice 1978). The critical stress is given by

$$\sigma = \sqrt{\frac{4\gamma\mu}{\pi(1-\nu)a}}. \tag{4.32}$$

A crack of length $2a$ subject to a stress greater than the critical stress given by equation (4.32) propagates unstably.

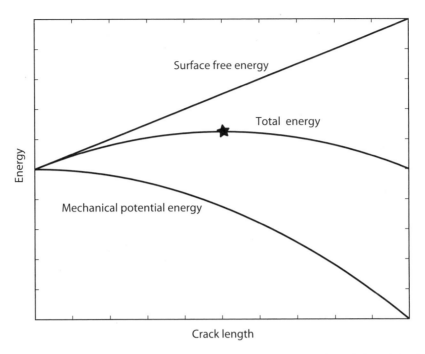

Figure 4.5. Variation in energy on increase in crack length. The critical crack length is shown by a star.

As noted by Griffith, the stress at the crack tip may be very large. For an ideally sharp crack tip (corresponding to the uniform stress drop crack considered in the previous section), the stress in some neighborhood of the crack tip is, in an r, θ coordinate system with origin at the crack tip, given by

$$\sigma_{ij} = \frac{K}{\sqrt{2r}} f_{ij}(\theta), \tag{4.33}$$

where $f_{ij}(\theta)$ are functions of the angular distance from the crack plane, and differ for the different crack modes (modes I, II, and III, corresponding to opening, inplane shear, and antiplane shear, respectively). K is the so-called *stress intensity factor*. The $1/\sqrt{r}$ singularity is universal (see problem 2) and applies to all crack geometries and loading configurations (In three dimensions, the tangent to the crack tip line must be smoothly varying.) Thus, all the information about the loading configuration is captured by the stress intensity factors K_I, K_{II}, and K_{III} corresponding to the three modes of deformation. Because the local stress is governed by K, the condition that local stress at the crack tip reach a critical value is equivalent to the stress intensity reaching a critical value, usually denoted K_{crit} and referred to as the *fracture toughness*.

On the plane in front of the crack tip, the near-tip stress (4.33) is

$$\sigma = \frac{K}{\sqrt{2r}}. \tag{4.34}$$

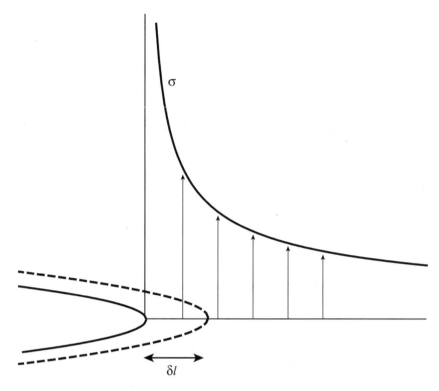

Figure 4.6. Crack tip stress and displacement. As the crack length increases by δl, the stress acting in front of the crack σ is reduced to zero. The work done during this increment of crack growth is given by equation (4.37).

(It should be noted that the stress intensity factor is sometimes defined in such a way that the denominator of [4.34] is replaced by $\sqrt{2\pi r}$.) The crack surface displacements measured behind the crack tip (figure 4.6) are

$$u = \frac{(1-v)K}{\mu}\sqrt{2r} \tag{4.35}$$

(e.g., Lawn and Wilshaw 1975), where here r measures distance from the crack tip *into* the crack. We can compute the work done as the crack tip propagates forward an increment δl. The additional crack surface displacements due to the increment of crack growth at fixed stress are

$$u = \frac{(1-v)K}{\mu}\sqrt{2(\delta l - r)}. \tag{4.36}$$

The work per unit surface of crack is given by equation (4.29). Making use of equations (4.34) and (4.36), we have

$$\delta U = -\frac{(1-v)K^2}{\mu}\int_0^{\delta l}\sqrt{\frac{\delta l - r}{r}}\,dr. \tag{4.37}$$

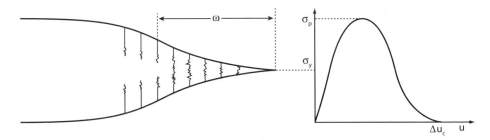

Figure 4.7. Detailed view of crack tip. Outside the crack tip region, the rock deforms elastically. Inside the crack, bonds deform inelastically until they eventually separate in this atomic-scale drawing. σ_y is the yield stress, σ_p is the peak stress, and u_c is the critical crack displacement at which point the stress acting across the crack walls drops to zero. The area under the stress-displacement curve is the work done to create new crack surface, or the fracture energy.

No work is done by displacement of the traction-free crack walls. The integral can be solved by making the substitution $r = \delta l \sin^2 \theta$. Doing so, we find that the integral yields $\pi \delta l / 2$. Equation (4.37) thus reduces to

$$\delta U = -\frac{\pi (1 - v) K^2}{2\mu} \delta l. \tag{4.38}$$

If, again, there is no work done on remote boundaries, equation (4.38) gives the total change in mechanical potential energy. The negative of the gradient in potential energy has the units of force and is referred to as the *energy release rate*, or *crack extension force*, often denoted as G (not to be confused with either shear modulus or the universal gravitation constant):

$$-\frac{\partial U_{mech}}{\partial l} \equiv G = \frac{\pi (1 - v) K^2}{2\mu}. \tag{4.39}$$

From an energetic perspective, we conclude that the crack grows when the rate of mechanical potential energy decrease exceeds the energy lost in creating new crack surface. From the perspective of the local stress acting at the tip of the crack, we say that the crack propagates when the stress intensity factor K reaches a critical value. Equation (4.39) shows that these perspectives are completely consistent. That is, there exists a critical energy release rate G_{crit} proportional to K_{crit}^2.

As an aside, it is worth noting that while the crack, with the $1/\sqrt{r}$ singularity, has a finite elastic strain energy in the neighborhood of the crack tip, the same is not true for the dislocation. Recall that the stress due to a single dislocation is proportional to $1/r$. The strain energy change is proportional to the integral of the stress times the strain, so the strain energy due to the dislocation scales as $\int_0^{2\pi} \int_{r_0}^{R} (1/r^2) r \, dr \, d\theta = 2\pi \ln(R/r_0)$. Thus, the strain energy in some volume around the dislocation is unbounded in the limit $r_0 \rightarrow 0$.

In actuality, the crack tip cannot be mathematically sharp in the sense of sustaining infinite stress immediately outside the crack. Rather, rock begins to yield under the great stress concentration acting at the crack tip. (Figure 4.7 illustrates the idealized case of a cleavage crack in a crystalline solid.) Once the stress exceeds the yield stress σ_y, bonds begin deforming inelastically. The stress increases to a peak value σ_p with increasing crack displacement.

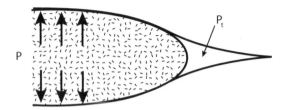

Figure 4.8. Schematic view of dike tip region. Magma (shown by stippled pattern) is at pressure p, which may vary along the dike. The excess pressure, $p - \sigma_n$, where σ_n is the far-field dike-normal stress, acts to dilate the dike. The dike tip region is filled with fluids at pressure p_t. If $p_t < \sigma_n$, then the excess pressure in the tip region is negative (a relative suction), and the crack walls close smoothly, such that the stress intensity is zero. The relative suction is analogous to the cohesive forces acting to pull the crack walls together in figure 4.7.

Eventually, the stress drops to zero at a critical crack displacement u_c. The region near the crack tip undergoing inelastic deformation is referred to as the *end zone*. If the dimension of the end zone is ω, then at distances $r \gg \omega$, the stress is well approximated by the purely elastic solution—that is, the stress decays with distance proportional to $K/\sqrt{2r}$.

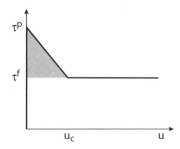

Figure 4.9. Slip-weakening model showing shear stress as a function of fault slip. τ^p is the peak stress, and τ^f is the residual frictional stress. In this simplified model, the strength is imagined to degrade linearly from τ^p to τ^f as slip increases to a critical value u_c. The shaded area is the work done in propagating the crack.

This picture can be usefully generalized to macroscopic opening mode fractures. In this case, the inelastic deformation may be dominated by grain boundary cracking in a wide zone around the fracture. Current models of dike propagation consider a dike tip region, which viscous magma cannot penetrate (e.g., Rubin 1995). The dike tip region may be filled with fluids exsolved from the melt or pore fluids from the surrounding rock mass. In order for the dike walls to dilate, the magma pressure p must exceed the remote dike-normal stress σ_n over much of the dike—that is, the excess magma pressure $p - \sigma_n$ is positive. If the fluid pressure in the dike tip cavity is less than the remote dike-normal stress, excess pressure is negative, and the dike walls close smoothly, as shown in figure 4.8.

The situation changes only slightly for a shear crack model of faulting. In the case of an open crack, the strength of the interface falls to zero behind the crack tip (figure 4.7). For faulting, a simplified slip-weakening model posits that the shear stress acting on the fault surface increases elastically up to a peak stress τ^p but then drops linearly to a constant residual friction stress τ^f (figure 4.9). The shaded area in figure 4.9 represents the inelastic work done in creating new fault surface, analogous to the fracture energy for the tensile crack.

Figure 4.10 shows the stress and displacement distributions corresponding to the *slip-weakening model* illustrated in figure 4.9. The fault surface is located on the x axis; for positive

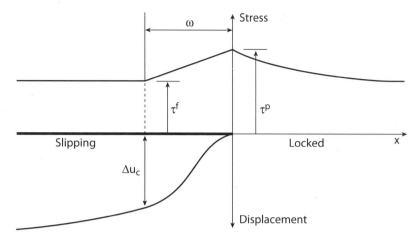

Figure 4.10. Stress in the neighborhood of a shear crack tip. Stress outside the crack increases to the peak stress, τ^p. Inside the crack tip region, the stress decays to the residual frictional stress, τ^f, over a distance scale ω. After Rice (1983).

values of x, the crack is locked. The stress increases toward the crack tip, where it reaches τ^p. Inside the crack tip, the strength degrades to τ^f over an end zone with dimension ω. As the strength degrades, the fault slip increases to Δu_c.

The slip-weakening model is widely used to describe faulting in the earth. If the dimensions of the end zone are small compared to the overall dimension of the slipping region ($\omega \ll l$), then the simplified crack model described in the previous section gives a good description of the elastic fields away from the end zone region. Furthermore, the slip-weakening model provides a physical connection between the critical energy release rate G_{crit} and the work done in reducing the strength between τ^p and τ^f.

Of course, we have simply asserted here that the resistive stress on the fault depends only on slip and remains constant after a critical displacement. In detail, the resistance to slip depends on the frictional properties of the fault as well as the pressure of pore fluids within the fault zone, as will be discussed in more detail in chapter 11.

4.4 Nonsingular Stress Distributions

It is useful to explicitly consider situations in which the stress is everywhere finite on the fault, and the end zone is not restricted to a very small region at the crack tip. It can be shown (Bilby and Eshelby 1968) that the condition that the stresses be everywhere finite, equivalent to the stress intensity factors vanishing, is given by

$$\int_{-a}^{a} \frac{\Delta \tau(\xi)}{\sqrt{a^2 - \xi^2}} \, \mathrm{d}\xi = 0,$$

$$\int_{-a}^{a} \frac{\xi \, \Delta \tau(\xi)}{\sqrt{a^2 - \xi^2}} \, \mathrm{d}\xi = 0. \tag{4.40}$$

If these equations are satisfied, and the net dislocation is zero, then equation (4.7) becomes

$$B(z) = \frac{-2\sqrt{a^2 - z^2}}{\mu \pi} \int_{-a}^{a} \frac{\Delta \tau(\xi)}{(z - \xi)\sqrt{a^2 - \xi^2}} \, \mathrm{d}\xi. \tag{4.41}$$

(see problem 5). Expanding the stress drop on the crack surface in terms of Chebyshev polynomials:

$$\Delta\tau(\xi) = \sum_{i=0}^{N} c_i T_i(z/a). \tag{4.42}$$

It can be shown (see problem 5) that the condition that equations (4.40) are satisfied is equivalent to $c_0 = c_1 = 0$. Thus, the sum in equation (4.42) is started at $i = 2$ for the finite stress case. From equations (4.41) and (4.42),

$$B(z) = -\frac{2\sqrt{a^2 - z^2}}{\mu\pi a} \sum_{i=2}^{N} c_i \int_{-1}^{1} \frac{T_i(z/a)}{(z/a - \xi)\sqrt{1 - \xi^2}} d\xi. \tag{4.43}$$

We can now make use of the integral relation (4.13) to write

$$B(z) = \frac{2\sqrt{a^2 - z^2}}{\mu a} \sum_{i=2}^{N} c_i U_{i-1}(z/a) \qquad |z/a| \le 1. \tag{4.44}$$

As an example, we shall consider the first nonsingular term in the expansion. Let

$$\Delta\tau = -c_2 T_2(z/a) = -c_2 \left[2(z/a)^2 - 1\right] \qquad |z| < a, \tag{4.45}$$

that is, the stress drop is quadratic within the rupture zone. The sign is chosen such that, for $c_2 > 0$, the stress drop is positive in the center of the crack. The beauty of this approach is that we can immediately see from equation (4.44) that the appropriate Burger's vector distribution is

$$B(z) = -\frac{2\sqrt{a^2 - z^2}}{\mu a} c_2 U_1(z/a) = -\frac{4c_2}{\mu}\frac{\sqrt{a^2 - z^2}}{a}(z/a). \tag{4.46}$$

To compute the slip, integrate the Burger's vector distribution (4.4):

$$s(z) = -\frac{4c_2}{\mu} \int_{-a}^{z} (\xi/a)\sqrt{1 - (\xi/a)^2} d\xi,$$

$$= \frac{4c_2 a}{3\mu} \left[1 - (z/a)^2\right]^{3/2}. \tag{4.47}$$

Figure 4.11 shows that the stress corresponding to the slip distribution (4.47) is quadratic within the slipping zone and everywhere finite on the fault.

To compute the displacements at the free surface, we use equation (4.23), taking into account the symmetry of the integrand:

$$u(x) = \frac{4c_2}{3\pi\mu}\frac{x}{a^2} \int_0^a \frac{(a^2 - z^2)^{3/2}}{x^2 + z^2} dz,$$

$$= \frac{4c_2}{3\pi\mu} \left[x \int_0^a \frac{(a^2 - z^2)^{1/2}}{x^2 + z^2} dz - \frac{x}{a^2} \int_0^a \frac{z^2(a^2 - z^2)^{1/2}}{x^2 + z^2} dz\right]. \tag{4.48}$$

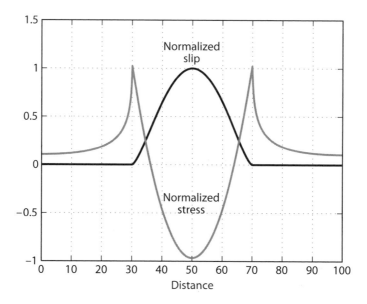

Figure 4.11. Normalized slip and stress corresponding to the slip distribution given by equation (4.47). The stress change within the crack is quadratic. Notice that the crack displacements taper smoothly to zero at the crack tips. This is always the case when the stress at the crack tip is finite.

The first of these integrals is examined in appendix C. The second can be computed using the same methods and is also given by Barnett and Freund (1975), who presented key ideas used in this section:

$$\frac{2}{a^2} \int_0^a \frac{\xi^2 \sqrt{a^2 - \xi^2}}{x^2 + \xi^2} d\xi = \frac{\pi}{2} - \pi \left(\frac{|x|}{a} \sqrt{\frac{x^2}{a^2} + 1} - \frac{x^2}{a^2} \right). \tag{4.49}$$

With these integrals, we find that the surface displacement is given by

$$u(x) = -\frac{c_2 a}{\mu} \left\{ \left(\frac{x}{a} \right) + \frac{2}{3} \left(\frac{x}{a} \right)^3 - \frac{2}{3} \text{sgn}(x) \left[1 + \left(\frac{x}{a} \right)^2 \right]^{3/2} \right\}. \tag{4.50}$$

4.5 Comparison of Slip Distributions and Surface Displacements

In order to compare the displacements for the various sources, we first normalize the slip distributions so that they have the same slip at the earth's surface, $s(z = 0)$. We then choose the depths at which the slip vanishes such that each slip distribution has the same seismic moment per unit length of fault—that is, the areas under the slip distributions as a function of depth are the same.

For the dislocation, the slip is s for $0 < z < a_d$, and zero elsewhere, where a_d is the depth of the dislocation. Let s_{max} be the maximum slip, which always occurs at the earth's surface and is therefore directly observable. For the uniform stress drop crack, we found (equation [4.18]) $s_{max} = 2 \Delta \tau a_c / \mu$, where a_c is the depth of the crack, so from equation (4.25),

$$s(z) = s_{max} \sqrt{1 - \left(\frac{z}{a_c} \right)^2}, \tag{4.51}$$

and

$$u(x) = \frac{s_{max}}{2}\left[\text{sgn}(x)\sqrt{1+\left(\frac{x}{a_c}\right)^2} - \left(\frac{x}{a_c}\right)\right].$$ (4.52)

The moment per unit length for the dislocation is $\mu s_{max}a_d$. For the uniform stress drop crack, it is one quarter the area of an ellipse with semi-axes s_{max} and a_c, multiplied by shear modulus—that is, $\mu\pi a_c s_{max}/4$. The requirement that the moment (per unit length) be equal for the crack and dislocation is $a_c = 4a_d/\pi$. Now consider the variable stress drop crack. From equation (4.47), the maximum slip is $4c_2 a_s/3\mu$, where a_s is the depth of the smoothly tapered crack. Requiring that the maximum slip be equal to that for the uniform stress drop crack:

$$c_2 = \frac{3}{2}\Delta\tau\frac{a_c}{a_s},$$ (4.53)

so that

$$s(z) = s_{max}\left[1-\left(\frac{z}{a_s}\right)^2\right]^{3/2},$$ (4.54)

and

$$u(x) = -\frac{3}{4}s_{max}\left\{\left(\frac{x}{a_s}\right) + \frac{2}{3}\left(\frac{x}{a_s}\right)^3 - \frac{2}{3}\text{sgn}(x)\left[1+\left(\frac{x}{a_s}\right)^2\right]^{3/2}\right\}.$$ (4.55)

Last, the ratio a_s/a_c is determined by the requirement that the moment per unit length is equal for each model. For the variable stress drop crack, the moment is found by integrating equation (4.54):

$$M_o = s_{max}\int_0^{a_s}\left[1-\left(\frac{z}{a}\right)^2\right]^{3/2}dz.$$ (4.56)

Making use of the integral

$$\int(1-\xi^2)^{3/2}\,d\xi = \left(\frac{5-2\xi^2}{8}\right)\xi\sqrt{1-\xi^2} + \frac{3}{8}\sin^{-1}(\xi)$$ (4.57)

(Petit–Bois 1961), we find that

$$M_o = \mu s_{max}a_s\frac{3\pi}{16}.$$ (4.58)

Matching this to the moment for the crack leads to $a_s = 4a_c/3$.

The three normalized slip distributions and their corresponding surface displacement fields are summarized in table 4.1 and figure 4.12. The results are striking. Despite the fact that the slip distributions are quite different, the surface displacement patterns are alarmingly similar. This suggests that it will be difficult to distinguish between different slip distributions with measurements of displacement at the earth's surface. But is it impossible? Figure 4.13 shows the differences between the constant stress drop crack and the uniform dislocation and between

TABLE 4.1

Three Normalized Slip Distributions and Corresponding Surface Displacement Fields

Model	Slip $s(z)$	Surface displacement $u(x)$
Uniform dislocation	$s(z) = \begin{cases} s & z \le a_d \\ 0 & z > a_d \end{cases}$	$u(x) = \dfrac{s}{\pi} \tan^{-1}\left(\dfrac{a_d}{x}\right)$
Constant-stress drop crack	$s(z) = s\sqrt{1 - \left(\dfrac{z}{a_c}\right)^2}$	$u(x) = \dfrac{s}{2}\left[\mathrm{sgn}(x)\sqrt{1 + \left(\dfrac{x}{a_c}\right)^2} - \left(\dfrac{x}{a_c}\right) \right]$
Tapered crack	$s(z) = s\left[1 - \left(\dfrac{z}{a_s}\right)^2\right]^{3/2}$	$u(x) = -\dfrac{3}{4}s\left\{ \left(\dfrac{x}{a_s}\right) + \dfrac{2}{3}\left(\dfrac{x}{a_s}\right)^3 \right.$ $\left. -\dfrac{2}{3}\mathrm{sgn}(x)\left[1 + \left(\dfrac{x}{a_s}\right)^2\right]^{3/2} \right\}$

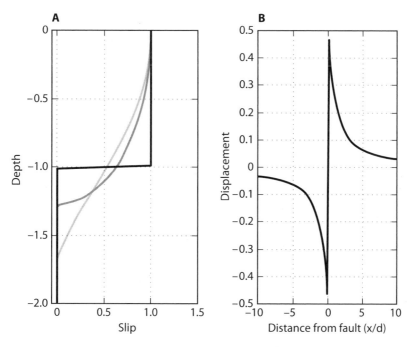

Figure 4.12. A: Three different slip distributions corresponding to a uniform-slip dislocation, a constant stress drop crack, and a tapered crack. B: The three displacement distributions corresponding to the slip distributions in A.

the nonsingular slip distribution and the uniform dislocation per meter of maximum slip. The difference between the constant stress drop crack and the uniform dislocation is at most 12 mm per meter of slip. For the nonsingular slip distribution and the uniform dislocation, the difference is about 20 mm per meter of maximum slip. For a large earthquake, with maximum slip of 4 or 5 meters, these differences (50 to 100 mm) are easily detected with current GPS and SAR instrumentation. Of course, we must not forget that other approximations that

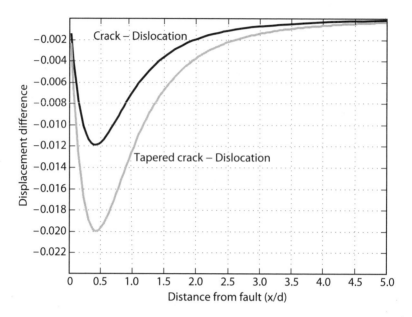

Figure 4.13. The differences between the slip distributions in figure 4.12 shown relative to the uniform slip dislocation.

have been made, such as uniform elastic properties, are also likely not to hold in detail. For smaller earthquakes with maximum slip of, say, 0.5 m, the differences between different slip distributions will be difficult to detect with confidence.

This example focused on variations in slip distribution with depth, normalized such that the significant differences between models are deep, where surface geodetic observations have limited resolution. In contrast, *along-strike* differences in slip distribution between uniform slip dislocations, constant stress drop cracks, and nonsingular cracks are far more easily recognized.

4.6 Boundary Element Methods

As we have seen, it can be difficult to find closed form expressions to the mixed boundary value problem. The plane strain problem in a half-space (for dip-slip faults) is more complex (Freund and Barnett 1976) than for antiplane strain, and solutions in three dimensions are quite limited.

One can, however, take a discrete approximation to the integral equation and use standard matrix inversion methods to obtain useful results. This is the idea behind the *boundary element method*. The key is to exploit the superposition principle and add up a series of solutions, each of which satisfies the governing equations off the crack surface. By choosing the right combination of these elemental solutions, we can (approximately) satisfy the boundary conditions and thus achieve a solution to the boundary value problem. The trade-off is that we lose the ability to inspect a closed form solution for the dependence on various parameters.

For the antiplane problem, one can use a pair of dislocations to represent the slip in the interval d_1 to d_2. Including image terms (to represent the free surface), we can write the stress on the plane $x_1 = 0$, at $x_2 = z_i$, due to slip in the interval d_{j-1} to d_j, as

$$\sigma_{13}(x_1 = 0, \, x_2 = z_i) = s_j g(z_i, d_j, d_{j-1}). \qquad (4.59)$$

The total stress at z_i is the *sum* of the contributions from all the N dislocation segments:

$$\sigma_i \equiv \sigma_{13}(x_1 = 0, \ x_2 = z_i) = \Sigma_{j=1}^{N} s_j g(z_i, d_j, d_{j-1}). \qquad (4.60)$$

The boundary conditions are given in terms of the stresses acting on the fault plane—namely, the σ_i at each element within the fault surface. (Note that there is automatically no displacement discontinuity outside the fault surface if there are no dislocation elements there.) In order to satisfy the boundary conditions at each of the fault elements, we must solve a set of N equations of the form

$$\sigma(z_i) = \Sigma_{j=1}^{N} s_j g(z_i, d_j, d_{j-1}) \qquad i = 1, 2, \dots N, \qquad (4.61)$$

for the N unknowns s_j. The usual procedure is to evaluate the stress at the center of each segment. In matrix form, equation (4.61) becomes

$$\sigma = Gs, \qquad (4.62)$$

with solution

$$s = G^{-1}\sigma. \qquad (4.63)$$

Given the estimated slip s_j within each element, it is straightforward to compute the surface displacements.

Sample results are shown in figure 4.14. Notice that with even 10 elements, the result is quite accurate.

4.7 Fourier Transform Methods

Boundary element methods are quite general, but they can be computationally slow. In some instances, it is possible to solve stress boundary value problems more efficiently using Fourier transform methods (see section A.1).

Combining equations (4.4) and (4.6), and letting the integral extend over the entire domain:

$$\sigma(z) = \frac{\mu}{2\pi} \int_{-\infty}^{\infty} \frac{-\partial s/\partial \xi}{z - \xi} \, d\xi, \qquad (4.64)$$

where $\sigma = -\Delta \tau$ represents the stress acting on the crack face due to slip. Note that the same equation applies to plane strain conditions if we replace μ with $\mu/(1 - \nu)$.

Equation (4.64) is in the form of a *Hilbert transform*. Recall, however, that we started this chapter by noting that crack problems involve mixed traction/displacement boundary conditions. We do not know a priori the stress acting on the fault plane outside the slipping zone. Thus, in general, it is not possible to couch the problem in terms of an integral over the entire fault plane as in equation (4.64). There are cases, however, where one knows the slip distribution and wants to compute the stress. In addition, when modeling the evolution of friction using rates and state-dependent constitutive laws (chapter 11), the fault is considered to be always slipping—albeit sometimes at an extremely low rate. In this case, we do not distinguish between inside the crack, where the stress is specified, and outside the crack, where the displacement is specified. In cases like these, it is appropriate to let the limits on the integral extend to $\pm\infty$.

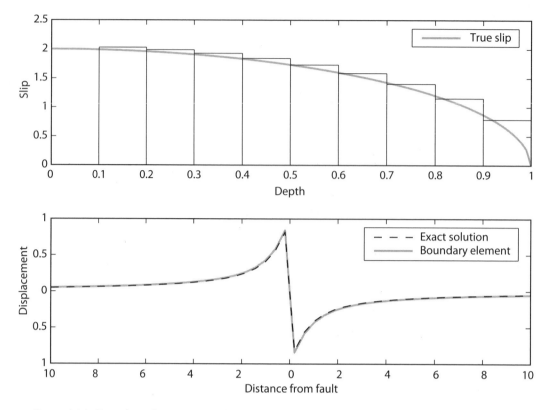

Figure 4.14. Boundary element approximation to constant stress drop crack. Top: Slip distribution as a function of depth. Bottom: Surface displacements.

Notice that equation (4.64) is in the form of a convolution (see equation [A.9]), $\int_{-\infty}^{\infty} f(\xi)g(z-\xi)d\xi = f \star g$,

$$\sigma = \frac{-\mu}{2\pi} s' \star \frac{1}{z}, \tag{4.65}$$

where s' denotes the spatial derivative of s. Convolution is equivalent to multiplication in the Fourier transform domain (A.10). Here we use the Fourier transform pair defined by equations (A.1) and (A.2). The Fourier transform of the derivative s' is simply (see equation [A.7])

$$\mathcal{F}(s') = ik\mathcal{F}(s), \tag{4.66}$$

where k is the spatial wavenumber, and $\mathcal{F}(s)$ indicates the Fourier transform of the slip s. The Fourier transform of $1/z$ is

$$\mathcal{F}(1/z) = -i\pi \operatorname{sgn}(k) \tag{4.67}$$

(Bracewell 1965, p. 130). The integral is evaluated using contour methods (see problem 6). Thus, the stress on the fault in the Fourier domain is

$$\mathcal{F}(\sigma) = \frac{-\mu}{2}|k|\mathcal{F}(s). \tag{4.68}$$

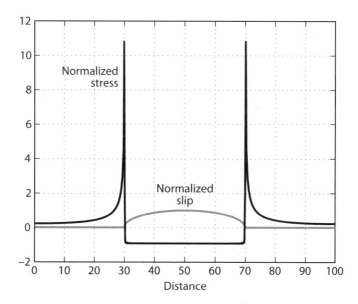

Figure 4.15. Fourier transform computation of stress change due to prescribed elliptical slip distribution. Notice that the stress is nearly constant inside the crack, with the exception of small oscillations, the so-called Gibbs effect, near the crack tip. $N = 512$.

The stiffness relating slip to stress in the Fourier domain is thus simply

$$-\frac{\mu}{2}|k|. \tag{4.69}$$

To compute the stress change due to any continuous slip distribution, one simply takes the transform of the slip multiplied by the stiffness and then computes the inverse transform. For a numerical implementation employing the discrete Fourier transform with N elements and maximum wavelength $\lambda = 2\pi/k_{\min}$, the stiffness in the transform domain is written as

$$\frac{-\pi\mu|n|}{\lambda} \qquad n = -\frac{N}{2} \text{ to } \frac{N}{2} - 1. \tag{4.70}$$

A sample calculation for a specified elliptical slip distribution with unit stress drop ($\Delta\sigma = 1$) is shown in figure 4.15. The final stress on the crack is nearly everywhere equal to -1 as expected. There are oscillations, known as the *Gibbs effect*, near the crack tip, which result from the steep slip gradients there.

4.8 Some Three-Dimensional Crack Results

Three-dimensional crack problems are generally beyond the scope of this text. We will consider one method for solving for the displacements due to a uniformly pressurized circular penny-shaped crack in chapter 7. These problems can also be addressed with Hankel transform methods (e.g., Sneddon 1995). Here, a few significant results are cited. Consider a flat crack with an elliptical plan shape:

$$\frac{x^2}{a^2} + \frac{y^2}{b^2} = 1, \tag{4.71}$$

where a and b are the semi-axes of the ellipse. Subject to a uniform internal pressure p, this crack could be taken as a model of a fluid-filled dike or sill. The displacement perpendicular to the crack plane is given by

$$u_z = \frac{b(1-v)p}{\mu E(k)}\sqrt{1 - \frac{x^2}{a^2} - \frac{y^2}{b^2}},\tag{4.72}$$

where $E(k)$ is the complete elliptic integral of the second kind with modulus $k = (1 - b^2/a^2)^{1/2}$, (e.g., Mura 1982). The net crack opening is twice the displacement. Notice that, as was the case for two-dimensional solutions, the displacement distribution is elliptical.

In the limit of a circular, penny-shaped crack, $a = b$, and $E(0) = \pi/2$, so

$$u_z = \frac{2a(1-v)p}{\mu\pi}\sqrt{1 - \frac{r^2}{a^2}},\tag{4.73}$$

where $r^2 = x^2 + y^2$.

In the case of a penny-shaped crack subject to shear stress τ, the displacement profile is also elliptical:

$$u = \frac{4(1-v)a\tau}{\pi(2-v)\mu}\sqrt{1 - \frac{r^2}{a^2}}\tag{4.74}$$

(Eshelby 1957). Expressions (4.72) and (4.74) can be used in an approximate fashion to scale three-dimensional dislocation solutions so that the average fault slip or dike opening is appropriate for a given stress drop and crack dimensions. It should be kept in mind, however, that these solutions are for cracks in a full-space. In a half-space, there will be effects due to the free surface that alter the displacement distribution on the cracks (e.g., Wu et al. 1991).

4.9 Summary and Perspective

Crack models provide a more realistic physical representation of faults and planar intrusions (dikes and sills) in which the tractions acting on the crack surface are specified, resulting in nonuniform distributions of slip or opening. For a planar intrusion filled with quiescent (nonflowing) magma, the shear traction acting on the crack walls must vanish. The normal tractions are given by the internal magma pressure. Except for some region near the crack tip, the static magma pressure is uniform along strike, and in "magma-static" equilibrium; otherwise, magma would flow from high to low pressure. The situation for faults is more complex; the degree to which shear tractions within the interior of a slipping fault remain constant depends on frictional properties, as discussed more fully in chapter 11. Nevertheless, simple crack models are physically more realistic than dislocation models and provide greater physical insight.

Linear elastic fracture mechanics provides a theoretical basis for considering whether a crack propagates. Propagation is favored when the rate of mechanical potential energy release equals the effective fracture energy. If the crack–surface tractions are nearly constant except for a small region near the crack tip, the energy-based fracture criterion is equivalent to one based on the strength of the stress intensity at the crack tip. These concepts have proven useful in understanding quasi-static nucleation of slip on faults characterized by rate- and state-dependent friction, as discussed in chapter 11.

Physical realism comes at a cost; crack problems are considerably more difficult to solve than comparable dislocation problems. A wide variety of analytical methods has been developed; here, we focused on a boundary integral representation that relates the unknown dislocation density (slip gradient) distribution to the specified crack–surface tractions. Expansion of the crack–surface tractions in terms of Chebyshev polynomials allows for direct computation of the corresponding dislocation density distribution, and hence crack–surface displacement. For an infinitely long strike-slip fault with uniform stress drop, the slip distribution is elliptical. This is generally true for uniform crack surface tractions as long as the crack is far from the free surface. If the crack is close to the earth's surface, interaction with the free surface distorts the crack–surface displacements so that they are no longer elliptical, even for a uniform traction distribution. This can be seen later in figure 7.27 for the case of a vertical dike growing toward the earth's surface.

For more complex problems, one can approximate the boundary integral with a finite sum and develop boundary element solutions. In some instances, it is possible to use highly efficient Fourier transform (FT) methods to solve crack problems. When should we use boundary element methods and when FT methods? For general crack problems, with specified stress change inside the crack, the boundary conditions are mixed displacement and traction. In this case, it is not possible to use FT. However, models of faults with rate- and state-dependent friction posit that faults are always slipping, albeit sometimes at extremely low rates. In this case, there is a direct relationship between fault slip rate and the time derivative of fault traction (see chapter 11). If the geometry is sufficiently symmetric that periodic replication of the fault is allowed, FT is the method of choice.

In section 4.5, the surface displacements due to a uniform-slip dislocation were compared to those resulting from a constant stress drop crack and a tapered nonsingular crack. We found that while the surface displacements are distinct, the differences are subtle. Is it worth the extra effort to compute displacements for more complex models, or should one stick with the simple uniform-slip dislocation? The answer depends on the question one is asking. If the goal is to estimate the first-order characteristics of an earthquake, then a simple uniform-slip dislocation model may be appropriate. Alternatively, it is equally valid to argue that while we have limited knowledge of the physics of faulting, we do know that the stress change associated with fault slip is everywhere finite. Thus, uniform-slip dislocations, with their associated stress singularities, can be excluded a priori. In many cases, we have independent constraints on the stress drop, or magma pressure, which favors the application of crack models over dislocation models. Furthermore, crack models lead naturally to smooth slip distributions, without the need for regularization that is generally required in distributed slip inversions.

4.10 Problems

1. Show that the critical stress for crack propagation is given by equation (4.32).

2. (a) Use an approximation to the Burger's vector distribution in equation (4.16), accurate near the crack tip, to show that the displacement near the crack tip scales with $\sqrt{r}$, where here r measures the distance away from the crack tip *into* the crack.

   (b) Use the same approximation to the Burger's vector distribution to show that the stress in front of the crack is proportional to $1/\sqrt{r}$, where r measures the distance *in front of* the crack tip. You may find the following integral helpful:

$$\int \frac{dx}{\sqrt{x}(r + x)} = \frac{2}{\sqrt{r}} \tan^{-1} \sqrt{\frac{x}{r}}. \tag{4.75}$$

   (c) By comparing with equation (4.34), show that the stress intensity factor must be proportional to $\Delta\tau$ and the square root of crack length. Use this result to show that the proportionality constant in part (a) is K/μ, as in equation (4.35). Note that the factor of $1 - \nu$ results from the difference between antiplane strain and plane strain.

3. Equation (4.30) gives the change in mechanical energy on introduction of a constant stress drop crack. Show that this gives the energy release rate G if the stress intensity in equation (4.39) is $\sigma\sqrt{a}$.

4. Show that the net dislocation is given by equation (4.8).
   Hint: Use the expansion for $B(\xi)$ given in equation (4.20) and the orthogonality relation for the T_n.

5. (a) Use the following identity and the equations (4.40) to show that (4.7) reduces to (4.41):

$$\frac{\sqrt{1 - \xi^2}}{\sqrt{1 - z^2}} - \frac{\sqrt{1 - z^2}}{\sqrt{1 - \xi^2}} = \frac{z^2 - \xi^2}{\sqrt{1 - \xi^2}\sqrt{1 - z^2}}. \tag{4.76}$$

   (b) Show using the orthogonality relations for the T_n that conditions (4.40) are satisfied only if $c_0 = c_1 = 0$.

6. Derive the Fourier transform in equation (4.67). Following Bracewell (1965), it is simpler to prove the inverse transform. The trick is to multiply the integrand by $\exp(-a|k|)$, solve the integral, and then take the limit $a \to 0$.
   Hint: Break the integral into two parts, one from $-\infty$ to 0 and the second from 0 to ∞.

7. In this problem, we will gain some insights into the boundary element method. Use boundary elements to solve the antiplane problem for strike slip on an infinitely long vertical fault from the surface to depth D (say, 10 km if you want to make it dimensional).
   Hint: When you compute the stress due to slip over an interval, the stress must be evaluated at the midpoint of the interval.

   (a) Find the (discrete) slip distribution needed to approximate constant stress drop (of magnitude -1) on the slipping fault surface. (Note that there is no displacement discontinuity for depths greater than D.) Compare with the elliptical distribution (4.17).

   (b) Compute the displacement at the earth's surface, and compare with the analytical solution (4.25).

   (c) Vary the number of elements used to approximate the continuous slip distribution. How many elements do you need to get a good approximation to the analytical result?

8. Compare the boundary element and Fourier transform methods for solving for the stress change given a known slip distribution. Assume an elliptical distribution of slip inside the fault and no slip outside, and use fast Fourier transform (FFT) to compute the stress everywhere. As a check on the solution accuracy, you should find uniform stress inside the crack.

   (a) How close to uniform is the stress varying the number of elements in FFT?

   (b) How does the FFT method compare with the boundary element method in terms of computation speed?

4.11 References

Abramowitz, M., and I. A. Stegun. 1972. *Handbook of mathematical functions*. New York: Dover Publications.

Barber, J. R. 1992. *Elasticity*. Boston: Kluwer, pp. 174–177.

Barnett, D. M., and L. B. Freund. 1975. An estimate of strike-slip fault friction stress and fault depth from surface displacement data. *Bulletin of the Seismological Society of America* **65**, 1259–1266.

Bilby, B. A., and J. D. Eshelby. 1968. Dislocations and the theory of fracture. In H. Liebowitz (Ed.), *A treatise on fracture*. New York: Wiley, pp. 99–182.

Bracewell, R. 1965. *The Fourier transform and its applications*. New York: McGraw-Hill.

Eshelby, J. D. 1957. The determination of the elastic field of an ellipsoidal inclusion and related problems. *Proceedings of the Royal Society of London A* **241**, 376–396.

Freund, L. B., and D. M. Barnett. 1976. A two-dimensional analysis of surface deformation due to dip-slip faulting. *Bulletin of the Seismological Society of America* **66**, 667–675.

Griffith, A. A. 1920. The theory of rupture and flow in solids. *Philosophical Transactions Royal Society of London A* **221**, 163–198.

Lawn, B. R., and T. R. Wilshaw. 1975. *Fracture of brittle solids*. London: Cambridge University Press.

Mason, J. C., and D. C. Handscomb. 2003. *Chebyshev polynomials*. Boca Raton, FL: Chapman & Hall/CRC.

Mura, T. 1982. *Micromechanics of defects in solids*. Norwell, MA: Kluwer.

Petit-Bois, G. 1961. *Tables of indefinite integrals*. New York: Dover Publications.

Rice, J. R. 1968. Mathematical analysis in the mechanics of fracture. In H. Liebowitz (Ed.), *Fracture: an advanced treatise*, vol. II. New York: Academic Press, pp. 191–311.

———. 1978. Thermodynamics of the quasi-static growth of Griffith cracks. *Journal of Mechanics and Physics of Solids* **26**, 61–78.

———. 1983. Constitutive relations for fault slip and earthquake instabilities. *Pure and Applied Geophysics* **121**, 443–475.

Rubin, A. M. 1995. Propagation of magma filled cracks. *Annual Review of Earth and Planetary Science* **23**, 287–336.

Sneddon, I. N. 1995. *Fourier transforms*. McGraw-Hill Books, 1951; republished New York: Dover Publications.

Wu, M., J. W. Rudnicki, C. H. Kuo, and L. M. Keer. 1991. Surface deformation and energy release rates for constant stress drop slip zones in an elastic half-space. *Journal of Geophysical Research* **96**(B10), 16509–16524, 10.1029/91JB01043.

5

Elastic Heterogeneity

The earth is not a homogeneous isotropic elastic half-space. We know from both geologic and geophysical observations that the crust is composed of different rock types, with different elastic wave speeds, densities, and elastic moduli. In this chapter, we will explore three different methods for computing dislocation solutions in elastically heterogeneous media. Image methods are useful when there are different material properties separated by planar boundaries. Images are relatively straightforward for antiplane problems but require more experience when solving plane strain and three-dimensional problems. Propagator matrices, on the other hand, are quite general and are widely used for horizontally stratified media in both two and three dimensions. The third method is a first-order perturbation approach that can treat arbitrarily complex variations in elastic properties but is only approximate. It provides accurate solutions when the variations in elastic moduli are modest but breaks down for strong variations in properties.

5.1 Long Strike-Slip Fault Bounding Two Media

To begin, we investigate the case of an infinitely long strike-slip fault with different shear modulus on either side of the fault. From the discussion in chapter 2, this problem is antiplane strain, so the displacement field on each side of the fault must satisfy Laplace's equation. Furthermore, the tractions must be continuous across the fault, and the displacements must be continuous except across the fault, where the difference in displacement equals the fault slip.

Denote the displacements on the side with modulus μ_1 as $u^{(1)}$ and those on the side with modulus μ_2 as $u^{(2)}$ (see figure 5.1). Recall that for a screw dislocation in a homogeneous material, the displacement is proportional to θ. We thus search for a solution of the form

$$u^{(1)} = A\theta, \qquad 0 \leq \theta \leq \pi, \tag{5.1}$$

$$u^{(2)} = B\theta, \qquad -\pi \leq \theta \leq 0. \tag{5.2}$$

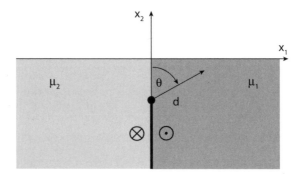

Figure 5.1. Strike-slip fault with differing shear modulus across the fault plane.

The angle θ is as specified in figure 5.1. The boundary conditions are thus

$$u^{(1)}(\theta = 0) = u^{(2)}(\theta = 0), \tag{5.3}$$

$$u^{(1)}(\theta = \pi) - u^{(2)}(\theta = -\pi) = s, \tag{5.4}$$

$$\sigma_{13}^{(1)} = \sigma_{13}^{(2)}, \quad on \quad x_1 = 0. \tag{5.5}$$

Note that the first boundary condition is satisfied automatically. The second boundary condition requires that

$$A + B = \frac{s}{\pi}, \tag{5.6}$$

and the boundary conditions for continuity of tractions are, writing $\theta = \tan^{-1}[x_1/(x_2 + d)]$,

$$\sigma_{13}^{(1)} = \mu_1 \frac{\partial u_3^{(1)}}{\partial x_1} = \mu_1 A \frac{x_2 + d}{x_1^2 + (x_2 + d)^2}, \tag{5.7}$$

$$\sigma_{13}^{(2)} = \mu_2 \frac{\partial u_3^{(2)}}{\partial x_1} = \mu_2 B \frac{x_2 + d}{x_1^2 + (x_2 + d)^2}. \tag{5.8}$$

Equating the tractions on the plane $x_1 = 0$ leads to

$$\mu_1 A = \mu_2 B. \tag{5.9}$$

Combining with equation (5.6) gives two equations in the unknowns A and B. Solving these yields

$$A = \frac{s}{\pi} \frac{\mu_2}{\mu_1 + \mu_2},$$

$$B = \frac{s}{\pi} \frac{\mu_1}{\mu_1 + \mu_2}. \tag{5.10}$$

This gives the solution in a full-space. We have yet to account for the traction-free condition on $x_2 = 0$. Following the procedure in chapter 2, we introduce image dislocations across the plane $x_2 = 0$ to render that surface traction free. This has the effect of doubling the displacements at $x_2 = 0$:

$$\boxed{\begin{aligned} u^{(1)}(x_2 = 0) &= \frac{2s}{\pi} \frac{\mu_2}{\mu_1 + \mu_2} \tan^{-1}\left(\frac{x_1}{d}\right), \\ u^{(2)}(x_2 = 0) &= \frac{2s}{\pi} \frac{\mu_1}{\mu_1 + \mu_2} \tan^{-1}\left(\frac{x_1}{d}\right). \end{aligned}} \tag{5.11}$$

Displacements are shown in figure 5.2 for a range of modulus contrasts. With increasing modulus contrast across the fault, the deformation becomes increasingly more concentrated on the more compliant side of the fault.

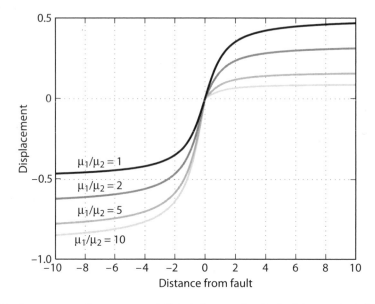

Figure 5.2. Displacement profiles across an infinitely long strike-slip fault in an elastic half-space with different shear modulus on either side of the fault. μ_1 refers to the shear modulus on the right side of the fault. See figure 5.1.

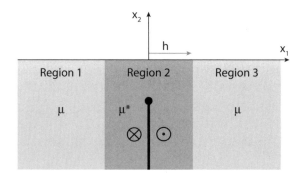

Figure 5.3. Fault zone of thickness $2h$ with differing shear modulus from the surrounding elastic half-space.

5.2 Strike-Slip Fault within a Compliant Fault Zone

Major faults are often located within broad damage zones. Seismic tomography as well as fault zone guided waves indicate zones of low seismic wave speeds, and by implication low stiffness. More recently, information about the fault zone compliance has come from SAR interferometry. Plate 5 illustrates a high-pass filtered interferogram covering the 1999 Hector Mine earthquake in southern California. The high-frequency fluctuations on faults adjacent to the Hector Mine rupture were interpreted by Fialko et al. (2002) as resulting from the coseismic stress changes acting on compliant fault zones. By comparing theoretical calculations with the observations, they infer that the fault zones are roughly two kilometers wide and that the stiffness is a factor of two or more less than the surrounding crust. The compliant fault zones must extend to considerable depth to explain the observations.

In this section, we will develop the solution for an infinitely long strike-slip fault in a compliant fault zone. The geometry is illustrated in figure 5.3. An infinitely long vertical

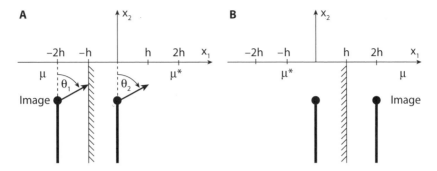

Figure 5.4. The first step in developing an image solution for a compliant fault zone.

strike-slip fault is located in the middle of a fault zone with thickness $2h$. The shear modulus of the fault zone is μ^*, while the modulus of the surrounding crust is μ.

Our procedure is to first account for the material interface forming the left boundary of the fault zone (figure 5.4A) and then account for the right boundary (figure 5.4B). The problem of a dislocation adjacent to a material boundary is very similar to that treated in the previous section, where the dislocation was located on the boundary. The fault is located at $x_1 = 0$. The left material boundary is located at $x_1 = -h$, so an image dislocation is placed at $x_1 = -2h$. Note that at this stage, we do not distinguish between region 2 and region 3 (figure 5.3); both are to the right of the boundary at $x_1 = -h$ and are treated as having modulus μ^*. We construct angles θ_1 measured clockwise from the image dislocation and θ_2 measured clockwise from the fault (figure 5.4A). Notice that θ_1 is continuous for $x_1 > -h$ and θ_2 is continuous for $x_1 < -h$. Thus, it makes sense to consider a solution of the form

$$u^{(2)} = \frac{s\theta_2}{2\pi} + A\theta_1, \tag{5.12}$$

$$u^{(1)} = B\theta_2. \tag{5.13}$$

The choice of $s/2\pi$ as a coefficient guarantees that the displacement increments by s across the fault, $x_1 = 0$. Note also that $u^{(1)}$ cannot depend on θ_1, as this would introduce a discontinuity in the displacement field in region 1. There are two unknown constants, A and B, which are determined by the boundary conditions—namely, continuity of displacement and traction across the material interface $x_1 = -h$. Specifically, the boundary conditions are

$$u^{(1)}(x_1 = -h) = u^{(2)}(x_1 = -h), \tag{5.14}$$

$$\sigma_{13}^{(1)}(x_1 = -h) = \sigma_{13}^{(2)}(x_1 = -h). \tag{5.15}$$

With reference to figure 5.4, the angles θ_1 and θ_2 are

$$\theta_1 = \tan^{-1}\left(\frac{x_1 + 2h}{x_2 + d}\right), \tag{5.16}$$

$$\theta_2 = \tan^{-1}\left(\frac{x_1}{x_2 + d}\right), \tag{5.17}$$

where d is the depth of the dislocation. Notice that on the boundary, $x_1 = -h$, $\theta_2 = -\theta_1$. Thus, the displacement boundary condition yields

$$A + B = \frac{s}{2\pi}. \tag{5.18}$$

Differentiating the displacements and making use of Hooke's law, the traction boundary condition becomes

$$\mu^* \left(\frac{s}{2\pi} + A \right) \frac{x_2 + d}{h^2 + (x_2 + d)^2} = B\mu \frac{x_2 + d}{h^2 + (x_2 + d)^2}, \tag{5.19}$$

which, when combined with equation (5.18), leads to

$$A = \frac{s}{2\pi} \frac{\mu - \mu^*}{\mu + \mu^*},$$

$$B = \frac{s}{2\pi} \frac{2\mu^*}{\mu + \mu^*}. \tag{5.20}$$

The displacements in region 1, denoted $u^{(1)}$, and in regions 2 and 3, denoted $u^{(2)} = u^{(3)}$ (see figure 5.3), are thus

$$u^{(1)} = \frac{s}{2\pi} \left[\tan^{-1} \left(\frac{x_1}{x_2 + d} \right) - \kappa \tan^{-1} \left(\frac{x_1}{x_2 + d} \right) \right],$$

$$u^{(2)} = u^{(3)} = \frac{s}{2\pi} \left[\tan^{-1} \left(\frac{x_1}{x_2 + d} \right) + \kappa \tan^{-1} \left(\frac{x_1 + 2h}{x_2 + d} \right) \right], \tag{5.21}$$

where

$$\kappa = \frac{\mu - \mu^*}{\mu + \mu^*}. \tag{5.22}$$

The reader can verify that equation (5.21) reduces to (5.11) in the limit that $h \to 0$, once image dislocations are added to account for the traction-free surface. Notice from equation (5.21) that the image dislocation on the stiff side of the boundary (assuming $\mu > \mu^*$) has strength $-\kappa s$, whereas the image dislocation on the compliant side has strength κs.

At this point, we have accounted for the leftmost material interface. We can summarize the result so far in tabular form, in which only the strength (amplitude) and the position of the dislocations, including image dislocations, are noted:

Region 1	source strength	s	$-\kappa s$
	source position $x_1 =$	0	0
Region 2	source strength	s	κs
	source position $x_1 =$	0	$-2h$
Region 3	source strength	s	κs
	source position $x_1 =$	0	$-2h$

At this stage, we have not yet distinguished between regions 2 and 3, so the solutions there are necessarily the same. We next satisfy the boundary condition between regions 2 and 3 by adding an image at $x_1 = 2h$ (figure 5.4B). By analogy with the previous solution, this requires adding an image for regions 1 and 2 located at $x_1 = 2h$ and an image for region 3 located an equal distance from the material interface at $x_1 = 0$. Regions 1 and 2 are now on the compliant side of the boundary, so the strength of the image for these regions is $+\kappa s$, while region 3 is on the stiff side of the boundary, so the image for this region has strength $-\kappa s$. After adding these terms, the solution is

Region 1	source strength	s	$-\kappa s$	κs
	source position $x_1 =$	0	0	$2h$
Region 2	source strength	s	κs	κs
	source position $x_1 =$	0	$-2h$	$2h$
Region 3	source strength	s	κs	$-\kappa s$
	source position $x_1 =$	0	$-2h$	0

At this stage, we have a first-order approximation to the solution. We have, however, neglected the effect of the image at $x_1 = -2h$ on the interface at $x_1 = h$, as well as the effect of the image at $x_1 = 2h$ on the interface at $x_1 = -h$. These effects are corrected by adding additional image dislocations. Consider first the effect of the image at $x_1 = -2h$ on the interface at $x_1 = h$ (figure 5.5A). The image dislocation is a distance $3h$ from the boundary at $x_1 = h$. We thus need to add an additional image dislocation for regions 1 and 2 at an equal distance on the other side of the boundary at $x_1 = 4h$. The new image is on the right side of the boundary, so the relative geometry is the same as in figure 5.4B. However, the magnitude of the image is κ times the strength of the original image. Thus, for regions 1 and 2, we add an image with strength $\kappa \times \kappa s$ at $x_1 = 4h$. In region 3, the strength of the dislocation is $-\kappa \times \kappa s$, and the image is located an equal distance on the opposite side of the boundary at $x_1 = -2h$.

Now consider the effect of the image at $x_1 = 2h$ on the interface at $x_1 = -h$ (figure 5.5B). The image is a distance $3h$ on the right side of the boundary, so an image for regions 2 and 3 will be located a distance $3h$ to the left of the boundary at $x_1 = -4h$. The relative geometry is the same as in figure 5.4A, so the strength of the image in regions 2 and 3 is $+\kappa \times \kappa s$. For region 1, the image is $3h$ to the right of the boundary at $x_1 = 2h$. The strength is $-\kappa \times \kappa s$. Considering the effect of these additional images, the solution is now

Figure				5.4A	5.4B	5.5A	5.5B
Region 1	source strength	s	$-\kappa s$	κs	$\kappa^2 s$	$-\kappa^2 s$
	source position $x_1 =$	0	0	$2h$	$4h$	$2h$
Region 2	source strength	s	κs	κs	$\kappa^2 s$	$\kappa^2 s$
	source position $x_1 =$	0	$-2h$	$2h$	$4h$	$-4h$
Region 3	source strength	s	κs	$-\kappa s$	$-\kappa^2 s$	$\kappa^2 s$
	source position $x_1 =$	0	$-2h$	0	$-2h$	$-4h$

Of course, the new images violate the boundary conditions on the alternate boundaries, so the process must be continued indefinitely. Fortunately, the sequence converges because successive images are increasingly distant from the compliant zone. From inspection, the

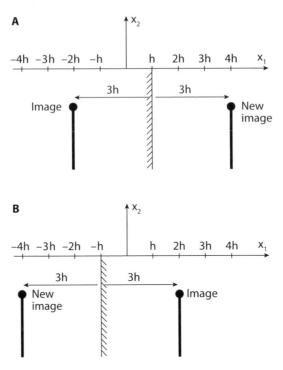

Figure 5.5. The second step in developing an image solution for a compliant fault zone. A: A new image must be introduced to account for the effect of the image at $x_1 = -2h$ on the right boundary $x_1 = h$. Notice that the relative geometry is the same as that in figure 5.4B. B: A new image must be introduced to account for the effect of the image at $x_1 = 2h$ on the left boundary $x_1 = -h$. Notice that the relative geometry is the same as that in figure 5.4A.

pattern can be recognized, and the solution is

$$
u^{(1)}(x_1 < -h) = \frac{(1-\kappa)s}{2\pi} \sum_{n=0}^{\infty} \kappa^n \tan^{-1} \left(\frac{x_1 - 2nh}{x_2 + d} \right) ,
$$

$$
u^{(2)}(-h < x_1 < h) = \frac{s}{2\pi} \left\{ \tan^{-1} \left(\frac{x_1}{x_2 + d} \right) + \sum_{n=1}^{\infty} \kappa^n \left[\tan^{-1} \left(\frac{x_1 - 2nh}{x_2 + d} \right) \right. \right.
$$

$$
\left. \left. + \tan^{-1} \left(\frac{x_1 + 2nh}{x_2 + d} \right) \right] \right\} ,
$$

$$
u^{(3)}(x_1 > h) = \frac{(1-\kappa)s}{2\pi} \sum_{n=0}^{\infty} \kappa^n \tan^{-1} \left(\frac{x_1 + 2nh}{x_2 + d} \right) . \tag{5.23}
$$

One should, of course, check that the displacements and tractions are indeed continuous across the boundaries (problem 1). The displacements at the free surface are easily found by adding image dislocations equidistant about the free surface and evaluating the resultant

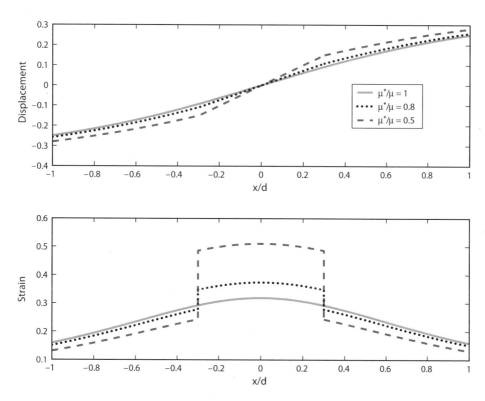

Figure 5.6. Displacements and shear strain for a strike-slip fault within a fault zone of thickness $2h$ with differing shear modulus from the surrounding crust. μ^* refers to the modulus of the fault zone. Notice that the strain is concentrated within the more compliant fault zone. Displacements are normalized by slip, s, while strain is normalized by s/d.

displacements at $x_2 = 0$, which has the effect of doubling the displacements. This solution reproduces one given by Rybicki and Kasahara (1977) and discussed by McHugh and Johnston (1977).

The displacements are shown in figure 5.6 for three different ratios of shear modulus μ^*/μ. As the fault zone becomes more compliant than the surroundings, the strain is increasingly concentrated within the fault zone. Chen and Freymueller (2002) analyzed repeated EDM and GPS surveys to determine the strain rate in four small-scale geodetic networks spanning the San Andreas fault. The networks extend one to two kilometers on either side of the fault trace. The strain rates observed in networks along the San Francisco segment of the fault are substantially larger than expected for slip below the seismogenically active fault at rates consonant with paleoseismic observations. Rather than appeal to anomalously shallow fault locking, Chen and Freymueller (2002) suggest that the data can be explained by a compliant fault zone. Their results suggest that within a several-kilometer-wide fault, the shear modulus is roughly a factor of two less than that of the surrounding crust.

5.3 Strike-Slip Fault beneath a Layer

We now consider the problem of a screw dislocation beneath a horizontal layer with different shear modulus (figure 5.7). The layer of thickness H has modulus μ_1 and overlies a half-space with modulus μ_2. The dislocation is at depth d below the layer boundary and depth $d + H$

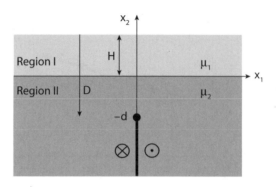

Figure 5.7. Strike-slip fault beneath a layer of thickness H. The screw dislocation is located at $x_2 = -d$, corresponding to depth $D = d + H$ beneath the free surface.

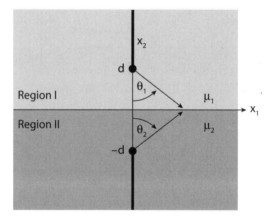

Figure 5.8. Screw dislocation perpendicular to a bimaterial interface. Dislocation is at $x_2 = -d$; image dislocation is at $x_2 = d$.

beneath the free surface. The first step is to develop the solution for a dislocation perpendicular to a bimaterial interface, as in figure 5.8. Following the image methods illustrated in the previous section, one can show (see problem 2):

$$u^{(2)} = \frac{s}{2\pi}\left[\tan^{-1}\left(\frac{x_1}{x_2 + d}\right) + \kappa \tan^{-1}\left(\frac{x_1}{x_2 - d}\right)\right],$$

$$u^{(1)} = \frac{s}{2\pi}(1 - \kappa)\tan^{-1}\left(\frac{x_1}{x_2 + d}\right), \tag{5.24}$$

where

$$\kappa = \frac{\mu_1 - \mu_2}{\mu_1 + \mu_2}. \tag{5.25}$$

Note that the material interface is located at $x_2 = 0$. We now build up the solution for the layer over half-space with a series of images to satisfy both the free-surface boundary conditions and the conditions on the material interface. Considering only the solution for the bimaterial, equations (5.24), the solution in region II consists of a dislocation of strength s

at $x_2 = -d$ and an image of strength κs at $x_2 = d$. The solution in region I consists of an image of strength $(1 - \kappa)s$ at $x_2 = -d$. We can write this in tabular form as

| Region II | source strength | | s | κs |
| | source position $x_2 =$ | | $-d$ | d |

| Region I | source strength | | $(1 - \kappa)s$ |
| | source position $x_2 =$ | | $-d$ |

This solution violates the free-surface boundary condition at $x_2 = H$, which is in region I. To fix the free-surface condition, we add an image dislocation opposite in sign to the dislocation in region I, symmetric about the free surface. The image is located at a distance $d + H$ from the free surface, at $x_2 = d + 2H$. Adding this term, our solution is

| Region II | source strength | | s | κs |
| | source position $x_2 =$ | | $-d$ | d |

| Region I | source strength | | $(1 - \kappa)s$ | $-(1 - \kappa)s$ |
| | source position $x_2 =$ | | $-d$ | $d + 2H$ |

At this point, the free-surface boundary condition is satisfied on $x_2 = H$, but the boundary condition at the material interface $x_2 = 0$ is violated. We can use the same scheme as before to fix the boundary condition on $x_2 = 0$. The image dislocation introduced at $x_2 = d + 2H$ is in region I, whereas the original dislocation is in region II. We can use the solution (5.24); however, the sign on κ changes to reflect the fact that the dislocation is on the opposite side of the material interface. Thus, the image in region II is located at $x_2 = d + 2H$, with strength

$$[1 - (-\kappa)] \times [-(1 - \kappa)s] = -(1 - \kappa^2)s,$$

where the second term in brackets is the strength of the source (equivalent to s in equation [5.24]). Similarly, the image in region I is located at $x_2 = -(d + 2H)$, with strength

$$-\kappa \times [-(1 - \kappa)s] = \kappa(1 - \kappa)s.$$

The total solution is now

| Region II | source strength | | s | κs | | $-(1 - \kappa^2)s$ |
| | source position $x_2 =$ | | $-d$ | d | | $d + 2H$ |

| Region I | source strength | | $(1 - \kappa)s$ | $-(1 - \kappa)s$ | $\kappa(1 - \kappa)s$ |
| | source position $x_2 =$ | | $-d$ | $d + 2H$ | $-(d + 2H)$ |

Of course, we have now violated the free-surface boundary condition at $x_2 = H$. Since the free surface is located in region I, we must add an image dislocation of strength $-\kappa(1 - \kappa)s$ to cancel the tractions there. The image is located at a distance $d + 3H$ from the surface, at $x_2 = d + 4H$. The total solution is now

| Region II | source strength | s | κs | | $-(1 - \kappa^2)s$ | |
| | source position $x_2 =$ | $-d$ | d | | $d + 2H$ | |

| Region I | source strength | $(1 - \kappa)s$ | $-(1 - \kappa)s$ | $\kappa(1 - \kappa)s$ | $-\kappa(1 - \kappa)s$ |
| | source position $x_2 =$ | $-d$ | $d + 2H$ | $-(d + 2H)$ | $d + 4H$ |

At this stage, we have to once again fix the boundary condition on the material interface at $x_2 = 0$. (Don't worry, we will add only one more term—by that time the pattern is clear.) The image dislocation we must correct for is again in region I, so the sign of κ is the same as it was in the previous iteration. Thus, the image in region II is located at $x_2 = d + 4H$, with strength

$$[1 - (-\kappa)] \times [-\kappa(1 - \kappa)s] = -\kappa(1 - \kappa^2)s.$$

Similarly, the image in region I is at $x_2 = -(d + 4H)$, with strength

$$[(-\kappa)] \times [-\kappa(1 - \kappa)s] = \kappa^2(1 - \kappa)s.$$

The solution at this stage is

Region II	source				
	strength	s	κs	$-(1 - \kappa^2)s$	$-\kappa(1-\kappa^2)s$
	source				
	position $x_2 =$	$-d$	d	$d + 2H$	$d + 4H$

Region I	source					
	strength		$(1 - \kappa)s$	$-(1 - \kappa)s$	$\kappa(1 - \kappa)s$	$-\kappa(1 - \kappa)s$	$\kappa^2(1 - \kappa)s$
	source						
	position $x_2 =$		$-d$	$d + 2H$	$-(d + 2H)$	$d + 4H$	$-(d + 4H)$

We can imagine repeating this sequence indefinitely. Since each of the image dislocations is increasingly far from the free surface, the solution converges.

If we restrict attention to displacements at the free surface, which is located in region I, then we have a series of dislocations at $x_2 = -d - 2mH$, $m = 0, 1, 2, 3, \ldots$, with strength $\kappa^m(1 - \kappa)s$, and a series of dislocations at $x_2 = d + 2nH$, $n = 1, 2, 3, \ldots$, with strength $-\kappa^{n-1}(1 - \kappa)s$.

The solution for the displacement field in region I is thus

$$u(x_1, x_2) = \frac{s}{2\pi}\left[(1 - \kappa)\sum_{m=0}^{\infty}\kappa^m\tan^{-1}\left(\frac{x_1}{x_2 + d + 2mH}\right) - (1 - \kappa)\sum_{n=1}^{\infty}\kappa^{n-1}\tan^{-1}\left(\frac{x_1}{x_2 - d - 2nH}\right)\right]. \tag{5.26}$$

Evaluating the displacements at the free surface $x_2 = H$ and noticing that the depth of the dislocation beneath the free surface is $D = d + H$, we have

$$u(x_1, x_2 = H) = \frac{s(1 - \kappa)}{2\pi}\left[\sum_{m=0}^{\infty}\kappa^m\tan^{-1}\left(\frac{x_1}{D + 2mH}\right) + \sum_{n=1}^{\infty}\kappa^{n-1}\tan^{-1}\left(\frac{x_1}{D + 2(n-1)H}\right)\right]. \tag{5.27}$$

The two sums are in fact the same, so equation (5.27) reduces to

$$u(x_1, x_2 = H) = \frac{s(1 - \kappa)}{\pi}\left[\sum_{m=0}^{\infty}\kappa^m\tan^{-1}\left(\frac{x_1}{D + 2mH}\right)\right]. \tag{5.28}$$

To write in the same form as Rybicki (1971), we define the ratio of moduli $\gamma = \mu_2/\mu_1$ so that

$$\kappa = \frac{1-\gamma}{1+\gamma}, \tag{5.29}$$

$$1 - \kappa = \frac{2\gamma}{1+\gamma}, \tag{5.30}$$

leading to

$$u(x_1, x_2 = H) = \frac{2s\gamma}{(1+\gamma)\pi} \left[\tan^{-1}\left(\frac{x_1}{D}\right) + \sum_{m=1}^{\infty} \left(\frac{1-\gamma}{1+\gamma}\right)^m \tan^{-1}\left(\frac{x_1}{D+2mH}\right) \right]. \tag{5.31}$$

Displacements are shown for three different modulus ratios in figure 5.9A. Notice that the deformation becomes increasingly localized near the fault when the layer is substantially more compliant than the underlying half-space.

5.4 Strike-Slip within a Layer over Half-Space

The same methods can be used to solve for the elastic fields when the slip is within the elastic layer overlying a half-space with differing shear modulus. In this example, the fault slips from the surface to depth D, where $D < H$. In the following expressions, the free surface is located at $x_2 = 0$. For x_2 within the surface layer—that is, $|x_2| < H$:

$$u(x_1, |x_2| < H) = \frac{s}{2\pi} \left\{ \tan^{-1}\left(\frac{x_2+D}{x_1}\right) - \tan^{-1}\left(\frac{x_2-D}{x_1}\right) \right.$$

$$+ \sum_{m=1}^{\infty} \left(\frac{1-\gamma}{1+\gamma}\right)^m \left[-\tan^{-1}\left(\frac{x_1}{x_2-2mH+D}\right) + \tan^{-1}\left(\frac{x_1}{x_2-2mH-D}\right) \right.$$

$$\left. \left. - \tan^{-1}\left(\frac{x_1}{x_2+2mH+D}\right) + \tan^{-1}\left(\frac{x_1}{x_2+2mH-D}\right) \right] \right\}. \tag{5.32}$$

For observation coordinates on the free surface, equation (5.32) reduces to

$$u(x_1, x_2 = 0) = \frac{s}{\pi} \left\{ \tan^{-1}\left(\frac{D}{x_1}\right) + \sum_{m=1}^{\infty} \left(\frac{1-\gamma}{1+\gamma}\right)^m \left[\tan^{-1}\left(\frac{D+2mH}{x_1}\right) \right. \right.$$

$$\left. \left. + \tan^{-1}\left(\frac{D-2mH}{x_1}\right) \right] \right\}. \tag{5.33}$$

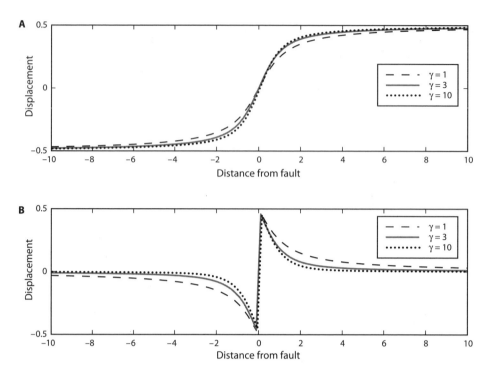

Figure 5.9. Effects of layering on displacements due to an infinitely long strike-slip fault. A: Uniform slip fault beneath a layer of thickness H extending to infinite depth. The top of the fault is at twice the layer thickness $D = 2H$. B: Fault within the layer. The fault slips from the free surface to the bottom of the layer $D = H$. γ is defined as the ratio of the moduli, where $\gamma = \mu_2/\mu_1$, and μ_1 refers to the shallow layer.

For x_2 below the surface layer—that is, $|x_2| > H$—the displacements are given by

$$u(x_1, |x_2| > H) = \frac{s}{\pi(1+\gamma)} \sum_{m=0}^{\infty} \left(\frac{1-\gamma}{1+\gamma}\right)^m \left[-\tan^{-1}\left(\frac{x_1}{x_2 - 2mH + D}\right) + \tan^{-1}\left(\frac{x_1}{x_2 - 2mH - D}\right)\right].$$

(5.34)

 Displacements are shown for three different modulus ratios in figure 5.9B. When the modulus contrast μ_2/μ_1 is significantly greater than unity, the displacements decay more rapidly with distance from the fault. In other words, the deformation is more concentrated near the fault when the layer is more compliant than the underlying half-space. Since the depth of faulting can be estimated by the decay of displacement with distance, failure to account for the effects of layering can bias the inferred fault depth. In particular, for a compliant surface layer, neglecting elastic heterogeneity will bias the inferred fault depth to be shallower than it actually is.

 Equations (5.31) and (5.33) indicate that solutions for a dislocation in a layered medium are mathematically equivalent to a series of dislocations in a homogeneous half-space with strength that depends on the elastic properties but generally decays with depth. This is illustrated in figure 5.10.

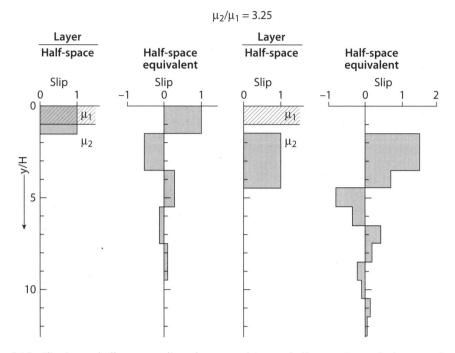

Figure 5.10. Slip in a shallow compliant layer overlying a half-space is equivalent to slip in a homogeneous half-space with equivalent dislocations. The example on the left side is for slip predominantly within the compliant layer, while the example on the right is for slip below the layer. After Savage (1987).

5.5 Propagator Matrix Methods

Propagator matrix methods are well suited for analyzing dislocations in horizontally stratified elastic earth models and are widely used in seismology to compute wave fields in layered media (e.g., Aki and Richards 1980). The first step is to rewrite the governing differential equations so as to isolate vertical derivatives. The kinematic, displacement-strain relations together with Hooke's law are used to eliminate stress components that do not exert tractions on horizontal boundaries. Stresses that result in tractions on horizontal surfaces are needed to enforce stress continuity on layer boundaries, as well as the stress-free boundary condition at the earth's surface. The remaining stress components can be reconstructed given the displacements and known stresses.

Fourier transforming in the horizontal directions reduces the partial differential equations to a system of ordinary differential equations in the vertical coordinate. Placing the displacements and the stress components of interest in a *stress-displacement vector* allows us to write the governing equations in matrix form. We next introduce the *propagator matrix*—a matrix that takes a stress-displacement vector at the bottom of a layer and propagates it to the top of that layer in such a way that the equations of elasticity are fully satisfied. By a series of matrix multiplications, we can then propagate a solution from depth through an arbitrary number of layers to the earth's surface, where the traction-free boundary condition is enforced. Last, introducing appropriate methods for including the dislocation source terms, we have the appropriate machinery for analyzing a variety of dislocation problems in layered earth models.

We begin with the simplest antiplane problem, following Ward (1985), writing the governing equations such that the vertical derivatives are separated. In antiplane strain, there is only one nontrivial equilibrium equation, which we write using the body force equivalent source representation (3.23), assuming slip on a vertical fault, as

$$
\frac{\partial \sigma_{23}}{\partial x_2} = -\frac{\partial \sigma_{13}}{\partial x_1} + M_{3j}\frac{\partial}{\partial x_j}\delta(\mathbf{x} - \boldsymbol{\xi}),
$$

$$
= -\frac{\partial \sigma_{13}}{\partial x_1} + \mu s_3 n_1 d\xi_2 \frac{\partial}{\partial x_1}\delta(\mathbf{x} - \boldsymbol{\xi}),
$$

$$
= -\mu \frac{\partial^2 u_3}{\partial x_1^2} + \mu s d\xi_2 \delta(x_2 - \xi_2)\frac{\partial}{\partial x_1}\delta(x_1 - \xi_1). \tag{5.35}
$$

The second term on the right-hand side is the body force equivalent of a line slip source, with slip amplitude s, located at $\boldsymbol{\xi}$ (see equation [3.19]). For slip on a vertical plane, the only nonzero component of the seismic moment (per unit length of fault in the along-strike direction) is, from equation (3.14), $M_{31} = \mu s_3 n_1 d\xi_2$, where $d\xi_2$ is the increment of fault height. The subscript on the slip is omitted following the second equation; for antiplane dislocations, it should be understood that $s = s_3$.

Similarly, the vertical gradient in displacement is written in terms of stress as

$$
\frac{\partial u_3}{\partial x_2} = \frac{1}{\mu}\sigma_{23}. \tag{5.36}
$$

Putting these two equations together,

$$
\frac{\partial}{\partial x_2}\begin{bmatrix} u_3 \\ \sigma_{23} \end{bmatrix} = \begin{bmatrix} 0 & \mu^{-1} \\ -\mu\dfrac{\partial^2}{\partial x_1^2} & 0 \end{bmatrix}\begin{bmatrix} u_3 \\ \sigma_{23} \end{bmatrix} + \begin{bmatrix} 0 \\ \mu s d\xi_2 \delta(x_2 - \xi_2)\dfrac{\partial\delta(x_1 - \xi_1)}{\partial x_1} \end{bmatrix}. \tag{5.37}
$$

Note that the terms we have separated on the left-hand side, u_3 and σ_{23}, are those that must be continuous across horizontal surfaces. The idea is then to propagate a valid solution at one depth upward, integrating the vertical derivatives. To simplify the equations, we take the Fourier transform of both sides with respect to x_1. We use the transform pair (A.1) and (A.2):

$$
\mathcal{F}[f(x)] = \bar{f}(k) = \int_{-\infty}^{+\infty} f(x)\exp(-ikx)dx, \tag{5.38}
$$

$$
\mathcal{F}^{-1}[\bar{f}(k)] = f(x) = \frac{1}{2\pi}\int_{-\infty}^{+\infty} \bar{f}(k)\exp(ikx)dk. \tag{5.39}
$$

To keep the notation simple, but to break with tradition in this text, we let the z axis be vertically down, while x remains the horizontal direction. This is accomplished by a rotation about the x_1 axis such that x_2 is positive downward. The x_3 axis now points into, rather than out of, the plane so that the sense of positive slip is reversed.

The Fourier transform of a derivative is equal to ik times the transform of the function (see equation [A.7]). Thus, the transform of the derivative of the delta function, making use of (A.3), is

$$
\mathcal{F}\left[\frac{\partial\delta(x_1 - \xi_1)}{\partial x_1}\right] = ik\mathcal{F}[\delta(x_1 - \xi_1)] = ike^{-ik\xi_1}, \tag{5.40}
$$

which can be set equal to ik, since without loss of generality, we may take the fault to be located at $\xi_1 = 0$. Thus, Fourier transforming equation (5.37) in x leads to

$$\frac{\partial}{\partial z} \begin{bmatrix} u_3 \\ \sigma_{23} \end{bmatrix} = \begin{bmatrix} 0 & \mu^{-1} \\ \mu k^2 & 0 \end{bmatrix} \begin{bmatrix} u_3 \\ \sigma_{23} \end{bmatrix} + \begin{bmatrix} 0 \\ ik\,\mu s\,\mathrm{d}z_s\,\delta(z - z_s) \end{bmatrix}. \tag{5.41}$$

Here z_s indicates source depth, and $\mathrm{d}z_s$ the increment of fault height. Note that the equation (5.41) is of the general form

$$\boxed{\frac{\mathrm{d}\mathbf{v}(z)}{\mathrm{d}z} = A\mathbf{v}(z) + \mathbf{f}(z),} \tag{5.42}$$

where $\mathbf{v}$ is the stress-displacement vector $\mathbf{v} = [u_3, \sigma_{23}]^T$, and $\mathbf{f}$ represents the source term.

We begin by considering the homogeneous equations corresponding to equation (5.42):

$$\frac{\mathrm{d}\mathbf{v}(z)}{\mathrm{d}z} = A\mathbf{v}(z). \tag{5.43}$$

Solutions of equation (5.43) are found from the eigenvectors and eigenvalues of the operator A (e.g., Strang 1976). If A has N eigenvalues λ_i and associated eigenvectors Ψ_i, then the solution to (5.43) is

$$\mathbf{v}(z) = c_1\Psi_1 \exp(\lambda_1 z) + c_2\Psi_2 \exp(\lambda_2 z) + \cdots + c_N\Psi_N \exp(\lambda_N z), \tag{5.44}$$

where the constants c_i are determined by the boundary conditions at $z = 0$. In the antiplane case under consideration, $N = 2$. If the eigen decomposition of A is $A = V\Lambda V^{-1}$, where V is a matrix of eigenvectors and Λ is a diagonal matrix of eigenvalues, then the solution to the homogeneous equations can be written in terms of the matrix exponential of Λ defined in equation (5.53),

$$\mathbf{v}(z) = Ve^{\Lambda z}\mathbf{c}, \tag{5.45}$$

where $\mathbf{c}$ is a vector containing the coefficients c_i. It is important to note that the eigenvectors are not unique solutions. Any eigenvector can be multiplied by a constant and still be a solution to the differential equation.

The propagator matrix $\mathbf{P}(z, z_0)$ is defined in such a way that it propagates the stresses and displacements from depth z_0 to depth z, $\mathbf{v}(z) = \mathbf{P}(z, z_0)\mathbf{v}(z_0)$. The propagator itself is a solution to the homogeneous equations (5.43) and is given by

$$\mathbf{P}(z, z_0) = \mathbf{I} + \int_{z_0}^{z} A(\zeta_1)\mathrm{d}\zeta_1 + \int_{z_0}^{z} A(\zeta_1) \left[\int_{z_0}^{\zeta_1} A(\zeta_2)\mathrm{d}\zeta_2 \right] \mathrm{d}\zeta_1 + \cdots, \tag{5.46}$$

where $\mathbf{I}$ is the identity matrix. Differentiating both sides of equation (5.46), we have

$$\frac{d\mathbf{P}(z, z_0)}{dz} = 0 + A(z) + A(z) \int_{z_0}^{z} A(\zeta_1)d\zeta_1 + \cdots,$$

$$= A(z) \left\{ \mathbf{I} + \int_{z_0}^{z} A(\zeta_1)d\zeta_1 + \int_{z_0}^{z} A(\zeta_1) \left[\int_{z_0}^{\zeta_1} A(\zeta_2)d\zeta_2 \right] d\zeta_1 + \cdots \right\},$$

$$= A(z)\mathbf{P}(z, z_0), \tag{5.47}$$

verifying that equation (5.46) solves (5.43).

Useful properties of propagator matrices $\mathbf{P}(z, z_0)$ are:

$$\mathbf{P}(z_0, z_0) = \mathbf{I}, \tag{5.48}$$

$$\mathbf{v}(z) = \mathbf{P}(z, z_0)\mathbf{v}(z_0), \tag{5.49}$$

$$\mathbf{P}(z, z_0) = [\mathbf{P}(z_0, z)]^{-1}, \tag{5.50}$$

(e.g., Aki and Richards 1980). The first property (5.48) follows directly from equation (5.46). The second property (5.49) follows from the fact that $\mathbf{P}(z, z_0)\mathbf{v}(z_0)$ satisfies the homogeneous equations (5.43). This is true because $\mathbf{P}(z, z_0)$ satisfies equation (5.43) and $\mathbf{v}(z_0)$ is independent of z. Thus, $\mathbf{P}(z, z_0)\mathbf{v}(z_0)$ satisfies equation (5.43) and equals $\mathbf{v}(z_0)$ when $z = z_0$ by equation (5.48). Since $\mathbf{P}(z, z_0)\mathbf{v}(z_0)$ satisfies equation (5.43) and equals $\mathbf{v}(z_0)$ at $z = z_0$, it must therefore be the solution, $\mathbf{v}(z) = \mathbf{P}(z, z_0)\mathbf{v}(z_0)$. Equation (5.49) is central in that it demonstrates that the effect of operating on vector $\mathbf{v}$ with $\mathbf{P}(z, z_0)$ is to transfer the stress and displacement at z_0 to z.

Notice that the propagators can be applied sequentially:

$$\mathbf{v}(z_2) = \mathbf{P}(z_2, z_1)\mathbf{v}(z_1), \tag{5.51}$$

$$= \mathbf{P}(z_2, z_1)\mathbf{P}(z_1, z_0)\mathbf{v}(z_0). \tag{5.52}$$

This is important in that it shows that the solution vector can be propagated through an arbitrary number of layers. Since this result may be applied to any $\mathbf{v}(z_2)$, choose it to be $\mathbf{v}(z_0)$, which proves the last property (5.50).

When the layers have constant properties so that A is constant within a layer, equation (5.46) reduces to

$$\mathbf{P}(z, z_0) = \mathbf{I} + A \int_{z_0}^{z} d\zeta_1 + AA \int_{z_0}^{z} \left[\int_{z_0}^{\zeta_1} d\zeta_2 \right] d\zeta_1 + \cdots,$$

$$= \mathbf{I} + A(z - z_0) + AA\frac{(z - z_0)^2}{2} + \cdots,$$

$$= \mathbf{e}^{[A(z-z_0)]}, \tag{5.53}$$

where the last equation defines the *matrix exponential*. Notice from the eigen decomposition $A = V\Lambda V^{-1}$, that $AA = (V\Lambda V^{-1})(V\Lambda V^{-1}) = V\Lambda^2 V^{-1}$. Extending this, $A^n = V\Lambda^n V^{-1}$. It follows then, from equation (5.53), that $\mathbf{e}^{[A(z-z_0)]} = V\mathbf{e}^{[\Lambda(z-z_0)]} V^{-1}$.

We now have the proper machinery to return to the inhomogeneous equations (5.42) and consider the general solution with a source term. The general solution is given by

$$\mathbf{v}(z) = \mathbf{P}(z, z_0)\mathbf{v}(z_0) + \int_{z_0}^{z} \mathbf{P}(z, \zeta)\mathbf{f}(\zeta)\mathrm{d}\zeta. \tag{5.54}$$

(e.g., Aki and Richards 1980). This can be verified by direct substitution into equation (5.42):

$$\frac{\mathrm{d}}{\mathrm{d}z} \left\{ \mathbf{P}(z, z_0)\mathbf{v}(z_0) + \int_{z_0}^{z} \mathbf{P}(z, \zeta)\mathbf{f}(\zeta)\mathrm{d}\zeta \right\} = A(z) \left\{ \mathbf{P}(z, z_0)\mathbf{v}(z_0) + \int_{z_0}^{z} \mathbf{P}(z, \zeta)\mathbf{f}(\zeta)\mathrm{d}\zeta \right\} + \mathbf{f}(z),$$

$$\frac{\mathrm{d}}{\mathrm{d}z} \int_{z_0}^{z} \mathbf{P}(z, \zeta)\mathbf{f}(\zeta)\mathrm{d}\zeta = A(z) \int_{z_0}^{z} \mathbf{P}(z, \zeta)\mathbf{f}(\zeta)\mathrm{d}\zeta + \mathbf{f}(z),$$

$$\frac{\mathrm{d}}{\mathrm{d}z} \int_{z_0}^{z} \mathbf{P}(z, \zeta)\mathbf{f}(\zeta)\mathrm{d}\zeta = \int_{z_0}^{z} \frac{\mathrm{d}}{\mathrm{d}z}\mathbf{P}(z, \zeta)\mathbf{f}(\zeta)\mathrm{d}\zeta + \mathbf{f}(z). \tag{5.55}$$

The first step follows from the fact that $\mathbf{P}(z, z_0)\mathbf{v}(z_0)$ satisfies the homogeneous equation as in equation (5.47). The second step is to take $A(z)$ inside the integral and apply (5.47). That the final equation represents an equality can be seen by taking the derivative inside the integral using Leibnitz's rule for differentiation of a definite integral:

$$\frac{\mathrm{d}}{\mathrm{d}c} \int_{a(c)}^{b(c)} f(x, c)\mathrm{d}x = \int_{a(c)}^{b(c)} \frac{\partial}{\partial c} f(x, c)\mathrm{d}x + \frac{\mathrm{d}b}{\mathrm{d}c} f(b, c) - \frac{\mathrm{d}a}{\mathrm{d}c} f(a, c). \tag{5.56}$$

Thus, we see that equation (5.54) is indeed a solution to (5.42).

5.5.1 The Propagator Matrix for Antiplane Deformation

The propagator matrix for the antiplane problem with layers of uniform properties can be computed from equation (5.53). With reference to equation (5.41), we have

$$A = \begin{bmatrix} 0 & \mu^{-1} \\ \mu k^2 & 0 \end{bmatrix}, \tag{5.57}$$

$$AA = \begin{bmatrix} 0 & \mu^{-1} \\ \mu k^2 & 0 \end{bmatrix} \begin{bmatrix} 0 & \mu^{-1} \\ \mu k^2 & 0 \end{bmatrix} = \begin{bmatrix} k^2 & 0 \\ 0 & k^2 \end{bmatrix}, \tag{5.58}$$

$$AAA = \begin{bmatrix} k^2 & 0 \\ 0 & k^2 \end{bmatrix} \begin{bmatrix} 0 & \mu^{-1} \\ \mu k^2 & 0 \end{bmatrix} = \begin{bmatrix} 0 & \mu^{-1}k^2 \\ \mu k^4 & 0 \end{bmatrix}, \tag{5.59}$$

so that eventually

$$\mathbf{P}(z, z_0) = \begin{bmatrix} 1 + \dfrac{(z - z_0)^2}{2}k^2 + \dfrac{(z - z_0)^4}{4!}k^4 \cdots & (z - z_0)\dfrac{1}{\mu} + \dfrac{(z - z_0)^3}{3!}\dfrac{k^2}{\mu} \cdots \\[2mm] \mu(z - z_0)k^2 + \mu\dfrac{(z - z_0)^3}{3!}k^4 \cdots & 1 + \dfrac{(z - z_0)^2}{2}k^2 + \dfrac{(z - z_0)^4}{4!}k^4 \cdots \end{bmatrix}. \tag{5.60}$$

The preceding simplifies to

$$\mathbf{P}(z, z_0) = \begin{bmatrix} \cosh[(z - z_0)|k|] & (\mu|k|)^{-1} \sinh[(z - z_0)|k|] \\[2mm] (\mu|k|) \sinh[(z - z_0)|k|] & \cosh[(z - z_0)|k|] \end{bmatrix}. \tag{5.61}$$

Note that the absolute value signs are not required in equation (5.61) since cosh is even, and while sinh is odd, it is premultiplied by k or $1/k$. However, I prefer to keep the absolute value signs on the $|k|$, to avoid confusion in subsequent derivations that depend on equation (5.61).

The general solution is given by equation (5.54), but how does one choose $\mathbf{v}(z_0)$? If we choose z_0 to be the shallowest point in the half-space below which there are no dislocation sources, then $\mathbf{v}(z)$ must satisfy the homogeneous differential equations (5.43) for $z \geq z_0$. In other words, we choose z_0 to be the shallowest point below which the problem is completely uniform, in the sense of there being neither layer boundaries nor dislocations below z_0. From equation (5.44), the homogeneous solution is of the form $\mathbf{v}(z) = C_1 \Psi_1 \exp(\lambda_1 z) + C_2 \Psi_2 \exp(\lambda_2 z)$, where λ_i and Ψ_i are the eigenvalues and eigenvectors of A, respectively. From equation (5.57), the eigenvalues are easily found to be $+k, -k$, with corresponding eigenvectors $[1/\mu, k]^T$ and $[1/\mu, -k]^T$.

The homogeneous solution is thus

$$\mathbf{v}(z) = C_1 \begin{bmatrix} 1/\mu \\ k \end{bmatrix} e^{kz} + C_2 \begin{bmatrix} 1/\mu \\ -k \end{bmatrix} e^{-kz}. \tag{5.62}$$

The solution must decay with depth as $z \to \infty$. Thus, the solution corresponding to the positive eigenvalue must vanish for positive wavenumbers, and vice versa. Equation (5.62) can thus be written in compact form, noting that the coefficient C is in general a function of wavenumber, as

$$\begin{bmatrix} u_3 \\ \sigma_{23} \end{bmatrix} = C(k) \begin{bmatrix} 1/\mu \\ -|k| \end{bmatrix} e^{-|k|z}. \tag{5.63}$$

The solution given by equation (5.63) is easily shown to satisfy (5.43) by direct substitution. The constant $C(k)$ is then determined by the boundary condition that the traction vanish at $z = 0$.

In order to develop an understanding of how to apply propagator methods, we explore several problems for which we have previously derived solutions based on other methods.

5.5.2 Vertical Fault in a Homogeneous Half-Space

There are two approaches to developing the propagator solution for a vertical dislocation in a homogeneous half-space. The first involves propagating the (as yet unknown) solution from the free surface down (positive z direction). Denote the solution vector at the free surface $\mathbf{v}(0)$.

For depths above any sources, we have

$$\mathbf{v}(z) = \mathbf{P}(z, 0)\mathbf{v}(0) \qquad z < z_s. \tag{5.64}$$

For simplicity, assume a point source at depth z_s (actually a line source extending infinitely in and out of the plane). For $z > z_s$, we pick up the source term, as in equation (5.54):

$$\mathbf{v}(z) = \mathbf{P}(z, 0)\mathbf{v}(0) + \int_0^z \mathbf{P}(z, \zeta) \begin{bmatrix} 0 \\ ik\,\mu s\,\mathrm{d}z_s\delta(\zeta - z_s) \end{bmatrix} \mathrm{d}\zeta,$$

$$= \mathbf{P}(z, 0)\mathbf{v}(0) + \mathbf{P}(z, z_s) \begin{bmatrix} 0 \\ ik\,\mu s\,\mathrm{d}z_s \end{bmatrix} \qquad z \geq z_s. \tag{5.65}$$

For a half-space, the homogeneous solution must apply for all $z \geq z_s$, so we are free to take $z = z_s$ in equation (5.65) and apply the homogeneous solution for $\mathbf{v}(z_s)$. Thus,

$$C(k) \begin{bmatrix} 1/\mu \\ -|k| \end{bmatrix} e^{-|k|z_s} = \mathbf{P}(z_s, 0) \begin{bmatrix} u_3(k, z = 0) \\ 0 \end{bmatrix} + \begin{bmatrix} 0 \\ ik\,\mu s\,\mathrm{d}z_s \end{bmatrix}, \tag{5.66}$$

since $\mathbf{P}(z_s, z_s) = I$. Note that $e^{-|k|z_s}$ is independent of depth and can be absorbed into the constant $C(k)$. Equation (5.66) represents two equations in two unknowns C and $u_3(z = 0)$ and can be solved for such in the present form. However, we can also premultiply both sides of the equation by $[\mathbf{P}(z_s, 0)]^{-1} = \mathbf{P}(0, z_s)$, yielding

$$\begin{bmatrix} u_3(k, z = 0) \\ 0 \end{bmatrix} = \mathbf{P}(0, z_s, \mu) \left\{ C(k) \begin{bmatrix} 1/\mu \\ -|k| \end{bmatrix} + \begin{bmatrix} 0 \\ -ik\,\mu s\,\mathrm{d}z_s \end{bmatrix} \right\}, \tag{5.67}$$

where we have introduced the notation $\mathbf{P}(0, z_s, \mu)$ to indicate that the region $0 \leq z \leq z_s$ has shear modulus μ.

The alternative approach is to start at the bottom and integrate up to the free surface. Since the deepest point where the exponentially decaying homogeneous solution no longer applies is z_s, we begin integrating from that point

$$\begin{bmatrix} u_3(k, z = 0) \\ 0 \end{bmatrix} = \mathbf{P}(0, z_s, \mu)C(k) \begin{bmatrix} 1/\mu \\ -|k| \end{bmatrix} + \int_{z_s}^0 \mathbf{P}(0, \zeta) \begin{bmatrix} 0 \\ ik\,\mu s\,\mathrm{d}z_s\delta(\zeta - z_s) \end{bmatrix} \mathrm{d}\zeta, \tag{5.68}$$

which yields equation (5.67) as long as we recognize that the integral in (5.68) is in the negative z direction, so we pick up a minus sign when integrating over the delta function.

The two equations in (5.67) can be solved for $C(k)$ and $u_3(k, z = 0)$. Specifically,

$$C(k) = -i\,s\,\mu\,\mathrm{d}z_s\,\mathrm{sgn}(k)\,\cosh(z_s|k|)e^{-z_s|k|}, \tag{5.69}$$

and

$$u_3(k, z = 0) = -i\,s\,\mathrm{d}z_s\,\mathrm{sgn}(k)\,e^{-z_s|k|}. \tag{5.70}$$

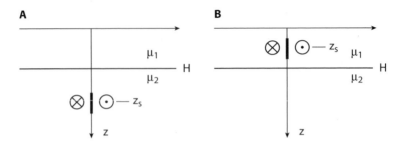

Figure 5.11. Geometry of the antiplane dislocation problem for (A) a dislocation in a half-space beneath a layer or (B) a dislocation in a layer overlying a half-space. In both cases, the layer thickness is H, and the line source is located at depth z_s.

To determine the displacement due to uniform slip from the surface to depth D, integrate

$$u_3(k, z=0) = -is\ \text{sgn}(k) \int_0^D e^{-z_s|k|} dz_s = \frac{-is}{k}\left(1 - e^{-|k|D}\right). \tag{5.71}$$

This transform can be inverted analytically (Ward 1985; see problem 7), yielding

$$u_3(x, z=0) = \frac{s}{\pi}\tan^{-1}(D/x), \tag{5.72}$$

recovering an earlier result (2.30).

5.5.3 Vertical Fault within Half-Space beneath a Layer

We will next consider the case of a strike-slip fault in a half-space with shear modulus μ_2 beneath a layer of thickness H of shear modulus μ_1 (figure 5.11A). We again assume a line source located at depth z_s, extending indefinitely in and out of the plane and ultimately integrate this solution to generate a fault with uniform slip. Working from the bottom up, the homogeneous half-space solution applies for all depths greater than z_s. We thus need to propagate both the half-space solution and the source terms from z_s to the free surface. This is achieved by applying $\mathbf{P}(0, H, \mu_1)\mathbf{P}(H, z_s, \mu_2)$ to both the source terms and the homogeneous half-space terms. The first propagator carries the solution from the source depth to the top of the half-space through a region with modulus μ_2. The second matrix carries the solution from the bottom of the layer to the free surface through a layer of modulus μ_1. Note that because the source is in the half-space, the appropriate shear modulus in the source term is μ_2. Furthermore, because we integrate in the negative z direction, the integral of the delta function picks up a minus sign, as in equation (5.68). This yields

$$\begin{bmatrix} u_3(k,0) \\ 0 \end{bmatrix} = \mathbf{P}(0, H, \mu_1)\mathbf{P}(H, z_s, \mu_2)\left\{ C(k)\begin{bmatrix} 1/\mu_2 \\ -|k| \end{bmatrix} + \begin{bmatrix} 0 \\ -ik\,\mu_2 s\,dz_s \end{bmatrix}\right\}. \tag{5.73}$$

Solving for $C(k)$,

$$C(k) = i\,s\mu_2 dz_s\,\text{sgn}(k)\frac{\{\mu_2\cosh(H|k|)\cosh[(H-z_s)|k|] - \mu_1\sinh(H|k|)\sinh[(H-z_s)|k|]\}}{\{\sinh[(H-z_s)|k|] - \cosh[(H-z_s)|k|]\}[\mu_2\cosh(H|k|) + \mu_1\sinh(H|k|)]}, \tag{5.74}$$

and

$$u_3(k, 0) = C(k) \left(\{\cosh[(H - z_s)|k|] - \sinh[(H - z_s)|k|]\} \left[\frac{1}{\mu_2} \cosh(H|k|) + \frac{1}{\mu_1} \sinh(H|k|) \right] \right)$$
$$- i s \mu_2 dz_s \ \text{sgn}(k) \left\{ \frac{1}{\mu_2} \cosh(H|k|) \sinh[(H - z_s)|k|] - \frac{1}{\mu_1} \sinh(H|k|) \cosh[(H - z_s)|k|] \right\}. \tag{5.75}$$

Combining equations (5.74) and (5.75) and making use of the following identity:

$$\cosh(z_1 - z_2) = \cosh(z_1) \cosh(z_2) - \sinh(z_1) \sinh(z_2), \tag{5.76}$$

leads to

$$u_3(k, 0) = -i s \ dz_s \ \text{sgn}(k) \ \frac{\exp[(H - z_s)|k|]}{\cosh(H|k|) + \dfrac{\mu_1}{\mu_2} \sinh(H|k|)} \qquad z_s > H. \tag{5.77}$$

Notice that in the limit that $\mu_1 = \mu_2$, $\cosh(H|k|) + \sinh(H|k|) = \exp(H|k|)$, so equation (5.77) reduces to (5.70). Last, integrating from D to ∞,

$$u_3(k, z = 0) = \frac{-i s \exp[(H - D)|k|]}{k[\cosh(H|k|) + \dfrac{\mu_1}{\mu_2} \sinh(H|k|)]} \qquad D > H. \tag{5.78}$$

A complication in numerically inverting the transform is that equation (5.78) is singular at $k = 0$. I have found that a good way to overcome this (following the approach for analytic inversion of problem 7) is to differentiate the displacement with respect to x to obtain the strain. This is achieved by multiplying by ik in the Fourier domain, thus removing the singularity. Following the inverse transformation, the displacement is recovered by integrating the strain. Figure 5.12 compares the solution computed by numerically inverting equation (5.78) with the infinite series image solution (5.31). The difference between the two methods is indistinguishable at the resolution of the plot. An alternative approach would be to take the inverse Fourier transform of equation (5.77), yielding the surface displacement due to a localized dislocation source, and then integrate this solution over source depth.

5.5.4 Vertical Fault in Layer over Half-Space

We next consider the case of a fault embedded in a layer with modulus μ_1 (figure 5.11B) over a homogeneous half-space with modulus μ_2. The layer thickness is again H. This is similar to the previous example except that the fault is located within the layer rather than the half-space.

Again working from the bottom up, we integrate the source terms from z_s to the surface. This is achieved by premultiplying the source term by $\mathbf{P}(0, z_s, \mu_1)$. Because there are no sources within the half-space, the homogeneous half-space solution applies for all depths greater H. Thus, the half-space solution is premultiplied by $\mathbf{P}(0, H, \mu_1)$. Again minding the sign on the source term, we have

$$\begin{bmatrix} u_3(z = 0) \\ 0 \end{bmatrix} = \mathbf{P}(0, H, \mu_1) \left\{ C(k) \begin{bmatrix} 1/\mu_2 \\ -|k| \end{bmatrix} \right\} + \mathbf{P}(0, z_s, \mu_1) \begin{bmatrix} 0 \\ -ik \mu_1 s \ dz_s \end{bmatrix}, \tag{5.79}$$

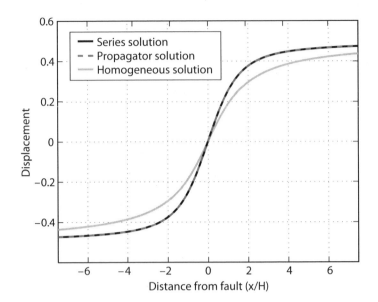

Figure 5.12. Displacements due to an infinitely long screw dislocation in a half-space below a layer of thickness H. The depth to the top of the dislocation is D; $H/D = 2/3$. The shear modulus of the half-space is μ_2, while that of the layer is μ_1 and $\mu_1/\mu_2 = 0.1$. The figure compares the series solution constructed using the method of images with the propagator matrix solution, using an FFT to compute the inverse transform. The solution for a homogeneous half-space is shown for reference.

where the fault is restricted to the depths $z_s < H$. (Note that the source term has modulus μ_1.) The first term propagates the decaying half-space term to the surface, while the second term propagates the source to the surface.

Solving for $C(k)$ yields

$$C(k) = \frac{-i\, s\, dz_s\, \text{sgn}(k)\, \mu_1 \mu_2\, \cosh(z_s|k|)}{\mu_2 \cosh(H|k|) + \mu_1 \sinh(H|k|)},\tag{5.80}$$

so that

$$u_3(k, 0) = C(k) \left[\frac{1}{\mu_2} \cosh(H|k|) + \frac{1}{\mu_1} \sinh(H|k|) \right] + is\, dz_s\, \text{sgn}(k) \sinh(z_s|k|)$$

$$= -is\, dz_s\, \text{sgn}(k) \left\{ \cosh(z_s|k|) \left[\frac{\mu_1 \cosh(H|k|) + \mu_2 \sinh(H|k|)}{\mu_2 \cosh(H|k|) + \mu_1 \sinh(H|k|)} \right] - \sinh(z_s|k|) \right\}.$$

$$\tag{5.81}$$

Note that this reproduces equation (5.70) in the limit that the shear modulus is the same in both the layer and the half-space.

Integrating the solution from 0 to D yields

$$u_3(k, 0) = \frac{-is}{k} \left\{ \sinh(D|k|) \left[\frac{\mu_1 \cosh(H|k|) + \mu_2 \sinh(H|k|)}{\mu_2 \cosh(H k) + \mu_1 \sinh(H|k|)} \right] - \cosh(D|k|) + 1 \right\},$$

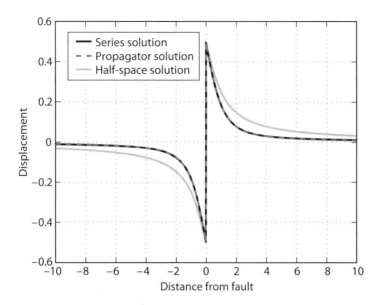

Figure 5.13. Displacements due to infinitely long screw dislocation in a layer over a half-space. Layer thickness is H, and depth of dislocation is D and $D = H$. The shear modulus of the half-space is three times that of the layer, $\mu_2/\mu_1 = 3$. The figure compares the series solution constructed using the method of images with the propagator matrix solution, using an FFT to compute the inverse transform. The solution in a homogeneous half-space is shown for reference.

$$u_3(k, 0) = \frac{-is}{k} \left(1 - \left\{ \frac{\mu_1 \sinh[(H - D)|k|] + \mu_2 \cosh[(H - D)|k|]}{\mu_2 \cosh(H|k|) + \mu_1 \sinh(H|k|)} \right\} \right) \qquad D < H. \qquad (5.82)$$

The last step following from application of the identities:

$$\sinh(z_1 + z_2) = \sinh(z_1) \cosh(z_2) + \cosh(z_1) \sinh(z_2), \qquad (5.83)$$

$$\cosh(z_1 + z_2) = \cosh(z_1) \cosh(z_2) + \sinh(z_1) \sinh(z_2). \qquad (5.84)$$

Figure 5.13 compares the solution computed by inverting equation (5.82) numerically, using an inverse FFT, with the infinite series image solution (5.33).

5.5.5 General Solution for an Arbitrary Number of Layers

It is now straightforward to generalize the method for an arbitrary number of layers. If there are additional layers between the source and the earth's surface, one simply premultiplies by the appropriate matrices to propagate the solution to the free surface. Note that the homogeneous solution applies to the half-space beneath the deepest source, where the medium is homogeneous and free of dislocations. We thus propagate the homogeneous solution from a depth corresponding to either the deepest source or the top of the half-space, *whichever is deepest*. There are two cases, as illustrated in figure 5.14: in one, the source is located

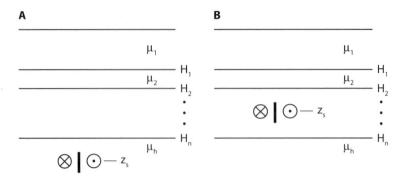

Figure 5.14. The structure of a propagator matrix solution. The earth model consists of n layers overlying a homogeneous half-space. The depth to the bottom of the nth layer is H_n. z_s refers to the depth of the dislocation point (line) source. A: The source is located in the half-space, $z_s > H_n$. B: The source is located somewhere within the stack of layers, $z_s < H_n$. In this case, H_s refers to the depth to the top of the layer containing the source.

in the half-space, and in the other, the source is located within any one of the layers. The most general formulation is thus

$$
\begin{bmatrix} u_3(z=0) \\ 0 \end{bmatrix} = \mathbf{P}(0,\, H_1,\, \mu_1)\mathbf{P}(H_1,\, H_2,\, \mu_2) \cdots \mathbf{P}(H_s,\, z_s,\, \mu_s) \begin{bmatrix} 0 \\ -ik\, s\mu_s dz_s \end{bmatrix}
$$
$$
+ \mathbf{P}(0,\, H_1,\, \mu_1)\mathbf{P}(H_1,\, H_2,\, \mu_2) \cdots
$$
$$
\cdots \mathbf{P}(H_{n-1},\, H_n,\, \mu_n) \begin{cases} \mathbf{P}(H_n,\, z_s,\, \mu_h) \begin{bmatrix} 1/\mu_h \\ -|k| \end{bmatrix} C(k) & z_s > H_n \\[2ex] \begin{bmatrix} 1/\mu_h \\ -|k| \end{bmatrix} C(k) & z_s < H_n \end{cases}
$$

$$(5.85)$$

Here H_1 is the depth of the first layer, with modulus μ_1, H_2 is the depth of the second layer, with modulus μ_2, and so on. H_s and z_s represent the depth to the top of the layer above the source and the source depth, respectively. μ_h is the shear modulus of the half-space, and μ_s is the shear modulus of the material containing the source. If the source is in the half-space, then $H_s = H_n$.

The first term propagates the source term from the source depth to the bottom of the overlying layer H_s, and from there to the earth's surface. The second term propagates the homogeneous solution to the free surface. Note that, as before, we have absorbed the term $e^{-|k|z_s}$ into the constant C. If the source is in the half-space (figure 5.14A), $z_s > H_n$, and equation (5.85) propagates the solution through the half-space to the bottom of the stack of layers and from there to the surface. If the source is above the half-space (figure 5.14B), $z_s < H_n$, and equation (5.85) propagates the solution from the top of the half-space through the stack of layers to the free surface.

5.5.6 Displacements and Stresses at Depth

In order to determine the stresses and displacements at depth, we propagate the known solution from the free surface into the body. For depths z shallower than any sources,

$$\mathbf{v}(z) = \mathbf{P}(z, 0)\mathbf{v}(0) \qquad z < z_s, \tag{5.86}$$

where $\mathbf{v}(0)$ represents the stress displacement vector at $z = 0$. Below the source depth, we pick up an additional source term so that

$$\mathbf{v}(z) = \mathbf{P}(z, 0)\mathbf{v}(0) + \int_0^z \mathbf{P}(z, \zeta)\mathbf{f}(\zeta)d\zeta,$$
$$= \mathbf{P}(z, 0)\mathbf{v}(0) + \mathbf{P}(z, z_s) \begin{bmatrix} 0 \\ ik\,\mu s dz_s \end{bmatrix} \qquad z \geq z_s. \tag{5.87}$$

For a homogeneous half-space, making use of equation (5.70), this leads to

$$u(z) = -is\,dz_s\,\text{sgn}(k)\,\cosh(z|k|)e^{-z_s|k|} \qquad z < z_s,$$
$$u(z) = -is\,dz_s\,\text{sgn}(k)\,\cosh(z_s|k|)e^{-z|k|} \qquad z > z_s. \tag{5.88}$$

As before, the displacements due to a finite fault are found by integrating the point sources. In this example, we consider a fault from the surface to depth D. In order to perform the integration, we need to distinguish between two cases. If we are evaluating the stress and displacement beneath the deepest source—that is, $z > D$—then $z > z_s$ for all z_s. In this case, the second equation in (5.88) applies for all z:

$$u(z) = -is\,\text{sgn}(k)e^{-z|k|} \int_0^D \cosh(z_s|k|)dz_s,$$
$$= -\frac{is}{k}e^{-z|k|}\sinh(D|k|) \qquad z > D. \tag{5.89}$$

If, on the other hand we evaluate the stress at $z < D$, we must break up the integration into two parts; from 0 to z and from z to D. In the first integral, $z > z_s$, so the second equation in (5.88) applies. In the second integral, $z < z_s$, so the first equation in (5.88) holds. Thus,

$$u(z) = -is\,\text{sgn}(k) \left[e^{-z|k|} \int_0^z \cosh(z_s|k|)dz_s + \cosh(z|k|) \int_z^D e^{-z_s|k|}dz_s \right],$$
$$= \frac{is}{k}\left[\cosh(z|k|)e^{-D|k|} - 1\right] \qquad z < D. \tag{5.90}$$

The Fourier transforms can be inverted analytically to obtain the familiar arctangent functions for displacement in the body (see problem 9). The same approach can readily be extended to determine the stress in the body and the elastic fields for more complex layered media.

5.5.7 Propagator Methods for Plane Strain

The equilibrium equations for plane strain were given in equation (3.1). Combining with Hooke's law and the strain-displacement relations, it is possible to separate the vertical derivatives in the equilibrium equations as

$$\frac{\partial \sigma_{12}}{\partial z} = -\frac{4\mu (\lambda + \mu)}{\lambda + 2\mu} \frac{\partial^2 u_1}{\partial x^2} - \frac{\lambda}{\lambda + 2\mu} \frac{\partial \sigma_{22}}{\partial x} - f_1,$$ (5.91)

$$\frac{\partial \sigma_{22}}{\partial z} = -\frac{\partial \sigma_{12}}{\partial x} - f_2.$$ (5.92)

We can also separate the vertical derivatives of the displacements

$$\frac{\partial u_1}{\partial z} = \frac{1}{\mu} \sigma_{12} - \frac{\partial u_2}{\partial x},$$ (5.93)

$$\frac{\partial u_2}{\partial z} = \frac{1}{\lambda + 2\mu} \sigma_{22} - \frac{\lambda}{\lambda + 2\mu} \frac{\partial u_1}{\partial x}.$$ (5.94)

The horizontal derivatives on the right sides of equations (5.91) to (5.94) are eliminated by Fourier transforming in x:

$$\frac{\partial \sigma_{12}}{\partial z} = \frac{4\mu (\lambda + \mu)}{\lambda + 2\mu} k^2 u_1 - \frac{\lambda}{\lambda + 2\mu} i k \sigma_{22} - f_1,$$ (5.95)

$$\frac{\partial \sigma_{22}}{\partial z} = -ik\sigma_{12} - f_2,$$ (5.96)

$$\frac{\partial u_1}{\partial z} = \frac{1}{\mu} \sigma_{12} - iku_2,$$ (5.97)

$$\frac{\partial u_2}{\partial z} = \frac{1}{\lambda + 2\mu} \sigma_{22} - \frac{\lambda}{\lambda + 2\mu} iku_1.$$ (5.98)

Last, equations (5.95) to (5.98) can be written in matrix form as

$$\begin{bmatrix} i\dfrac{\partial u_1}{\partial z} \\[2ex] \dfrac{\partial u_2}{\partial z} \\[2ex] i\dfrac{\partial \sigma_{12}}{\partial z} \\[2ex] \dfrac{\partial \sigma_{22}}{\partial z} \end{bmatrix} = \begin{bmatrix} 0 & k & 1/\mu & 0 \\[2ex] \dfrac{-k\lambda}{\lambda + 2\mu} & 0 & 0 & \dfrac{1}{\lambda + 2\mu} \\[2ex] \dfrac{4\mu k^2 (\lambda + \mu)}{\lambda + 2\mu} & 0 & 0 & \dfrac{k\lambda}{\lambda + 2\mu} \\[2ex] 0 & 0 & -k & 0 \end{bmatrix} \begin{bmatrix} iu_1 \\[2ex] u_2 \\[2ex] i\sigma_{12} \\[2ex] \sigma_{22} \end{bmatrix} + \begin{bmatrix} 0 \\[2ex] 0 \\[2ex] -if_1 \\[2ex] -f_2 \end{bmatrix}.$$ (5.99)

The propagator matrix is given by the matrix exponential of A defined in equation (5.99). Because the result is rather long, we will give the propagator only for $\lambda = \mu$:

$$\mathbf{P}(z, z_0) =$$

$$\begin{bmatrix} \mathcal{C} + \frac{2}{3}k(z-z_0)\mathcal{S} & \frac{2}{3}k(z-z_0)\mathcal{C} + \frac{1}{3}\mathcal{S} & \frac{1}{3k\mu}[k(z-z_0)\mathcal{C} + 2\mathcal{S}] & \frac{1}{3\mu}(z-z_0)\mathcal{S} \\[2mm] -\frac{2}{3}k(z-z_0)\mathcal{C} + \frac{1}{3}\mathcal{S} & \mathcal{C} - \frac{2}{3}k(z-z_0)\mathcal{S} & -\frac{1}{3\mu}(z-z_0)\mathcal{S} & -\frac{1}{3k\mu}[k(z-z_0)\mathcal{C} - 2\mathcal{S}] \\[2mm] \frac{4}{3}k\mu[k(z-z_0)\mathcal{C} + \mathcal{S}] & \frac{4}{3}k^2(z-z_0)\mu\mathcal{S} & \mathcal{C} + \frac{2}{3}k(z-z_0)\mathcal{S} & \frac{1}{3}[2k(z-z_0)\mathcal{C} - \mathcal{S}] \\[2mm] -\frac{4}{3}k^2(z-z_0)\mu\mathcal{S} & -\frac{4}{3}k\mu[k(z-z_0)\mathcal{C} - \mathcal{S}] & -\frac{1}{3}[2k(z-z_0)\mathcal{C} + \mathcal{S}] & \mathcal{C} - \frac{2}{3}k(z-z_0)\mathcal{S} \end{bmatrix},$$

$$(5.100)$$

where

$$\mathcal{C} = \cosh[k(z-z_0)],$$
$$\mathcal{S} = \sinh[k(z-z_0)]. \tag{5.101}$$

For a point source with moment tensor M_{ij}, the equivalent body forces are, from equation (3.19),

$$f_i = -M_{ij}\frac{\partial\delta(\mathbf{x}-\mathbf{x}_s)}{\partial x_j}, \tag{5.102}$$

where $\mathbf{x}_s$ is the source coordinate. The delta function can be expanded as the product of delta functions in x and z, $\delta(\mathbf{x}-\mathbf{x}_s) = \delta(x-x_s)\delta(z-z_s)$. Thus,

$$f_1 = -M_{11}\delta(z-z_s)\frac{\partial\delta(x-x_s)}{\partial x} - M_{12}\delta(x-x_s)\frac{\partial\delta(z-z_s)}{\partial z},$$

$$f_2 = -M_{21}\delta(z-z_s)\frac{\partial\delta(x-x_s)}{\partial x} - M_{22}\delta(x-x_s)\frac{\partial\delta(z-z_s)}{\partial z}. \tag{5.103}$$

Taking the Fourier transform with respect to x eliminates the horizontal derivatives:

$$f_1 = -ikM_{11}\delta(z-z_s)e^{-ikx_s} - M_{12}e^{-ikx_s}\frac{\partial\delta(z-z_s)}{\partial z},$$

$$f_2 = -ikM_{21}\delta(z-z_s)e^{-ikx_s} - M_{22}e^{-ikx_s}\frac{\partial\delta(z-z_s)}{\partial z}. \tag{5.104}$$

The general solution is of the form (5.54). To evaluate the integral of the source term, introduce the following vectors:

$$F_1 = [0\ 0\ -if_1\ 0]^T,\ F_2 = [0\ 0\ 0\ -f_2]^T. \tag{5.105}$$

Thus, the general solution (5.54) for the plane strain problem can be written as

$$\mathbf{v}(z) = \mathbf{P}(z, z_0)\mathbf{v}(z_0) + \int_{z_0}^{z} \mathbf{P}(z, \zeta) \left\{ F_1(\zeta) + F_2(\zeta) \right\} d\zeta,$$

$$= \mathbf{P}(z, z_0)\mathbf{v}(z_0) + \int_{z_0}^{z} \mathbf{P}(z, \zeta)F_1(\zeta)d\zeta + \int_{z_0}^{z} \mathbf{P}(z, \zeta)F_2(\zeta)d\zeta. \tag{5.106}$$

Introducing equation (5.104), the first integral becomes

$$\int_{z_0}^{z} \mathbf{P}(z, \zeta)F_1(\zeta)d\zeta = i \left[\mathbf{P}(z, z_s)ikM_{11}e^{-ikx_s} + M_{12}e^{-ikx_s} \int_{z_0}^{z} \mathbf{P}(z, \zeta)\frac{\partial \delta(\zeta - z_s)}{\partial \zeta}d\zeta \right] E_3,$$

$$= i \left[\mathbf{P}(z, z_s)ikM_{11}e^{-ikx_s} - M_{12}e^{-ikx_s} \int_{z_0}^{z} \delta(\zeta - z_s)\frac{\partial \mathbf{P}(z, \zeta)}{\partial \zeta}d\zeta \right] E_3,$$

$$= i \left[\mathbf{P}(z, z_s)ikM_{11}e^{-ikx_s} + M_{12}e^{-ikx_s} \int_{z_0}^{z} \delta(\zeta - z_s)A\mathbf{P}(z, \zeta)d\zeta \right] E_3,$$

$$= \mathbf{P}(z, z_s) \left[-kM_{11}E_3 + iM_{12}AE_3 \right] e^{-ikx_s}, \tag{5.107}$$

where we have introduced the unit basis vectors:

$$E_3 = [0\ 0\ 1\ 0]^T, \; E_4 = [0\ 0\ 0\ 1]^T. \tag{5.108}$$

The first step in equation (5.107) is to integrate by parts. The second makes use of (5.47)—specifically,

$$\frac{\partial \mathbf{P}(z, \zeta)}{\partial \zeta} = -Ae^{A(z-\zeta)}. \tag{5.109}$$

The last step also makes use of the fact the $A\mathbf{P}(z, z_0) = \mathbf{P}(z, z_0)A$, which follows from equation (5.53). In a similar fashion, the second integral reduces to

$$\int_{z_0}^{z} \mathbf{P}(z, \zeta)F_2(\zeta)d\zeta = \mathbf{P}(z, z_s) \left[ikM_{12}E_4 + M_{22}AE_4 \right] e^{-ix_s k}. \tag{5.110}$$

The solution is thus of the form

$$\mathbf{v}(z = 0) = \mathbf{P}(0, z_0)\mathbf{v}(z_0) - \mathbf{P}(0, z_s) \begin{bmatrix} iM_{12}/\mu \\ M_{22}/(\lambda + 2\mu) \\ -kM_{11} + k\lambda M_{22}/(\lambda + 2\mu) \\ 0 \end{bmatrix} e^{-ix_s k}. \tag{5.111}$$

In this expression, the term $\mathbf{P}(0, z_0)$ may actually consist of several matrices multiplied together that propagate the solution at z_0 to the free surface. Similarly, the term $\mathbf{P}(0, z_s)$ may consist of a number of matrices that propagate the solution from the source depth z_s to the free surface. Note again that the negative sign in front of the $\mathbf{P}(0, z_s)$ term arises because we are propagating the solution from depth to the surface in the negative z direction.

The final step is to derive solutions to the homogeneous equations. It turns out that in plane strain, the matrix A has only two independent eigenvalues: $\lambda_1 = k$ and $\lambda_2 = -k$. The eigenvectors thus provide only two linearly independent solutions to the homogeneous equations. In cases where the eigenvalues are degenerate, we find four linearly independent solutions from the Jordan decomposition of A (e.g., Strang 1976). The Jordan decomposition is also of the form $A = U J U^{-1}$, where the matrix U contains the eigenvectors and the generalized eigenvectors; however, in this case J is not diagonal. The solution to the homogeneous equations parallels equation (5.45) and is given by

$$\mathbf{v}(z) = U e^{Jz} \mathbf{c} \tag{5.112}$$

(e.g., Strang 1976), where $\mathbf{c}$ are coefficients determined by the boundary conditions. For the matrix A defined by equation (5.99), the Jordan decomposition is given by

$$J = \begin{bmatrix} -k & 1 & 0 & 0 \\ 0 & -k & 0 & 0 \\ 0 & 0 & k & 1 \\ 0 & 0 & 0 & k \end{bmatrix}, \tag{5.113}$$

and

$$U = \begin{bmatrix} -\dfrac{1}{2\,k\mu} & \dfrac{\lambda + 2\mu}{2\,k^2\mu\,(\lambda + \mu)} & -\dfrac{1}{2\,k\mu} & \dfrac{-\lambda - 2\mu}{2\,k^2\mu\,(\lambda + \mu)} \\[2ex] -\dfrac{1}{2\,k\mu} & \dfrac{-1}{2\,k^2\,(\lambda + \mu)} & \dfrac{1}{2\,k\mu} & \dfrac{-1}{2\,k^2\,(\lambda + \mu)} \\[2ex] 1 & -\dfrac{1}{k} & -1 & -\dfrac{1}{k} \\[2ex] 1 & 0 & 1 & 0 \end{bmatrix}. \tag{5.114}$$

Thus, the solution (5.112) can be expressed as

$$\mathbf{v}(z) = (c_1 \Psi_1 + c_2 \Psi_2) e^{-kz} + (c_3 \Psi_3 + c_4 \Psi_4) e^{kz}. \tag{5.115}$$

Here, Ψ_1 and Ψ_3 are given by the first and third columns of U in equation (5.114), and Ψ_2 and Ψ_4 are given by

$$
\Psi_2 = \begin{bmatrix} \dfrac{\lambda - kz\,\lambda + (2-kz)\,\mu}{2\,k^2\mu\,(\lambda+\mu)} \\[2ex] \dfrac{-\,[\mu + kz\,(\lambda+\mu)]}{2\,k^2\mu\,(\lambda+\mu)} \\[2ex] \dfrac{-1+kz}{k} \\[2ex] z \end{bmatrix}
\qquad
\Psi_4 = \begin{bmatrix} -\,\dfrac{\lambda + kz\,\lambda + (2+kz)\,\mu}{2k^2\mu\,(\lambda+\mu)} \\[2ex] \dfrac{-\mu + kz\,(\lambda+\mu)}{2k^2\mu(\lambda+\mu)} \\[2ex] -\,\dfrac{1+kz}{k} \\[2ex] z \end{bmatrix}.
\qquad (5.116)
$$

The complete solution for the stress displacement vector is thus, from equations (5.111) and (5.115), of the form

$$
\mathbf{v}(z) = \mathbf{P}(z, z_0)\left[(c_1\Psi_1 + c_2\Psi_2)e^{-kz_0} + (c_3\Psi_3 + c_4\Psi_4)e^{kz_0}\right] + \mathbf{P}(z, z_s)\mathbf{f_s}, \qquad (5.117)
$$

where the eigenvectors are evaluated at $z = z_0$, and the source term is

$$
\mathbf{f_s} = \begin{bmatrix} i\,M_{12}/\mu \\[1.5ex] M_{22}/(\lambda + 2\mu) \\[1.5ex] -kM_{11} + k\lambda M_{22}/(\lambda + 2\mu) \\[1.5ex] 0 \end{bmatrix} e^{-i x_s k}. \qquad (5.118)
$$

It is worth noting that the eigenvectors determined by the Jordan decomposition are not unique solutions to the homogeneous equations. A change of coordinates of the form $\mathbf{v} = M\mathbf{v}'$ leads to a first-order system of equations, with the matrix A replaced by $M^{-1}AM$. This is known as a *similarity transformation*. While the eigenvalues of $M^{-1}AM$ are the same as the eigenvalues of A, the eigenvectors are not (see Strang 1976). Thus, while the solution can always be expressed by equations (5.117), the particular choice of vectors Ψ_i is not unique. As long as the Ψ_i solve the homogeneous system (equations [5.99] without the body force terms) and are linearly independent, they form a basis of the solution space. Different authors may choose a different basis resulting in different coefficients c_i but an equivalent solution for the displacements.

The constants c_i are determined by the boundary conditions. The solution must decay with depth as $z \to \infty$, which means that the solution corresponding to the positive eigenvalue must vanish for positive wavenumbers, and vice versa. Thus, $c_1 = c_2 = 0$ for negative wavenumbers, and $c_3 = c_4 = 0$ for positive wavenumbers. The solution can be written more compactly, in a form valid for all wavenumbers k,

$$\mathbf{v}(z) = \mathbf{P}(z, z_0) \left\{ c_1 \begin{bmatrix} -\dfrac{1}{2k\mu} \\[2mm] -\dfrac{1}{2|k|\mu} \\[2mm] \mathrm{sgn}(k) \\[2mm] 1 \end{bmatrix} e^{-|k|z_0} + c_2 \begin{bmatrix} \mathrm{sgn}(k)\dfrac{\lambda(1 - |k|z) + \mu(2 - |k|z)}{2\,k^2\,\mu\,(\lambda + \mu)} \\[2mm] -\dfrac{\mu + |k|\,z\,(\lambda + \mu)}{2\,k^2\,\mu\,(\lambda + \mu)} \\[2mm] \dfrac{-1 + |k|z}{k} \\[2mm] z \end{bmatrix} e^{-|k|z_0} \right\}$$

$$+ \mathbf{P}(z, z_s) \begin{bmatrix} i\,M_{12}/\mu \\[2mm] M_{22}/(\lambda + 2\mu) \\[2mm] -kM_{11} + k\lambda M_{22}/(\lambda + 2\mu) \\[2mm] 0 \end{bmatrix} e^{-ix_s k}, \tag{5.119}$$

where the constants c_1 and c_2 are of course not the same as those in equations (5.117). Notice that for any wavenumber k, the $\exp(-|k|z_0)$ terms are constant and can be absorbed into the c_i. Equation (5.119) can be written more compactly as

$$\begin{bmatrix} iu_1(z = 0) \\ u_2(z = 0) \\ 0 \\ 0 \end{bmatrix} = \mathbf{P}(0, z_0) V \mathbf{c} - \mathbf{P}(0, z_s) \mathbf{f_s}, \tag{5.120}$$

where V is a matrix containing the two generalized eigenvectors in equation (5.119), $\mathbf{c}$ is a two-vector containing the coefficients c_1 and c_2, and $\mathbf{f_s}$ is the source term defined in equation (5.118). Equation (5.120) is symbolic in the sense that both $\mathbf{P}(0, z_0)$ and $\mathbf{P}(0, z_s)$ may actually be the product of a series of propagator matrices. Recall that z_0 is the source depth or the depth to the top of the half-space, whichever is deeper. The constants $\mathbf{c}$ are determined from the boundary condition that the shear and normal traction vanish at the free surface. The shear and normal tractions are contained in the third and fourth elements of the stress-displacement vector. Solving the lower two equations for the coefficients $\mathbf{c}$ yields $\mathbf{c} = [\mathbf{P}_2(0, z_0)V]^{-1}\mathbf{P}_2(0, z_s)\mathbf{f_s}$, where the subscript $_2$ refers to the lower two rows of the matrix. Substituting into the first two equations gives the displacements at the free surface:

$$\begin{bmatrix} iu_1(z = 0) \\ u_2(z = 0) \end{bmatrix} = \mathbf{P}_1(0, z_0) V[\mathbf{P}_2(0, z_0)V]^{-1}\mathbf{P}_2(0, z_s)\mathbf{f_s} - \mathbf{P}_1(0, z_s)\mathbf{f_s}, \tag{5.121}$$

where the subscript $_1$ indicates the upper two rows of the matrix. With multiple layers, one simply premultiplies by the appropriate propagator matrices, as in equation (5.85).

Are we done? Essentially. We now have the solution for the surface displacements due to a point source anywhere in a layered medium. If we followed the prescription in the previous section on antiplane deformation, we would finish by integrating the point-source solutions over the fault. The details become unwieldy but can be accomplished with symbolic integrators. Alternatively, one can numerically integrate the point sources using, for example,

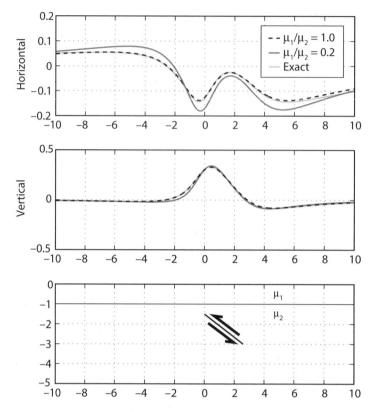

Figure 5.15. Displacements due to a dipping thrust fault beneath a layer with different shear modulus. The layer has shear modulus μ_1, while the half-space has shear modulus μ_2. Note that when the two moduli are equal, the propagator matrix solution agrees well with the analytical solution from chapter 3. A compliant surface layer alters the horizontal displacements more than the vertical displacements. $\nu = 0.25$ in both the layer and the half-space.

Gaussian quadrature (see section 3.6.6). A final point on numerical implementations is warranted: the elements in the propagator matrices can become extremely large for large wavenumber k. The matrix subtraction in equation (5.121) thus can lead to significant loss of precision and even numerical instability. Gilbert and Backus (1966) present matrix minor methods that address this problem. Alternatively, appropriate filtering of large $|k|$ components in the wavenumber domain prior to application of the inverse FFT can achieve good results.

Figure 5.15 shows the displacements due to a dipping thrust fault beneath a more compliant layer. Note that for equal moduli, the propagator solution agrees well with the analytical solution from chapter 3. However, when the near-surface layer is more compliant, the deformation is more localized near the fault, as it was for strike-slip faulting. The horizontal displacements are more affected than the vertical displacements; this appears to hold true for a range of fault geometries (see Savage 1998).

5.6 Propagator Solutions in Three Dimensions

In order to compare theoretical predictions to data, we generally require solutions for finite faults in three dimensions. To this point, we have limited discussion to antiplane and plane strain solutions. Interestingly, with modest additional work, we can construct three-dimensional solutions for dislocations in layered earth models from the two-dimensional solutions already obtained.

Early results in this area were obtained by expanding the displacement field in *cylindrical vector harmonics*, following the approach of Ben-Meneham and Singh (1968) and Singh (1970) and expanded on by Jovanovich et al. (1974). Similar results were presented by Sato (1971) and Sato and Matsu'ura (1973). A recent review and discussion of both static and dynamic sources is given by Zhu and Rivera (2002). An alternative derivation, following Barker (1976), is presented here.

In three dimensions, we first Fourier transform in both horizontal coordinates:

$$\bar{f} = \int_{-\infty}^{+\infty} \int_{-\infty}^{+\infty} f(x_1, x_2) e^{-i(k_1 x_1 + k_2 x_2)} \mathrm{d}x_1 \mathrm{d}x_2, \tag{5.122}$$

where k_1 and k_2 are horizontal wavenumbers. Transforming Hooke's law written in terms of displacement gradients and separating the vertical derivatives yields

$$\frac{\partial \bar{u}_1}{\partial z} = \frac{\bar{\sigma}_{13}}{\mu} - ik_1 \bar{u}_3,$$

$$\frac{\partial \bar{u}_2}{\partial z} = \frac{\bar{\sigma}_{23}}{\mu} - ik_2 \bar{u}_3,$$

$$\frac{\partial \bar{u}_3}{\partial z} = \frac{\bar{\sigma}_{33}}{\lambda + 2\mu} - \frac{i\lambda}{\lambda + 2\mu}(k_1 \bar{u}_1 + k_2 \bar{u}_2), \tag{5.123}$$

where the overbar indicates Fourier transform. Next, transform the equilibrium equations

$$-[\mu k^2 + (\lambda + \mu)k_1^2]\bar{u}_1 - (\lambda + \mu)k_1 k_2 \bar{u}_2 + \frac{\partial \bar{\sigma}_{13}}{\partial z} + i\lambda k_1 \frac{\partial \bar{u}_3}{\partial z} + \bar{f}_1 = 0,$$

$$-[\mu k^2 + (\lambda + \mu)k_2^2]\bar{u}_2 - (\lambda + \mu)k_1 k_2 \bar{u}_1 + \frac{\partial \bar{\sigma}_{23}}{\partial z} + i\lambda k_2 \frac{\partial \bar{u}_3}{\partial z} + \bar{f}_2 = 0,$$

$$-\mu k^2 \bar{u}_3 + i\mu \left(k_1 \frac{\partial \bar{u}_1}{\partial z} + k_2 \frac{\partial \bar{u}_2}{\partial z} \right) + \frac{\partial \bar{\sigma}_{33}}{\partial z} + \bar{f}_3 = 0, \tag{5.124}$$

where $k^2 = k_1^2 + k_2^2$. Isolating the vertical derivatives in equations (5.124) and combining with equation (5.123) allows us to write the system of equations in matrix form in terms of the displacement stress vector $[\bar{u}_1, \bar{u}_2, \bar{u}_3, \bar{\sigma}_{13}, \bar{\sigma}_{23}, \bar{\sigma}_{33}]$. This yields a coupled 6×6 system of first-order differential equations. Rather than take this approach, we simplify this system considerably by transforming from the x_1, x_2 system into radial and transverse coordinates. To do so, we follow Barker (1976) and introduce the functions $\mathbf{U}^{(2)}$ and $\mathbf{U}^{(4)}$ defined by the following relations:

$$-kU_1^{(2)} = ik_2 \bar{u}_1 - ik_1 \bar{u}_2,$$

$$-kU_2^{(2)} = ik_2 \bar{\sigma}_{13} - ik_1 \bar{\sigma}_{23}, \tag{5.125}$$

and

$$
\begin{aligned}
-kU_1^{(4)} &= ik_1\bar{u}_1 + ik_2\bar{u}_2, \\
U_2^{(4)} &= \bar{u}_3, \\
-kU_3^{(4)} &= ik_1\bar{\sigma}_{13} + ik_2\bar{\sigma}_{23}, \\
U_4^{(4)} &= \bar{\sigma}_{33}.
\end{aligned}
\tag{5.126}
$$

Introducing the expressions (5.125) and (5.126) into the equilibrium equations (5.124) and Hooke's law (5.123), we find that the 6×6 system decouples into a 2×2 system and a 4×4 system:

$$
\frac{d}{dz}
\begin{bmatrix} U_1^{(2)} \\ U_2^{(2)} \end{bmatrix}
=
\begin{bmatrix} 0 & 1/\mu \\ \mu k^2 & 0 \end{bmatrix}
\begin{bmatrix} U_1^{(2)} \\ U_2^{(2)} \end{bmatrix}
+ \bar{F}^{(2)},
\tag{5.127}
$$

$$
\frac{d}{dz}
\begin{bmatrix} U_1^{(4)} \\ U_2^{(4)} \\ U_3^{(4)} \\ U_4^{(4)} \end{bmatrix}
=
\begin{bmatrix}
0 & -k & 1/\mu & 0 \\
\dfrac{k\lambda}{\lambda + 2\mu} & 0 & 0 & \dfrac{1}{\lambda + 2\mu} \\
\dfrac{4\mu k^2 (\lambda + \mu)}{\lambda + 2\mu} & 0 & 0 & -\dfrac{k\lambda}{\lambda + 2\mu} \\
0 & 0 & k & 0
\end{bmatrix}
\begin{bmatrix} U_1^{(4)} \\ U_2^{(4)} \\ U_3^{(4)} \\ U_4^{(4)} \end{bmatrix}
+ \bar{F}^{(4)},
\tag{5.128}
$$

where $\bar{F}^{(2)}$, $\bar{F}^{(4)}$ are vectors of body forces. What is remarkable is that the 2×2 and 4×4 systems are exactly the antiplane (compare to equation [5.41]) and plane strain (compare to equation [5.99]) problems we have previously analyzed! In seismological parlance, these are the SH and P-SV systems, respectively. The superscript $^{(2)}$ denotes the 2×2 system, while the superscript $^{(4)}$ refers to the 4×4 system. In retrospect, this important result should not be surprising. The Fourier transform represents the displacement and stress fields as a superposition of plane waves; for each wavenumber $\mathbf{k}$, equations (5.125) and (5.126) rotate the horizontal displacements and stresses into wavenumber parallel and perpendicular coordinates. For example, $U_1^{(4)}$ is (to within a constant) the projection of the horizontal displacements onto the wavenumber vector, whereas $U_1^{(2)}$ is the orthogonal component of the horizontal displacements. Thus, the $U_j^{(4)}$ contain the inplane horizontal and vertical displacements and tractions, whereas the $U_j^{(2)}$ contain the antiplane displacements and tractions.

The derivation of equations (5.127) and (5.128) proceeds as follows. The first equation in (5.127) is found by combining the second equation in (5.125) with the first two equations (5.123). The second equation in (5.127) is found by multiplying the first equilibrium equation (5.124) by $-ik_2$, the second by ik_1 and adding, and so on.

The dislocation source is represented by body force double couples in $\bar{F}^{(2)}$ and $\bar{F}^{(4)}$. We do not anticipate that the body force representations will be the same as they were in the two-dimensional cases—after all, in two dimensions, they were line sources, and in three dimensions, they are point sources. Rather,

$$
\bar{F}^{(2)} =
\begin{bmatrix}
0 \\
ik^{-1}(k_2 \bar{f}_1 - k_1 \bar{f}_2)
\end{bmatrix},
\tag{5.129}
$$

and

$$
\bar{F}^{(4)} = \begin{bmatrix} 0 \\ 0 \\ ik^{-1}(k_1 \bar{f}_1 + k_2 \bar{f}_2) \\ -\bar{f}_3 \end{bmatrix}. \tag{5.130}
$$

Notice that the 2×2 and 4×4 systems are coupled through the body force terms.

Following the procedure outlined for the plane strain case in equations (5.102) to (5.104), we find that the appropriate body force terms are

$$
\begin{bmatrix} \bar{f}_1 \\ \bar{f}_2 \\ \bar{f}_3 \end{bmatrix} = - \begin{bmatrix} ik_1 M_{11}\delta(z-z_s) + ik_2 M_{12}\delta(z-z_s) + M_{13}\dfrac{\partial \delta(z-z_s)}{\partial z} \\ ik_1 M_{21}\delta(z-z_s) + ik_2 M_{22}\delta(z-z_s) + M_{23}\dfrac{\partial \delta(z-z_s)}{\partial z} \\ ik_1 M_{31}\delta(z-z_s) + ik_2 M_{32}\delta(z-z_s) + M_{33}\dfrac{\partial \delta(z-z_s)}{\partial z} \end{bmatrix} e^{-i(k_1 x_1^{(s)} + k_2 x_2^{(s)})}, \tag{5.131}
$$

where in a rather mixed notation, $[x_1^{(s)}, x_2^{(s)}, z_s]$ represents the coordinates of the point source. To write the solution vectors in the form $\mathbf{v} = \mathbf{P}(z, z_0)V\mathbf{c} + \mathbf{P}(z, z_s)\mathbf{f}_s$, we need to carry out the integrations of the point sources following the procedure in (5.107). Combining equations (5.129) and (5.130) with (5.131) and carrying out the integration yields

$$
\mathbf{U}^{(j)} = \mathbf{P}(z, z_0)^{(j)} V^{(j)}\mathbf{c} + \mathbf{P}^{(j)}(z, z_s)\mathbf{g}^{(j)} \qquad j = 2, 4, \tag{5.132}
$$

and

$$
\begin{aligned}
\mathbf{g}^{(2)} &= \left\{ [k_1 k_2 (M_{11} - M_{22}) + k_2^2 M_{12} - k_1^2 M_{21}]E_2 - i(k_2 M_{13} - k_1 M_{23})A^{(2)}E_2 \right\} k^{-1} e^{-i(k_1 x_1^{(s)} + k_2 x_2^{(s)})}, \\
\mathbf{g}^{(4)} &= \left\{ [k_1^2 M_{11} + k_2^2 M_{22} + k_1 k_2 (M_{12} + M_{21})]\, E_3 k^{-1} - i(k_1 M_{13} + k_2 M_{23})A^{(4)}E_3 k^{-1} \right. \\
&\quad \left. + \left[ik_1 M_{31} + ik_2 M_{32} + M_{33}A^{(4)} \right] E_4 \right\} e^{-i(k_1 x_1^{(s)} + k_2 x_2^{(s)})},
\end{aligned} \tag{5.133}
$$

where $E_2 = [0, 1]^T$, $E_3 = [0, 0, 1, 0]^T$, and $E_4 = [0, 0, 0, 1]^T$ are basis vectors, and $A^{(2)}$ and $A^{(4)}$ refer to the 2×2 and 4×4 systems, respectively. At this point, the system is fully described. Note that the boundary conditions that shear and normal traction vanish at the free surface require that $U_2^{(2)} = U_3^{(4)} = U_4^{(4)} = 0$ on $z = 0$, so the machinary set up for the two-dimensional solutions can be employed to solve for the $\mathbf{U}^{(j)}$. Given the $\mathbf{U}^{(j)}$, the Fourier transformed displacements are found by rotating back to x_1, x_2 coordinates, by inverting the transformations (5.125) and (5.126). Displacements in physical space are finally obtained by numerically inverting the Fourier transforms.

Because the solutions are invariant to translations in the horizontal coordinates, it is straightforward to integrate the sources over a finite length in a horizontal direction prior to inverse Fourier transforming (see problem 11). Integration in the vertical direction is more involved but can be accomplished with numerical quadrature.

Figure 5.16 shows the surface displacements due to a strike-slip point source computed using propagator matrix methods described in this section. Agreement (not shown) with analytic solutions given in chapter 3, equations (3.98), is excellent for homogeneous half-spaces.

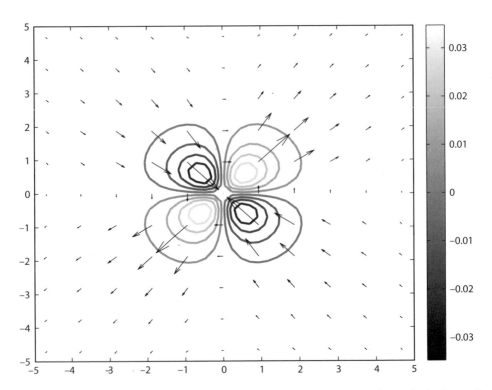

Figure 5.16. Surface displacements due to a point strike-slip source in a layered elastic medium computed with propagator matrix methods. Horizontal distance scales are normalized by source depth. Vertical displacements are indicated by contour intervals, with dark indicating subsidence.

5.7 Approximate Solutions for Arbitrary Variations in Properties

The previously described method of images and propagator matrix methods work well for horizontally stratified media. You have also seen that image methods can be used to construct solutions for slip along a material interface. More general solutions can become quite tedious. However, it is possible to construct approximate solutions using a perturbation expansion approach. The general formulation of the moduli perturbation approach is given in Du et al. (1994; note corrections in Cervelli et al. 1999). Three-dimensional examples are given in Du et al. (1997).

We begin with the generalized form of Hooke's law

$$\sigma_{ij} = C_{ijkl}\epsilon_{kl} = C_{ijkl}u_{k,l},\tag{5.134}$$

and the equilibrium equations

$$\sigma_{ij,j} + f_i = 0.\tag{5.135}$$

The notation $_{,j} \equiv \partial/\partial x_j$ is used liberally in this section. To these governing equations, we add boundary conditions, either traction or displacement:

$$\sigma_{ij}n_j = T_i \quad \text{on } S_T,\tag{5.136}$$

$$u_i = \Omega_i \quad \text{on } S_\Omega,\tag{5.137}$$

where S_T is the part of the boundary (possibly all or none) where traction boundary conditions apply, and S_Ω is the part of the boundary (possibly all or none) where displacement boundary conditions apply.

From equations (5.134) and (5.135), the equilibrium conditions can be written as

$$C_{ijkl}u_{k,lj} + u_{k,l}C_{ijkl,j} + f_i = 0. \tag{5.138}$$

Note that because the moduli C_{ijkl} are general functions of position, the spatial derivatives do not vanish. The concept of the *perturbation expansion* is that we assume that the $C_{ijkl}(\mathbf{x})$ can be written as the sum of a spatially uniform part, $C_{ijkl}^{(0)}$, and a small spatially varying perturbation, $\delta C_{ijkl}(\mathbf{x})$:

$$C_{ijkl}(\mathbf{x}) = C_{ijkl}^{(0)} + \varepsilon\delta C_{ijkl}(\mathbf{x}), \tag{5.139}$$

where $\varepsilon \ll 1$. Similarly, the displacements are expanded in a power series in ε:

$$u_i(\mathbf{x}) = u_i^{(0)}(\mathbf{x}) + \varepsilon u_i^{(1)}(\mathbf{x}) + \varepsilon^2 u_i^{(2)}(\mathbf{x}) + \dots . \tag{5.140}$$

Substituting equations (5.139) and (5.140), into (5.138), we obtain the equilibrium equations:

$$\left(C_{ijkl}^{(0)} + \varepsilon\delta C_{ijkl}\right)\left(u_{k,lj}^{(0)} + \varepsilon u_{k,lj}^{(1)} + \dots\right) + \left(u_{k,l}^{(0)} + \varepsilon u_{k,l}^{(1)} + \dots\right)\varepsilon\delta C_{ijkl,j} + f_i = 0. \tag{5.141}$$

Similarly, the stresses are written as

$$\sigma_{ij} = \left(C_{ijkl}^{(0)} + \varepsilon\delta C_{ijkl}\right)\left(u_{k,l}^{(0)} + \varepsilon u_{k,l}^{(1)} + \dots\right). \tag{5.142}$$

Notice that equation (5.141) must be satisfied for all values of $\varepsilon \ll 1$. Thus, equation (5.141) must hold for all powers of ε *independently*. Collecting all of the zero-order terms—that is, terms independent of ε—leads to the following system:

$$C_{ijkl}^{(0)}u_{k,lj}^{(0)} + f_i = 0, \tag{5.143}$$

with boundary conditions

$$C_{ijkl}^{(0)}u_{k,l}^{(0)}n_j = T_i \text{ on } S_T, \tag{5.144}$$

$$u_i^{(0)} = \Omega_i \text{ on } S_\Omega. \tag{5.145}$$

Note that this is precisely the statement of the dislocation problem in a homogenous space. The zero-order approximation is to ignore material heterogeneity completely. Next, collect all terms of order ε, which leads to the following system:

$$C_{ijkl}^{(0)}u_{k,lj}^{(1)} + \delta C_{ijkl}u_{k,lj}^{(0)} + u_{k,l}^{(0)}\delta C_{ijkl,j} = 0,$$
$$= C_{ijkl}^{(0)}u_{k,lj}^{(1)} + \underbrace{(u_{k,l}^{(0)}\delta C_{ijkl})_{,j}}_{\hat{f}_i} = 0, \tag{5.146}$$

where the second equation follows from an application of the chain rule. The first-order system has boundary conditions:

$$C_{ijkl}^{(0)}u_{k,l}^{(1)}n_j + \underbrace{\delta C_{ijkl}u_{k,l}^{(0)}n_j}_{-\hat{T}_i} = 0 \quad \text{on } S_T, \tag{5.147}$$

$$u_i^{(1)} = 0 \quad \text{on } S_\Omega. \tag{5.148}$$

There are several important points to notice about these equations. First, consider the term $(u_{k,l}^{(0)}\delta C_{ijkl})_{,j}$ in equation (5.146). Notice that all the terms in the parentheses are known: the modulus perturbation δC_{ijkl} is known by definition of the problem, and $u_k^{(0)}$ is the solution to the zero-order problem. Second, note that this term is equivalent to an effective body force in the equilibrium equation for $u_k^{(1)}$ (compare to equation [5.143]). So the solution for $u_k^{(1)}$ is equivalent to a solution in a *homogeneous* body with a specified body force distribution:

$$\hat{f}_i = (u_{k,l}^{(0)}\delta C_{ijkl})_{,j}. \tag{5.149}$$

The next thing to notice is that the displacement boundary condition—for example, on the dislocation surface—is solved fully in the zero-order problem. The traction boundary conditions (5.147) also contain a term $\delta C_{ijkl}u_{k,l}^{(0)}n_j$ that is known after solution of the zero-order problem. This term acts like a specified traction

$$\hat{T}_i = -u_{k,l}^{(0)}\delta C_{ijkl}n_j, \tag{5.150}$$

acting on S_T. In sum, the first-order correction is composed of specified tractions and body forces in a homogeneous body with strength depending on the zero-order solution.

If we follow the same procedure to order n, we obtain the system

$$C_{ijkl}^{(0)}u_{k,lj}^{(n)} + \left(\delta C_{ijkl}u_{k,l}^{(n-1)}\right)_{,j} = 0, \tag{5.151}$$

$$C_{ijkl}^{(0)}u_{k,l}^{(n)}n_j + \left(\delta C_{ijkl}u_{k,l}^{(n-1)}\right)n_j = 0 \quad \text{on } S_T, \tag{5.152}$$

$$u_i = 0 \quad \text{on } S_\Omega. \tag{5.153}$$

The problem can thus be solved iteratively. The zero-order solution is a dislocation in a homogenous half-space. From this, we obtain $u_i^{(0)}$ and $\delta C_{ijkl}(\mathbf{x})$, which allows us to compute equivalent body force and traction distributions to generate a boundary value problem to be solved for the first-order correction, $u_i^{(1)}$. In principle, we could then use this solution to pose a boundary value problem for a second correction, and so on. In practice, the algebra often becomes too complex to obtain a second-order correction. Fortunately, as you shall see, the first-order correction is often reasonably accurate.

The only thing needed then to obtain a first-order correction for the effects of heterogeneity is the elastostatic Green's function $g_m^i(\mathbf{x}, \boldsymbol{\xi})$ for a *homogenous* half-space (see chapter 3). Recall that the Green's function $g_m^i(\mathbf{x}, \boldsymbol{\xi})$ gives the displacement in the m direction at $\mathbf{x}$ due to a point force in the ith direction at $\boldsymbol{\xi}$. Given this, we can write the displacements due to distributed forces $\hat{f}_i(\boldsymbol{\xi})$ as

$$u_m(\mathbf{x}) = \int_V \hat{f}_i(\boldsymbol{\xi})g_m^i(\mathbf{x}, \boldsymbol{\xi})dv(\boldsymbol{\xi}). \tag{5.154}$$

Substituting equations (5.149) and (5.150) into (5.154) yields

$$u_m^{(1)}(\mathbf{x}) = \int_V \left(\delta C_{ijkl}(\boldsymbol{\xi}) u_{k,l}^{(0)}(\boldsymbol{\xi}) \right)_{,j} g_m^i(\mathbf{x}, \boldsymbol{\xi}) \mathrm{d}v(\boldsymbol{\xi}) - \int_S \delta C_{ijkl}(\boldsymbol{\xi}) u_{k,l}^{(0)}(\boldsymbol{\xi}) n_j g_m^i(\mathbf{x}, \boldsymbol{\xi}) \mathrm{d}s(\boldsymbol{\xi}). \qquad (5.155)$$

These two integrals can be reduced to a single volume integral. First, use the chain rule to write

$$u_m^{(1)}(\mathbf{x}) = \int_V (\delta C_{ijkl} u_{k,l}^{(0)} g_m^i)_{,j} \mathrm{d}v - \int_V \delta C_{ijkl} u_{k,l}^{(0)} g_{m,j}^i (\mathbf{x}, \boldsymbol{\xi}) \mathrm{d}v - \int_S \delta C_{ijkl} u_{k,l}^{(0)} n_j g_m^i \mathrm{d}s(\boldsymbol{\xi}). \qquad (5.156)$$

The first integral can be transformed to a surface integral using the divergence theorem

$$\int_V (\delta C_{ijkl} u_{k,l}^{(0)} g_m^i)_{,j} \mathrm{d}v = \int_S \delta C_{ijkl} u_{k,l}^{(0)} n_j g_m^i \mathrm{d}s(\boldsymbol{\xi}), \qquad (5.157)$$

so that the first and third integrals exactly cancel. We are left with

$$\boxed{u_m^{(1)}(\mathbf{x}) = - \int_V \delta C_{ijkl} u_{k,l}^{(0)} g_{m,j}^i (\mathbf{x}, \boldsymbol{\xi}) \mathrm{d}v.} \qquad (5.158)$$

The same procedure applies at order n, so we may write

$$u_m^{(n)}(\mathbf{x}) = - \int_V \delta C_{ijkl}(\boldsymbol{\xi}) u_{k,l}^{(n-1)}(\boldsymbol{\xi}) g_{m,j}^i (\mathbf{x}, \boldsymbol{\xi}) \mathrm{d}v(\boldsymbol{\xi}). \qquad (5.159)$$

Notice that the correction is in terms of a volume integral over known quantities: the known modulus perturbation, $\delta C_{ijkl}(\boldsymbol{\xi})$; the solution at the previous iteration, $u_k^{(n-1)}$; and the elastic Green's tensors for the *homogenous* body, $g_m^i(\mathbf{x}, \boldsymbol{\xi})$.

5.7.1 Variations in Shear Modulus

The range of Poisson's ratios for typical rocks is modest. The principal effect of heterogeneous elastic properties is thus likely to be in the shear modulus. For the special case where there are variations only in shear modulus, we may write

$$\delta C_{ijkl} = \frac{\delta \mu}{\mu_0} C_{ijkl}^{(0)}, \qquad (5.160)$$

where $\delta \mu$ is the difference between the shear modulus and that in the reference state, μ_0. In order to guarantee convergence of the perturbation expansion, we choose the reference modulus μ_0 so that $\delta \mu / \mu_0 < 1$. In this case, the general solution for the first-order correction $u_m^{(1)}(\mathbf{x})$, given by equation (5.158) reduces to

$$\begin{aligned} u_m^{(1)}(\mathbf{x}) &= -\frac{C_{ijkl}^{(0)}}{\mu_0} \int_V \delta \mu(\boldsymbol{\xi}) u_{k,l}^{(0)}(\boldsymbol{\xi}) g_{m,j}^i (\mathbf{x}, \boldsymbol{\xi}) \mathrm{d}v(\boldsymbol{\xi}), \\ &= -\frac{1}{\mu_0} \int_V \delta \mu(\boldsymbol{\xi}) \sigma_{ij}^{(0)}(\boldsymbol{\xi}) g_{m,j}^i (\mathbf{x}, \boldsymbol{\xi}) \mathrm{d}v(\boldsymbol{\xi}). \end{aligned} \qquad (5.161)$$

Here we have taken advantage of the fact that $C_{ijkl}^{(0)} u_{k,l}^{(0)} = \sigma_{ij}^{(0)}$ is the stress in the zero-order solution—that is, the stress due to a dislocation in a homogeneous half-space.

For piece-wise constant variations in shear modulus, the volume integral in equation (5.161) can be reduced to a more easily computed surface integral. Begin by noting that

$$[\sigma_{ij}^{(0)} g_m^i(\mathbf{x}, \xi)]_{,j} = \sigma_{ij,j}^{(0)} g_m^i(\mathbf{x}, \xi) + \sigma_{ij}^{(0)} g_{m,j}^i(\mathbf{x}, \xi),$$
$$= \sigma_{ij}^{(0)} g_{m,j}^i(\mathbf{x}, \xi), \tag{5.162}$$

where the first term on the right-hand side is zero, by equilibrium. Thus, if $\delta\mu$ is constant over the domain of integration, equation (5.161) becomes

$$u_m^{(1)}(\mathbf{x}) = -\frac{\delta\mu}{\mu_0} \int_V \sigma_{ij}^{(0)} g_{m,j}^i(\mathbf{x}, \xi) \mathrm{d}v(\xi),$$
$$= -\frac{\delta\mu}{\mu_0} \int_V \left[\sigma_{ij}^{(0)} g_m^i(\mathbf{x}, \xi)\right]_{,j} \mathrm{d}v(\xi),$$

$$\boxed{u_m^{(1)}(\mathbf{x}) = -\frac{\delta\mu}{\mu_0} \int_S \sigma_{ij}^{(0)}(\xi)\, n_j\, g_m^i(\mathbf{x}, \xi) \mathrm{d}s(\xi),} \tag{5.163}$$

where the last step is by application of the divergence theorem.

Notice that the term $\sigma_{ij}^{(0)} n_j$ is simply the traction induced by the zero-order solution and the integral is over the interface between different materials.

5.7.2 Screw Dislocation

We revisit the ubiquitous screw dislocation problem, because it is one for which we have exact solutions, and therefore it provides a good check on the perturbation approach. Note that the stress due to a screw dislocation was given in chapter 2, while the antiplane Green's function was derived in chapter 3. In many cases, the integrations in equation (5.163) will be done numerically. For the screw dislocation, the integrations are tedious but can be done exactly. Du et al. (1994) give the following first-order correction to the free-surface displacements for a screw dislocation at depth d beneath a layer of thickness H:

$$u_3^{(1)}(x_2 = 0) = \frac{s(\gamma - 1)}{2\pi\gamma} \left[\tan^{-1}\left(\frac{x_1}{d}\right) - \tan^{-1}\left(\frac{x_1}{d + 2H},\right)\right], \tag{5.164}$$

where as before $\gamma = \mu_2/\mu_1$. For the case where the dislocation lies in the shallow layer and breaks from the surface to depth d, the first-order correction is given by

$$u_3^{(1)}(x_2 = 0) = \frac{s(\gamma - 1)}{2\pi\gamma} \left[\tan^{-1}\left(\frac{x_1}{d - 2H}\right) + \tan^{-1}\left(\frac{x_1}{d + 2H}\right)\right]. \tag{5.165}$$

Figure 5.17 compares the perturbation expansion with the image solution for an infinitely long screw dislocation in a layer over a half-space. The modulus of the half-space is four times that of the layer, $\mu_2 = 4\mu_1$. The first-order approximate solution is remarkably accurate, especially considering that a factor of four contrast is hardly small. The error in the approximate solution can be quantified by

$$\varepsilon \equiv \frac{u_{\text{series}} - u_{\text{perturb}}}{u_{\text{half-space}}}. \tag{5.166}$$

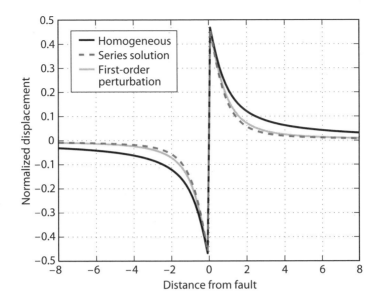

Figure 5.17. Displacements due to infinitely long screw dislocation in a layer over a half-space, comparing first-order perturbation solution to the infinite series solution. Layer thickness is H, and depth of dislocation is $D = 0.8H$. The shear modulus of the half-space is four times that of the layer $\mu_2 = 4\mu_1$. The displacements in a homogeneous half-space are given for reference.

Results shown in figure 5.18 demonstrate that the first-order solution is accurate to better than 15% even for modulus contrasts, $\gamma = \mu_2/\mu_1$, of a factor of four. The second-order correction is accurate to better than a few percent for a similar range of shear moduli.

5.7.3 Edge Dislocation

Figure 5.19 compares several methods, including a first-order perturbation calculation for a vertical fault underlying a compliant surface layer. For a factor of 4 contrast in modulus, the first-order perturbation is reasonably accurate in predicting the vertical displacements, but less so for the horizontal displacements.

While it is difficult to draw general conclusions from these examples, it does appear that the first-order perturbation result is likely to be highly accurate for modest variations in elastic properties and may be reasonably accurate for contrasts in shear modulus of as much as a factor of two. The advantage of this method is that one can treat arbitrarily complex geometries. The method is not expected to yield accurate results with strong contrasts in properties, however. For horizontally stratified variations in properties, propagator matrix methods are superior to perturbation methods.

5.8 Summary and Perspective

In this chapter, we return to dislocation sources but relax the previous assumption of elastic homogeneity. A number of approaches are explored. For relatively simple geometries, image methods are expedient. Antiplane dislocations at, or near, planar material interfaces across which there is a change in shear modulus are analyzed by simply extension of methods from chapter 2. Exploiting the principle of superposition allows us to sum multiple elemental solutions and thus to satisfy boundary conditions on multiple surfaces. In this way, we generate a series solution for a fault within a compliant fault zone. Superposition of images

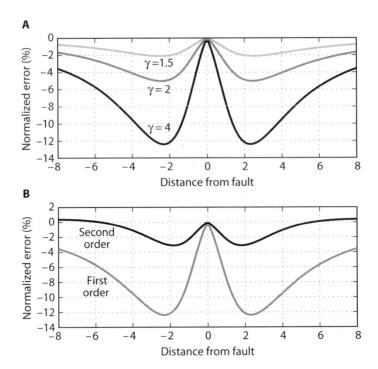

Figure 5.18. Percent error in perturbation expansion solution for infinitely long screw dislocation in a layer over a half-space. Layer thickness is H, and depth of dislocation is $D = 0.8H$. A: Error in first-order approximation for differing values of $\gamma = \mu_2/\mu_1$. B: Error in first- and second-order approximations when the shear modulus of the half-space is $\mu_2 = 4\mu_1$.

that account for material interfaces with those that enforce the free-surface condition yields series solutions for infinitely long strike-slip faults, either within or below a layer overlying a half-space with different shear modulus.

Rather than extend image methods to plane strain and three-dimensional problems, we investigate propagator matrix methods for horizontally stratified earth models. The approach is to isolate vertical derivatives and components of the stress tensor that result in tractions on horizontal interfaces. The propagator matrix carries a solution from one depth to another via simple matrix multiplication. Appropriate consideration of the dislocation source terms and matching the free-surface boundary condition yields the displacement fields of interest. A significant advantage of the propagator matrix approach is that, other than bookkeeping, it is no more difficult to compute a solution for 10 layers than it is for a single layer.

The previously described methods are not well suited to heterogeneity with irregular geometry. Perturbation methods can be used to approximate the effects of arbitrarily shaped heterogeneity but are guaranteed to be accurate only if the variation in elastic properties is modest. Surprisingly, we have found that, at least in some cases, a first-order perturbation solution is accurate even for contrasts in shear modulus of several hundred percent.

Is it worth the additional effort required to compute solutions for dislocations in elastically heterogeneous earth models? The answer is, in some circumstances, yes. Unlike elastodynamics, where variations in elastic properties give rise to all the reflected and refracted waves that many seismologists spend their careers studying, with quasi-static deformations, we can often ignore the effects of earth structure to first order. If, for example, one is interested in obtaining a rough estimate of the source properties, homogenous half-space models may be

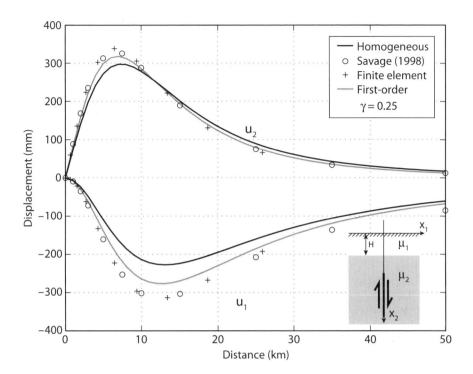

Figure 5.19. Comparison of different solutions for the plane-strain problem of a vertical fault beneath a compliant layer. The shear modulus of the surface layer is 25% of the underlying half-space. Solutions given by Savage (1998) following Lee and Dunders (1973) agree quite well with finite element calculations. A first-order perturbation solution is quite accurate in predicting the vertical displacements but underestimates the correction on the horizontal displacements by as much as 50%. Notice that the x_2 axis is down. After Cervelli et al. (1999).

adequate. However, if one is attempting to derive accurate quantitative estimates of earthquake or volcanic source parameters, particularly the depth of the source or details of the slip distribution on the fault, then effects of heterogeneity may be significant.

A general result is that for faults buried beneath a more compliant layer, the effect is to cause the deformation to be more localized (figures 5.12 and 5.13). This has the effect of biasing inversions to shallower depths (Du et al. 1994; Savage 1987). Figure 5.20 illustrates this effect for a dipping fault beneath a compliant surface layer. Ignoring the effect of layering biases the estimated source depth to be shallower than it actually is. The bias increases with increasing modulus contrast and thickness of the compliant surface layer.

Savage (1998) also investigated this effect for both vertical and horizontal faults. He finds that with a compliant near-surface layer, the depth of faulting is systematically underestimated and the slip overestimated. For example, with a modulus contrast of a factor of three, the depth of a horizontal fault was underestimated by 11% when inverting the horizontal displacements and by 17% when inverting the vertical displacements. The magnitude of the slip and length of the fault were also biased. For a vertical fault with modulus contrast of a factor of 4, the upper fault edge was biased shallow by 4 to 5%, while the lower edge was biased by 6 to 16%, depending on whether the vertical or horizontal displacements were inverted.

The techniques developed in this chapter have utility beyond the analysis of dislocations in heterogeneous elastic earth models. In the next chapter, we will study transient deformations

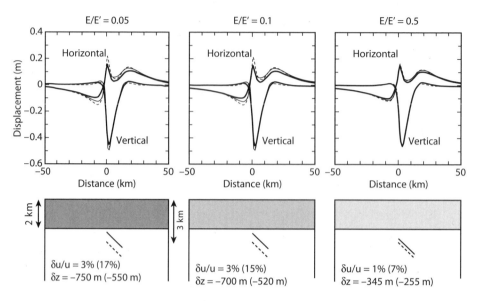

Figure 5.20. Effect of a compliant surface layer on inversion for a uniform-slip dislocation. E is Young's modulus of the compliant layer, E' that of the half-space. The fault dips 40 degrees, and the thickness of the layer is 2 km. Top: The dashed line shows the displacements due to the dislocation in the layered medium; the heavy solid line shows the displacements for the same dislocation in a half-space. The thin solid lines show the displacements for the best fitting "equivalent" source in a homogeneous half-space. Bottom: The dashed line shows the actual source location, and the solid line shows the best fitting "equivalent" source in a homogeneous half-space. For each case, the bias in fault depth δz and estimated slip $\delta u/u$ due to neglect of layering are indicated. The values in parentheses indicate biases on inverting only the horizontal displacements. Note that failure to account for the effect of layering biases the estimated depth of the dislocation, causing the fault to appear shallower than it really is. The effect is more severe for large modulus contrast. After Cattin et al. (1999).

induced by relaxation of stress in viscously deforming layers within the earth. As you will see, a key step in solving viscoelastic problems is to analyze corresponding problems with dislocations in heterogeneous earth models. We will therefore make extensive use of solutions already developed in this chapter.

5.9 Problems

1. Verify that equations (5.23) satisfy the displacement and traction boundary conditions on the boundaries of the compliant fault zone.

2. Derive the displacement field due to an infinitely long strike-slip fault perpendicular to a bimaterial interface (equations [5.24]).

3. Equation (5.26) gives the displacements within the shallow layer (region I) due to an infinitely long strike-slip fault within a half-space beneath a layer with differing shear modulus. Find the displacements within the half-space $|x_2| > H$ (region II). Show that when the depth of faulting is equal to the thickness of the layer—that is, $D = H$—the

displacements are

$$u(x_1, |x_2| > H) = \frac{-s}{2\pi} \left(\tan^{-1} \left(\frac{H - x_2}{x_1} \right) - \tan^{-1} \left(\frac{H + x_2}{x_1} \right) + \sum_{m=1}^{\infty} \left(\frac{1 - \gamma}{1 + \gamma} \right)^m \right.$$
$$\left. \left\{ \tan^{-1} \left[\frac{(2m + 1)H - x_2}{x_1} \right] - \tan^{-1} \left[\frac{(2m - 3)H - x_2}{x_1} \right] \right\} \right). \qquad (5.167)$$

4. Derive the solution for an infinitely long strike-slip fault embedded within an elastic layer overlying a half-space with different shear modulus (equations [5.32] and [5.34]) using image sources.

5. Show that for the antiplane strain system, the propgator matrix given by equation (5.61) satisfies the reciprocal relation given in (5.50).

6. Write out the propagator solution in *symbolic* form, as in equation (5.85), for the case of four layers overlying a half-space. Layer 1 is shallowest with shear modulus μ_1, layer 2 below that with shear modulus μ_2, and so forth. Denote the shear modulus of the half-space μ_h. Consider a point source dislocation located in

  (a) The half-space beneath the layers.

  (b) Layer 3.

  (c) Layer 1.

7. Use the propagator matrix method to derive the solution for an infinitely long strike-slip fault with uniform slip from the free surface to depth D. Show that the inverse Fourier transform of $(s/ik)(1 - e^{-|k|D})$ is given by $s/\pi \tan^{-1}(D/x)$.
   Hint: In chapter 4, we found that $\mathcal{F}(1/z) = -i\pi \, \text{sgn}(k)$. You can use the reciprocal property of the Fourier transform to get the inverse transform of $1/k$. To solve the remaining part, differentiate both sides of the equation with respect to x, giving the transform of the strain. Do the inverse transform and then integrate to get back to displacement.

8. Use the propagator matrix method to derive the solution for a strike-slip fault in a layer overlying a uniform half-space (equation [5.82]).

9. Invert the Fourier transforms in equations (5.89) and (5.90) to show that the displacement in the body can be written as

$$u(x, z) = \frac{s}{2\pi} \left[\tan^{-1} \left(\frac{x}{z - D} \right) - \tan^{-1} \left(\frac{x}{z + D} \right) \right] \qquad z > D,$$
$$u(x, z) = \frac{s}{2\pi} \left[\tan^{-1} \left(\frac{z + D}{x} \right) - \tan^{-1} \left(\frac{z - D}{x} \right) \right] \qquad z < D. \qquad (5.168)$$

(Use the hint in problem 7.)

10. Use propagator methods for antiplane strain dislocations to determine both components of stress in the body for a dislocation from the free surface to depth D in a homogeneous half-space.

11. Consider the three-dimensional propagator matrix representation of a source in a plane-layered earth model. Show that a dislocation source that extends in the x_1 direction over the range $-L \leq x_1 \leq L$ can be described by multiplying the point-source solution in the Fourier domain by

$$
\begin{aligned}
&\frac{2}{k_1} \sin(k_1 L) \quad k_1 \neq 0, \\
&2L \qquad\qquad k_1 = 0.
\end{aligned}
\tag{5.169}
$$

5.10 References

Aki, K., and P. G. Richards. 1980. *Quantitative seismology: theory and methods.* New York: W. H. Freeman.

Barker, T. 1976. Quasi-static motions near the San Andreas Fault zone. *Geophysical Journal of the Royal Society* **45**, 689–705.

Ben-Meneham, A., and S. J. Singh. 1968. Multipolar elastic fields in a layered half-space. *Bulletin of the Seismological Society of America* **58**, 1519–1572.

Cattin, R., P. Briole, H. Lyon-Caen, P. Bernard, and P. Pinettes. 1999. Effects of superficial layers on coseismic displacements for a dip-slip fault and geophysical implications. *Geophysical Journal International* **137**, 149–158.

Cervelli, P., S. Kenner, and P. Segall. 1999. Correction to "Dislocations in inhomogeneous media via a moduli-perturbation approach: general formulation and 2-D solutions," by Y. Du, P. Segall, and H. Gao. *Journal of Geophysical Research* **104**, 23,271–23,277.

Chen, Q., and J. T. Freymueller. 2002. Geodetic evidence for a near-fault compliant zone along the San Andreas fault in the San Francisco Bay area. *Bulletin of the Seismological Society of America* **92**, 656–671.

Du, Y., P. Segall, and H. Gao. 1994. Dislocations in inhomogeneous media via a moduli-perturbation approach: general formulation and 2-D solutions. *Journal of Geophysical Research* **99**, 13,767–13,779.

———. 1997. Quasi-static dislocations in three-dimensional inhomogeneous media. *Geophysical Research Letters* **24**, 2347–2350.

Fialko, Y., D. Sandwell, D. Agnew, M. Simons, P. Shearer, and B. Minster. 2002. Deformation on nearby faults induced by the 1999 Hector Mine earthquake. *Science* **297**, 1858–1862.

Gilbert, F., and G. E. Backus. 1966. Propagator matrices in elastic wave and vibration problems. *Geophysics* **31**, 326–332.

Jovanovich, D. B., M. I. Husseini, and M. A. Chinnery. 1974. Elastic dislocations in a layered half-space—I. Basic theory and numerical methods. *Geophysical Journal of the Royal Astronomical Society* **39**, 205–217.

Lee, M.-A., and J. Dundurs. 1973. Edge dislocation in a surface layer. *International Journal of Engineering Science* **11**, 87–94.

McHugh, S., and M. Johnston. 1977. Surface shear stress, strain, and shear displacement for screw dislocations in a vertical slab with shear modulus contrast. *Geophysical Journal of the Royal Astronomical Society* **49**, 715–722.

Rybicki, K. 1971. The elastic field of a very long strike-slip fault in the presence of a discontinuity. *Bulletin of the Seismological Society of America* **61**, 79–92.

Rybicki, K., and K. Kasahara. 1977. A strike-slip fault in a laterally inhomogeneous medium. *Tectonophysics* **42**, 127–138.

Sato, R. 1971. Crustal deformation due to dislocation in a multi-layered medium. *Journal of Physics of the Earth* **19**, 32–46.

Sato, R., and M. Matsu'ura. 1973. Static deformations due to the fault spreading over several layers in a multi-layered medium. Part I: Displacement. *Journal of Physics of the Earth* **21**, 227–249.

Savage, J. C. 1987. Effect of crustal layering on dislocation modeling. *Journal of Geophysical Research* **92**, 10,595–10,600.

———. 1998. Displacement field for an edge dislocation in a layered half-space. *Journal of Geophysical Research* **102**, 2439–2446.

Singh, S. J. 1970. Static deformation of a multilayered half-space by internal sources. *Journal of Geophysical Research* **75**, 3257–3263.

Strang, G. 1976. *Linear algebra and its applications*. New York: Academic Press.

Ward, S. N. 1985. Quasi-static propagator matrices: creep on strike-slip faults. *Tectonophysics* **120**, 83–106.

Zhu, L., and L. A. Rivera. 2002. A note on the dynamic and static displacements from a point source in a layered media. *Geophysical Journal International* **148**, 619–627.

6

Postseismic Relaxation

Up to this point, we have considered only time-independent processes, except for simple models of interseismic strain accumulation in which the material behavior is elastic and the only parameter that varies with time is the amount of slip on the fault (chapter 2). There is considerable evidence that such descriptions are inadequate. First of all, we expect a priori that with increasing depth, and hence temperature, rock will behave in a ductile fashion. Ductile flow in response to stress changes induced by earthquakes, among other forcings, can lead to transient deformation at the earth's surface. We also recognize that rock is porous and liquid saturated. Flow of pore fluids induced by coseismic stress changes may also cause transient deformations that could be observed at the earth's surface. Furthermore, there is good observational evidence of time-dependent deformation following large earthquakes. Such deformation is referred to as *postseismic*, although the timescale may vary from hours to decades depending on the dominant process and the data set in question.

Some of the earliest evidence of transient postseismic deformation came from triangulation data collected after the 1906 San Francisco earthquake. Thatcher (1975) first showed that post-1906 strain rates, derived from measurements centered over the San Andreas fault, were a factor of 4 higher than contemporary strain rates. Figure 6.1 summarizes peak shear strain rate as a function of time since the most recent great earthquake in California. For northern California, the most recent great event was the 1906 San Francisco earthquake; for southern California, the 1857 Fort Tejon earthquake. A significant decrease in strain rate with a timescale of several decades is clearly seen in the data.

Transient postseismic deformation is not limited to strike-slip earthquakes. Thatcher (1984) examined leveling data in Japan adjacent to the Nankai trough and showed that the pattern of uplift and subsidence varied with time following the great 1944 Tonankai and 1946 Nankaido earthquakes (figure 6.2). More accurate and frequently sampled GPS measurements following the 1994 Sanriku-Haruka-Oki earthquake off Hokaido exhibit a spectacular postseismic transient. The displacement in the year following the earthquake alone is comparable to the coseismic displacement (figure 6.3). Heki and others (1997) noted that the cumulative moment of the postseismic signal, if interpreted as afterslip along the subduction interface, had a moment (M_W 7.7) that exceeded that of the earthquake (M_W 7.6). This has clear implications for the slip budget at the plate boundary and thus for forecasts of seismic hazard.

Post-earthquake GPS and InSAR measurements are becoming much more readily available, are far more accurate than historical measurements, and are very likely to rewrite our understanding of postseismic deformation. Peltzer and others (1996) show spectacular interferograms of postseismic motions following the 1992 Landers earthquake. Localized deformations within the steps along the faults that ruptured in the earthquake were interpreted by these authors as resulting from transient pore-fluid flow induced by the earthquake (see chapter 10). Longer wavelength components of these and similar data sets have also been modeled by a number of research groups as resulting from distributed viscoelastic relaxation. Plate 6 illustrates the transient horizontal displacements determined from continuous GPS networks in the 7 years following the M 7.1 1999 Hector Mine earthquake in southern California, 7 to 14 years after the nearby M 7.3 1992 Landers earthquake. Significant transient deformation

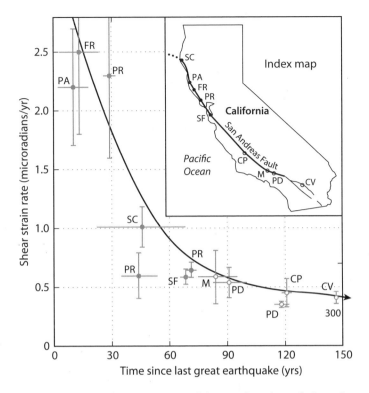

Figure 6.1. Peak shear strain rate (engineering strain) as a function of time since the last great earthquake on the San Andreas fault, after Thatcher (1990). The two-letter codes refer to the geodetic network; locations are shown on inset map. Solid symbols refer to northern California, and open symbols refer to southern California.

extends more than 200 km from the ruptures and is interpreted by Freed and others (2007) as requiring distributed flow of the upper mantle at depths below 40 km in the earth.

As is readily apparent from this discussion, there are a variety of mechanical processes that could give rise to postseismic deformation, including viscous relaxation of rock below the seismogenic zone; transient fault slip, either within the rupture zone or below the rupture on the downdip extension of the fault; and flow of pore fluids induced by the earthquake. This chapter focuses on viscoelastic relaxation; chapter 10 describes poroelastic effects, including post-earthquake adjustments, and chapter 11 covers frictional properties of faults, including afterslip.

Thatcher (1983) examined afterslip and distributed viscoelastic deformation, nicely differentiating between what he referred to as *thick lithosphere* and *thin lithosphere* models of faulting (figure 6.4). In the thick lithosphere model, rock adjacent to the fault behaves elastically to depths substantially greater than the maximum depth of slip during earthquakes, D. Postseismic deformation occurs due to transient slip on the downdip extent of the fault. Employing a thick lithosphere model, Thatcher found that the post-1906 triangulation data could be fit with slip in the depth interval of 15 to 25 km that decayed exponentially with a time constant of ~ 30 years.

In the thin lithosphere model, elastic behavior is limited to a relatively thin elastic plate with thickness $H \geq D$, which overlies a *viscoelastic* asthenosphere, in which the shear stress relaxes with time. This leads to a decaying strain transient at the surface. Thatcher concluded

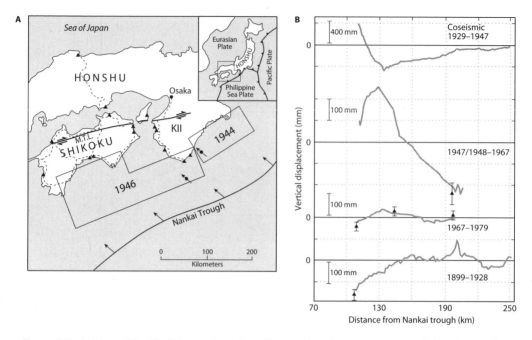

Figure 6.2. A: Map of the Nankai trough region of Japan showing rupture areas of the 1944 and 1946 earthquakes and position of leveling lines. B: Height changes as a function of distance from the Nankai Trough for different time periods, after Thatcher (1984). Triangles mark elevation changes determined by tide gauge data. Thatcher assumes that the pre-earthquake time period, 1899–1928, represents deformation late in the cycle and thus is reasonably shown after the 1967–1979 period.

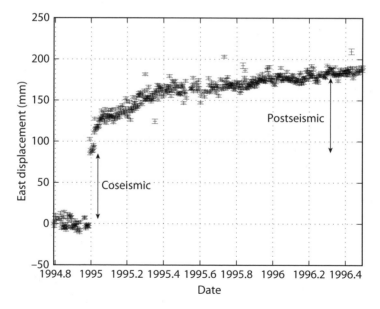

Figure 6.3. Variations in the east component of a GPS station position with time along the northern coast of Honshu, Japan, showing postseismic deformation following the 1994 Sanriku-Haruka-Oki earthquake, after Heki and others (1997). Data courtesy of Shin'ichi Miyazaki.

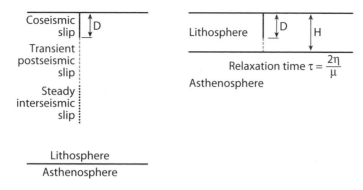

Figure 6.4. Thick (left) and thin (right) lithosphere models, after Thatcher (1983).

that he could not differentiate between the thick and thin lithosphere models based on their fits to the data from California. Barker (1976) appears to have been the first to point out that deep fault slip and distributed viscoelastic flow yield similar deformation patterns at the earth's surface. This will be discussed considerably more in detail later in this chapter.

6.1 Elastic Layer over Viscous Channel

We will begin by considering the simple model illustrated in figure 6.5 of an elastic plate with thickness H overlying a viscous channel with thickness h, originally due to Elsasser (1969).

The deformation is again antiplane. Let

$$u(x_1) \equiv \frac{1}{H} \int_{-H}^{0} u_3(x_1, x_2) dx_2 \tag{6.1}$$

be the displacement averaged over the thickness of the elastic plate, and let

$$\sigma(x_1) \equiv \frac{1}{H} \int_{-H}^{0} \sigma_{13}(x_1, x_2) dx_2 \tag{6.2}$$

be the thickness-averaged stress acting on vertical planes. Last, let $\tau \equiv \sigma_{23}(x_1, x_2 = -H)$ be the shear traction acting in the x_3 direction, exerted on the base of the elastic plate due to viscous flow. Note that all variables, u, σ, and τ, are functions of lateral distance x_1 only, so in what follows, we will drop the subscript. Equilibrium requires that the sum of the forces in the x_3 direction balance. If we examine an element of the elastic plate of dimensions H by δx, then the force on the left vertical face is σH, whereas the force on the right vertical face is $H[\sigma + (\partial \sigma / \partial x)\delta x]$ (figure 6.5). The gradient in the stress must be balanced by the force exerted by the basal shear traction $\tau \delta x$. Thus,

$$H \frac{\partial \sigma}{\partial x} = \tau. \tag{6.3}$$

The plate is elastic, by assumption, so the thickness-averaged stress is proportional to the thickness-averaged strain:

$$\sigma = \mu \frac{\partial u}{\partial x}. \tag{6.4}$$

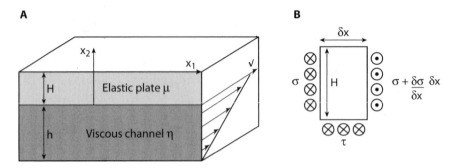

Figure 6.5. A: Geometry of an elastic plate overlying a viscous channel. B: Element of the elastic plate.

The material in the asthenospheric channel is assumed to be linearly viscous—that is, the shear stress is proportional to the strain rate, $\tau = 2\eta\dot{\epsilon}$. The strain rate in the channel is $v/2h$ since the velocity increases from zero at the base of the channel to v at the top (figure 6.5B). Thus, the basal traction is given by

$$\tau = \frac{\eta}{h}\frac{\partial u}{\partial t}, \tag{6.5}$$

where η is the viscosity of the asthenosphere. Substituting equations (6.5) and (6.4) into (6.3) yields

$$\frac{hH\mu}{\eta}\frac{\partial^2 u}{\partial x^2} = \frac{\partial u}{\partial t}. \tag{6.6}$$

The displacement u follows a *diffusion equation* with diffusivity κ:

$$\kappa = \frac{hH\mu}{\eta}, \tag{6.7}$$

with units of length squared per unit time. Knowing that the displacement satisfies a diffusion equation tells us immediately that following an earthquake, displacement diffuses into the elastic plate to a distance that scales with $\sqrt{\kappa t}$ (see appendix B).

The initial conditions and boundary conditions specify that there is no displacement prior to the earthquake. At time $t = 0$, the fault (located at $x = 0$) suddenly displaces by an amount Δu but is thereafter locked:

$$u(x, t = 0) = 0, \tag{6.8}$$

$$u(x = 0, t) = \Delta u \qquad t > 0. \tag{6.9}$$

The solution to the diffusion equation for these boundary and initial conditions is well known (a derivation is given in appendix B):

$$u(x, t) = \Delta u \operatorname{erfc}\left(\frac{x}{2\sqrt{\kappa t}}\right), \tag{6.10}$$

where erfc is the complementary error function and is given by

$$\text{erfc}(z) = 1 - \text{erf}(z) = 1 - \frac{2}{\sqrt{\pi}} \int_0^z e^{-y^2} \mathrm{d}y. \tag{6.11}$$

It is left as an exercise (problem 1) to show that equation (6.10) is indeed the solution to equation (6.6).

It is useful to put the solution (6.10) into dimensionless form. Scale the horizontal distance by the elastic layer thickness:

$$\tilde{x} = x/H, \tag{6.12}$$

and u by the displacement at the fault surface:

$$\tilde{u} = u/\Delta u. \tag{6.13}$$

This leads to

$$\tilde{u} = \text{erfc}\left(\frac{\tilde{x}}{2\sqrt{\tilde{t}}}\right), \tag{6.14}$$

where time is normalized by the characteristic relaxation time $\tilde{t} = t/t_R$:

$$t_R = \frac{H\eta}{h\mu}. \tag{6.15}$$

To compute velocity, we differentiate equation (6.14) with respect to dimensionless time. Note from equation (6.11) that time appears in the integration limits, so that we must apply Leibnitz's rule (5.56). The dimensionless velocity and shear strain rate are thus given by

$$\tilde{v} = \frac{1}{\tilde{t}} \frac{\tilde{x}}{2\sqrt{\pi\tilde{t}}} \exp\left[-\frac{\tilde{x}^2}{4\tilde{t}}\right], \tag{6.16}$$

and

$$\frac{\partial\tilde{\gamma}}{\partial\tilde{t}} = \frac{\partial\tilde{v}}{\partial\tilde{x}} = \frac{1}{\tilde{t}} \frac{1}{2\sqrt{\pi\tilde{t}}} \left(1 - \frac{\tilde{x}^2}{2\tilde{t}}\right) \exp\left[-\frac{\tilde{x}^2}{4\tilde{t}}\right]. \tag{6.17}$$

The normalized displacement, velocity, and strain rate are shown in figure 6.6. Notice that the shear strain rate changes sign at $\tilde{x} = \sqrt{2\tilde{t}}$. This is reminiscent of the result for a buried screw dislocation solution. Recall from chapter 2 that slip in a confined depth interval generates a positive strain rate over the fault but a negative strain rate away from the fault (see figure 2.12). This feature is exhibited by all solutions where postseismic deformation is confined to a finite depth interval, and while diagnostic of the effect, cannot necessarily be used to discriminate between models.

While the simple Elsasser model is useful in that it provides considerable insight into the physics of postseismic stress diffusion, it suffers from two serious limitations. First, the asthenosphere cannot behave as a viscous fluid. At short timescales, it must behave elastically. The asthenosphere transmits shear waves, and coseismic displacements are well described by earth models that are elastic at all depths. A better description of the asthenosphere would be a viscoelastic medium, which behaves elastically at short timescales yet flows over long periods

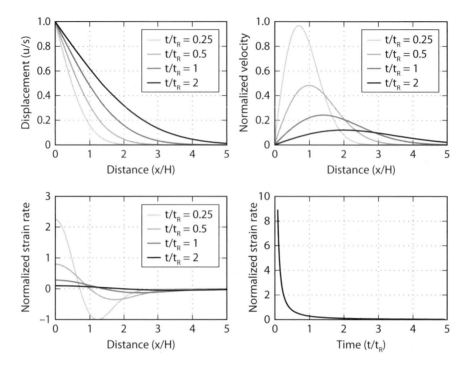

Figure 6.6. Results of the Elsasser model. Upper left: Displacement as a function of distance from the fault. Upper right: Velocity as a function of distance from the fault. Lower left: Strain rate as a function of distance from the fault. All are shown for different normalized times. Lower right: Strain rate at the fault trace as a function of time.

of time. The second problem with the Elsasser model is that it ignores vertical gradients in displacement and stress. Lehner et al. (1981) explored a generalization of the Elsasser model, including viscoelastic response of the asthenosphere, to suddenly imposed dislocations or stress drop. Li and Rice (1983) extend this work to consider approximate three-dimensional behavior as a coupling of a plane stress plate problem that describes the along-strike behavior and an antiplane relation between the fault stress and depth-dependent slip distribution. Last, Li and Rice (1987) consider an antiplane model of an infinite sequence of earthquakes that rupture an elastic plate overlying a Maxwell foundation, including stress-driven creep on a downdip extension of the fault in the elastic plate. Earthquake cycle models such as this are described in more detail in chapter 12.

For now, we will consider models in which the earth can be described as an elastic plate overlying a viscoelastic half-space, as in the thin lithosphere model of figure 6.4.

6.2 Viscoelasticity

Viscoelastic models of earthquakes have generally been restricted to simple constitutive laws that can be thought of as combinations of elastic and viscous elements (e.g., Flugge 1975). In figure 6.7, springs represent elastic elements, whereas dashpots represent viscous elements. We consider three different materials. A spring and dashpot in series is known as a *Maxwell material* or *Maxwell fluid*, because at infinite times (or low frequencies), it behaves as a viscous fluid. A spring and a dashpot in parallel is known as a *Kelvin material* or a *Kelvin solid*, as it

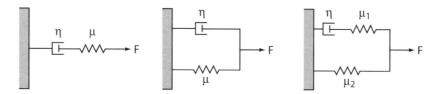

Figure 6.7. Simple viscoelastic representations. Left: Maxwell material. Center: Kelvin solid. Right: Standard linear solid.

behaves like an elastic solid at long times and low frequencies. At short times, the Kelvin material is rigid and does not deform. A Maxwell element in parallel with a spring is known as a *standard linear solid* (figure 6.7), although the standard linear solid is sometimes represented as a spring in series with a Kelvin element. More complex combinations can be easily generated and analyzed based on the principles outlined here.

The different material behaviors can be analyzed after first formulating their constitutive equations. For example, consider the Maxwell material. The total strain rate is the sum of the strain rate in the spring and the dashpot. Force balance, however, requires that the stress must be the same in each element; thus,

$$
\dot{\epsilon}_{\text{total}} = \dot{\epsilon}_{\text{spring}} + \dot{\epsilon}_{\text{dashpot}},
$$
$$
= \frac{\dot{\sigma}}{2\mu} + \frac{\sigma}{2\eta}, \tag{6.18}
$$

where the superimposed dot signifies time derivative, and the factors of two are introduced to retain consistency with tensor forms of the constitutive laws as discussed later.

For the Kelvin material, the stress is the sum of the stress in the spring and the dashpot. The strain, however, must be the same in each; thus,

$$
\sigma_{\text{total}} = \sigma_{\text{spring}} + \sigma_{\text{dashpot}},
$$
$$
= 2\mu\epsilon + 2\eta\dot{\epsilon}. \tag{6.19}
$$

Last, the constitutive equation for the standard linear solid (problem 4) is

$$
\sigma + \frac{\eta}{\mu_1}\dot{\sigma} = 2\mu_2\epsilon + 2\eta\left(1 + \frac{\mu_2}{\mu_1}\right)\dot{\epsilon}. \tag{6.20}
$$

Note that if the spring in series with the dashpot becomes infinitely stiff, $\mu_1 \rightarrow \infty$, equation (6.20) reduces to that of a Kelvin solid. Similarly, if $\mu_2 \rightarrow 0$, equation (6.20) reduces to that of a Maxwell fluid.

Consider the behavior of the Maxwell fluid, from equation (6.18). For a step change in stress, the material first deforms elastically by an amount $\sigma/2\mu$, since the dashpot has no time to relax. Thereafter, the material strains at a constant rate $\sigma/2\eta$ (figure 6.8). At short times, the material behaves elastically, but at long times, it behaves like a fluid. On the other hand, if we apply a sudden strain ϵ_0, which is thereafter held constant, the stress will instantaneously rise by an amount $2\mu\epsilon_0$. After that, the total strain is constant, so we must solve the differential

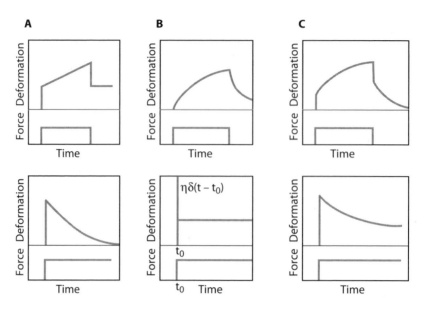

Figure 6.8. Creep (top) and relaxation functions (bottom) for simple viscoelastic rheologies. A: Maxwell. B: Kelvin. C: Standard linear solid. After Fung (1965).

equation (6.18) with zero left-hand side. This yields the so-called *relaxation function*:

$$\sigma = 2\mu\epsilon_0 e^{-\frac{\mu}{\eta}t} \qquad t > 0 \tag{6.21}$$

(figure 6.8), demonstrating that the Maxwell material has a characteristic relaxation time of $t_R = \eta/\mu$.

For the Kelvin solid subject to a suddenly applied stress σ_0, which is thereafter held constant, the strain is found by solving equation (6.19) with $\sigma_{\text{total}} = \sigma_0$:

$$\epsilon = \frac{\sigma_0}{2\mu}\left(1 - e^{-\frac{\mu}{\eta}t}\right) \qquad t \geq 0. \tag{6.22}$$

The Kelvin material exhibits an exponentially increasing strain, again with characteristic time $t_R = \eta/\mu$. The strain increases from zero at $t = 0$ to $\sigma_0/2\mu$ at infinite time. This *creep function* for the Kelvin material (6.22) is shown in figure 6.8B. The dashpot in the Kelvin solid resists rapid deformation, and thus there is a singularity in stress when subject to an instantaneous increase in applied strain. The creep and relaxation functions for a standard linear solid are left for exercises (problem 5).

The *Burger's rheology*, sometimes referred to as the *Burger's body*, has also been used to describe postseismic deformation (Pollitz 2003; Hetland and Hager 2005). The Burger's rheology consists of Maxwell and Kelvin elements in series (figure 6.9). This rheology is referred to as *biviscous* in that it possesses two relaxation modes, a recoverable deformation associated with the Kelvin element and a nonrecoverable deformation associated with the Maxwell element. It should be emphasized that the biviscous rheology is linear and should be distinguished from nonlinear creep laws in which the strain rate is proportional to stress to a power $n > 1$. With two relaxation times, $\tau_1 = \eta_1/\mu_1$ and $\tau_2 = \eta_2/\mu_2$, the biviscous rheology can be

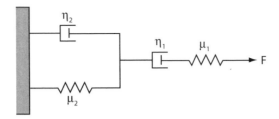

Figure 6.9. Spring-dashpot representation of the Burger's rheology, which consists of a Maxwell element in series with a Kelvin element.

fit to postseismic deformation measurements that exhibit a short-term transient followed by a longer term relaxation. Whether these observations are best explained by distributed deformation with multiple relaxation times, by a nonlinear stress-dependent rheology, or by a combination of post-earthquake fault creep (chapter 11) followed by a more distributed relaxation remains to be determined.

Before continuing with solution methods, we will consider the appropriate tensor extensions of the simple spring and dashpot models. Ductile creep of rock is primarily in shear; low-porosity rocks exhibit limited volumetric relaxation. We thus take the volumetric strain to be elastic, while for a Maxwell medium, we take the deviatoric strain rate to be the sum of elastic and viscous components. For a Newtonian fluid, the deviatoric strain rate follows $2\eta\dot{\epsilon}'_{ij} = \sigma_{ij} - \sigma_{kk}\delta_{ij}/3$, where the prime indicates deviatoric strain. For the elastic component, we separate the deviatoric strain as

$$\sigma_{ij} = 2\mu\left(\epsilon_{ij} - \frac{\epsilon_{kk}}{3}\delta_{ij}\right) + \left(\lambda + \frac{2}{3}\mu\right)\epsilon_{kk}\delta_{ij},$$

$$= 2\mu\epsilon'_{ij} + K\epsilon_{kk}\delta_{ij}. \tag{6.23}$$

Differentiating Hooke's law with respect to time, the governing equation for a Maxwell material can thus be shown to be

$$\dot{\sigma}_{ij} + \frac{\mu}{\eta}\left(\sigma_{ij} - \frac{\sigma_{kk}}{3}\delta_{ij}\right) = 2\mu\dot{\epsilon}_{ij} + \lambda\dot{\epsilon}_{kk}\delta_{ij}. \tag{6.24}$$

To apply these ideas to models of faulting, we have to solve a time-dependent boundary value problem. Note, however, that the limiting behavior at short time, the so-called *unrelaxed* response, is a completely elastic solution. Similarly, in the infinite time, or fully *relaxed* state, all viscous stresses have dissipated, and the state of the system corresponds to a (different) elastic solution. Thus, both the relaxed and unrelaxed states can be found by solving an appropriate elasticity problem. Interestingly, the fully time-dependent solution can also be found from a corresponding elastic solution, as discussed in the next section.

6.2.1 Correspondence Principle

Compare the viscoelastic problem to an elastic problem. The equilibrium equations are the same, as are the kinematic (strain-displacement) equations. The only differences are the constitutive equations. Note that the constitutive equation for a Maxwell material, from equation (6.18), can be written as

$$\frac{\partial}{\partial t}\epsilon = \frac{1}{2}\left(\mu^{-1}\frac{\partial}{\partial t} + \eta^{-1}\right)\sigma. \tag{6.25}$$

This is still a linear relationship between stress and strain, but one in which the relationship is governed by linear differential operators. Indeed, any linear viscoelastic constitutive relation can be written in this form. This suggests that we make use of the *Laplace transform* (section A.2), defined as

$$\mathcal{L}\{f(t)\} \equiv \bar{f}(s) = \int_0^\infty f(t)e^{-st}dt \tag{6.26}$$

(equation [A.13]). If the function vanishes at $t = 0$, the transform of the derivative of that function (A.15) is $\mathcal{L}\{\partial f(t)/\partial t\} = s\bar{f}(s)$. Assuming that the stress and strain vanish at $t = 0$, equation (6.25) can thus be written as

$$\bar{\sigma} = 2\bar{\mu}\bar{\epsilon}, \tag{6.27}$$

where

$$\boxed{\bar{\mu}(s) = \frac{s}{\mu^{-1}s + \eta^{-1}}.} \tag{6.28}$$

We see that the constitutive law for a viscoelastic material in the Laplace transform domain exactly parallels Hooke's law. Since time does not enter the quasi-static equilibrium or compatibility equations, it follows that the Laplace transform of a viscoelastic problem is equivalent to an elastic problem. This is known as the *correspondence principle*. It means that if we know the solution to the corresponding elastic problem, we need only perform the inverse transform to find the solution to the viscoelastic problem.

The correspondence principle applies to the general, tensor form of the viscoelastic constitutive equations. For example, the Laplace transform of equation (6.24) can be written in the form of Hooke's law, noting that the corresponding elastic parameters are given by

$$\bar{\mu} = \frac{s\mu}{s + \dfrac{\mu}{\eta}},$$

$$\bar{\lambda} = \frac{s\lambda + K\dfrac{\mu}{\eta}}{s + \dfrac{\mu}{\eta}}. \tag{6.29}$$

6.3 Strike-Slip Fault in an Elastic Plate Overlying a Viscoelastic Half-Space

We will consider first an infinitely long strike-slip fault in an elastic layer overlying a viscoelastic half-space, first presented by Nur and Mavko (1974). The problem is antiplane strain, and for simplicity, we take the elastic shear modulus to be the same in the layer and the half-space. The corresponding elastic problem is one of a screw dislocation in an elastic layer over an elastic half-space with different shear modulus, a problem previously treated in chapter 5. Slip on the fault is given by

$$\Delta u(t) = \Delta u H(t), \tag{6.30}$$

where $H(t)$ is the Heavyside function—that is, zero for $t < 0$ and unity for $t > 0$. The Laplace transform of the Heavyside function (A.14) is simply $\mathcal{L}\{H(t)\} = 1/s$. Note that in this chapter, Δu is used to denote the magnitude of the fault slip, since s is reserved for the Laplace transform variable. We see immediately from the elastic solution for a screw dislocation in a layer overlying a half-space with different shear modulus (5.33) that the solution in the Laplace transform domain is given by

$$\bar{u}(x, s) = \frac{\Delta u}{s\pi} \left[\tan^{-1} \left(\frac{D}{x} \right) + \sum_{n=1}^{\infty} \left(\frac{\mu_1 - \bar{\mu}_2}{\mu_1 + \bar{\mu}_2} \right)^n F_n(x, D, H) \right], \tag{6.31}$$

where

$$
\begin{aligned}
F_n(x, D, H) &= \left[\tan^{-1} \left(\frac{D + 2nH}{x} \right) + \tan^{-1} \left(\frac{D - 2nH}{x} \right) \right], \\
&= \tan^{-1} \left[\frac{2xD}{x^2 + (2nH)^2 - D^2} \right],
\end{aligned}
\tag{6.32}
$$

and $\bar{\mu}_2$ given by equation (6.28). The relation in equation (6.32) comes from application of the addition formula for the inverse tangent. Notice that because the layer with modulus μ_1 is elastic, and thus independent of time, it does not change in the Laplace transform domain.

Note that for a homogeneous viscoelastic half-space, equation (6.31) reduces to $(\Delta u/s\pi) \tan^{-1}(D/x)$. We find the rather surprising result that the displacements resulting from a dislocation in a homogeneous Maxwell viscoelastic half-space are independent of time and equal to the elastic displacements. Why? Mathematically, the elastic half-space solution is independent of μ. Thus, the only term that transforms is $\Delta u(t)$, and inverse transformation recovers the elastic solution. The stresses, however, depend on μ and are time dependent. Physically, as the stresses relax, elastic strain is transformed to viscous (inelastic) strain, the total strain staying the same. The same will not be true if the fault is driven by a specified stress, in which case, the displacements are functions of μ. For fixed stress, the displacements will thus increase with time.

Returning to equation (6.31), define

$$\beta = \frac{\mu_2}{2\eta} = \frac{1}{t_R}, \tag{6.33}$$

where t_R is the Maxwell relaxation time. Note that in the Maxwell constitutive law (6.21), the material relaxation time was taken to be $t_R = \eta/\mu$. One should be careful when comparing results from different authors to clarify whether the factor of two is included in the definition of the relaxation time. From equation (6.31), the displacement in the Laplace domain is

$$\bar{u}(x, s) = \frac{\Delta u}{s\pi} \left[\tan^{-1} \left(\frac{D}{x} \right) + \sum_{n=1}^{\infty} \left(\frac{\beta}{s + \beta} \right)^n F_n(x, D, H) \right]. \tag{6.34}$$

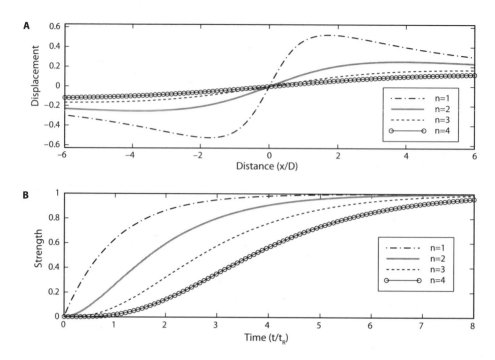

Figure 6.10. Modes for the antiplane model of an infinitely long strike-slip fault overlying a Maxwell viscoelastic half-space. A: First four spatial modes. B: First four temporal coefficients.

Recall that for simplicity, the elastic shear modulus in the layer and the half-space are assumed to be equal, $\mu_1 = \mu_2$. Inverting the Laplace transform is discussed in section A.2. First, note that the term proportional to $\tan^{-1}(D/x)$ transforms to the elastic solution, as discussed earlier. The problem thus reduces to finding the inverse Laplace transform:

$$\mathcal{L}^{-1}\left\{ \frac{1}{s} \left(\frac{\beta}{s+\beta} \right)^n \right\}. \tag{6.35}$$

It is shown in section A.2, equation (A.32), that the inverse transform is given by

$$\mathcal{L}^{-1}\left\{ \frac{1}{s} \left(\frac{\beta}{s+\beta} \right)^n \right\} = 1 - e^{-\beta t} \sum_{m=1}^{n} \frac{(\beta t)^{n-m}}{(n-m)!}, \tag{6.36}$$

so that the time varying displacements are

$$u(x, t) = \frac{\Delta u}{\pi}\left\{ \tan^{-1}(D/x) + \sum_{n=1}^{\infty} \left[1 - e^{-t/t_R} \sum_{m=1}^{n} \frac{(t/t_R)^{n-m}}{(n-m)!} \right] F_n(x, D, H) \right\} \qquad t \geq 0. \tag{6.37}$$

Figure 6.10 shows both the spatial terms $F_n(x, D, H)$ at different order n and the corresponding time dependence, given by equation (6.36). Notice that the higher order terms (larger n) are smaller in amplitude and correspond to longer spatial wavelengths. The higher order terms thus contribute less to the deformation, except at great distance from the fault.

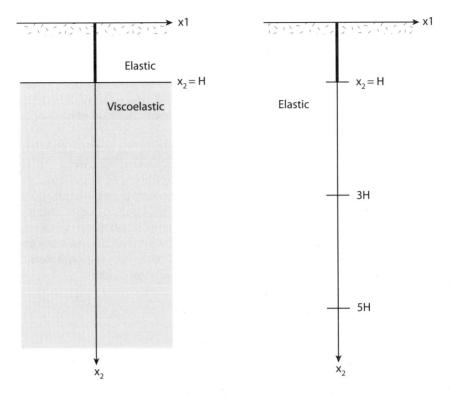

Figure 6.11. Comparison of a viscoelastic model to an equivalent dislocation solution. Left: Lithosphere–asthenosphere model. Right: Equivalent elastic half-space model.

Figure 6.10B illustrates that the higher order terms grow more slowly in time. This means that the postseismic displacements close to the fault grow quickly, whereas the displacements far from the fault grow more slowly. In this way, strain diffuses away from the fault as in the Elsasser model.

Savage (1990) noted the correspondence between the viscoelastic solution and an equivalent elastic half-space solution with specified slip (figure 6.11). The solution (6.37) is mathematically equivalent to fault slip within an elastic half-space in a series of strips below the coseismic rupture. For the case where coseismic slip breaks the full elastic plate, $D = H$, the first strip, corresponding to the first term in the sum, extends from depth D to $3D$, the second strip from depth $3D$ to $5D$, and so on. Slip is spatially uniform within each strip. The time dependence of slip within each strip is given by equation (6.36), the first few terms of which are

$$n = 1: \qquad 1 - e^{-t/t_R},$$

$$n = 2: \qquad 1 - e^{-t/t_R}\left[t/t_R + 1\right],$$

$$n = 3: \qquad 1 - e^{-t/t_R}\left[\frac{(t/t_R)^2}{2} + t/t_R + 1\right],$$

$$n = 4: \qquad 1 - e^{-t/t_R}\left[\frac{(t/t_R)^3}{3!} + \frac{(t/t_R)^2}{2} + t/t_R + 1\right]. \qquad (6.38)$$

Notice that each term is zero when $t = 0$ and evolves to unity as $t \to \infty$. The result (6.37) shows that prescribed slip within an elastic full-space, by the amount given by equation (6.38) in each strip, is precisely equivalent to the viscoelastic solution. In this sense, measurement of surface velocity, or strain, cannot distinguish between the thick and thin lithosphere models. Because slip in the first strip closest to the surface tends to dominate the surface deformation, the $n = 1$ term in equation (6.38) explains why Thatcher was able to find equivalent fits to the data with a layer over a viscoelastic half-space and one with exponentially decreasing postseismic slip rate on a downdip extension of the fault.

While this result emphasizes the difficulty in distinguishing between different models of faulting, all hope is not lost. First, the mathematical equivalence between the two models is restricted to infinitely long strike-slip faults. It does not apply to finite faults, or to dip-slip faults. Second, it is far from obvious that a piece-wise constant-with-depth slip distribution, such as implied by this analogy, could ever arise in nature. Last, we may be able to bring other geophysical and geologic information to bear on distinguishing between models.

We complete this analysis by computing the velocity and strain rate. The velocity is obtained by differentiating equation (6.37) with respect to time; the only time dependence coming from the term in equation (6.36). Differentiating (6.36) gives

$$\beta e^{-\beta t} \left[\sum_{m=1}^{n} \frac{(\beta t)^{n-m}}{(n-m)!} - \sum_{m=1}^{n-1} \frac{(\beta t)^{n-m-1}}{(n-m-1)!} \right].$$

Letting $q = m + 1$ in the second sum, we notice that all terms cancel except for the $m = 1$ term in the first sum, so that the velocity becomes simply

$$v(x, t) = \frac{\Delta u}{\pi t_R} e^{-t/t_R} \sum_{n=1}^{\infty} \frac{(t/t_R)^{n-1}}{(n-1)!} F_n(x, D, H). \qquad (6.39)$$

To compute the strain rate, differentiate the velocity (6.39) with respect to x:

$$\dot{\gamma}(x, t) = \frac{\Delta u}{\pi t_R} e^{-t/t_R} \sum_{n=1}^{\infty} \frac{(t/t_R)^{n-1}}{(n-1)!} G_n(x, D, H),$$

$$G_n(x, D, H) = \frac{-(2nH + D)}{x^2 + (2nH + D)^2} + \frac{(2nH - D)}{x^2 + (2nH - D)^2}. \qquad (6.40)$$

Results for $D/H = 1$ are shown in figure 6.12. Qualitatively, the behavior is similar to that of the Elsasser model. However, in the case of the elastic layer over a viscoelastic half-space, the postseismic velocities are lower and the deformation is more diffuse. This is a consequence of the instantaneous elastic behavior of the half-space. Figure 6.13 compares the behavior for the case where the earthquake ruptures only half of the lithospheric thickness.

It is instructive to compare the post-1906 triangulation data with the prediction of the layer over viscoelastic half-space solution. The triangulation data best constrain the maximum strain rate directly over the fault. For this analysis, I have supplemented these data with more recent strain-rate determinations following Kenner and Segall (2000). Notice from figure 6.14B that the observed strain rates decay to some nominally steady value after approximately

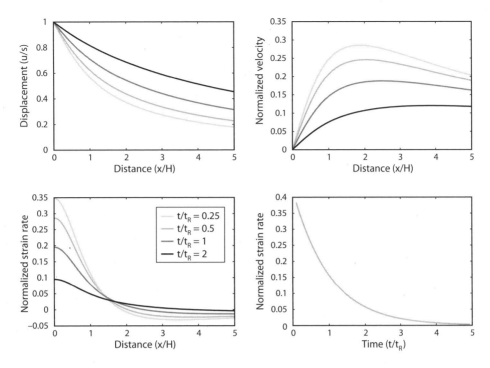

Figure 6.12. Results for an antiplane viscoelastic model, with $D/H = 1$. Upper left: Displacement. Upper right: Velocity. Lower left: Strain rate. Lower right: Peak strain rate as a function of time.

80 years. Since the Nur-Mavko model does not account for strain accumulation, I have added the observed strain rate for the last observing period to the model predictions from equation (6.40), evaluated at $x = 0$. In this way, I attempt to fit only the transient strains. A better approach is developed in chapter 12, where we will investigate interseismic deformation, including the long-term deformation associated with far-field plate motions. Also for simplicity, I take the average coseismic slip in 1906 to be 5.5 meters and assume that the earthquake ruptured the entire elastic layer. This leaves two parameters to be estimated from the data: relaxation time t_R and elastic layer thickness H. I estimated the best fitting values using a simple grid search (figure 6.14). The optimal elastic thickness is roughly 15 km, while the preferred relaxation time is ~22 years, corresponding to a viscosity of 7.2×10^{18} Pa-s for a reasonable shear modulus.

6.3.1 Stress in Plate and Half-Space

Before leaving this solution, it is worth noting that we have developed all the machinery for computing the stresses as a function of space and time, within both the elastic layer and the Maxwell half-space. Equations (5.32) and (5.34) give the displacements in the two regions for the corresponding elastic solution, from which the stresses are readily derived.

The stress is shown at six different times in figure 6.15 for a vertical strike-slip fault that cuts the entire elastic plate. Notice that immediately after the earthquake, the stress has decreased in the elastic plate in the vicinity of the fault but increased below. With time, the stresses in the half-space relax, and stress is transferred back to the overlying elastic plate. As time increases, the net stress perturbation, coseismic plus postseismic, decreases in both the elastic plate and the half-space.

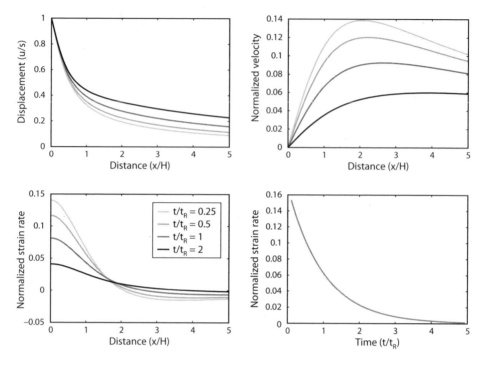

Figure 6.13. Results for an antiplane viscoelastic model, with $D/H = 0.5$. Upper left: Displacement. Upper right: Velocity. Lower left: Strain rate. Lower right: Peak strain rate as a function of time.

6.4 Strike-Slip Fault in Elastic Layer Overlying a Viscoelastic Channel

The previous section employed the infinite series solution for a dislocation in an elastic layer over an elastic half-space to solve the corresponding viscoelastic solution. Chapter 5 also developed propagator matrix methods for solving layered elastic dislocation problems. For example, equation (5.82) gives the solution for an infinitely long fault in an elastic layer over a half-space with differing modulus. Applying the correspondence principle, we have

$$\bar{u}_3(k, s) = \frac{-i \Delta u}{ks} \left\{ \sinh(D|k|) \left[\frac{(s + 2\beta) \cosh(H|k|) + s \sinh(H|k|)}{s \cosh(H|k|) + (s + 2\beta) \sinh(H|k|)} \right] - \cosh(D|k|) + 1 \right\}, \tag{6.41}$$

where Δu is fault slip, and the notation $\bar{u}_3(k, s)$ indicates that the displacement is a function of the Fourier wavenumber k and the Laplace transform variable s. The term in brackets can be written as

$$[\ \ ] = \left[\frac{s \exp(H|k|) + 2\beta \cosh(H|k|)}{s \exp(H|k|) + 2\beta \sinh(H|k|)} \right],$$

$$= \left\{ \frac{s + [1 + \exp(-2H|k|)]/t_R}{s + [1 - \exp(-2H|k|)]/t_R} \right\}, \tag{6.42}$$

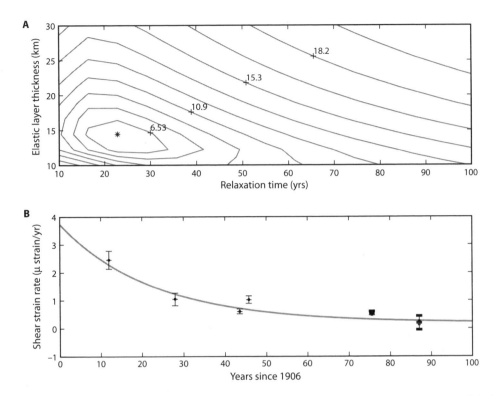

Figure 6.14. Comparison of post-1906 strain-rate determinations with predictions from a model of a faulted elastic layer overlying a Maxwell viscoelastic half-space. A: The misfit to the data is contoured as a function of the relaxation time and elastic layer thickness. The star marks the best fitting parameters. B: Comparison between the model prediction and observations.

so that

$$\bar{u}_3(k, s) = \frac{-i\Delta u}{ks} \left(\sinh(D|k|) \left\{ \frac{s + [1 + \exp(-2H|k|)]/t_R}{s + [1 - \exp(-2H|k|)]/t_R} \right\} - \cosh(D|k|) + 1 \right). \qquad (6.43)$$

The last two terms are proportional to $1/s$, so invert directly to the Heavyside function $H(t)$. The first term is of the form $(s + b)/[s(s + a)]$, where

$$b = [1 + \exp(-2H|k|)]/t_R,$$
$$a = [1 - \exp(-2H|k|)]/t_R. \qquad (6.44)$$

The transform is inverted as (see section A.2)

$$\mathcal{L}^{-1} \left\{ \frac{1}{s} \left(\frac{s+b}{s+a} \right) \right\} = \frac{b}{a}(1 - e^{-at}) + e^{-at}, \qquad (6.45)$$

so that the time-dependent displacements in the Fourier domain are

$$u_3(k, t) = \frac{-i\Delta u}{k} \left\{ \sinh(D|k|) \left[e^{-at} + \coth(H|k|)(1 - e^{-at}) \right] - \cosh(D|k|) + 1 \right\}. \qquad (6.46)$$

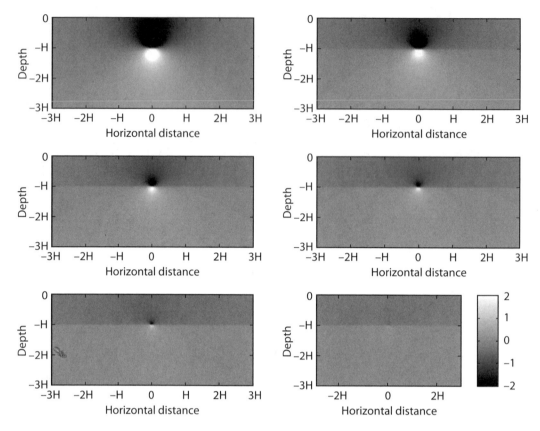

Figure 6.15. Total stress, coseismic plus postseismic, in an elastic plate overlying a Maxwell viscoelastic half-space. Each panel shows the shear stress on vertical planes (σ_{13}) at a different time, where time is normalized by the Maxwell relaxation time. At time zero, a uniform-slip dislocation is introduced that cuts the entire elastic plate, $D = H$. Panels show $t/t_R = 0.0, 0.6, 1.2, 1.8, 2.4$, and 4.0.

Note that the effective relaxation time, which is a function of wavenumber, is given by

$$a^{-1} = \frac{t_R}{1 - e^{-2H|k|}}, \qquad (6.47)$$

so that only the largest wavenumbers (smallest wavelengths) decay with characteristic time t_R. At the longest wavelengths, as $k \to 0$, the effective relaxation time becomes $(t_R/4\pi)(\lambda/H)$ and is unbounded. Here the wavelength λ is given by $2\pi/k$.

Figure 6.16 shows that an inverse FFT of equation (6.46) yields a solution that is in good agreement with the infinite series solution.

There are a number of interesting limiting cases of equation (6.46). The limit as $t \to 0$ recovers the elastic solution for a dislocation in a homogeneous half-space (5.71), as it must. The limit as $t \to \infty$ gives the solution for a dislocated plate:

$$\lim_{t \to \infty} u_3(k, t) = \frac{-i\Delta u}{k} \left[\sinh(D|k|) \coth(H|k|) - \cosh(D|k|) + 1 \right] . \qquad (6.48)$$

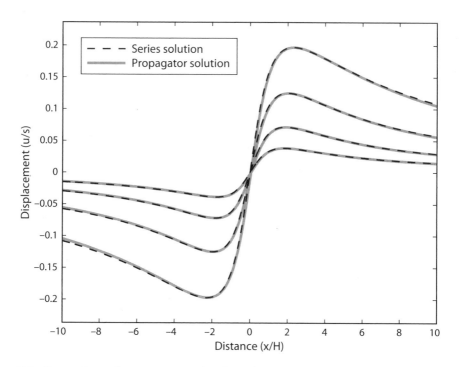

Figure 6.16. Comparison of propagator and series solutions to the problem of an infinitely long strike-slip fault in an elastic layer overlying a Maxwell viscoelastic half-space, $D = H$. Postseismic displacements are shown for $t/t_R = 0.25, 0.5, 1$, and 2.

If the depth of faulting is equal to the elastic layer thickness ($D = H$), then equation (6.46) reduces to

$$u_3(k, t) = \frac{-i\Delta u}{k} \left[1 - e^{-a(k)t}e^{-D|k|}\right].$$
(6.49)

The limit of equation (6.49) as $t \to \infty$ (or equation [6.48] for $D = H$) is simply $-i/k$, which is the transform of a step function. That is, if the earthquake ruptures the full elastic plate, as the viscous stresses dissipate, the plate relaxes to a rigid body offset.

The form of equation (6.49) suggests an asymptotic limit for $t/t_R \ll 1$. It is possible to invert the resulting expression analytically (problem 9), which yields

$$u_3(x, t) \cong \frac{\Delta u}{\pi} \left\{\tan^{-1}\left(\frac{D}{x}\right) + \frac{t}{t_R}\left[\tan^{-1}\left(\frac{x}{D}\right) - \tan^{-1}\left(\frac{x}{3D}\right)\right]\right\}.$$
(6.50)

Equation (6.50) provides a quite good approximation for $t/t_R < 0.5$, as can be seen in figure 6.17.

These methods can be extended to develop the solution for a fault in an elastic layer of thickness H overlying a viscoelastic channel of thickness h. This solution provides some insight into transient deformation behavior in the case that the earth's effective viscosity increases substantially with depth below the asthenosphere. Furthermore, the approach can be generalized to include an arbitrary number of layers with differing viscosities and elastic properties.

The procedure, by now familiar, is to solve the corresponding elastic problem for two layers of differing shear modulus, overlying an elastic half-space. We take the lower layer to be

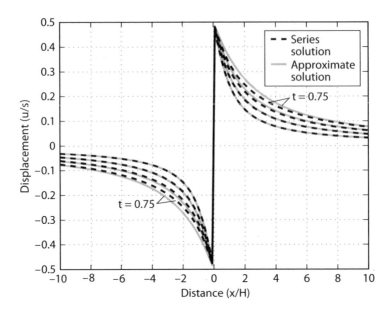

Figure 6.17. Comparison of the series solution and an approximate solution given by equation (6.50), at $t/t_R = 0$, 0.25, 0.5, and 0.75. Notice that the approximation is reasonably good for $t/t_R < 0.5$.

viscoelastic and apply the correspondence principle. After some considerable algebra, we find that

$$\bar{u}_3(k, s) = \frac{-i\Delta u}{ks} \left[\sinh(D|k|) A(H, h, k, s) - \cosh(D|k|) + 1 \right], \tag{6.51}$$

where

$$A(H, h, k, s) = \frac{s^2 + \beta s(2 + e^{-2H|k|} - e^{-2(h+H)|k|}) + \beta^2(1 - e^{-2h|k|} + e^{-2H|k|} - e^{-2(h+H)|k|})}{s^2 + \beta s(2 - e^{-2H|k|} + e^{-2(h+H)|k|}) + \beta^2(1 - e^{-2h|k|} - e^{-2H|k|} + e^{-2(h+H)|k|})}. \tag{6.52}$$

Notice that the denominator of $A(H, h, k, s)$ is quadratic in s, so there are two distinct relaxation times. In contrast, the infinitely long fault in a layer over a half-space had only a single relaxation time given by equation (6.47). The relaxation times are shown as a function of wavenumber in figure 6.18. For wavelengths short compared to the elastic layer thickness H, the effective relaxation times both equal the material relaxation time t_R. For longer wavelengths, there are two branches, one decays more slowly than the half-space solution, and the other has an effective relaxation time that tends toward half the material relaxation time in the limit $|k| \to 0$. Thus, we expect the long wavelength components to decay much more quickly than they do in the half-space solution.

We can write equation (6.51) in the following form:

$$\bar{u}_3(k, s) = \frac{-i\Delta u}{ks} \left[\sinh(D|k|) \frac{\mathcal{N}(s)}{a(s - s_1)(s - s_2)} - \cosh(D|k|) + 1 \right], \tag{6.53}$$

where s_1 and s_2 are the two roots of the denominator of A, and $\mathcal{N}$ represents the numerator of A. Thus, using the contour integration approach described in section A.2, the solution in the

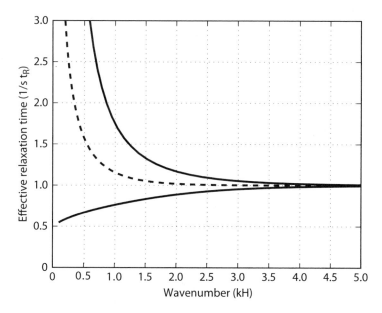

Figure 6.18. Effective relaxation times for an antiplane fault in a layer overlying a viscoelastic channel for the case when the channel thickness equals that of the elastic layer, $h = H$ (solid curves). Dashed curve gives the relaxation time for the layer over a half-space. Relaxation time is normalized by t_R, and wavenumber is normalized by H.

time domain is immediately found as

$$u_3(k, t) = \frac{-i\Delta u}{k} \left\{ \sinh(D|k|) \left[\frac{\mathcal{N}(s_1)}{a s_1 (s_1 - s_2)} e^{s_1 t} - \frac{\mathcal{N}(s_2)}{a s_2 (s_1 - s_2)} e^{s_2 t} + \frac{\mathcal{N}(0)}{a s_1 s_2} \right] - \cosh(D|k|) + 1 \right\}.$$

(6.54)

For a viscoelastic channel with thickness equal to the elastic layer thickness (figure 6.19, bottom), we find much smaller far-field displacements, in comparison with those resulting from a dislocation in an elastic layer over a half-space (figure 6.16), or equivalently a very thick viscoelastic channel (figure 6.19, top). The maximum postseismic displacement is also smaller in amplitude than it is for a viscoelastic half-space.

6.5 Dip-Slip Faulting

The analysis in the previous section suggests a procedure for dip-slip faulting in a two-dimensional elastic layer overlying a viscoelastic medium. We will write the elastic solution for a point source in a layer of modulus μ_1 overlying a half-space with modulus μ_2, following equation (5.117), as

$$\begin{bmatrix} i u_1 (z = 0) \\ u_2 (z = 0) \\ 0 \\ 0 \end{bmatrix} = \mathbf{P}(0, H, k, \mu_1) \left[c_1(k) \mathbf{v}_1(\mu_2, k) + c_2(k) \mathbf{v}_2(\mu_2, k) \right] - \mathbf{P}(0, z_s, k, \mu_1) \mathbf{f}_s(M, \mu_1),$$

(6.55)

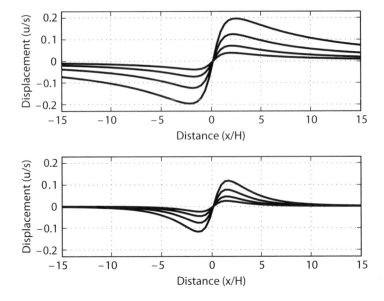

Figure 6.19. Postseismic displacement for an antiplane fault in a layer overlying a viscoelastic channel. The top panel shows the case when the viscoelastic channel is very large, $h/H = 100$, corresponding to the half-space result. The lower panel gives the result when the viscoelastic channel thickness h equals the elastic layer thickness H. Curves are drawn for several times: $t/t_R = 0.25, 0.5, 1$, and 2.

where k is the wavenumber, and $\mathbf{v}_1(\mu_2, k)$ and $\mathbf{v}_2(\mu_2, k)$ are the generalized eigenvectors of the homogeneous solution. The coefficients c_1 and c_2 are specified to satisfy the boundary conditions, and $\mathbf{f}_s$ is the source vector that depends on the moment tensor M and elastic properties. The term $\mathbf{P}(0, H, k, \mu_1)$ propagates the half-space solution to the surface, whereas the term $\mathbf{P}(0, z_s, k, \mu_1)$ propagates the source term to the surface. Since the source is within the elastic layer, only the half-space terms become functions of the transform variable s upon application of the correspondence principle. For an elastic layer over a half-space, only $\mathbf{v}_1$ and $\mathbf{v}_2$ are functions of μ_2 and therefore s. Combining the two generalized eigenvectors in a matrix $V(s)$ and defining $\mathbf{c} = [c_1, c_2]^T$, equation (6.55) can then be written compactly in the form of equation (5.120) as

$$\begin{bmatrix} iu_1(z=0) \\ u_2(z=0) \\ 0 \\ 0 \end{bmatrix} = \mathbf{P}(0, H)V(s)\mathbf{c} - \mathbf{P}(0, z_s)\frac{\mathbf{f}_s}{s}, \qquad (6.56)$$

where we have assumed a Heavyside source term, corresponding to an earthquake at time $t = 0$. $\mathbf{P}(0, H)$ and $\mathbf{P}(0, z_s)$ are functions only of μ_1 and therefore independent of s. The coefficients $\mathbf{c}$ are found by solving the last two equations. Thus,

$$\mathbf{c} = [\mathbf{P}_2(0, H)V(s)]^{-1}\mathbf{P}_2(0, z_s)\frac{\mathbf{f}_s}{s}, \qquad (6.57)$$

where the subscript notation $(_2)$ indicates the last two rows of the matrix. The displacements at the free surface are thus written in the form

$$
\begin{bmatrix} iu_1(z=0) \\ u_2(z=0) \end{bmatrix} = \mathbf{P}_1(0,\,H)V(s)[\mathbf{P}_2(0,\,H)V(s)]^{-1}\mathbf{P}_2(0,\,z_s)\frac{\mathbf{f}_s}{s} - \mathbf{P}_1(0,\,z_s)\frac{\mathbf{f}_s}{s},
\tag{6.58}
$$

where subscript $(_1)$ indicates the first two rows of the matrix.

The inverse of a two-by-two matrix can be written as

$$
A^{-1} = \frac{Adj(A)}{|A|} = \frac{1}{|A|}\begin{bmatrix} A_{22} & -A_{12} \\ -A_{21} & A_{11} \end{bmatrix},
\tag{6.59}
$$

where $|A|$ is the determinant, and $Adj(A)$ is known as the *adjugate*. This suggests that if we write $V(s)$ such that all terms are nonsingular in s, the only poles in equation (6.58), other than that due to the source term, arise from the determinant $|\mathbf{P}_2(0,\,H)V(s)|$. Since the coefficients c_1 and c_2 are arbitrary functions of wavenumber and elastic constants, it is possible to scale each of the generalized eigenvectors so that they are nonsingular. For the plane strain case, this results in the following form for $V(s)$:

$$
V = \begin{bmatrix} -(s+\beta) & -(s+\beta)[(1+Hk)(K\beta+s\lambda)+(2+Hk)s\mu] \\ s+\beta & (s+\beta)\{-s\mu+Hk[K\beta+s(\lambda+\mu)]\} \\ -2ks\mu & -2ks\mu(1+Hk)[K\beta+s(\lambda+\mu)] \\ 2ks\mu & 2Hk^2s\mu[K\beta+s(\lambda+\mu)] \end{bmatrix} \quad k<0,
\tag{6.60}
$$

and

$$
V = \begin{bmatrix} -(s+\beta) & (s+\beta)[(1-Hk)(K\beta+s\lambda)+(2-Hk)s\mu] \\ -(s+\beta) & -(s+\beta)\{s\mu+Hk[K\beta+s(\lambda+\mu)]\} \\ 2ks\mu & 2ks\mu(Hk-1)[K\beta+s(\lambda+\mu)] \\ 2ks\mu & 2Hk^2s\mu[K\beta+s(\lambda+\mu)] \end{bmatrix} \quad k>0.
\tag{6.61}
$$

Note that K is the bulk modulus $K = \lambda + 2\mu/3$, and k is the wavenumber. As before, β is the inverse Maxwell relaxation time. It should be noted, recalling the elastic case, that equations (6.60) and (6.61) are not a unique representation. In detail, I multiplied the first eigenvector by $2\mu ks$ and the second by $2\,k^2s\mu\,[K\,\beta+s\,(\lambda+\mu)]$. It is not difficult to show that in this case, the determinant of $[\mathbf{P}_2(0,\,H)V(s)]$ is cubic in s; hence

$$
\begin{bmatrix} iu_1(k,\,z=0,\,s) \\ u_2(k,\,z=0,\,s) \end{bmatrix} = \frac{\mathbf{P}_1(0,\,H)V(s)Adj[\mathbf{P}_2(0,\,H)V(s)]\mathbf{P}_2(0,\,z_s)\mathbf{f}_s}{as(s-s_1)(s-s_2)(s-s_3)} - \mathbf{P}_1(0,\,z_s)\frac{\mathbf{f}_s}{s},
\tag{6.62}
$$

where the s_i, $i = 1, 2, 3$ are the roots of the determinant. The roots are found *exactly* by factoring the determinant in powers of s and then using the analytical formula for the roots of a cubic. These calculations are quite tedious and best done with a symbolic manipulator. However, written in this form, it is straightforward to invert the Laplace transform analytically using the residue theorem as discussed in section A.2. The displacements in the Fourier domain are

$$
\begin{bmatrix} iu_1(k, z = 0, t) \\ u_2(k, z = 0, t) \end{bmatrix} =
$$

$$
\frac{\mathbf{P}_1(0, H)V(s_1)Adj[\mathbf{P}_2(0, H)V(s_1)]\mathbf{P}_2(0, z_s)\mathbf{f}_s}{as_1(s_1 - s_2)(s_1 - s_3)}e^{s_1 t} + \frac{\mathbf{P}_1(0, H)V(s_2)Adj[\mathbf{P}_2(0, H)V(s_2)]\mathbf{P}_2(0, z_s)\mathbf{f}_s}{as_2(s_2 - s_1)(s_2 - s_3)}e^{s_2 t}
$$

$$
+\frac{\mathbf{P}_1(0, H)V(s_3)Adj[\mathbf{P}_2(0, H)V(s_3)]\mathbf{P}_2(0, z_s)\mathbf{f}_s}{as_3(s_3 - s_1)(s_3 - s_2)}e^{s_3 t} - \frac{\mathbf{P}_1(0, H)V(0)Adj[\mathbf{P}_2(0, H)V(0)]\mathbf{P}_2(0, z_s)}{as_1 s_2 s_3}
$$

$$
-\mathbf{P}_1(0, z_s)\mathbf{f}_s. \tag{6.63}
$$

Note that the preceding can be written in compact form:

$$
\boxed{u_i(k, t) = u_i^e(k)\mathcal{H}(t) + \sum_{n=1}^{3} \Upsilon_i^{(n)}(H, k, \mathbf{x}_s)(e^{s_n t} - 1),} \tag{6.64}
$$

where u_i^e are the elastic displacements, and $\Upsilon_i^{(n)}(H, k, \mathbf{x}_s)$ are the rather involved functions found by carrying out the matrix operations in equation (6.63). H is the elastic layer thickness, $\mathcal{H}(t)$ is the Heaviside function, k is the wavenumber, $\mathbf{x}_s$ is the source location, and s_n are the inverse relaxation times. The elastic response is found by setting $t = 0$ in equation (6.63); we see that the $\Upsilon_i^{(n)}$ are thus simply the expressions in equation (6.63) subtracting the elastic component.

At this point, one can generate solutions, either by numerically summing point-source solutions or by analytically integrating the previous point-source results over the fault surface. The latter is again best achieved with a symbolic mathematics package. Figure 6.20 shows the relaxation times corresponding to the three roots of the determinant in equation (6.63). Notice that while two of the relaxation times increase with increasing wavelength, the short wavelength limits are not equal to the material relaxation time. It is not difficult to show that adding an additional Maxwell viscoelastic layer causes the determinant to become a fifth-order polynomial in s, so there are two additional relaxation times (see problem 10).

6.5.1 Examples
Some examples are now warranted. Figure 6.21 shows the postseismic displacements due to a 30-degree dipping thrust fault that cuts fully through the elastic layer. Notice that the horizontal displacements are directed inward toward the bottom edge of the fault, with the sign change occurring at approximately one elastic layer thickness from the fault trace. There is post-earthquake uplift above the hanging wall of the fault at distances $x \geq H$, with lesser subsidence for $x \leq H$. Figure 6.22 shows the same information in vector form.

The vertical pattern of deformation changes dramatically when the fault breaks only the upper half of the elastic plate (figure 6.23). In this case, there is subsidence broadly symmetric about the hanging wall with lesser amounts of uplift in the flanking regions. The horizontal displacements are again directed inward toward the lower edge of the fault, the sign change occurring at roughly $x = 0.8H$.

Surprisingly, when the fault cuts only the lower part of the elastic plate, the vertical displacements change sign (figure 6.24). In this case, there is broad uplift in the vicinity of the

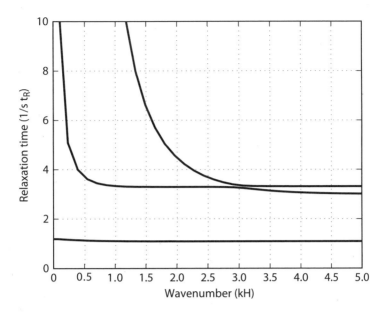

Figure 6.20. Relaxation times for dip-slip faulting in an elastic layer overlying a Maxwell half-space. Relaxation time is normalized by t_R, and wavenumber is normalized by H. Both layer and half-space have $\lambda = \mu$.

fault, with lesser amounts of subsidence on either side. Again, the horizontal displacements are directed toward the fault, but in this instance, the sign change occurs at roughly $x = 2H$.

The preceding figures show only the transient postseismic displacements, corresponding to those that would be measured geodetically following a large earthquake. To compute the total displacement, postseismic plus coseismic, one simply adds in the elastic displacements due to the earthquake. For an elastically homogeneous half-space, this can be accomplished using the analytic expressions for the surface displacements derived in chapter 3. This may be advantageous for faults that break the earth's surface, since the discontinuous coseismic displacements are not well modeled by a finite Fourier expansion. The postseismic deformations, however, are continuous at the earth's surface and well modeled with a finite Fourier transform.

Figure 6.25 shows the total cumulative displacement when a thrust fault cuts the entire elastic layer. As expected, as $t \to \infty$, the displacement approaches rigid body block motion, since stresses must decay to zero in the Maxwell half-space. Notice that because we have not yet included the effects of gravity, the fully relaxed state has the entire foot wall subsiding and the entire hanging wall uplifting. This of course cannot occur in nature, where residual elastic stresses in the relaxed configuration are balanced by gravitational forces, an effect that is treated in chapter 9.

6.6 Three-Dimensional Calculations

The approach to solving viscoelastic relaxation problems in three dimensions should now be clear. The first step is to solve for a point-source dislocation in a three-dimensional layered elastic medium, as outlined in chapter 5. As discussed there, the problem in three dimensions contains two parts, one equivalent to the antiplane strain problem and one equivalent to the plane strain problem, that are coupled through the dislocation source terms. Given the elastic solution, one then applies the correspondence principle and inverts the resulting Laplace transforms, either semianalytically as we have done here, or using an appropriate numerical

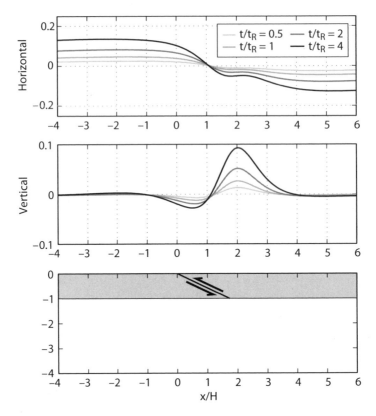

Figure 6.21. Postseismic displacements due to a 30-degree dipping thrust fault that cuts an elastic layer overlying a Maxwell half-space. Displacements are scaled by the fault slip and do not include the coseismic contribution. Lower panel illustrates fault geometry.

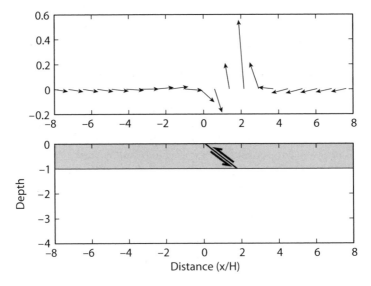

Figure 6.22. Postseismic displacements in vector form for a 30-degree dipping thrust fault that cuts an elastic layer overlying a Maxwell half-space. $t/t_R = 4$. Lower panel illustrates fault geometry.

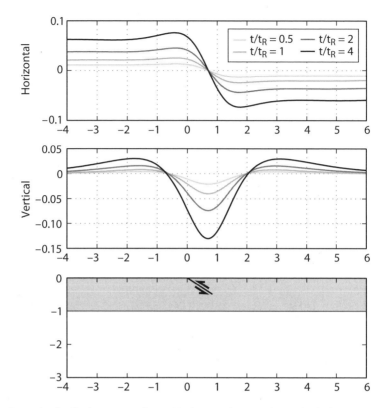

Figure 6.23. Postseismic displacements for a 30-degree dipping thrust fault that cuts the upper half of an elastic layer overlying a Maxwell half-space. Distance scale normalized by elastic plate thickness, and displacements normalized by the coseismic slip. Lower panel illustrates fault geometry.

procedure. The displacements are then reconstructed with a inverse Fourier transform. In the case of an elastic layer overlying a Maxwell viscoelastic half-space with N intervening Maxwell viscoelastic layers, there will be $3 + 2N$ plane strain relaxation modes and $1 + N$ antiplane relaxation modes, for a total of $4 + 3N$ modes.

The first results following this approach appear to have been presented by Barker (1976). Following this, Rundle (1978) and subsequent work used the solution for a point dislocation in a layered half-space by expanding the elastic fields in cylindrical vector harmonics. He then used an approximate technique to invert the Laplace transform. Matsu'ura and others (1981) extended these methods to a layered system and consider the effects of a confined viscoelastic layer. Figure 6.26 shows vertical displacements for a thrust fault breaking the upper part of an elastic layer overlying a Maxwell viscoelastic half-space. The results compare qualitatively to the two-dimensional solution shown in Figure 6.23.

6.7 Summary and Perspective

Considering that temperatures increase with depth in the earth and that rocks become increasingly ductile with increasing temperature leads to the expectation of time-dependent rheology below the shallow brittle crust. This expectation is broadly consistent with observations of time-dependent deformation following large earthquakes, as well as changes in surface loads including the melting of continental ice sheets. Linear viscoelastic constitutive

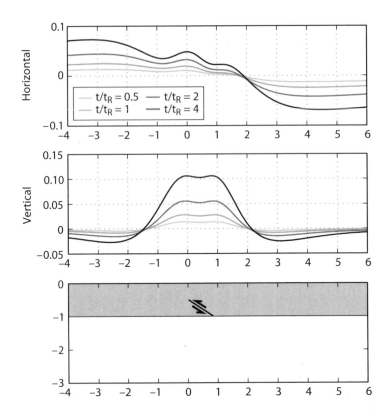

Figure 6.24. Postseismic displacements for a 30-degree dipping thrust fault that cuts the lower half of an elastic layer overlying a Maxwell half-space. Distance scale normalized by elastic plate thickness, and displacements normalized by the coseismic slip. Lower panel illustrates fault geometry.

laws exhibit instantaneous elastic response as well as longer term stress relaxation. While many studies have focused on the Maxwell rheology, arbitrarily complex rheologies can be visualized as combinations of elastic springs and viscous dashpots. Relaxation of rock is dominantly in shear, so the tensor form of the constitutive laws attributes viscous relaxation only to the deviatoric components, while the isotropic response is perfectly elastic.

The correspondence principle associates a viscoelastic boundary value problem with a corresponding inhomogeneous, elastic problem in the Laplace domain. If the solution to the elastic problem can be found—for example, using image or propagator matrix methods (as discussed in chapter 5)—the solution to the viscoelastic problem is reduced to inverting the resultant Laplace transform. In many cases, this can be accomplished by simply using the residue theorem.

In this chapter, we explore solutions for infinitely long strike-slip and dip-slip faults in an elastic layer overlying a Maxwell viscoelastic half-space. Extension of these results to include multiple viscoelastic layers, or more complex rheologies (for example, the standard linear solid or Burger's rheology) is in principle straightforward, although the algebra may become unwieldy. Poles in the Laplace domain correspond to independent relaxation times, each a function of spatial wavenumber. For a layer overlying a Maxwell half-space, it is possible to determine the poles analytically, greatly facilitating computations. Solutions in three dimensions are found from the corresponding layered elastic solutions discussed in chapter 5, although details are not presented here.

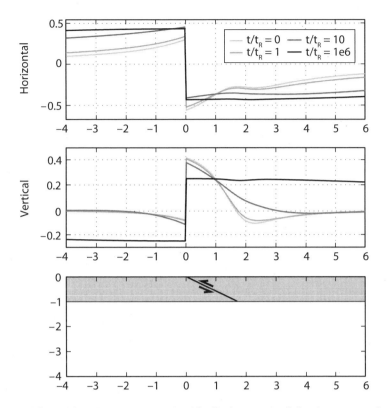

Figure 6.25. Cumulative (coseismic + postseismic) displacements following an earthquake on a 30-degree dipping thrust fault that cuts through an elastic layer overlying a Maxwell half-space. Gravitational effects are not included (see chapter 9). Distance scale normalized by elastic plate thickness, and displacements normalized by the coseismic slip. Lower panel illustrates fault geometry.

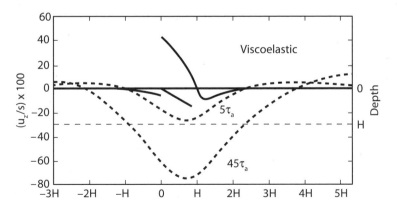

Figure 6.26. Reverse faulting in the upper part of a three-dimensional elastic layer overlying a Maxwell viscoelastic half-space. Displacements are shown along the symmetry axis of the fault. Fault geometry shown on right axis. After Rundle (1982).

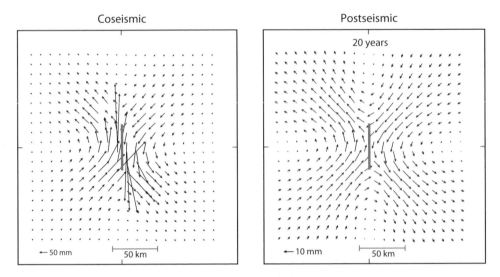

Figure 6.27. Coseismic and postseismic deformation due to slip on a vertical strike-slip fault. Elastic layer thickness is 16 km; underlying Maxwell medium has viscosity 10^{19} Pa-s, $\mu = 30$ Gpa. One meter of slip on a 50-km-long fault cuts the entire elastic layer. Postseismic displacements are computed 20 years after the earthquake. Calculation courtesy of Fred Pollitz.

The principal limitation of the presentation in this chapter is that we have neglected the effects of gravity, a discussion of which is postponed until chapter 9. There we find that gravitational effects become increasingly important with time as elastic stresses relax. Thus, solutions ignoring gravity may be approximately valid at short times, relative to the material relaxation time, following an earthquake. At longer times, the errors in these solutions become increasingly severe.

An alternative approach to the three-dimensional viscoelastic problem has been to use spherical harmonic basis functions to find solutions for elastic deformation in a radially symmetric spherical earth, followed by application of the correspondence principle. This was first done by Pollitz (1992) and later by Piersanti and others (1995). The principal differences between these approaches are in how gravitational and compressibility effects are treated.

Figure 6.27 shows horizontal displacements due to a vertical strike-slip fault in a spherical earth model. As expected, the postseismic displacements are much smaller and more widely distributed than the coseismic displacements. The coseismic displacements are discontinuous across the fault, whereas the postseismic displacements vanish at the fault trace.

The discussion in this chapter has been limited to linear viscoelastic rheologies, in large part because linear constitutive laws are far more amenable to analytic treatment. Laboratory experiments on rocks at temperatures that are a significant fraction of their melting point exhibit temperature-dependent creep. Steady-state shear strain rates in these experiments can be related to the deviatoric stress τ_{ij} through flow laws of the form

$$\dot{\epsilon}'_{ij} = A\tau_s^{n-1} \exp\left[\frac{-(Q + pV)}{RT}\right] \tau_{ij} \quad \tau_s = \sqrt{J_2}, \tag{6.65}$$

where $\dot{\epsilon}'_{ij}$ is the deviatoric strain rate, and τ_s is the maximum shear stress given by the square root of the second deviatoric stress invariant J_2; see equation (1.48). Here, Q is an activation energy, p is pressure, V is the activation volume, R is the gas constant, T is the absolute temperature, and n is an empirical stress exponent. The factor A depends on composition, grain size, and activity of H_2O, among other factors.

Ductile creep of crystalline materials occurs due to the motion of crystal dislocations, stress-controlled diffusion of lattice point defects, and grain boundary sliding. Increased dislocation density with ongoing deformation results in strain hardening. Hardening can be counteracted by recovery processes such as dislocation climb, a process enabled by thermally activated intracrystalline diffusion. Steady-state creep (sometimes referred to as *secondary creep*) occurs when recovery balances strain hardening, leading to an equilibrium microstructure.

For diffusion-controlled creep, the observed stress exponents n are close to 1, and the strain rate is linear in stress. However, for creep controlled by dislocation climb, the exponents n tend to be in the range of 3 to 6. In this case, the effective viscosity, defined as the ratio of stress to twice the strain rate, can be written as

$$\eta_{\text{eff}} = \frac{1}{2A} \tau_s^{1-n} \exp\left[\frac{(Q + pV)}{RT}\right]. \tag{6.66}$$

The effective viscosity decreases with increasing temperature, hence depth, for a given composition. For $n > 1$, the effective viscosity decreases with increasing stress magnitude. For rocks directly below the coseismic rupture, this could lead to a low effective viscosity immediately following an earthquake followed by an increasing effective viscosity as shear stresses diminish. Such an effect has been proposed to explain changes in deformation rates following large earthquakes, although the applicability of steady-state creep laws must be considered if the stress changes are large. Other workers have explained dramatic changes in deformation rates using a linear biviscous rheology (e.g., see equation [6.68]).

Pore-fluid flow induced by stress changes associated with earthquakes (see chapter 10) can also cause measurable postseismic deformation. There has been a tendency for individual studies to focus on a single mechanism for postseismic deformation. There is no reason to rule out the possibility that multiple processes are operative simultaneously and indeed interacting with one another. However, it may be that some processes dominate at particular timescales following an earthquake or in particular geologic environments. For example, rapid postseismic deformation may be explained by a combination of rapid afterslip (see chapter 11) and longer duration, distributed viscoelastic relaxation (see also chapter 12). More accurate and complete data sets following recent large earthquakes will undoubtedly shed light on these questions in the coming years.

6.8 Problems

1. Show that equation (6.10) is indeed the solution to diffusion equation (6.6). Show that the dimensionless velocity is given by equation (6.16) and the dimensionless strain rate by (6.17).

2. Using the Elsasser model for postseismic deformation, and assuming an elastic layer of thickness 15 km, an asthenospheric channel of 15 km, a shear modulus of 3×10^{10} Pa, and a viscosity of 10^{18} Pa-s, what is the predicted displacement at 15 km from the fault after 1 year? At 30 km from the fault after 4 years? Why does the displacement behave in this way?

3. A prediction of the Elsasser model is that the shear strain rate changes sign some distance from the fault. Ten years after an earthquake, you find that the shear strain rate changes sign 30 km from the fault. What is your estimate of the asthenospheric viscosity? (You may assume that all the other parameters are the same as in the previous problem.)

4. Derive the constitutive equation (6.20) for the standard linear solid.

5. Compute the creep and relaxation functions for a standard linear solid.

6. Derive the generalized tensor form of the Maxwell constitutive equation (6.24).

7. Show that the governing differential equation for the Burger's rheology is

$$2\mu_2\dot{\epsilon} + 2\eta_2\ddot{\epsilon} = \frac{\eta_2}{\mu_1}\ddot{\sigma} + \left(\frac{\mu_2}{\mu_1} + \frac{\eta_2}{\eta_1} + 1\right)\dot{\sigma} + \frac{\mu_2}{\eta_1}\sigma. \tag{6.67}$$

From this differential equation, show that the strain response to a step change in stress is

$$\epsilon(t) = \frac{\sigma_0}{2\mu_1} + \frac{\sigma_0}{2\mu_2}\left(1 - e^{-\frac{t}{\tau_2}}\right) + \frac{\sigma_0}{2\mu_1}\frac{t}{\tau_1}, \tag{6.68}$$

where $\tau_1 = \eta_1/\mu_1$ and $\tau_2 = \eta_2/\mu_2$. Note that the creep response consists of an instantaneous elastic response followed by a decaying transient with time constant τ_2 superimposed on a steady-state viscous creep.

8. Show that the stress in a Maxwell half-space beneath an infinitely long strike-slip fault in an elastic layer follows

$$\sigma_{13}(x_1, |x_2| > H, t) = \frac{\mu \Delta u}{2\pi}\sum_{m=0}^{\infty}\frac{(t/t_R)^m e^{-t/t_R}}{m!}\left[-\frac{x_2 - 2mH + D}{x_1^2 + (x_2 - 2mH + D)^2} + \frac{x_2 - 2mH - D}{x_1^2 + (x_2 - 2mH - D)^2}\right]. \tag{6.69}$$

9. Derive the approximate closed form viscoelastic solution for a strike-slip fault appropriate for times short compared to the material relaxation time, equation (6.50).

10. Show that for plane strain deformation, the response of a dip-slip fault in an elastic layer overlying a Maxwell viscoelastic layer overlying a Maxwell half-space has five independent relaxation modes—that is, show that the determinant is a fifth-order polynomial in s.

    Hint: One need not do all of the matrix computations. It suffices to write all terms in the appropriate powers of the Laplace transform variable s and then carry out the matrix operations to determine the order of the determinant.

6.9 References

Barker, T. 1976. Quasi-static motions near the San Andreas fault zone. *Geophysical Journal of the Royal Society* **45**, 689–705.

Elsasser, W. M. 1969. Convection and stress propagation in the upper mantle. In S. K. Runcorn (Ed.), *The application of modern physics to the earth and planetary interiors*. New York: Wiley, pp. 223–246.

Flugge, W. 1975. *Viscoelasticity*, 2nd ed. New York: Springer-Verlag.

Freed, A. M., R. Burgmann, and T. Herring. 2007. Far-reaching transient motions after Mojave earthquakes require broad mantle flow beneath a strong crust. *Geophysical Research Letters* **34**, L19302, doi:10.1029/2007GL030959.

Fung, Y. C. 1965. *Foundations of solid mechanics*. Englewood Cliffs, NJ: Prentice Hall.

Heki, K., S. Miyazaki, and H. Tsuji. 1997. Silent fault slip following an interplate thrust earthquake at the Japan Trench. *Nature* **386**(6625), 595–598.

Hetland, E. A., and B. H. Hager. 2005. Postseismic and interseismic displacements near a strike-slip fault: a two-dimensional theory for general linear viscoelastic rheologies. *Journal of Geophysical Research* **110**, B10401, doi:10.1029/2005JB003689.

Kenner, S., and P. Segall. 2000. Postseismic deformation following the 1906 San Francisco earthquake. *Journal of Geophysical Research* **105**, 13,195–13,209

Lehner, F. K., V. C. Li, and J. R. Rice. 1981. Stress diffusion along rupturing plate boundaries. *Journal of Geophysical Research* **86**, 6155–6169.

Li, V. C., and J. R. Rice. 1983. Preseismic rupture progression and great earthquake instabilities at plate boundaries. *Journal of Geophysical Research* **88**, 4231–4246.

———. 1987. Crustal deformation in great California earthquake cycles. *Journal of Geophysical Research* **92**, 11,533–11,551.

Matsu'ura, M., T. Tanimoto, and T. Iwasaki. 1981. Quasi-static displacements due to faulting in a layered half-space with an intervenient viscoelastic layer. *Journal of Physics of the Earth* **29**, 23–54.

Nur, A., and G. Mavko. 1974. Postseismic viscoelastic rebound. *Science* **183**, 204–206.

Peltzer, G., P. Rosen, F. Rogez, and K. Hudnut. 1996. Postseismic rebound in fault step-overs caused by pore fluid flow. *Science* **273**, 1202–1204.

Piersanti, A., G. Spada, R. Sabadini, and M. Bonafede. 1995. Global postseismic deformation. *Geophysical Journal International* **120**, 544–566.

Pollitz, F. 1992. Postseismic relaxation theory on the spherical earth. *Bulletin of the Seismological Society of America* **82**, 422–453.

———. 2003. Transient rheology of the uppermost mantle beneath the Mojave Desert, California. *Earth and Planetary Science Letters* **215**, 89–104.

Rundle, J. B. 1978. Viscoelastic crustal deformation by finite quasi-static sources. *Journal of Geophysical Research* **83**, 5937–5945.

———. 1982. Viscoelastic-gravitational deformation by a rectangular thrust fault in a layered earth. *Journal of Geophysical Research* **87**, 7787–7796.

Savage, J. C. 1990. Equivalent strike-slip earthquake cycles in half-space and lithosphere-asthenosphere earth models. *Journal of Geophysical Research* **95**, 4873–4879.

Thatcher, W. 1975. Strain accumulation and release mechanism of the 1906 San Francisco earthquake. *Journal of Geophysical Research* **80**, 4862–4872.

———. 1983. Nonlinear strain buildup and the earthquake cycle on the San Andreas fault. *Journal of Geophysical Research* **88**, 5893–5902.

———. 1984. The earthquake deformation cycle on the Nankai trough, southwest Japan. *Journal of Geophysical Research* **89**, 5674–5680.

———. 1990. Present-day crustal movements and the mechanics of cyclic deformation. In R. E. Wallace (Ed.), *The San Andreas fault system*, USGS Professional Paper 1515. Washington, DC: U.S. Government Printing Office, pp. 189–205.

7

Volcano Deformation

Measurements of deformation are one of the most important means for studying magmatic processes and monitoring active volcanoes. Indeed, deformation along with seismic monitoring is one of the principal means of assessing the potential for future eruptive activity. The reason for this is straightforward. As magma migrates toward the earth's surface, it forces aside the surrounding crust. This inevitably causes deformation that can be detected by a variety of modern techniques. Because the shallow crust is brittle, the deformation usually results in earthquakes that are also easily detected. In some cases, there is evidence to suggest that the onset of measurable deformation precedes the onset of volcanic seismicity (Langbein et al. 1993).

Figure 7.1 shows distance changes prior to the 18 May 1980, eruption of Mount St. Helens made with a geodimeter, a laser distance measuring device. Note that station North Point displaced 27 meters in 18 days, a rate of 1.5 meters/day. These extreme rates were localized to the north flank of the volcano, the area that eventually failed in a massive landslide that unleashed a northward-directed lateral blast. The massive bulging of the north flank, which was easily detected even in photographs of the volcano, was caused by the intrusion of a subsurface lava dome, or *cryptodome*, into the edifice of the volcano itself. The line length changes were roughly linear in time; there was no rate increase prior to the eruption, at least over the time period for which measurements were made. Instead, a shallow earthquake triggered the landslide, which unloaded the magma body and the overlying hydrothermal system, initiating the eruption.

A very different example is illustrated in figure 7.2, which shows transient strains recorded by a borehole strainmeter some 30 minutes prior to an eruption of Hekla volcano in Iceland. The strain rate changed at the same time as a sequence of small earthquakes, presumably coincident with the initiation of dike propagation from depth to the surface. The much lower viscosity basalt, relative to the Mount St. Helens dacite, resulted in fast propagation to the surface. A similar pattern of deformation and seismicity preceded a subsequent eruption of Hekla in 2000, leading to a short-term prediction by the Civil Defense of Iceland (Agustsson et al. 2000).

The objectives of volcano deformation studies are to (1) determine the geometry of subsurface magma bodies—i.e., whether the source of deformation is a dike, a sill, a roughly equidimensional chamber, or a hybrid source; (2) to quantify parameters of the source, for example its depth, dimensions, volume, and internal magma pressure; and (3) to better understand the physics of magma transport and eruption dynamics.

Observations of uplift and tilt from basaltic shield volcanoes, particularly on the Big Island of Hawaii, have given rise to the concept of *inflation and deflation cycles*. The observations— for example, as illustrated in figure 7.3—indicate relatively long periods of volcanic uplift and tilting away from the volcano summit, punctuated by rapid periods of subsidence and tilting toward the volcano. The subsidence, or deflation episodes, are accompanied either by eruptions or by dike intrusion into the flanks of the volcano. The evidence for intrusion consists of earthquake swarms as well as deformation patterns consistent with emplacement of dikes. From these and other data, a conceptual model has been developed with a shallow, central magma chamber that is supplied with melt from the mantle. As the pressure in the

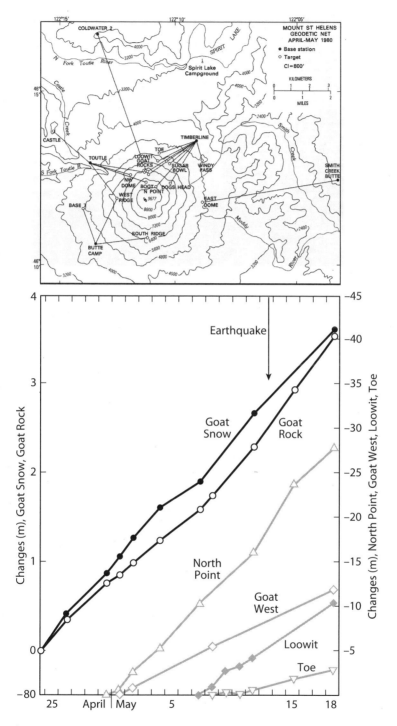

Figure 7.1. Deformation at Mount St. Helens prior to the catastrophic 18 May 1980 eruption. Top: A map view of the volcano and measured geodimeter lines. The change in distance along these lines was measured repeatedly prior to the eruption. Bottom: Changes in line length as a function of time. Note that although the rates of line-length change were extremely high, in some cases meters per day, the rates did not change significantly prior to the 18 May eruption. After Lipman and others (1981).

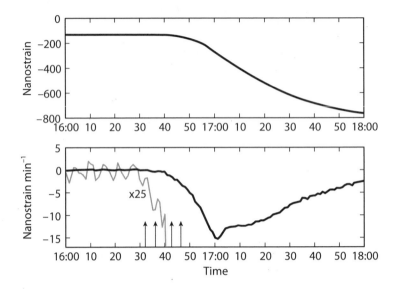

Figure 7.2. Borehole strain meter data, showing transient deformation beginning 30 minutes prior to the 1991 eruption of Hekla volcano, Iceland. Top: The volumetric strain. Bottom: The strain rate, also amplified by a factor of 25. The arrows mark the times of small earthquakes. The eruption occurred at roughly 17:00. After Linde and others (1993).

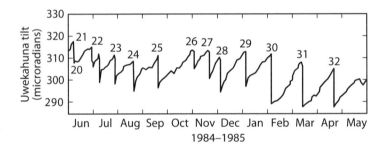

Figure 7.3. Tilt cycles at Kilauea summit. Positive tilts correspond to inflation; negative tilts to deflation. Each of the deflation episodes corresponds to a high fountaining phase during the early years of the Pu'u O'o eruption, which began in 1983. Courtesy Hawaiian Volcano Observatory website, http://hvo.wr.usgs.gov/.

chamber increases, the ground bulges upward, and the volcano expands, as in plate 7 (left). After some time, the increasing pressure causes the walls of the chamber to fracture, and a dike propagates, carrying magma either to the surface or into the volcano flanks, as illustrated in plate 7 (right).

As has been our custom, we will begin with a simple model of a magma chamber. Specifically, we investigate a pressurized spherical cavity in a homogeneous elastic half-space. We relate this solution to a point-source solution involving the elastic Green's functions used for elementary dislocation solutions in chapter 3. Next, we will investigate different magma chamber geometries, including ellipsoidal cavities, dikes, and sills. Needless to say, these models only approximate realistic magma chambers. In most cases, the boundary conditions imposed on the chamber walls only approximate those that act in the earth. For example, we typically approximate the boundary of the magma body as a liquid-elastic solid interface, although in a later section, we will include the effects of viscoelastic response in a zone around

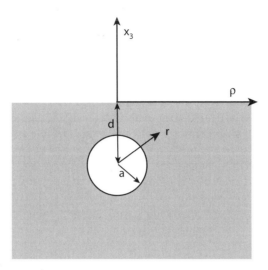

Figure 7.4. Geometry of a spherical magma chamber. The chamber has radius a, with center at depth d beneath the free surface, $x_3 = 0$. r denotes the radial distance from the center of the magma chamber, whereas ρ marks the distance from the center of symmetry along the free surface.

a spherical magma body. We do not analyze here the effects of elastic heterogeneity, which can be explored using methods described in chapter 5. The effect of nonplanar surface topography, which can be prominent with stratovolcanoes, is treated in chapter 8. Last, it is common to ignore thermoelastic and poroelastic effects; these are treated in chapter 10.

We begin by exploring the simplest model, that of a spherical magma chamber subjected to uniform internal pressure in a homogeneous elastic half-space.

7.1 Spherical Magma Chamber

The problem geometry is specified in figure 7.4. We consider a spherical cavity of radius a, with center at depth d in a uniform half-space subjected to uniform internal pressure p. We require two coordinate systems: one a spherical system with origin at the center of the spherical cavity, where r denotes radial distance from the center; the second a cylindrical coordinate system centered above the chamber, where ρ measures the radial distance from the center of symmetry on the earth's surface. The boundary conditions are

$$\sigma_{i3} = 0 \quad \text{on} \quad x_3 = 0, \; i = 1, 2, 3,$$

$$\sigma_{rr} = -p \quad \text{on} \quad r = a,$$

$$\sigma_{r\theta} = \sigma_{r\phi} = 0 \quad \text{on} \quad r = a. \tag{7.1}$$

The first boundary condition specifies that the plane $x_3 = 0$ is traction free, the other two conditions specify that the radial stress equals a uniform pressure p on the chamber walls, where the shear tractions vanish.

The solution to this problem is far from trivial. The most useful solution is an approximate one given by McTigue (1987), and we will outline the derivation here, omitting substantial detail. The idea is as follows: We begin with the solution for a pressurized spherical cavity in an elastic full-space. This solution is straightforward but violates the boundary conditions on the plane $x_3 = 0$. We cancel these stresses by applying equal and opposite tractions on the plane $x_3 = 0$, causing it to be traction free. We now have an approximate solution that satisfies the free-surface boundary condition (first of equations [7.1]), but only approximately satisfies the boundary conditions on the spherical cavity (second and third of equations [7.1]). We show that this solution is equivalent to a point-source approximation, widely known as the *Mogi model* after Mogi (1958), who applied the mathematical solution given by Yamakawa (1955) to measurements of volcano subsidence and uplift. The same solution, for a point center of dilation in an elastic half-space, was found earlier by T. Sezawa (1931), who included a spherical plastic shell with constant yield stress surrounding the magma chamber, and by Anderson (1936), who used the predicted stress trajectories to rationalize the orientations of planar intrusions surrounding magmatic centers.

McTigue (1987) continues by removing the induced tractions on $r = a$ so that they once again satisfy the boundary condition on the chamber walls. This then violates the free-surface condition, which can be again corrected by applying equal and opposite tractions on $x_3 = 0$. The procedure could be followed indefinitely, leading to evermore accurate solutions. While this procedure of "reflections" is easy to state, it is far from simple to carry out.

The first step is to solve the problem of a pressurized spherical cavity in a full-space. The full-space solution possesses radial symmetry. From equation (1.99), stress equilibrium for a purely radial displacement field in the absence of body forces reduces to

$$\frac{\partial \sigma_{rr}}{\partial r} + \frac{1}{r}(2\sigma_{rr} - \sigma_{\theta\theta} - \sigma_{\phi\phi}) = 0. \tag{7.2}$$

By symmetry, $\sigma_{\theta\theta} = \sigma_{\phi\phi}$. From equation (1.24), the radial strain is $\epsilon_{rr} = \partial u_r / \partial r$, while with radial symmetry, the hoop strains are given by $\epsilon_{\theta\theta} = \epsilon_{\phi\phi} = u_r / r$. Combining with Hooke's law yields the equilibrium equation in terms of displacement:

$$\frac{d}{dr}\left(\frac{du_r}{dr} + 2\frac{u_r}{r}\right) = 0, \tag{7.3}$$

assuming uniform elastic properties (problem 1). Integrating once leads to a first-order equation, which is solved as follows. Multiply each term by $e^{\Phi(r)}$:

$$e^{\Phi(r)}\frac{du_r}{dr} + \frac{2u_r}{r}e^{\Phi(r)} = Ae^{\Phi(r)}.$$

Now let $y = u_r e^{\Phi(r)}$, which leads to

$$\frac{dy}{dr} + \left(\frac{2}{r} - \frac{d\Phi}{dr}\right)u_r e^{\Phi(r)} = Ae^{\Phi(r)}. \tag{7.4}$$

The solution is to let the term in parentheses vanish, so that $\Phi = \log r^2$, and thus equation (7.4) reduces to $dy/dr = Ar^2$. This leads directly to

$$u_r = A\frac{r}{3} + \frac{B}{r^2}. \tag{7.5}$$

The requirement that the displacements vanish far from the magma chamber implies that $A = 0$. The constant B is determined by the boundary condition on the chamber wall. The radial strain is $\epsilon_{rr} = \partial u_r / \partial r$, while the hoop strains are given by $\epsilon_{\theta\theta} = \epsilon_{\phi\phi} = u_r / r$. Thus, the volumetric strain ϵ_{kk} vanishes, and the radial stress is simply

$$\sigma_{rr} = 2\mu\epsilon_{rr} = \frac{-4\mu B}{r^3}, \tag{7.6}$$

so that $B = pa^3/4\mu$. Thus, the displacements and stresses are

$$u_r = \frac{pa^3}{4\mu r^2},$$

$$\sigma_{rr} = \frac{-pa^3}{r^3},$$

$$\sigma_{\phi\phi} = \sigma_{\theta\theta} = 2\mu\frac{u_r}{r} = \frac{pa^3}{2r^3}. \tag{7.7}$$

Notice, perhaps surprisingly, that the state of stress is pure shear, $\sigma_{kk} = 0$. The radial stress is compressive, as expected, but the circumferential ("hoop") stresses are tensile, and exactly cancel the radial stress. You will see shortly that the mean normal stress due to a pressurized spherical cavity in a half-space does not vanish.

Notice also that the stresses decay with distance as $(a/r)^3$, so that on the plane $x_3 = 0$, the free-surface condition is violated at order $(a/d)^3$. We next transform the displacements and stresses to cylindrical polar coordinates, where ρ is the radial coordinate along the free surface. The radial displacements can be decomposed into vertical and horizontal components, where $u_z = u_r d/r$ and $u_\rho = u_r \rho/r$, so that

$$u_z(\rho, x_3 = 0) = \frac{pa^3}{4\mu}\frac{d}{(\rho^2 + d^2)^{3/2}},$$

$$u_\rho(\rho, x_3 = 0) = \frac{pa^3}{4\mu}\frac{\rho}{(\rho^2 + d^2)^{3/2}}. \tag{7.8}$$

To transform the stresses to cylindrical polar coordinates, using transformation rules from chapter 1, we note that

$$\sigma_{\rho z} = (\sigma_{rr} - \sigma_{\theta\theta})\sin\theta\cos\theta,$$

$$\sigma_{zz} = \sigma_{rr}\cos^2\theta + \sigma_{\theta\theta}\sin^2\theta, \tag{7.9}$$

where $\sin\theta = \rho/r$, $\cos\theta = d/r$. This then leads to

$$\sigma_{zz}(\rho, x_3 = 0) = -\frac{pa^3}{2}\left[\frac{3d^2}{(\rho^2 + d^2)^{5/2}} - \frac{1}{(\rho^2 + d^2)^{3/2}}\right],$$

$$\sigma_{\rho z}(\rho, x_3 = 0) = -\frac{3pa^3}{2}\frac{\rho d}{(\rho^2 + d^2)^{5/2}}. \tag{7.10}$$

To remove the tractions on $x_3 = 0$, we apply equal and opposite tractions to the surface of an elastic half-space. This can be done using Hankel transforms (McTigue 1987), which we omit for brevity. The displacements generated by the applied surface tractions are

$$u_z(\rho, x_3 = 0) = \frac{(3 - 4v)pa^3}{4\mu}\frac{d}{(\rho^2 + d^2)^{3/2}},$$

$$u_\rho(\rho, x_3 = 0) = \frac{(3 - 4v)pa^3}{4\mu}\frac{\rho}{(\rho^2 + d^2)^{3/2}}, \tag{7.11}$$

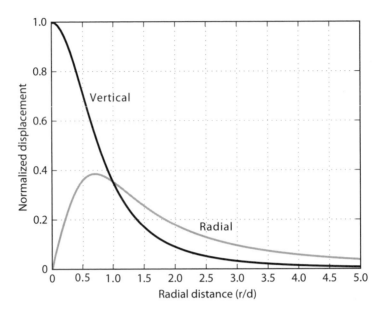

Figure 7.5. First-order approximation to surface displacements from a pressurized spherical magma chamber; equations (7.12) and (7.14).

so that the total displacement field, equations (7.8) plus (7.11), at this level of approximation, is

$$
\begin{aligned}
u_z &= \frac{(1-v)pa^3}{\mu}\frac{d}{(\rho^2+d^2)^{3/2}}, \\
u_\rho &= \frac{(1-v)pa^3}{\mu}\frac{\rho}{(\rho^2+d^2)^{3/2}}.
\end{aligned}
\tag{7.12}
$$

Note that the displacements on the free surface are a factor of $4(1-v)$ times the displacements evaluated on that plane in the full space (7.8). (Davies [2003] has shown that this holds for arbitrarily shaped inclusions that undergo a uniform expansion.)

Notice also that the ratio of horizontal to vertical displacement is $u_\rho/u_z = \rho/d$, which means that the displacements at the free surface are directed radially away from the center of the spherical source, as they were in the full-space. This is not true everywhere in the half-space; in general the displacements are not radial except near the chamber walls for small a/d, and at the free surface (as shown later in figure 7.9).

The imposed surface tractions have a length scale of d. The stresses due to the applied tractions decay from the surface with $(d/x_3)^3$. Near the cavity boundary, $x_3 \sim d$, the induced stresses are thus of the same order as the surface tractions, $(a/d)^3$. Thus, the solution is accurate everywhere to order $(a/d)^2$. The boundary conditions on $x_3 = 0$ are met exactly, but those on the cavity boundary are met only approximately. This approximation to the problem of a pressurized sphere in a half-space is widely known as the *Mogi model*, but perhaps should more properly be referred to as the *Sezawa-Anderson-Mogi-Yamakawa model*.

The pressure change p occurs with the cavity radius in the term pa^3. Thus, at least at this level of approximation, it is not possible to estimate the size of the cavity and the pressure change independently. It is possible, however, to estimate the *volume change* associated with

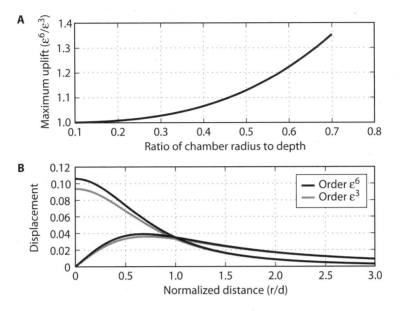

Figure 7.6. First- and second-order approximations to surface displacements from a pressurized spherical magma chamber. A: Maximum uplift normalized by the uplift predicted by the Mogi solution as a function of a/d. B: Vertical and radial displacements at order ϵ^3 and ϵ^6 for $a/d = 0.5$.

the deformation. The volume change ΔV is given by the integral of the radial displacement over the surface of the cavity. If the cavity radius is small compared to the depth of burial, then we may approximate the displacements on the cavity boundary by the full-space solution (7.7). By symmetry, $\Delta V = 4\pi a^2 u_r (r = a)$. From equation (7.7), $u_r (r = a) = pa/4\mu$, so that

$$\Delta V = \pi p a^3 / \mu. \tag{7.13}$$

We can thus rewrite equation (7.12) as

$$
\begin{aligned}
u_z &= \frac{(1-v)\Delta V}{\pi} \frac{d}{(\rho^2 + d^2)^{3/2}}, \\[2ex]
u_\rho &= \frac{(1-v)\Delta V}{\pi} \frac{\rho}{(\rho^2 + d^2)^{3/2}}.
\end{aligned}
\tag{7.14}
$$

The surface displacements due to the Mogi source are shown in figure 7.6. Notice that the uplift falls to half the maximum amount at a radial distance of $d(2^{2/3} - 1)^{1/2} \simeq 0.77d$. One can use vertical displacement data to estimate the depth of the source, simply by noting the radial distance at which the peak uplift drops by one-half. The maximum uplift can then be used to estimate the change in volume within the source, or the pressure change on the cavity boundary assuming that the radius of the chamber is known.

McTigue (1987) continues the analysis looking for higher order terms that might allow one to separately estimate the chamber radius and pressure change. The solution is sought in terms of a series expansion in the parameter $\epsilon = a/d$. The next step is to transform the order ϵ^3 stresses due the free-surface correction back into spherical coordinates and compute the tractions on the cavity wall. Expanding the *spatially variable* surface tractions in powers of ϵ

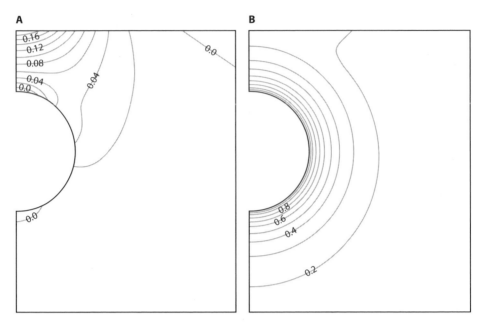

Figure 7.7. Stress due to pressurization of a spherical chamber. A: Mean normal stress. B: Deviatoric stress. Stresses are accurate to order $(a/d)^5$. After McTigue (1987).

leads to further corrections of order ϵ^3, ϵ^4, etc. The stresses due to these tractions on the cavity wall again decay with distance from the cavity as $(a/r)^3$. Thus, the newly applied tractions on the cavity violate the free-surface condition on $x_3 = 0$ at order ϵ^6, so that the next correction for the displacements is of order ϵ^6 where displacements are scaled by pd/μ. The free-surface displacement includes no contribution due to corrections at order ϵ^4 and ϵ^5, which indicates that unless the chamber is very close to the free surface, the simple approximation will be very accurate. Indeed, the effects of a finite cavity are sufficiently small that if one is going to consider them in modeling volcanic deformation, one should most likely consider other approximations (the spherical geometry of the chamber, the perfectly elastic behavior outside the chamber, homogeneous and isotropic material response, and neglect of surface topography) that may have equal or greater impact on the predicted deformation fields.

Figure 7.6 shows the vertical and radial displacements accurate to order ϵ^3 and ϵ^6, for $a/d = 0.5$. The upper figure shows the effect of the finite chamber radius on the maximum uplift. When the radius is 70% of the depth to the center of the cavity, the peak uplift is 35% greater than for the standard Mogi solution. The stresses in the neighborhood of the spherical cavity are shown in figure 7.7, accurate to order ϵ^5. Notice that the contours of deviatoric stress are nearly concentric around the chamber and that changes in mean normal stress are localized near the free surface. This reflects the fact that the spherical chamber in a full-space is a pure shear source and that the changes in mean stress occur only due to interactions with the free surface.

7.1.1 Center of Dilatation

While McTigue's (1987) analysis elegantly illustrates why the simplest approximation holds for such large ratios of chamber radius to depth, it is not the simplest way to generate the Mogi solution. A more direct method is to make use of the elastic Green's tensors to generate an isotropically expanding point source, known as a *center of dilatation*.

Figure 7.8. A center of dilatation consists of three mutually orthogonal double forces. Outside the neighborhood of the point source, the displacements and stresses are equivalent to those of a pressurized spherical cavity.

Recall that the elastostatic Green's tensors give the displacement due to a concentrated force as

$$u_i(x) = F_j g_i^j(x, \xi). \qquad (7.15)$$

A double force without moment is written as $F_j[g_i^j(x, \xi + d\xi_j) - g_i^j(x, \xi)]$, in the limit as $d\xi$ goes to zero. This is equivalent to differentiating the Green's function with respect to the source coordinate in the direction that the force acts. The displacement due to a double force $u_i^{DF}(x)$ is thus

$$u_i^{DF}(x) = F_j d\xi_j \frac{\partial g_i^j(x, \xi)}{\partial \xi_j} \qquad \text{no sum on } j. \qquad (7.16)$$

Taking three pairs of double forces, all without moment, acting in three mutually orthogonal directions leads to a center of dilatation (figure 7.8):

$$u_i^{CD}(x) = (F d\xi) \sum_{j=1}^{3} \frac{\partial g_i^j(x, \xi)}{\partial \xi_j}. \qquad (7.17)$$

Recall that the Green's function for the full-space is given by (equation [3.91])

$$g_i^j(x, \xi) = \frac{1}{4\pi\mu} \left[\frac{\delta_{ij}}{r} - \frac{1}{4(1-v)} \frac{\partial}{\partial x_j} \frac{\partial r}{\partial x_i} \right],$$

$$r = \sqrt{(x_1 - \xi_1)^2 + (x_3 - \xi_2)^2 + (x_3 - \xi_3)^2}. \qquad (7.18)$$

Substituting equation (7.18) into (7.17) and noting that by symmetry in the full-space (not so for half-space!), $\partial/\partial\xi_i = -\partial/\partial x_i$ gives

$$u_i^{CD}(x) = -\frac{(F d\xi)}{4\pi\mu} \left[\frac{\partial}{\partial x_i} \left(\frac{1}{r} \right) - \frac{1}{4(1-v)} \frac{\partial}{\partial x_i} \nabla^2 r \right],$$

$$= -\frac{(F d\xi)}{8\pi\mu} \left(\frac{1-2v}{1-v} \right) \frac{\partial}{\partial x_i} \left(\frac{1}{r} \right),$$

$$= \frac{(F d\xi)}{8\pi\mu} \left(\frac{1-2v}{1-v} \right) \frac{(x_i - \xi_i)}{r^3}. \qquad (7.19)$$

Notice that the motion is everywhere radially outward:

$$u_r^{CD} = \frac{(F d\xi)}{8\pi\mu} \left(\frac{1-2v}{1-v} \right) \frac{1}{r^2}. \qquad (7.20)$$

The radial stress is given by

$$\sigma_{rr}^{CD} = -\frac{(F\,d\xi)}{2\pi}\left(\frac{1-2\nu}{1-\nu}\right)\frac{1}{r^3}. \tag{7.21}$$

Comparing to equation (7.7), we note that the displacements and the stresses are of the same form as the previously derived solution for a pressurized cavity in a full-space for $r > a$. The equivalence can be made more precise by considering the following operations. Cut out a sphere of radius a around the center of dilatation. Replace the elastic material with a fluid at pressure

$$p = \frac{(F\,d\xi)}{2\pi a^3}\left(\frac{1-2\nu}{1-\nu}\right). \tag{7.22}$$

By comparison to equation (7.7), the displacements and stresses are the same for $r > a$, so the center of dilatation and the pressurized spheroid are equivalent in the limit that the chamber radius is small compared to the depth to the center of the chamber.

The equivalence is completed by noting that the volume change is, as before, $\Delta V = 4\pi a^2 u_r(r = a)$, which leads to

$$F\,d\xi = 2\mu\left(\frac{1-\nu}{1-2\nu}\right)\Delta V. \tag{7.23}$$

From equation (3.12), this implies that the moment tensor equivalent for a volumetric point source is given by

$$M_{ij} = 2\mu\left(\frac{1-\nu}{1-2\nu}\right)\Delta V\delta_{ij}. \tag{7.24}$$

To find the solution for a center of dilatation in a half-space, as opposed to a full-space, we simply use the half-space (Mindlin) Green's functions (3.93) and (3.94) in equation (7.17). By the equivalence discussed earlier, this is a first-order approximation to a pressurized spherical cavity in a full-space. The boundary conditions on the earth's surface are met exactly, because the Green's functions satisfy these boundary conditions, whereas the boundary conditions on the cavity wall are met only approximately. Thus, this approximation is equivalent to the first two steps in McTigue's (1987) analysis. The displacements in the body are shown in figure 7.9. Near the source, the displacements are radially symmetric as expected. At the free surface, the displacements are also radial to the source; however, in between they do not have this symmetry. The horizontal strains are extensional above and below the source and compressional off to the sides (figure 7.10).

The point center of dilatation model is very widely used in comparing model predictions to surface deformation data in volcanic terrains. This is partly due to the simplicity of the expressions for the surface deformation, but also due to the fact that a considerable amount of data is well fit by this model. Plate 8 shows one example from the Galápagos, where spatially extensive InSAR data are well fit by a point center of dilatation.

Figure 7.11 shows the fit of a center of dilatation model to cumulative uplift at Long Valley caldera, in eastern California, after Battaglia et al. (2003a). The uplift was computed by comparing elevations determined by GPS in the late 1990s with previous leveling surveys in the early 1980s. The uplift data is well fit by a center of expansion at a depth of 11.4 km, with a volume change of 0.24 km³. As will be shown, the horizontal displacements at Long Valley, however, are not well fit by this model. This emphasizes the importance of using both vertical and horizontal deformation data when inferring source shape. We will reexamine these data after introducing an ellipsoidal source model later in this chapter. The elliptical source is able to fit both the horizonal and the vertical data.

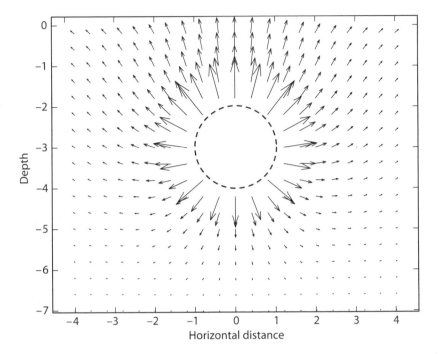

Figure 7.9. Displacements within the half-space due to a center of dilatation. Vectors within some radius of the source are not plotted.

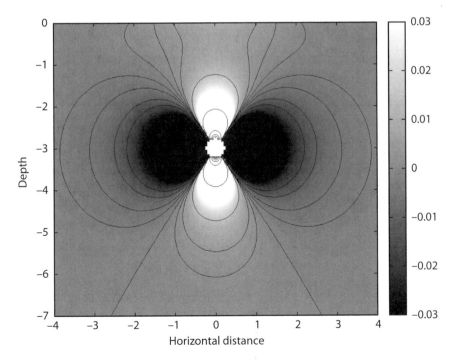

Figure 7.10. Horizontal strain due to center of dilation. Values within some radius of the source are not plotted. Strain is normalized by $\Delta V/d^3$.

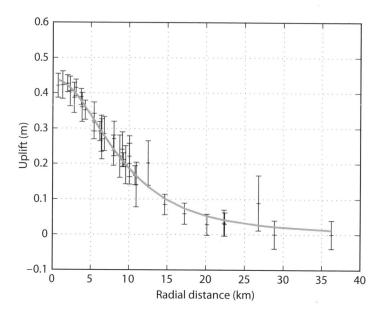

Figure 7.11. Uplift at Long Valley caldera, California, between 1982 and 1999 as a function of radial distance from the center of uplift. Fit is to a center of dilatation at a depth of 11.4 km with a volume change of 0.24 km^3. After Battaglia et al. (2003a).

7.1.2 Volume of the Uplift, Magma Chamber, and Magma

It is straightforward to show that the volume of the uplift is not the same as the volume change in a spherical magma chamber unless the surrounding crust is incompressible ($\nu = 0.5$). Integrating the vertical displacement over the free surface, the volume of the uplift is

$$V_{uplift} = 2(1 - \nu)\Delta V, \tag{7.25}$$

where ΔV is the volume change in the magma chamber. Two things may seem strange about this result. First, the volume of uplift is actually greater than the volume change of the magma chamber for $\nu < 0.5$. Second, as Poisson's ratio decreases and the material becomes more compressible, the discrepancy actually increases. It should be noted that this result holds only for a spherical magma chamber; for a horizontal circular sill, $V_{uplift} = \Delta V$ (Fialko et al. 2001; see section 7.4.1).

It is also interesting to ask how the change in the volume of the magma chamber relates to the volume of intruded magma (Johnson 1992; Delaney and McTigue 1994). Consider the change in mass of the magma $m = \rho_m V_m$:

$$\delta m = \rho_m \delta V_m + V_m \delta \rho_m, \tag{7.26}$$

where V_m and ρ_m are the volume and density of the magma. The compressibility of the magma is defined as

$$\beta_m = \frac{1}{\rho_m} \frac{\partial \rho_m}{\partial p}, \tag{7.27}$$

where p is pressure. Thus,

$$\delta m = \rho_m \delta V_m + \rho_m \beta_m V_m \delta p. \tag{7.28}$$

For a spherical magma chamber, $\delta V_m = \pi a^3 \delta p/\mu$ (equation [7.13]). Thus,

$$\delta m = \rho_m \pi a^3 \left(\frac{4}{3}\beta_m + \frac{1}{\mu} \right) \delta p. \tag{7.29}$$

One might have been tempted to estimate the mass change as simply the volume change of the chamber times the magma density $\rho_m \pi a^3 \delta p/\mu$. To do so, however, underestimates the mass change in the magma chamber by a factor of

$$\left[\frac{4\beta_m\mu}{3} + 1 \right]. \tag{7.30}$$

If we take the shear modulus of a somewhat fractured crust to be 1×10^{10} Pa and the compressibility of a gas-poor basalt of $10^{-10}\,\text{Pa}^{-1}$ (Murase and McBirney 1973), then ignoring compressibility of the melt could lead to an underestimate of the mass of added magma by a factor of 2.3. If, on the other hand, the magma is saturated with respect to a volatile phase, the compressibility of the magma can increase dramatically. The density of magma with a separate gas phase is given by

$$\rho_m = \left[\frac{\chi_g}{\rho_g} + \frac{1-\chi_g}{\rho_l} \right]^{-1}, \tag{7.31}$$

where χ_g is the mass fraction of volatile constituent in the gas phase, and ρ_g and ρ_l are the gas and liquid densities, respectively. It is typically assumed that the gas density is given by the ideal gas law $\rho_g = pM/RT$, where M is the molar mass, R is the gas constant, and T is temperature. (Note that the isothermal compressibility of an ideal gas is $1/p$, whereas the adiabatic compressibility is $[c_v/c_p]/p$, where c_v and c_p are the specific heat capacities at constant volume and pressure respectively, and c_v/c_p is of order unity.) For example, a basalt magma at a depth of 3 km that is saturated with respect to CO_2 at a depth of 15 km will contain 0.25 weight percent CO_2, have a bubble volume fraction of 1.8%, and a compressibility of 3.6×10^{-10} 1/Pa. The simplified calculation here ignores the effect of increased gas exsolution as the pressure decreases, which increases the compressibility, as well as the influence of multiple volatile phases. Employing the same shear modulus earlier, ignoring the magma compressibility, would underestimate the intruded mass by a factor of 5.8.

These considerations need to be kept in mind when comparing volume changes associated with magma withdrawal from a source reservoir with the volume change of intruded or erupted magma. It is after all the mass that is conserved, not the volume. If the mass change in the source reservoir is δm^{source} and the mass change at the sink is δm^{sink}, then mass conservation requires $\delta m^{source} + \delta m^{sink} = 0$. Define a magma chamber compressibility as $\beta_c = (1/V)\partial V/\partial p$. Equation (7.28) becomes

$$\delta m = \rho_m \delta V \left(1 + \frac{\beta_m}{\beta_c} \right). \tag{7.32}$$

Only in the limit that the magma is incompressible relative to the magma chamber is the mass change given by $\delta m = \rho_m \delta V$. Conserving the mass between the source and the sink then leads to

$$\delta V^{source} = -\delta V^{sink} \left(\frac{\rho_m^{sink}}{\rho_m^{source}} \right) \left(\frac{1 + \beta_m^{sink}/\beta_{sink}}{1 + \beta_m^{source}/\beta_{source}} \right), \tag{7.33}$$

where ρ_m^{source} and β_m^{source} are the magma density and compressibility in the source region, and ρ_m^{sink} and β_m^{sink} refer to the same parameters in the magma sink. β_{source} and β_{sink} refer to the compressibility of the source and sink reservoirs.

If the magma erupts, then $\beta_{sink} = \infty$. Similarly, because planar bodies such as dikes and sills are effectively compliant, $1 + \beta_m^{sink}/\beta_{sink} \sim 1$. This suggests that the volume decrease of a spheroidal magma chamber might be considerably less than the volume increase of a dike fed by that chamber. In addition, as magma moves from deep reservoirs into the shallow crust, volatile constituents will exsolve from solution. This causes $\rho_m^{sink}/\rho_m^{source}$ to be less than one and may help to explain why broadscale subsidence is not always observed in association with uplift and shallow intrusion (Rivalta and Segall 2008).

7.2 Ellipsoidal Magma Chambers

Actual magma chambers are unlikely to be perfectly spherical. An early analysis of surface deformation due to a variety of magma chamber shapes was conducted by Dieterich and Decker (1975). They examined a number of axially symmetric magma chamber geometries using the finite element method, including a sphere, a circular sill, a "pill-shaped" stock, and a point source. The important result is that the vertical displacements at the earth's surface are very similar for all models, if one scales the depth of the source appropriately (figure 7.12). Since the source depth is usually not known a priori—this is one of the things we hope to determine from the data—it is apparent that we cannot determine the source shape (and hence depth) with vertical displacement data alone.

On the other hand, the horizontal displacements are quite different for different magma chamber shapes (figure 7.12). As one would expect, the horizontal sill produces relatively little horizontal displacement, whereas the stock produces relatively more horizontal displacement. Prior to the advent of GPS, it was rare to have accurate measurements of vertical and horizontal displacements at the same time and place. With accurate three-dimensional deformation measurements, however, we can determine some features of the magma chamber geometry. It is worth remembering that forward models may be biased by the effects of heterogeneous earth structure, topography, and so on.

An obvious extension of the spherical cavity would be to a general three-dimensional ellipsoidal cavity subjected to uniform internal pressure. Given the difficulty of finding an exact solution for the spherical cavity, we anticipate that a complete treatment of the ellipsoid will be quite challenging. Hence, efforts have focused on various approximate solutions. For completeness, we state the boundary conditions here:

$$\sigma_{i3} = 0 \quad \text{on} \quad x_3 = 0, \quad i = 1, 2, 3,$$
$$\sigma_{ij}n_j n_i = -p \quad \text{on} \quad \mathcal{S}, \tag{7.34}$$

where $\mathcal{S}$ is the surface of the ellipsoid, and $\mathbf{n}$ is the unit normal to the surface.

Approximate solutions for a pressurized ellipsoidal cavity build on Eshelby's (1957) solution for an ellipsoidal inclusion in a elastic medium. Eshelby's method is based on a hypothetical "cut and weld" thought experiment. The first step is to cut out an ellipsoidal region from an elastic full-space (figure 7.13). The matrix is not subjected to stress, so it undergoes no deformation due to the removal of the inclusion. The inclusion is then subjected to a spatially uniform stress-free *transformation* strain, ϵ_{ij}^T. An example of a transformation strain would be a thermoelastic strain, or a strain due to a phase change. Note that at this stage, the stress in both the inclusion and the matrix is zero. The second step is to apply tractions to the boundary of the inclusion such that it is elastically deformed back to its initial size and shape. The required tractions are

$$T_i = -C_{ijkl}\epsilon_{kl}^T n_j, \tag{7.35}$$

such that the inclusion is now subjected to a uniform elastic stress $\sigma_{ij} = -C_{ijkl}\epsilon_{kl}^T$.

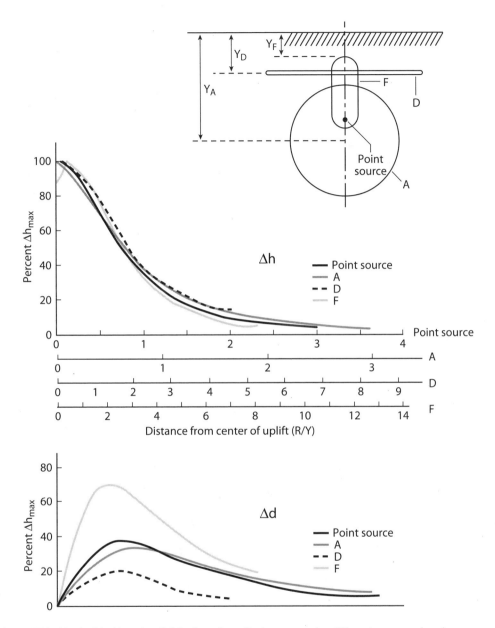

Figure 7.12. Vertical (Δh) and radial (Δd) surface displacements for different magma chamber geometries. Displacements are normalized by maximum uplift. Radial distance scale is normalized by different depths y, for different source shapes. After Dieterich and Decker (1975).

The inclusion now fits precisely back into the matrix. The next step is to "weld" the inclusion in place, maintaining the surface tractions. At this stage, the matrix remains stress free, the total strain in the inclusion (the sum of inelastic transformation strain and elastic strain) is zero, whereas the stress in the inclusion, proportional to the *elastic* strain is $\sigma_{ij} = -C_{ijkl}\epsilon_{kl}^T$.

The final step is to relax the tractions acting on the boundary of the ellipsoid and allow the inclusion to deform. The inclusion cannot fully relax because it is constrained by the

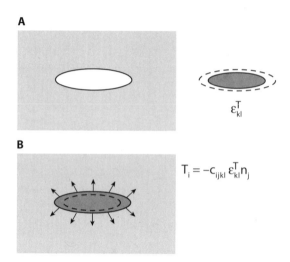

Figure 7.13. Eshelby cut-and-weld procedure. A: An ellipsoidal region is cut from an unstressed elastic medium. The inclusion is then subject to a stress-free "transformation" strain, ϵ_{kl}^T. B: Tractions $T_i = -C_{ijkl}\epsilon_{kl}^T n_j$ applied to the boundaries of the inclusion restore it to its initial shape so that it fits back into the matrix.

surrounding matrix. Mathematically, this step is equivalent to adding a layer of point forces to the boundary between the inclusion and the matrix that are equal and opposite to the tractions T_i. Given elastic Green's functions $g_i^k(\mathbf{x}, \boldsymbol{\xi})$, the induced displacements in both the inclusion and matrix are given by

$$u_i^c(\mathbf{x}) = -\int_S T_k(\boldsymbol{\xi})g_i^k(\mathbf{x}, \boldsymbol{\xi})\mathrm{d}S, \qquad (7.36)$$

where the superscript c in u_i^c denotes *constraint*, in that it represents the elastic constraint of the surroundings. The integration is over the surface of the ellipsoid. Substituting equation (7.35) into (7.36) yields

$$u_i^c(\mathbf{x}) = \int_S \sigma_{kj}^T(\boldsymbol{\xi})n_j g_i^k(\mathbf{x}, \boldsymbol{\xi})\mathrm{d}S, \qquad (7.37)$$

where $\sigma_{ij}^T = C_{ijkl}\epsilon_{kl}^T$. Now, use the divergence theorem to transform to a volume integral:

$$u_i^c(\mathbf{x}) = \int_V \sigma_{kj}^T(\boldsymbol{\xi})\frac{\partial g_i^k}{\partial \xi_j}(\mathbf{x}, \boldsymbol{\xi})\mathrm{d}V, \qquad (7.38)$$

where the integral is over the volume of the ellipsoid. Thus, the strain in the matrix is given by $\epsilon_{ij}^M(\mathbf{x}) = [u_{i,j}^c(\mathbf{x}) + u_{j,i}^c(\mathbf{x})]/2$ for $\mathbf{x}$ outside V, and the stress in the matrix is $\sigma_{ij}^M = C_{ijkl}\epsilon_{kl}^M$.

In the inclusion, the strain is given by the spatial derivatives of the u_i^c, this time evaluated inside V. However, the *elastic* strain is the sum of ϵ_{ij}^c and $-\epsilon_{ij}^T$, and thus the stress in the inclusion is

$$\sigma_{ij}^I = C_{ijkl}(\epsilon_{kl}^c - \epsilon_{kl}^T). \qquad (7.39)$$

Introducing the full-space Green's functions (3.90) into equation (7.38) yields, after some analysis,

$$u_i^c(\mathbf{x}) = \frac{\sigma_{kj}^T}{16\pi\mu(1-v)} \int_V \frac{f_{ijk}(\mathbf{l})}{r^2} dV, \qquad (7.40)$$

where the function f_{ijk} depends on the unit vector $\mathbf{l}$ pointing from $\boldsymbol{\xi}$ to $\mathbf{x}$—that is, $l_i = (x_i - \xi_i)/r$:

$$f_{ijk}(\mathbf{l}) = (1-2v)(l_j\delta_{ik} + l_k\delta_{ij}) - l_i\delta_{jk} + 3l_il_jl_k, \qquad (7.41)$$

and we have made use of the fact that, because of the symmetry in the stress tensor, $f_{ijk} = f_{ikj}$. The integral (7.40) can be recast in terms of the transformation strain:

$$u_i^c(\mathbf{x}) = \frac{\epsilon_{kj}^T}{8\pi(1-v)} \int_V \frac{q_{ijk}(\mathbf{l})}{r^2} dV, \qquad (7.42)$$

where

$$q_{ijk}(\mathbf{l}) = (1-2v)(l_j\delta_{ik} + l_k\delta_{ij} - l_i\delta_{jk}) + 3l_il_jl_k. \qquad (7.43)$$

Note that the transformation strain, and hence stress, are uniform, so they can be taken out of the volume integral in equations (7.40) and (7.42). Eshelby (1957) showed that for an observation point in the interior of an ellipsoidal inclusion with semi-axes (a_1, a_2, a_3), the integral (7.42) reduces to (see problem 6)

$$u_i^c(\mathbf{x}) = -\frac{\epsilon_{kj}^T}{8\pi(1-v)} \int_\Sigma q_{ijk}(\mathbf{l})r(\mathbf{l})d\omega, \qquad (7.44)$$

where $d\omega$ is an area element on the surface of a unit sphere centered on the observation point $\mathbf{x}$, and the integral is taken over the surface of the unit sphere, Σ. $r(\mathbf{l})$ is the distance from the observation point, with coordinates $\mathbf{x} = (x_1, x_2, x_3)$, to the surface of the ellipsoid and is thus given by the positive root (since $r \geq 0$) of

$$\frac{(x_1 + rl_1)^2}{a_1^2} + \frac{(x_2 + rl_2)^2}{a_2^2} + \frac{(x_3 + rl_3)^2}{a_3^2} = 1. \qquad (7.45)$$

The appropriate root is

$$r = -\frac{\beta}{\alpha} + \sqrt{\frac{\beta^2}{\alpha^2} + \frac{\gamma}{\alpha}}, \qquad (7.46)$$

where

$$\beta = \sum_{i=1}^{3} l_i x_i / a_i^2,$$

$$\alpha = \sum_{i=1}^{3} l_i^2 / a_i^2,$$

$$\gamma = 1 - \sum_{i=1}^{3} x_i^2 / a_i^2. \qquad (7.47)$$

Eshelby pointed out that the square root term vanishes in the integral because it is even in $\mathbf{l}$ while the q_{ijk} are odd, so that simply $r = -\beta/\alpha$. Defining $\psi_i = l_i/a_i^2$ (no sum on i), such that

$\beta_i = \psi_i x_i$, the integral (7.44) then reduces to

$$u_i^c(\mathbf{x}) = \frac{\epsilon_{kj}^T x_m}{8\pi(1-v)} \int_\Sigma \frac{\psi_m q_{ijk}}{\alpha} d\omega. \tag{7.48}$$

Thus, finally, the constraint strain is given by

$$\epsilon_{il}^c(\mathbf{x}) = \frac{\epsilon_{kj}^T}{16\pi(1-v)} \int_\Sigma \frac{\psi_l q_{ijk} + \psi_i q_{ljk}}{\alpha} d\omega. \tag{7.49}$$

The important result is that the constraint strain is uniform, independent of $\mathbf{x}$ within the ellipsoid.

Recognizing that the integral in equation (7.49) is a fourth-rank tensor, the constraint strain within the inclusion can be written more compactly as

$$\boxed{\epsilon_{ij}^c = S_{ijmn}\epsilon_{mn}^T,} \tag{7.50}$$

where S_{ijmn} is known as the *shape tensor*, since the various components depend on the shape of the ellipsoid. These are determined by integrals of the form given by equation (7.49). The components of the shape tensor have the following symmetry: $S_{ijmn} = S_{jimn} = S_{ijnm}$; however, $S_{ijmn} \neq S_{mnij}$. For an isotropic medium, the only nonzero elements are S_{1111}, S_{1122}, S_{1133}, S_{2211}, S_{2222}, S_{2233}, S_{3311}, S_{3322}, S_{3333}, S_{1212}, S_{1313}, and S_{2323} and are given by

$$S_{1111} = \frac{3}{8\pi(1-v)}a_1^2 I_{11} + \frac{1-2v}{8\pi(1-v)}I_1,$$

$$S_{1122} = \frac{1}{8\pi(1-v)}a_2^2 I_{12} - \frac{1-2v}{8\pi(1-v)}I_1,$$

$$S_{1133} = \frac{1}{8\pi(1-v)}a_3^2 I_{13} - \frac{1-2v}{8\pi(1-v)}I_1,$$

$$S_{1212} = \frac{a_1^2+a_2^2}{16\pi(1-v)}I_{12} + \frac{1-2v}{16\pi(1-v)}(I_1+I_2), \tag{7.51}$$

where all other nonzero components are determined by cyclic permutation of the indices (Eshelby 1957). Eshelby (1957) further showed, following the work of Routh (1892), that the surface integrals arising in equation (7.49) can be reduced to standard forms:

$$I_1 = \int_\Sigma \frac{l_1^2 d\omega}{\alpha a_1^2} = 2\pi a_1 a_2 a_3 \int_0^\infty \frac{ds}{(a_1^2+s)\Delta(s)},$$

$$I_{11} = \int_\Sigma \frac{l_1^4 d\omega}{\alpha a_1^4} = 2\pi a_1 a_2 a_3 \int_0^\infty \frac{ds}{(a_1^2+s)^2\Delta(s)},$$

$$I_{12} = \int_\Sigma \frac{l_1^2 l_2^2 d\omega}{\alpha a_1^2 a_2^2} = \frac{2}{3}\pi a_1 a_2 a_3 \int_0^\infty \frac{ds}{(a_1^2+s)(a_2^2+s)\Delta(s)}, \tag{7.52}$$

where $\Delta(s) = (a_1^2+s)^{1/2}(a_2^2+s)^{1/2}(a_3^2+s)^{1/2}$. The remaining terms are found by cyclic permutation of $(1,2,3)$, (a_1, a_2, a_3), and (l_1, l_2, l_3). The integrals I_k can be reduced to more convenient forms involving elliptic integrals (Kellog 1929):

$$I_1 = \frac{4\pi a_1 a_2 a_3}{(a_1^2 - a_2^2)(a_1^2 - a_3^2)^{1/2}}[F(\theta, k) - E(\theta, k)],$$

$$I_3 = \frac{4\pi a_1 a_2 a_3}{(a_2^2 - a_3^2)(a_1^2 - a_3^2)^{1/2}}\left[\frac{a_2(a_1^2 - a_3^2)^{1/2}}{a_1 a_3} - E(\theta, k)\right], \tag{7.53}$$

where the $F(\theta, k)$ and $E(\theta, k)$ are elliptic integrals

$$F(\theta, k) = \int_0^\theta \frac{dt}{(1 - k^2 \sin^2 t)^{1/2}},$$

$$E(\theta, k) = \int_0^\theta (1 - k^2 \sin^2 t)^{1/2} dt, \tag{7.54}$$

and

$$\theta = \sin^{-1}(1 - a_3^2/a_1^2)^{1/2},$$

$$k^2 = \frac{a_1^2 - a_2^2}{a_1^2 - a_3^2}. \tag{7.55}$$

The following useful expressions can be derived from the integrals (7.52) (see Mura 1987):

$$I_1 + I_2 + I_3 = 4\pi,$$

$$3I_{11} + I_{12} + I_{13} = 4\pi/a_1^2,$$

$$I_{12} = (I_2 - I_1)/(a_1^2 - a_2^2),$$

$$3a_1^2 I_{11} + a_2^2 I_{12} + a_3^2 I_{13} = 3I_1. \tag{7.56}$$

To now return to the pressurized magma chamber, we need to choose a value of the transformation strain such that the stress in the inclusion is isotropic. The stress in the inclusion is given by equation (7.39). Setting this equal to an isotropic pressure and making use of (7.50) yields

$$-p\delta_{ij} = C_{ijkl}(S_{klmn} - \delta_{km}\delta_{ln})\epsilon_{mn}^T. \tag{7.57}$$

Making use of equation (1.102) for the stiffness tensor in an isotropic medium, and taking advantage of the symmetry of the shape tensor, leads to

$$-p\delta_{ij} = \lambda\delta_{ij}(S_{kkmn} - \delta_{nn})\epsilon_{mn}^T + 2\mu(S_{ijmn} - \delta_{im}\delta_{jn})\epsilon_{mn}^T. \tag{7.58}$$

The contraction of equation (7.58) is

$$-3p = (3\lambda + 2\mu)(S_{kkmn} - \delta_{mn})\epsilon_{mn}^T. \tag{7.59}$$

Notice that equation (7.59) is of the form $-p = K(\epsilon_{kk}^c - \epsilon_{kk}^T)$, reflecting the fact that the *elastic strain* in the inclusion is the difference between the constraint strain and the transformation

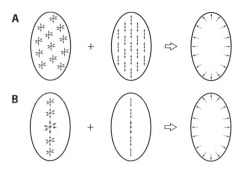

Figure 7.14. Equivalent source distributions for the ellipsoidal cavity. A: Equation (7.62) shows that a pressurized ellipsoidal cavity can be represented by a uniform distribution of centers of dilatation and double forces. B: Yang and others (1988) suggested that a quadratic distribution of sources along the line connecting the foci of the ellipsoid also yield a uniform pressure on the walls of the cavity. After Yang and others (1988).

strain. Substituting equation (7.59) into (7.58) yields

$$\frac{-p\delta_{ij}}{(3\lambda + 2\mu)} = (S_{ijmn} - \delta_{im}\delta_{jn})\epsilon_{mn}^{T}, \tag{7.60}$$

which provides a relationship between the pressure in the magma chamber and the transformation strains. First note that if $i \neq j$, equation (7.60) requires that the shear strains, ϵ_{12}^{T}, ϵ_{13}^{T}, and ϵ_{23}^{T}, vanish. For example, set $i = 1$, $j = 2$. Considering the nonzero elements of the shape tensors, equation (7.60) reduces to $0 = (S_{1212} - 1)\epsilon_{12}^{T}$, which implies that $\epsilon_{12}^{T} = 0$. Similar arguments apply to the other shear strains. The remaining three equations, which provide constraints on the normal strains, can be written in matrix form:

$$\frac{-1}{3K}\begin{bmatrix} p \\ p \\ p \end{bmatrix} = \begin{bmatrix} S_{1111} - 1 & S_{1122} & S_{1133} \\ S_{2211} & S_{2222} - 1 & S_{2233} \\ S_{3311} & S_{3322} & S_{3333} - 1 \end{bmatrix}\begin{bmatrix} \epsilon_{11}^{T} \\ \epsilon_{22}^{T} \\ \epsilon_{33}^{T} \end{bmatrix}. \tag{7.61}$$

The procedure is as follows. From the shape of the ellipsoid, determine the components of the shape tensor needed in equation (7.61). Solve (7.61) for the transformation strains ϵ_{11}^{T}, ϵ_{22}^{T}, and ϵ_{33}^{T}. Compute the corresponding stresses from $\sigma_{ij}^{T} = C_{ijkl}\epsilon_{kl}^{T}$, and last, compute the displacements in the matrix by evaluating integrals (7.37) or (7.38).

Consider now the case where the x_3 axis of the ellipsoid is vertical and the two horizontal axes are equal. This implies that $\sigma_{11}^{T} = \sigma_{22}^{T} \neq \sigma_{33}^{T}$. From equation (7.38), the displacements in this case are given by

$$u_i^c(\mathbf{x}) = \int_V \left\{ \sigma_{11}^T(\xi)\frac{\partial g_i^k}{\partial \xi_k} + [\sigma_{33}^T(\xi) - \sigma_{11}^T(\xi)]\frac{\partial g_i^3}{\partial \xi_3} \right\} dV. \tag{7.62}$$

We recognize the first term in the integrand as a center of dilatation and the second as a vertical force dipole. Thus, equation (7.62) shows that a pressurized ellipsoidal cavity can be represented by a *uniform* distribution of centers of dilatation and force dipoles (figure 7.14), or equivalently by a uniform distribution of three orthogonal force dipoles. The force dipoles are greatest parallel to the minimum principal axes of the ellipsoid. For example, for a vertical prolate ellipsoid, the vertical double forces are compressive such that the force couples in the horizontal direction are greatest. This makes sense in that the magma pressure acts on a larger vertical surface area, giving rise to a greater horizontal force. For an oblate ellipsoid, the double forces are tensile, so the vertical force couples are maximal.

We complete this analysis by relating the volume change to the magma pressure inside the chamber. The change in volume is given by $\delta V = \epsilon_{kk}^c V$, since the total strain in the inclusion is equal to the constraint strain, the transformation strain having been cancelled by the application of the surface tractions. The constraint strain is related to the transformation strain through equation (7.50). Making use of equation (7.59), we derive

$$\frac{\delta V}{V} - \epsilon_{kk}^T = -\frac{3p}{(3\lambda + 2\mu)} = -\frac{p}{K}. \tag{7.63}$$

The transformation strains are found from inversion of equations (7.61). Substituting the trace of the transformation strains, ϵ_{kk}^T, into equation (7.63) yields the volume change.

As an example, consider the spherical case, for which the transformation strain is isotropic, $\epsilon_{mn}^T = e\delta_{mn}$, where e is a scalar. In this case, equation (7.59) can be written as

$$-9p = (3\lambda + 2\mu)(S_{kkmn} - 3)\epsilon_{kk}^T. \tag{7.64}$$

From equations (7.51) and (7.56), it can be shown that

$$S_{kkmn} = \frac{1 + \nu}{1 - \nu}. \tag{7.65}$$

Combining equations (7.63), (7.64), and (7.65) yields

$$\delta V = \frac{3pV}{4\mu}, \tag{7.66}$$

which recovers equation (7.13) for the spherical magma chamber. Equations (7.63) and (7.61) can be combined to give the volume change for general ellipsoidal geometries. Results for penny-shaped cracks and elongated pipelike ellipsoids are given by Amoruso and Crescentini (2009) and can be derived from results given in sections 7.3 and 7.4.

Davis (1986) used the Eshelby solution, which holds for the full-space, to generate an approximate solution for a pressurized ellipsoidal cavity in a half-space. The first approximation is to replace the full-space Green's functions with the half-space Green's functions. Recall that we made the same approximation when deriving the approximate solution for a spherical magma chamber in section 7.1.1. The second approximation is to assume that the cavity dimensions are small compared to the distance from the source to the free surface so that the Green's functions are essentially constant throughout the inclusion volume. This allows the integral in equation (7.38) to be replaced by

$$u_i^c(\mathbf{x}) = \sigma_{kj}^T V \frac{\partial g_i^k}{\partial \xi_j}, \tag{7.67}$$

where V is the volume of the ellipsoid. Comparing with Volterra's formula, equations (3.12) and (3.13), it is clear that the equivalent moment tensor for a point ellipsoidal

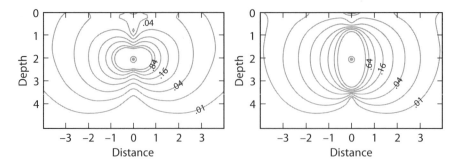

Figure 7.15. A: Deviatoric stress for a triaxial point source approximating a vertical prolate ellipsoid (Davis 1986). B: Deviatoric stress for a finite source approximation to a vertical prolate ellipsoid. After Yang and others 1988.

cavity is

$$M_{ij} = \sigma_{ij}^T V, \tag{7.68}$$

and that the transformation stresses are thus equivalent to a volumetric moment density (units of $N - m/m^3$). For a thin ellipsoid of thickness h, such that $dV = h\,dA$, where dA is the area increment, then $\sigma_{kj}^T h$ is equivalent to the areal moment tensor density in Volterra's formula.

Davis (1986) proposed an inversion strategy that works as follows. First, invert for the components of the moment tensor and the location of a generalized point source. The eigenvectors of the moment tensor give the orientation of the ellipsoidal source and the eigenvalues give the principal stress components σ_{ij}^T. Next, determine the shape of the ellipsoid through equation (7.61), where the S_{ijkl} are functions of the aspect ratios of the ellipsoid. Determination of the aspect ratios a/c and b/c is a nonlinear optimization problem that could be accomplished by a number of well-established methods.

Notice that by replacing the uniform distribution of sources with a single point source, Davis (1986) no longer fits the boundary conditions on the surface of the magma chamber *even in the limit that the chamber is far from the free surface*. In other words, the point-source representation does not accurately approximate the stress field in the neighborhood of the cavity (see figure 7.15). Yang and others (1988) sought a more accurate solution for a prolate ellipsoid that would fit the boundary conditions in the limit that the magma body is far from the free surface, analogous to the Mogi approximation to a spherical magma chamber. They hypothesized that for a prolate ellipsoid, the uniform distribution of centers of dilatation and double forces throughout the inclusion, as in equation (7.62), could be represented by some distribution of point sources on the line joining the two foci of the ellipse (figure 7.14). They looked for a polynomial distribution of sources and found that a quadratic distribution of sources could satisfy the boundary conditions.

As in the previous analysis, instead of the full-space Green's functions, Yang and others (1988) use the half-space Green's functions. In this way, the boundary conditions on $x_3 = 0$ are met exactly. However, the boundary conditions on the cavity wall are satisfied only approximately, the approximation equivalent to the first two steps of McTigue's analysis of the pressurized spherical cavity. In this respect, the Yang et al. (1988) solution is equivalent to the Mogi approximation. The integration of the sources along the line is done analytically, and the displacements are given in Yang and others (1988) and not repeated here.

Figure 7.16 shows the displacements due to vertical prolate ellipsoidal magma chambers with different aspect ratios. In all cases, $a/d = 0.3$, where a is the major semi-axis and d is the

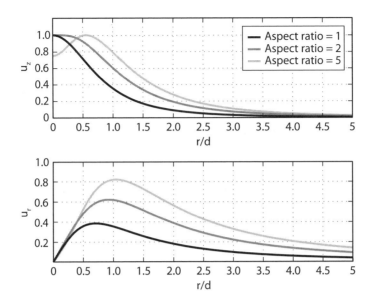

Figure 7.16. Effect of aspect ratio on surface displacements for an ellipsoidal cavity. The displacements are normalized by the maximum vertical displacement, u_z(max), found for each value of the aspect ratio. $a/d = 0.3$.

depth to the center. As the aspect ratio increases, the horizontal displacements become more significant relative to the vertical displacements. Notice that the plots are normalized by the maximum vertical displacement found for each value of the aspect ratio. (The unnormalized displacements decrease with increasing aspect ratio.) Interestingly, as the ellipsoid becomes more prolate, the vertical displacement is no longer maximal over the center of the chamber. The slight dip in the top of the uplift is caused by the strong radial extension over the cavity. Poisson's effect induces a slight subsidence associated with this extension.

Figure 7.17 shows the displacements for various source depths for an ellipsoidal chamber with aspect ratio of 3. As anticipated, the displacements are more localized when the source depth is shallow. The peak vertical displacement moves closer to the origin as the source moves toward the surface.

The two ellipsoid models we have considered are both approximate. The Davis (1986) model is a point-source approximation, whereas the Yang and others (1988) solution is finite. Both models fit the boundary conditions on the free surface, but neither fits the boundary condition on the walls of the cavity. The finite-source model does, however, in the limit that the depth of burial is much greater than the dimensions of the cavity. How different are the predictions of the two models? Figure 7.18 compares the vertical displacements for the finite-source model of Yang and others (1988) with the point-source representation of Davis (1986) for a vertical prolate ellipsoid. We see that the predictions are similar when the semimajor axis a is less than half the depth to the center of the chamber. For $a/d > 0.5$, the point-source model underpredicts the uplift generated by the finite source.

Figure 7.11 compared the fit of a center of dilatation to vertical displacement data from Long Valley caldera, California, determined by a combination of leveling and GPS. Not shown was the fit of this model to horizontal displacements determined by two frequency laser EDM. In fact, the center of dilatation cannot fit both data sets together (figure 7.19). The two-frequency laser data are considerably more precise than the uplift measurements. Because of this, a weighted inversion fits the horizontal displacements quite well (comparable to that shown in

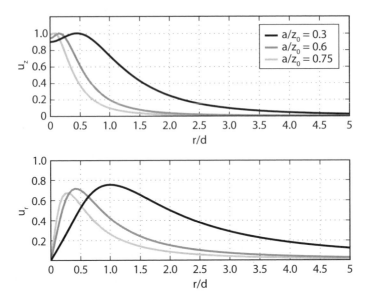

Figure 7.17. Effect of source depth on surface displacements for a vertical prolate ellipsoidal cavity. The displacements are normalized by the maximum vertical displacement, u_z(max), found for each value of the source depth. The aspect ratio is 3 in all cases. a is the semimajor axis, which is vertical, and z_0 is the depth to the center of the ellipsoid.

figure 7.20). However, the optimal model does not fit the vertical uplift data satisfactorily. The best fitting center of dilatation source is located at a depth of 8 km and has a volume change, between 1985 and 1999, of 0.12 cubic kilometers. On the other hand, both data sets can be well fit by a prolate ellipsoidal magma chamber (figure 7.20). The best fitting ellipsoid has a

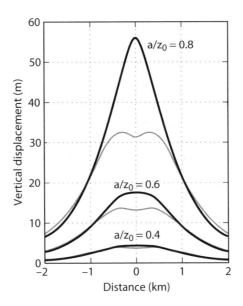

Figure 7.18. Comparison of the surface vertical displacements due to a triaxial point source (Davis 1986; light lines) and the ellipsoidal model of Yang and others (1988; dark lines). From Yang and others (1988).

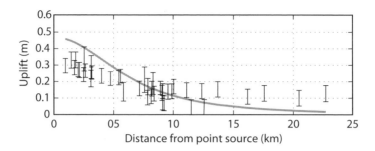

Figure 7.19. Comparison of the Long Valley uplift data from 1985 to 1999 to that predicted by a center of dilatation that is fit to both the horizontal laser ranging data and the vertical uplift data. After Battaglia et al. (2003b).

centroid depth of 5.9 km and an aspect ratio (the ratio of the horizontal to vertical semi-axes) of 0.475, with the long axis vertical. The estimated volume change between 1985 and 1999 is 0.09 cubic kilometers.

The aspect ratio of the best fitting ellipsoid is 0.47, which might not seem very far from spherical. However, this results in a substantial difference in the source parameters determined from the two models. Fitting the spherically symmetric source to the uplift data alone yielded a source depth of 11.4 km, whereas the best fitting ellipsoidal source to both the vertical and horizontal deformation data has a centroid depth of 5.9 km. Not surprisingly, the estimated volume changes are quite different. It takes far less volume change to fit the data for a source at 6 km than it does for a source at 11 km. This example once again emphasizes how important it is to use both horizontal and vertical deformation observations when estimating the shape of magma chambers.

7.3 Magmatic Pipes and Conduits

So far, we have considered only roughly equidimensional magma chambers: spheres or ellipsoids. Long-lived magma conduits tend to develop into cylindrical, pipelike structures, and the deformation associated with these conduits is of considerable interest on active volcanoes.

The first effort in this regard was due to Walsh and Decker (1971). They simply integrated the half-space solution for a center of dilatation along a line. For the case where the conduit is vertical and extends from depth d to infinity, the vertical and radials surface displacements are given by

$$u_z = \frac{(1-v)pa^2}{\mu} \frac{1}{\sqrt{\rho^2 + d^2}},$$

$$u_\rho = \frac{(1-v)pa^2}{\mu} \frac{\sqrt{\rho^2 + d^2} - d}{\rho \sqrt{\rho^2 + d^2}}. \tag{7.69}$$

As expected, these displacements decay more slowly with distance from the source than the displacements due to a center of dilatation (figure 7.21).

The Walsh and Decker (1971) solution is kinematic in that it does not attempt to satisfy a pressure boundary condition on the conduit walls. A better approach, developed by Bonaccorso and Davis (1999), is to consider the limiting case of a very narrow prolate ellipsoid. In the limit as the semimajor axis becomes much greater than the other two semi-axes

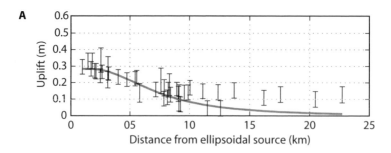

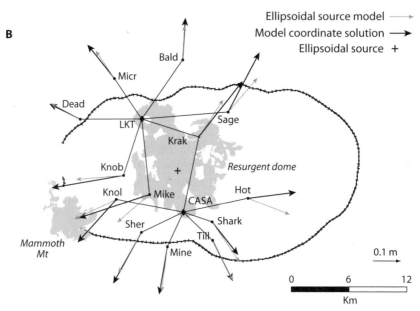

Figure 7.20. Fit of the Long Valley uplift data from 1985 to 1999 to a pressurized prolate ellipsoid. A: Fit to vertical displacement data determined by a combination of leveling and GPS. B: Fit to two-color laser geodimeter data. + indicates horizontal location of source. After Battaglia et al. (2003b).

(equal to a), equation (7.62) becomes

$$u_i(\mathbf{x}) = \pi a^2 \int_{d_1}^{d_2} \left[\sigma_{11}^T(\xi) \frac{\partial g_i^k}{\partial \xi_k} + (\sigma_{33}^T - \sigma_{11}^T) \frac{\partial g_i^3}{\partial \xi_3} \right] \, d\xi_3. \tag{7.70}$$

Mura (1987) gives the components of the shape tensor for this case as:

$$S_{1111} = S_{2222} = \frac{5 - 4\nu}{8(1 - \nu)},$$

$$S_{1122} = S_{2211} = \frac{-1 + 4\nu}{8(1 - \nu)},$$

$$S_{1133} = S_{2233} = \frac{4\nu}{8(1 - \nu)},$$

$$S_{3311} = S_{3322} = S_{3333} = 0. \tag{7.71}$$

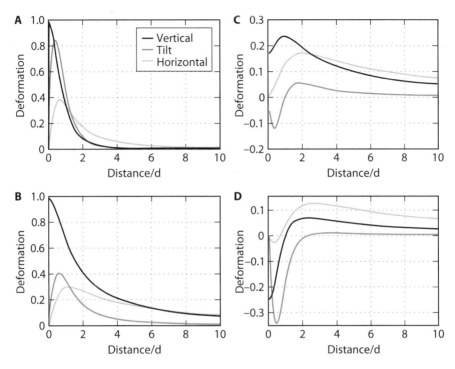

Figure 7.21. Comparison of different pipe models. A: Center of dilatation. B: Walsh and Decker semi-infinite line of centers of dilatation. C: Vertical prolate ellipsoid. D: Radial expansion of a cylindrical pipe. For A, the distance scale is normalized by the source depth; for the others, it is normalized by the depth to the top of the pipe, which extends to infinite depth. For A, displacements are normalized by $(1 - v)pa^3/\mu d_2$. For B, displacements are normalized by $(1 - v)pa^2/\mu d_1$. For C, displacements are normalized by $3pa^2/4\mu d_1$. For D, displacements are normalized by sa/d_1. After Bonaccorso and Davis (1999).

Solving for the transformation strains corresponding to a uniform pressure from equation (7.61) leads to

$$\sigma_{33}^T = \frac{p}{(1 - 2v)},$$

$$\sigma_{11}^T = \frac{2p(1 - v)}{(1 - 2v)}. \tag{7.72}$$

For the special case of $v = 1/4$, this gives $\sigma_{33}^T = 2p$ and $\sigma_{11}^T = 3p$, in agreement with the results of Bonaccorso and Davis (1999). Using equations (7.72), equation (7.70) becomes

$$u_i(\mathbf{x}) = \pi a^2 p \int_{d_1}^{d_2} \left[\frac{2(1 - v)}{(1 - 2v)} \frac{\partial g_i^k}{\partial \xi_k} - \frac{\partial g_i^3}{\partial \xi_3} \right] d\xi_3. \tag{7.73}$$

Thus, the pressurized pipe is equivalent to the sum of a line of centers of dilatation and a line of *compressive* (note the negative sign) double forces. For $v = 1/4$, the ratio of the strengths of the centers of dilatation and the double forces is 3 : 1. This contrasts with the Walsh and Decker (1971) solution, which is simply a line of centers of dilatation.

Both of these models assume closed conduits. For an erupting volcano, however, the conduit is open to the surface. For low-viscosity magmas, the effect of a magma-filled conduit will mainly be the excess pressure exerted by the magma on the conduit walls. Unlike closed conduits, there is no force exerted at the top of the pipe. For higher viscosity magmas, there will also be viscous shear stresses exerted on the conduit walls when magma is flowing. Bonaccorso and Davis (1999) presented a simple kinematic model of an open conduit, in which the walls of a cylindrical conduit displace outward by an amount s. From Volterra's formula (3.15), we can write the displacements as

$$u_k(\mathbf{x}) = a \int_{d_1}^{d_2} \int_0^{2\pi} \left[\mu s_i n_j \left(\frac{\partial g_k^i}{\partial \xi_j} + \frac{\partial g_k^j}{\partial \xi_i} \right) + \lambda s_i n_i \frac{\partial g_k^m}{\partial \xi_m} \right] d\theta d\xi_3, \qquad (7.74)$$

where a is again the pipe radius. For the last term, $s_i n_i = s$, and the integration over θ is trivial. The integrations of the other terms over θ are also elementary, so the problem reduces to an integral over depth:

$$u_i(\mathbf{x}) = 2\pi s \mu a \int_{d_1}^{d_2} \left[\frac{1}{(1-2\nu)} \frac{\partial g_i^k}{\partial \xi_k} - \frac{\partial g_i^3}{\partial \xi_3} \right] d\xi_3. \qquad (7.75)$$

Remarkably, in this case, the displacements are also a combination of centers of dilatation and compressive double forces. For $\nu = 1/4$, the ratio of strengths is $2:1$, whereas in the closed pressurized conduit, the ratio is $3:1$. This makes sense in that we expect the closed conduit to generate greater vertical deformation.

To compute the integrals, we need the Green's functions for centers of dilatation and vertical double forces. The former are readily obtained from equations (7.14) and (7.23):

$$\frac{\partial g_3^k}{\partial \xi_k} = -\frac{(1-2\nu)}{2\pi\mu} \frac{\xi_3}{r^3},$$

$$\frac{\partial g_\rho^k}{\partial \xi_k} = \frac{(1-2\nu)}{2\pi\mu} \frac{\rho}{r^3} \qquad r^2 = \rho^2 + \xi_3^2, \qquad (7.76)$$

where here the source coordinate ξ_3 is negative when the source is below the earth's surface. We can deduce the Green's functions for vertical double forces from Mindlin's solution (3.93). This yields

$$\frac{\partial g_3^3}{\partial \xi_3} = \frac{\xi_3}{4\pi\mu} \left[\frac{2\nu}{r^3} - \frac{3\xi_3^2}{r^5} \right],$$

$$\frac{\partial g_\rho^3}{\partial \xi_3} = \frac{\rho}{4\pi\mu} \left[\frac{3\xi_3^2}{r^5} - \frac{2\nu}{r^3} \right]. \qquad (7.77)$$

Integrating equations (7.73) using (7.76) and (7.77) gives, for the *closed pipe*,

$$u_3(\mathbf{x}) = \frac{a^2 p}{4\mu} \left[\frac{(1-2\nu)}{r} + \frac{\rho^2}{r^3} \right]_{d_1}^{d_2},$$

$$u_\rho(\mathbf{x}) = \frac{a^2 p\rho}{4\mu} \left[\frac{(3-2\nu)\xi_3}{\rho^2 r} + \frac{\xi_3}{r^3} \right]_{d_1}^{d_2}. \qquad (7.78)$$

Note that this corrects an error in Bonaccorso and Davis (1999), in that they use a form of equation (7.73) that is appropriate only for $\nu = 1/4$, combined with the general Green's functions for arbitrary value of Poisson's ratio. The results do agree for $\nu = 1/4$.

For the open pipe, fixed displacement model, equation (7.75) with the Green's functions from (7.76) and (7.77) yields

$$
\begin{aligned}
u_3(\mathbf{x}) &= sa \left[\frac{(1-2v)}{2r} - \frac{\rho^2}{2r^3} \right]_{d_1}^{d_2}, \\
u_\rho(\mathbf{x}) &= sa \frac{\xi_3}{\rho} \left[\frac{(1+v)}{r} - \frac{\xi_3^2}{2r^3} \right]_{d_1}^{d_2}.
\end{aligned}
\tag{7.79}
$$

Figure 7.21 compares the line of centers of dilatation with the closed and open pipe models. Both of the pipe models generate considerably more horizontal displacement, relative to the peak uplift, in comparison to the Walsh and Decker line of centers of dilatation. Mathematically, this results from the compressive vertical double forces. In particular, the open pipe generates little uplift, and indeed subsidence over the source due to a Poisson contraction. For conduits that are plugged at depths of a few hundred meters, we would expect deformation that decays rapidly with distance from the volcanic vent.

It is important to distinguish between deformation caused by inflation of buried magma bodies from that caused by erupting volcanoes that are venting to the surface. For closed chambers in an elastically deforming crust, the deformation is usually spread out over a broad area, depending on the depth, and to some extent the shape, of the magma body. When volcanoes are open to the surface, the deformation tends to be much more localized about the vent, although one expects broad subsidence due to depressurization of a deep magma reservoir feeding the eruption.

7.4 Dikes and Sills

The simplest models of dikes and sills are those based on elastic dislocations with prescribed uniform opening. In this way, Volterra's formula can be utilized, as in chapter 3, to compute the surface displacements from planar sources.

The surface displacement field for a vertical dike is very characteristic and easily distinguished from more equidimensional magma bodies. The vertical displacements for steeply dipping buried dikes exhibits two zones of uplift on either side of the dike plane with an intervening zone of subsidence above the dike (figure 7.22). For dikes that are very long compared to their height, such that the deformation is effectively two-dimensional plane strain, there is no net displacement immediately above a vertical dike (as seen later in figure 7.28). For insight into why this must be so, see problem 7 and Rubin (1992). As the dip decreases, the uplift increases on the side closest to the earth's surface and diminishes on the opposite side (figure 7.22). For nearly horizontal dips, the secondary uplift disappears, leaving only a single region of uplift. The result is that it is generally easy to distinguish near vertical dikes from other possible magmatic source geometries based solely on the pattern of vertical deformation. The dip of the dike can also usually be well determined. On the other hand, sills are difficult to distinguish from other more equidimensional bodies given only vertical deformation data. Recall that sills are relatively inefficient at generating horizontal deformation (figure 7.12), so it is possible to distinguish sills from more equidimensional sources using a combination of vertical and horizontal deformation data.

For a near vertical dike, the horizontal displacements are directed largely perpendicular to and away from the dike plane. Figure 7.23 shows horizontal components of the displacements accompanying an intrusion and eruption on Kilauea volcano, Hawaii. The data are well

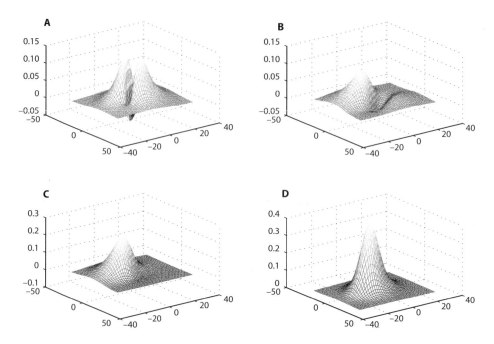

Figure 7.22. Vertical surface displacements for different dike dips. A: 90 degrees. B: 75 degrees. C: 60 degrees. D: Zero. Notice that the vertical scale is different in C and D.

modeled by a simple uniform opening rectangular dislocation. Notice the motion perpendicular to the dike plane and how it decays with distance from the dike. Also note the much weaker inward motions off the ends of the dike. A number of features suggests that this simple kinematic model actually describes the intrusion within the earth. First, the best fitting dike source aligns with the surface rifts shown by the heavy line. The model dike extends to the west of the surface breaks, but it is reasonable to expect that the dike did not break the surface along its entire length. Second, the estimated dike opening of 2 meters agrees well with independent measurements of the opening across the surface fractures (Owen et al. 2000). Third, the depth to the bottom of the dike, which is constrained by the rate at which the displacements decay with distance from the dike, is consistent with the depths of microearthquakes that accompany shallow intrusion events on Kilauea.

As indicated by the previous example, the surface displacement fields yield important information about dike depth. Figure 7.24 shows the vertical and horizontal displacements along a profile across the midpoint of a vertical opening mode dislocation with the top at different depths. Notice that for a buried dike, the maximum vertical displacement moves toward the dike plane as the top of the dike propagates vertically. Considering the tilt perpendicular to the dike plane (the horizontal derivative of the vertical displacement), one expects that a tiltmeter placed close to the dike plane would first tilt toward the dike and then, as the dike propagates upward, tilt away from the dike. Very repeatable tilt signals of this type have been observed accompanying dike intrusions off the east coast of the Izu peninsula, Japan (Okada et al. 2000). Nearly all of the tilt episodes are associated with an initial tilt toward the dike plane. Interestingly, the tilt signals begin prior to the onset of seismicity accompanying the intrusion. In this way, monitoring of tilt signals can be a useful predictor of vertical dike propagation and ensuing eruptions.

The maximum horizontal displacements also occur at a distance from the dike plane that depends on the dike depth. Thus, the ground above a buried dike is driven into extension

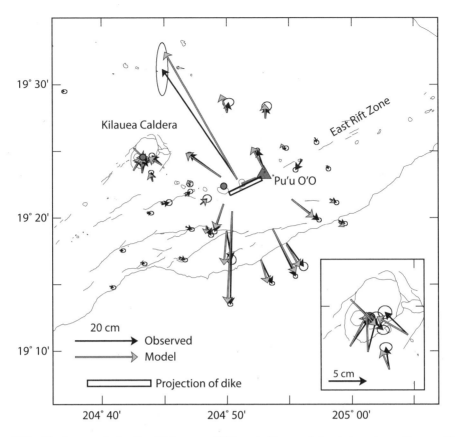

Figure 7.23. Displacements for the 30 January 1997 east rift zone intrusion and eruption on Kilauea volcano determined from GPS measurements. The vectors with error ellipses mark the observations; those without error ellipses indicate displacements predicted by a uniform opening dislocation model in an elastic half-space. The surface projection of the dike is shown as a rectangle. Two point centers of dilation are also included to model subsidence of summit and rift zone magma reservoirs. After Owen and others (2000).

perpendicular to the dike, whereas it is driven into compression off to the sides. As the dike propagates upward, the boundary between the zones of extension and compression moves toward the dike. Once the dike breaches the surface, the maximum vertical and horizontal displacements occur at the dike, and the strain is everywhere compressive perpendicular to the dike.

7.4.1 Crack Models of Dikes and Sills

Dislocation models with an imposed displacement discontinuity do not accurately represent the boundary conditions on dike walls. Neglecting viscous stresses due to flowing magma, the appropriate boundary condition on the dike wall is the pressure exerted by the melt. As in chapter 4, figure 4.2, we can separate the uniform far-field stress and model the deformation due to the dike as resulting from a crack subject to an *excess pressure* given by the pressure in excess of the normal stress in the rock acting perpendicular to the plane of the dike, $\Delta p = p - \sigma_n$. The excess pressure is often written simply as p in this chapter, although it should be understood to represent $p - \sigma_n$. An arbitrary, nonuniform opening distribution can be represented by a sequence of edge dislocations, as in figure 4.4.

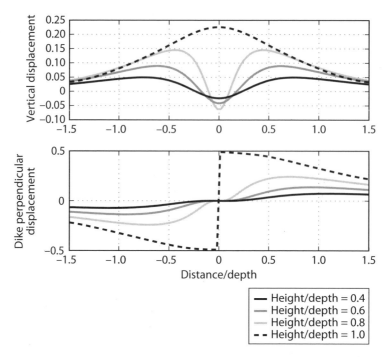

Figure 7.24. Surface displacements resulting from a rectangular dislocation model of a vertical dike, showing the evolution of the deformation field as the dike approaches the earth's surface. Depth is to the bottom edge of the dike, and height is the vertical extent of the dike. When the height/depth = 1, the dike has breached the earth's surface.

Recall that for a three-dimensional penny-shaped crack in a full-space, the displacements on the crack wall are given by equation (4.73):

$$u_z = \frac{2a(1-v)p}{\mu\pi}\sqrt{1 - \frac{r^2}{a^2}}. \tag{7.80}$$

We can now use the Eshelby methods described in this chapter to derive this result. Note that since the strain within the inclusion is spatially uniform, the displacement of the crack wall is

$$u_3 = \epsilon_{33}^c x_3. \tag{7.81}$$

For an oblate ellipsoid $a_1 = a_2 \gg a_3$, the boundary of the ellipsoid is described by

$$x_3 = a_3\sqrt{1 - \frac{r^2}{a_1^2}}. \tag{7.82}$$

Given the shape factors for a penny-shaped ellipsoid (see problem 5), it is possible to show that in the limit of $a_3/a_1 \to 0$,

$$\epsilon_{33}^c = \frac{2(1-v)p}{\mu\pi}\frac{a_1}{a_3}, \tag{7.83}$$

which when combined with equations (7.81) and (7.82) yields (7.80).

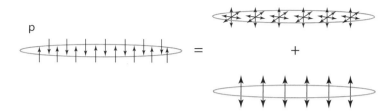

Figure 7.25. Deformation due to a thin, penny-shaped sill can be approximated by a disk of centers of dilatation and double forces.

Following problem 5, it can also be shown that in the limit $a_3/a_1 \to 0$,

$$\sigma_{11}^T = \frac{4\nu(1-\nu)}{\pi(1-2\nu)} \frac{a_1}{a_3} p, \qquad (7.84)$$

and

$$\sigma_{33}^T - \sigma_{11}^T = \frac{4(1-\nu)}{\pi} \frac{a_1}{a_3} p. \qquad (7.85)$$

If we take the 3-direction to be vertical, then the crack represents a penny-shaped sill. According to equation (7.62), the source can be represented as the sum of centers of dilation and vertical double forces (figure 7.25). For this geometry, the double forces are extensional; the sill generates relatively greater vertical displacements. Also note that the ratio of strengths of the double forces to centers of dilatations is $(1 - 2\nu)/\nu$, which for a Poisson solid is $2:1$.

Following the procedure established earlier, we can generate an approximate solution for the displacements due to a circular sill in a half-space using the half-space Green's functions. If the depth to the sill is substantially greater than its radius, a point-source approximation (after Davis [1986]) should be adequate. Figure 7.26 compares the point-source approximation with a quasi-analytical solution for a pressurized penny-shaped crack in a half-space due to Fialko et al. (2001). The Fialko et al. procedure is to relate the stresses and displacements to Neuber-Papkovich potentials. These functions are then expanded in Hankel transforms. The boundary conditions on the crack faces lead to a pair of coupled integral equations (analogous to the integral equation [4.6] for the antiplane crack problem in chapter 4). Fialko et al. (2001) present efficient numerical procedures for solving the dual integral equations. From figure 7.26, we see that the point-source approximation is fully adequate for $d/a = 5$, where d is the source depth, and a is the sill radius. The approximation begins to break down for $d/a = 2.5$ and is poor for $d/a = 1$, as expected.

The mixed boundary value problem for pressurized cracks in a half-space can become quite involved. One efficient strategy already discussed in chapter 4 is the boundary element method. Problem 8 develops the needed expressions for these calculations in two dimensions. Figure 7.27 shows the opening distribution of a two-dimensional dike growing toward the free surface. Because of the constant pressure boundary condition, the dike opening increases as the dike extends. The effect of the free surface is to increase the dike opening over what it would be in a full-space. This is seen most clearly when the upper dike tip gets close to the free surface.

Figure 7.28 shows the surface deformation caused by the growing dike in figure 7.27. The first-order observation is that the amplitude of the uplift increases dramatically as the dike grows. This is due in part to the fact that the dike is getting closer to the surface, but more so because the dike opening increases as the dike grows (see figure 7.27). Comparing figure 7.28

234

Chapter 7

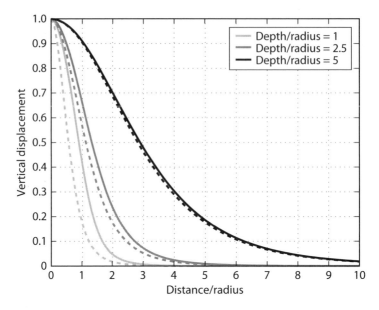

Figure 7.26. Normalized vertical displacement due to a horizontal circular sill. Solid curves are displacements calculated from the quasi-analytic solution of Fialko et al. (2001); dashed lines are for a point-source approximation.

to the corresponding calculation for a uniform dislocation (figure 7.24), in which the opening does not increase as the dike propagates, illustrates the importance of the latter effect. For both models, the maximum flanking uplift moves toward the dike plane as the dike nears the free surface.

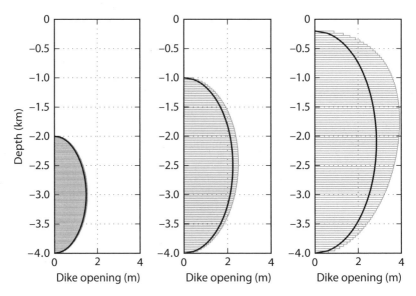

Figure 7.27. Boundary element calculation of a two-dimensional vertical dike growing under constant internal pressure toward the free surface. The bottom edge is held fixed. The solid curves represent the analytical solution for a two-dimensional pressurized crack in a full-space, so departures from this opening distribution represent the effect of the free surface on the dike opening.

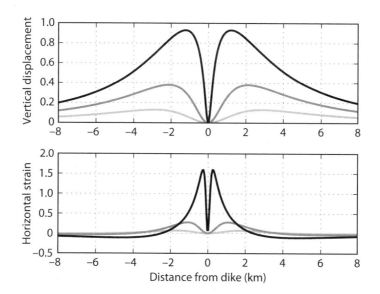

Figure 7.28. Deformation pattern corresponding to the vertically propagating dike shown in figure 7.27. Top: Vertical displacement. Bottom: Horizontal strain perpendicular to the dike at the earth's surface. The three curves correspond to the three panels in figure 7.27. Displacement in meters; strain in units of 10^{-3}.

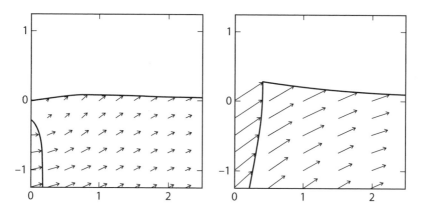

Figure 7.29. Change in displacement pattern when a buried dike breaches the free surface. After Pollard et al. (1983).

Inflation of the dike causes horizontal compression of the crust to either side of the dike (figure 7.28, bottom). Intuitively, we expect extensional strain above the dike, and this is in fact observed. What is perhaps surprising is that the surface-parallel horizontal strain is not peaked over the dike. Rather, the strain distribution is bimodal, with maximum strains off to either side of the dike and a minimum in extension immediately over the dike.

As the dike breaches the surface, the ligament of elastic material at the top of the dike is broken, and the displacement pattern changes dramatically (figure 7.29). When this happens, both the vertical and horizontal displacements increase substantially. The subsidence and extension over the buried dike disappear; in two dimensions, the crust is everywhere uplifted and compressed.

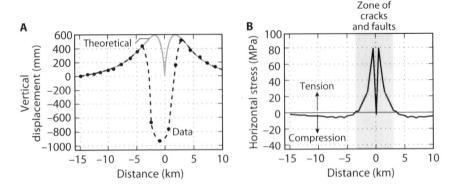

Figure 7.30. A: Observed and predicted uplift pattern over an intruded dike, Krafla volcano, Iceland. Notice that the elastic model (marked theoretical) does a good job of predicting the data on the flanks of the dike but does not predict the subsidence over the dike. B: Horizontal normal stress at the earth's surface. After Pollard et al. (1983).

7.4.2 Surface Fracturing and Dike Intrusion

Extensive fracturing and faulting is observed above dikes that approach the earth's surface. Often a graben with inward-dipping normal faults forms above the dike. Leveling observations above intruded dikes show that elastic models that ignore the fracturing and faulting can fit the measured deformation far from the dike but cannot fit the subsidence observed immediately above the dike.

Figure 7.30 shows one example using data from the 7 January 1976 intrusion event on Krafla volcano, Iceland. During the late 1970s, Krafla was the site of a dramatic rifting episode that involved the intrusion of numerous dikes both to the south and to the north of the central volcano. The leveling data in figure 7.30 shows uplift of 40 to 50 cm on the flanks of the intrusion and subsidence in the central graben of roughly 100 cm. Pollard et al. (1983) fit the data from the flanking uplift, ignoring the data from the graben, and estimate a vertical dike 5.5 km tall with a depth to center of 3 km. Assuming a shear modulus of 4×10^4 MPa, they estimate a magma pressure in excess of the least principal compressive stress of roughly 20 MPa and a dike thickness of 2 meters, not atypical for basaltic dikes.

Because both dislocation (figure 7.24) and crack models (figure 7.28) predict little or no absolute subsidence over the dike, the graben subsidence observed above the dike in figure 7.30 must reflect inelastic deformation. Pollard et al. (1983) present several other examples of this phenomenon from Kilauea volcano, and similar observations have been made in volcanic rift zones in the Afar region of Africa and elsewhere.

Tensile fractures are observed in front of some recently emplaced dikes, as illustrated in figure 7.31. Pollard et al. (1983) noted that the fractures tended to form two zones in front of the eruptive fissures. These fracture zones are believed to form on either side of the lateral extension of the dike at depth. They explained the growth of these surface fractures as resulting from the two zones of horizontal extension and relative tensile stress above the dike tip (figure 7.30). The stress maxima are located at a distance away from the trace of the dike plane comparable to the depth to the top of the dike. There is no stress change immediately above the dike. In some cases, the surface fractures coalesce to form normal faults, which downdrop the block above the dike with respect to the surroundings.

Field observations indicate that fracturing and normal faulting are common above shallowly emplaced dikes. As illustrated by figure 7.30, elastic models that ignore the faulting above the dike can fit the deformation far from the dike plane but do a poor job of describing

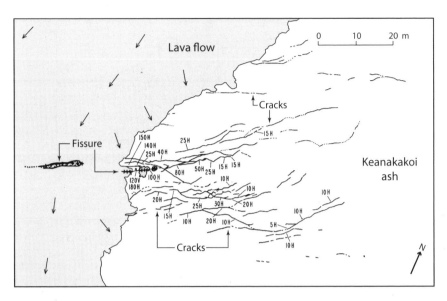

Figure 7.31. Surface fractures associated with the 31 December 1974 dike intrusion and eruption in the southwest rift zone of Kilauea volcano, Hawaii. Numbers indicate horizontal (*H*) and vertical (*V*) separation in millimeters. After Pollard et al. (1983).

the deformation in the vicinity of the dike. Rubin and Pollard (1988) considered the growth of normal faults due to the stresses induced by dike intrusion. They noted that induced compressive stresses off to the side of the dike inhibit normal fault slip adjacent to the dike once the dike has formed. On the other hand, limiting the faulting to the small region of extension over the dike cannot explain the graben subsidence. They suggested that the fault slip occurred *in front of* the propagating dike in a region of extension. This allows fault slip to substantial depth before the arrival of the dike and the associated horizontal compression. This interpretation permits a graben to develop over the dike and does a much better job of describing the observed vertical deformation patterns determined by leveling (figure 7.32). There is some evidence to suggest that during the September 1977 intrusion at Krafla volcano, normal fault slip occurred prior to the arrival of the dike. This intrusion is well known because the dike actually intersected a kilometer-deep geothermal borehole, erupting a few cubic meters of basaltic tephra from the well (Larsen et al. 1979). According to Rubin and Pollard (1988), fault slip closed the main highway 1 hour prior to the eruption, suggesting that slip preceded the dike.

7.5 Other Magma Chamber Geometries

Most other magma chamber geometries are not readily amenable to analytical treatment but can be analyzed with numerical methods. Yun et al. (2006) used three-dimensional boundary element methods to compare the surface deformation due to a sill and a diapirlike body, modeled as a truncated cone (figure 7.33). For shallow burial depths, the surface deformation is essentially identical to that of a sill at the same depth as the top of the diapir (figure 7.34). For sources at shallow depths, it is the upper surface of the magma body that drives surface deformation; the sides and lower edges do not displace significantly.

Fialko et al. (2001) determined the ratio of the volume change of the lower half of the crack to the total volume change for the circular penny-shaped crack. For a deeply buried sill, this ratio is, by symmetry, 1.0. As the radius of the sill increases relative to the depth, more and

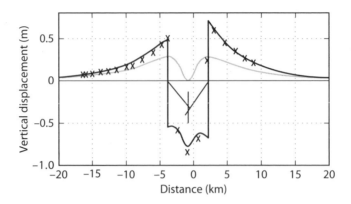

Figure 7.32. Elevation changes accompanying the 1978 eruption of Krafla volcano, Iceland. Data (Xs) are from the Kelduhverfi region 40 km north of the Krafla caldera. Model fit to a prediction (black curve) from an inflating dike accompanied by fault slip (inset). The dike extends from 1.5 to 6 km depth and has 1.5 m of opening. The faults on either side of the dike are inferred to have slipped prior to the arrival of the dike tip and slipped from the surface to depths of 5 to 6 km, accumulating slip of 1.3 to 1.7 meters, respectively. Displacement due to dike alone shown with gray curve. From Rubin and Pollard (1988).

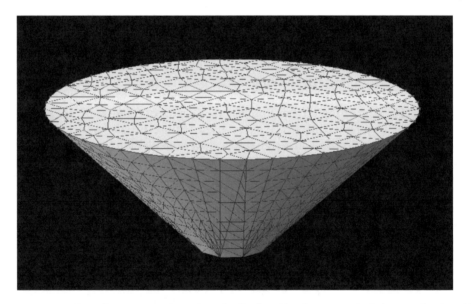

Figure 7.33. Boundary element representation of a diapir, with a flat top and 45-degree dipping sides. After Yun et al. (2006).

more of the opening, and therefore volume change, occurs in the upper half of the crack. For a radius equal to the depth, $\Delta V_{bottom}/\Delta V_{total} \sim 0.3$. By the time the sill radius is 2.5 times the depth, only 10% of the volume change is in the lower half of the crack. At this point, the intrusion has taken on the shape of a laccolith (figure 7.35). The transition occurs at a depth to radius ratio near unity.

For very shallow horizontal intrusions, one can approximate the displacement of the overlying rock using expressions for a bending elastic plate clamped at the crack ends

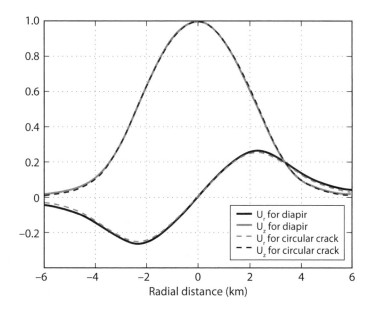

Figure 7.34. Comparison of the vertical and radial displacements due to the diapir shown in figure 7.33 and a sill at the same depth. The radius of the sill and diapir top is 3.0 km, and the depth to the top is 1.9 km. After Yun et al. (2006).

(e.g., Jackson and Pollard 1988). The elastic plate approximation has deflection

$$w(r) = \frac{3(1-\nu)p}{32\mu} \frac{a^4}{d^3} \left(1 - \frac{r^2}{a^2}\right)^2,$$ (7.86)

where a is the radius of the sill, and d is the depth of burial. For small d/a, the displacements at the earth's surface should be reasonably close to the displacements given by equation (7.86). Figure 7.35 compares the scaled displacements from equation (7.86) with those given by the Fialko et al. (2001) solution for the penny-shaped crack. The two distributions are remarkably similar; however, there are important differences. The plate model underpredicts the absolute displacements (see of Fialko et al. 2001, figure 2). Furthermore, the scaling of the maximum displacement with crack radius is different. For a deeply buried crack, the maximum opening increases linearly with the crack radius (7.80), and the volume increases as the cube of the radius. The numerical results of Fialko et al. (2001) for circular sills with radius greater than the burial depth show that volume change scales roughly as a^5, which implies that the maximum opening scales as a^3. The bending plate model (7.86) predicts the the opening scales with a^4.

Last, while the uplift distributions in figure 7.35 look similar to a laccolith (see inset), it is important to recall that the *undeformed* shape prior to pressurization is a thin circular slit. The surface deformation due to pressurization of a preexisting laccolith could be substantially different. Inelastic deformation in the hinge zones of the overlying plate could also alter the deformation field at the earth's surface.

Solving for very general magma chamber shapes subjected to uniform pressure boundary conditions is, as we have seen, a difficult problem. An alternative approach, which is well suited for inversion, but perhaps not well motivated on physical grounds, is to assume that the deformation can be represented by some spatial distribution of volume change. In this

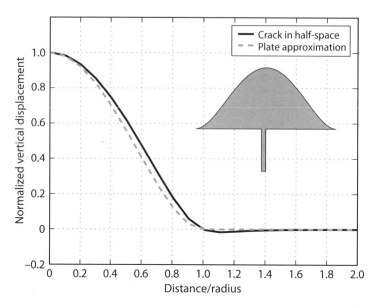

Figure 7.35. Displacements predicted by the penny-shaped crack model for $d/a = 0.2$ compared to those predicted by the deflection of an elastic plate over the intrusion as given by equation (7.86). Inset shows the shape of a laccolith.

way, we can write the displacement at any surface point $\mathbf{x}$, from equation (7.14), as

$$u_i(\mathbf{x}) = \frac{1-\nu}{\pi} \int_V \Delta V(\xi) \frac{x_i - \xi_i}{r^3} dV_\xi, \tag{7.87}$$

where ΔV is the volume change at point ξ, and $r = \|\mathbf{x} - \xi\|$ is the Euclidean distance between $\mathbf{x}$ and ξ (e.g., Vasco et al. 1988). While this approach leads to a displacement field that is linear in ΔV and therefore amenable to inversion, the interpretation of the resulting volume change distribution is not clear. The simplest models of magma chambers involve pressurized cavities in an elastic earth. Except for the case of a point volume change, as in the center of dilatation, a distributed volume change will not correspond to a constant pressure boundary condition on the surface of a cavity in an elastic medium. For example, we have seen that in the case of an ellipsoidal cavity, there is an equivalence to a uniform distribution of centers of dilatation and double forces.

7.6 Viscoelastic Relaxation around Magma Chambers

Bonafede et al. (1986) considered two types of volumetric sources in a viscoelastic half-space: a pressure source and a center of dilatation. While we have seen that these solutions are equivalent in an elastic half-space, they are different in a viscoelastic medium. The first corresponds to a constant pressure boundary condition, whereas the second corresponds to a fixed displacement at the wall of the magma chamber. Bonafede et al. (1986) also consider two types of material response: Maxwell and standard linear solid, the latter having finite long-term elastic stiffness. For the fixed displacement boundary condition, the displacements at any point asymptotically approach a constant value for either rheology. On the other hand, the constant pressure source in a Maxwell half-space produces displacements that monotonically increase with time, as the long-term response is one of a Newtonian fluid. While we might

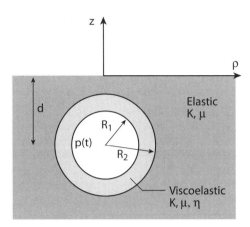

Figure 7.36. Geometry of a spherical magma chamber with a Maxwell viscoelastic shell. R_1 is the radius of the magma chamber. R_2 is the radius of the shell.

imagine a viscoelastic-like response at depth or near the hot magma body, it is unlikely that the entire crust behaves as a viscoelastic medium, at least one with the same relaxation time as hotter rocks at depth.

Dragoni and Magnanensi (1989) presented an interesting model of a pressurized spherical chamber surrounded by a viscoelastic shell. This model is geologically plausible in that we expect the rock surrounding the magma chamber to be hotter than ambient and therefore to relax under the applied stresses. Dragoni and Magnanensi (1989) present results for a Maxwell shell in a full-space. We show here how their results are simply, albeit approximately, extended to the half-space.

Consider the geometry shown in figure 7.36. The magma chamber is represented by a spherical chamber with radius R_1. Surrounding this is a viscoelastic shell with radius R_2. The depth to the center of the chamber is d, and the pressure acting on the wall of the chamber ($r = R_1$) is given by $p(t)$. We begin with an elastic solution for a spherical cavity surrounded by an elastic shell with different moduli, in a full-space. Following the methods outlined in chapter 6, we then use the correspondence principle to obtain the solution of the viscoelastic problem in the Laplace transform domain. The inverse transform yields the viscoelastic solution in a full-space. We then follow the method in section 7.1 to obtain an approximate solution for the half-space, valid for $R_2/d \ll 1$.

Before proceeding with the analysis, we can predict some features of the solution. For times very short compared to the relaxation time of the viscoelastic shell, the displacements are equivalent to those due to a pressurized sphere of radius R_1 in an elastic half-space. At very long time compared to the Maxwell relaxation time, the material in the shell completely relaxes, transmitting the magma pressure to the outer wall of the shell. In this limit, the displacements are equivalent to those due to a pressurized sphere of radius R_2. If $R_1 < R_2 \ll d$, then the geometry does not change with time, so the shape of the surface displacement field does not change with time. The amplitude, however, will increase as the apparent radius of the source increases from R_1 to R_2.

The corresponding elastic solution proceeds as follows: the shell $R_1 \leq r \leq R_2$ has shear modulus μ_1 and will be referred to as region 1, and the exterior region (region 2), $r \geq R_2$, has shear modulus μ_2. Because of the radial symmetry, we return to the general solution given in equation (7.5). The solution must remain finite in the limit that $r \to \infty$, which implies that the term proportional to r must be zero in region 2, and thus

$$u_r^{(1)} = A\frac{r}{3} + \frac{B}{r^2},$$

$$u_r^{(2)} = \frac{C}{r^2}. \tag{7.88}$$

Note that the volumetric strain is zero for $r \geq R_2$, as was the case of the pressurized cavity in a homogeneous full-space, but within the shell $R_1 \leq r \leq R_2$, the volumetric strain is A. The stresses, from Hooke's law, are

$$\sigma_{rr}^{(1)} = K_1 A - \frac{4\mu_1 B}{r^3},$$

$$\sigma_{rr}^{(2)} = -\frac{4\mu_2 C}{r^3}, \tag{7.89}$$

where K_1 is the bulk modulus in region 1. The boundary conditions require that both the displacements and stresses match at the interface $r = R_2$ and that the radial stress at the inner boundary equal the applied pressure:

$$u_r^{(1)}(r = R_2) = u_r^{(2)}(r = R_2),$$

$$\sigma_{rr}^{(1)}(r = R_2) = \sigma_{rr}^{(2)}(r = R_2),$$

$$\sigma_{rr}^{(1)}(r = R_1) = -p. \tag{7.90}$$

This leads to three equations in the three unknowns A, B, and C, which after some algebra yields

$$A = -3pR_1^3(\mu_1 - \mu_2)/D,$$

$$B = -pR_1^3 R_2^3(3K_1 + 4\mu_2)/4D,$$

$$C = -pR_1^3 R_2^3(3K_1 + 4\mu_1)/4D, \tag{7.91}$$

where

$$D = 3K_1 R_1^3(\mu_1 - \mu_2) - \mu_1 R_2^3(3K_1 + 4\mu_2) \tag{7.92}$$

(Dragoni and Magnanensi 1989). To find the viscoelastic solution, we appeal to the correspondence principle (section 6.2.1) and replace μ_1 with its Laplace transform (section A.2). For a Maxwell material,

$$\bar{\mu}(s) = \frac{s}{\mu^{-1}s + \eta^{-1}} \tag{7.93}$$

(equation [6.28]). Because there is no bulk relaxation, the bulk modulus is unchanged in the Laplace domain. Furthermore, we assume no relaxation external to the shell (region 2), so μ_2 is unchanged. To keep the algebra as simple as possible, we assume that the intrinsic elastic stiffness is the same in both regions; this will be labeled simply as μ.

The solutions are thus of the form

$$\bar{u}_r^{(1)}(r,\, s) = \frac{\bar{A}(s)r}{3} + \frac{\bar{B}(s)}{r^2},$$

$$\bar{u}_r^{(2)}(r,\, s) = \frac{\bar{C}(s)}{r^2}, \tag{7.94}$$

where the overbar indicates the Laplace transform and s is the transform variable. Some tedious algebra leads to

$$\bar{A} = -\frac{\bar{p}}{2\eta}\left(\frac{1-2v}{1-v}\right)\left(\frac{R_1}{R_2}\right)^3 \bar{D}^{-1},$$

$$\bar{B} = \frac{\bar{p}R_1^3}{4\mu}(s+\mu/\eta)\bar{D}^{-1},$$

$$\bar{C} = \frac{\bar{p}R_1^3}{4\mu}\left[s + \frac{\mu(1+v)}{3\eta(1-v)}\right]\bar{D}^{-1}, \tag{7.95}$$

where

$$\bar{D} = s + \frac{\mu(1+v)R_1^3}{3\eta(1-v)R_2^3}. \tag{7.96}$$

We first consider an instantaneous step increase in magma pressure, $p(t) = p_0 H(t)$, where $H(t)$ is the Heavyside function, such that $\bar{p}(s) = 1/s$. In this case, each of the terms in equation (7.95) is of the form

$$\frac{as+b}{s(s+t_R^{-1})}, \tag{7.97}$$

where the characteristic relaxation time is

$$t_R = \frac{3\eta(1-v)R_2^3}{\mu(1+v)R_1^3}. \tag{7.98}$$

Notice that the characteristic relaxation time depends on the material relaxation time η/μ and geometric factors. The inverse Laplace transform of equation (7.97) is considered in section A.2, and takes the form

$$ae^{-t/t_R} + bt_R(1 - e^{-t/t_R}). \tag{7.99}$$

Using the general form (7.99), the inverse transforms of equation (7.95) are

$$\mathcal{L}^{-1}(\bar{A}) = -\frac{3(1-2v)p_0}{2\mu(1+v)}(1 - e^{-t/t_R}),$$

$$\mathcal{L}^{-1}(\bar{B}) = \frac{p_0 R_1^3}{4\mu}\left[e^{-t/t_R} + \frac{3(1-v)R_2^3}{(1+v)R_1^3}(1 - e^{-t/t_R})\right],$$

$$\mathcal{L}^{-1}(\bar{C}) = \frac{p_0 R_1^3}{4\mu}\left[e^{-t/t_R} + \frac{R_2^3}{R_1^3}(1 - e^{-t/t_R})\right]. \tag{7.100}$$

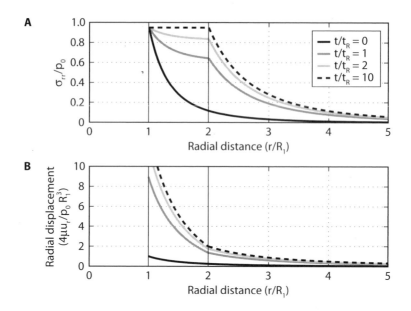

Figure 7.37. Stresses and displacements due to a pressurized sphere in a full-space surrounded by a Maxwell viscoelastic shell. A: Radial stress normalized by magma pressure. B: Normalized radial displacement. The vertical lines at $R_1 = 1$ mark the radius of the magma chamber and at $R_2 = 2$ mark the radius of the viscoelastic shell for this particular example.

Substituting equation (7.100) into (7.94) yields expressions for the displacements as a function of time and radial distance:

$$u_r^{(1)}(r, t) = \frac{p_0 R_1^3}{4\mu} \left\{ \frac{e^{-t/t_R}}{r^2} + \left[\frac{3(1-\nu)}{r^2} \left(\frac{R_2}{R_1} \right)^3 - \frac{2(1-2\nu)r}{R_1^3} \right] \frac{(1 - e^{-t/t_R})}{1+\nu} \right\} \qquad R_1 < r < R_2,$$

$$u_r^{(2)}(r, t) = \frac{p_0 R_1^3}{4\mu r^2} \left[e^{-t/t_R} + \frac{R_2^3}{R_1^3}(1 - e^{-t/t_R}) \right], \qquad r > R_2. \tag{7.101}$$

Notice that the displacements outside the relaxing shell look exactly like those of a source in an elastic body with time varying strength. Initially, the source strength is proportional to $p_0 R_1^3$. With time, the displacements increase, at an exponentially decaying rate, after the instantaneous elastic response (figure 7.37). For $t \gg t_R$, the strength is proportional to $p_0 R_2^3$.

From equation (7.89), the radial stresses in the Laplace transform domain are

$$\bar{\sigma}_{rr}^{(1)} = K_1 \bar{A}(s) - 4\bar{\mu}(s) \frac{\bar{B}(s)}{r^3},$$

$$\bar{\sigma}_{rr}^{(2)} = -4\mu \frac{\bar{C}(s)}{r^3}, \tag{7.102}$$

where

$$\bar{\mu}(s)\bar{B}(s) = \frac{p_0 R_1^3}{4} \frac{1}{s + t_R^{-1}}. \tag{7.103}$$

Making use of equation (7.100), the stresses are given by

$$\sigma_{rr}^{(1)}(r, t) = \frac{-p_0 R_1^3}{r^3} e^{-t/t_R} - p_0(1 - e^{-t/t_R}) \qquad R_1 < r < R_2,$$

$$\sigma_{rr}^{(2)}(r, t) = \frac{-p_0 R_1^3}{r^3} \left[e^{-t/t_R} + \frac{R_2^3}{R_1^3}(1 - e^{-t/t_R}) \right] \qquad r > R_2. \tag{7.104}$$

As expected, the spatial dependence of the radial stresses outside the relaxing shell are the same as those due to a center of dilatation in an elastic body. For $t \ll t_R$, the effective source radius is R_1, whereas for $t \gg t_R$, the effective source radius is R_2. The stresses in the shell decay so that in the fully relaxed state $\sigma_{rr}^{(1)}(r, t \to \infty) = -p_0$; in this limit, the shell behaves as a fluid at pressure p_0 (figure 7.37).

We now proceed to consider the displacements at the free surface. An exact treatment is not likely or useful, as we have seen in the case of the pressurized cavity in an elastic medium. However, we can find an approximate solution that is valid in the limit that the depth to the center of the chamber is large compared to the radius of the cavity and its viscoelastic aureole.

Notice from equations (7.101) and (7.104) that the displacements and stresses outside the viscoelastic shell are of the same form as the displacements and stresses for a pressurized cavity in an elastic body. Specifically, the displacements decay with distance as $1/r^2$, and the stresses as $1/r^3$. Thus, if we follow the procedure in section 7.1 and compute the displacements and stresses on the plane $z = 0$, they turn out to be scaled, time-dependent versions of equations (7.8) and (7.10), respectively. Following the procedure in section 7.1, we can remove the shear and normal tractions on $z = 0$ by adding equal and opposite stresses. The displacements associated with these surface tractions are scaled versions of equations (7.11). At this stage of the analysis, the boundary conditions are satisfied on the plane $z = 0$ exactly, while the boundary conditions on the cavity boundary $r = R_1$ and the constitutive equations within the viscoelastic shell are met only approximately. Specifically, these conditions are violated at order $(R_2/d)^3$. As long as R_2 is sufficiently small compared to the depth to the center of the chamber, the approximate solution should be reasonably accurate. Indeed, by analogy with the elastic case, we can anticipate that the approximate solution is accurate even for only modestly small R_2/d.

The approximate displacements at the free surface are thus scaled, time-dependent versions of the Mogi solution in an elastic half-space (equations [7.12]):

$$u_z(\rho, z = 0, t) = \frac{(1 - v)p_0 R_1^3}{\mu d^2} \left[e^{-t/t_R} + \frac{R_2^3}{R_1^3}(1 - e^{-t/t_R}) \right] \frac{1}{(1 + \rho^2)^{3/2}},$$

$$u_\rho(\rho, z = 0, t) = \frac{(1 - v)p_0 R_1^3}{\mu d^2} \left[e^{-t/t_R} + \frac{R_2^3}{R_1^3}(1 - e^{-t/t_R}) \right] \frac{\rho}{(1 + \rho^2)^{3/2}}, \tag{7.105}$$

where ρ is the radial distance from the center of the source normalized by the source depth.

The displacements have the same spatial distribution as for the pressurized cavity in a completely elastic half-space. Following the instantaneous elastic response, the displacements continue to increase, at an exponentially decreasing rate that scales with t_R. The postintrusion displacements scale with $(R_2/R_1)^3 - 1$ and are thus potentially very significant. For example, a viscoelastic shell of only 20% the radius of the magma chamber (figure 7.38) leads to time-dependent displacements that are 70% of the instantaneous elastic displacements.

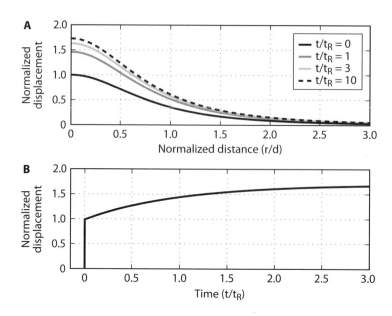

Figure 7.38. Vertical displacements due to a pressurized sphere surrounded by a Maxwell viscoelastic shell in a half-space. Displacements are normalized by $(1 - \nu)p_0 R_1^3/\mu d^2$. A: Vertical displacement as a function of radial distance along the free surface at four different nondimensional times. B: Displacement as a function of nondimensional time immediately above the magma chamber. $R_2/R_1 = 1.2$.

Although actual intrusions are not step functions, equations (7.105) should be accurate as long as the duration of the intrusion is much shorter than the characteristic relaxation time. Relaxation times from days to decades are expected for viscosities in the range of 10^{15} to 10^{18} Pa-s. We can, however, easily extend the analysis to consider gradual, rather than instantaneous, increases in magma pressure—for example, let

$$p(t) = p_0 \left(1 - e^{-t/t_s}\right).\tag{7.106}$$

In this pressure history (7.106), the rate of pressurization decreases with time (figure 7.39). This is reasonable if intrusion of melt into the shallow magma chamber is driven by the pressure difference between the chamber and a deeper reservoir. From equation (7.94), we see that the effect of a gradual pressurization of the magma chamber is only in the time dependence of the deformation. The spatial dependence is unchanged.

The Laplace transform of equation (7.106) is

$$\bar{p}(s) = p_0 \left(\frac{1}{s} - \frac{1}{s + t_s^{-1}}\right).\tag{7.107}$$

Considering only the displacements outside the viscoelastic shell, we need only invert $\bar{C}$, which in this case is given by

$$\bar{C}(s) = \frac{p_0 R_1^3}{4\mu} \left[\frac{s+a}{s(s+t_R^{-1})} - \frac{s+a}{(s+t_s^{-1})(s+t_R^{-1})}\right],\tag{7.108}$$

where

$$a = \frac{\mu(1+\nu)}{3\eta(1-\nu)}.$$

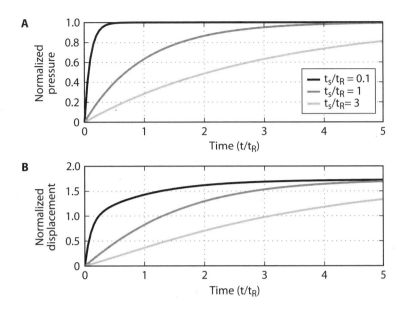

Figure 7.39. Vertical displacements due to a pressurized sphere surrounded by a Maxwell viscoelastic shell in a half-space. A: Source pressure history, normalized by p_0, given by equation (7.106). B: Maximum vertical displacement at $\rho = 0$, $z = 0$ as a function of time. Displacements are normalized by $(1 - v)p_0 R_1^3/\mu d^2$. Results are shown for three different ratios of the source duration t_s to the relaxation time t_R.

We require the new inverse transform

$$\mathcal{L}^{-1}\left\{ \frac{s+a}{(s+t_s^{-1})(s+t_R^{-1})} \right\} = \frac{1}{t_R^{-1} - t_s^{-1}} \left[(a - t_s^{-1})e^{-t/t_s} - (a - t_R^{-1})e^{-t/t_R} \right], \qquad (7.109)$$

which is valid except for $t_R^{-1} = t_s^{-1}$. In the latter case, the relevant transform is

$$\mathcal{L}^{-1}\left\{ \frac{s+a}{(s+t_R^{-1})^2} \right\} = ate^{-t/t_R} + (1 - t/t_R)e^{-t/t_R}. \qquad (7.110)$$

Combining equations (7.94), (7.108), (7.99), (7.109), and (7.110) yields the displacement field outside the relaxing shell:

$$u_r^{(2)}(r, t) = \frac{p_0 R_1^3}{4\mu r^2}\left[\Lambda(e^{-t/t_R} - e^{-t/t_s}) + \frac{R_2^3}{R_1^3}(1 - e^{-t/t_R}) \right] \qquad r > R_2,\ t_s \neq t_R,$$

$$u_r^{(2)}(r, t) = \frac{p_0 R_1^3}{4\mu r^2}\left[\left(1 - \frac{R_2^3}{R_1^3}\right)\frac{t}{t_R}e^{-t/t_R} + \frac{R_2^3}{R_1^3}(1 - e^{-t/t_R}) \right] \qquad r > R_2,\ t_s = t_R, \qquad (7.111)$$

where

$$\Lambda = \frac{(t_s/t_R)(R_2/R_1)^3 - 1}{(t_s/t_R) - 1}. \qquad (7.112)$$

Following the same procedure as before, the displacements at the surface of an elastic half-space are given by (see figure 7.39)

$$u_z(\rho, z = 0, t) = \frac{(1 - \nu)p_0 R_1^3}{\mu d^2} \frac{1}{(1 + \rho^2)^{3/2}} f(t, t_s, t_R, R_2/R_1),$$

$$u_\rho(\rho, z = 0, t) = \frac{(1 - \nu)p_0 R_1^3}{\mu d^2} \frac{\rho}{(1 + \rho^2)^{3/2}} f(t, t_s, t_R, R_2/R_1),$$

$$f(t, t_s, t_R, R_2/R_1) = \Lambda e^{-t/t_R} - \Lambda e^{-t/t_s} + \frac{R_2^3}{R_1^3}(1 - e^{-t/t_R}) \qquad t_s \neq t_R,$$

$$f(t, t_s, t_R, R_2/R_1) = \left(1 - \frac{R_2^3}{R_1^3}\right) t e^{-t/t_R} + \frac{R_2^3}{R_1^3}(1 - e^{-t/t_R}) \qquad t_s = t_R. \qquad (7.113)$$

The limiting behavior for both forms of $f(t, t_s, t_R, R_2/R_1)$ is that $f(0, t_s, t_R, R_2/R_1) = 0$, since there has been no pressure change at time zero. Also, $f(t \to \infty, t_s, t_R, R_2/R_1) = R_2^3/R_1^3$, so that in the fully pressurized and relaxed state, the displacements are those due to a chamber of radius R_2. Last, note that if $t_s/t_R \ll (R_1/R_2)^3 < 1$, then $\Lambda = 1$, and equation (7.113) reduces to (7.105). If on the other hand, the source time is long compared to the relaxation time $t_s \gg t_R$, then $\Lambda = (R_2/R_1)^3$, and equation (7.113) reduces to

$$u_r^{(2)}(r, t) = \frac{p_0 R_2^3}{4\mu r^2}\left(1 - e^{-t/t_s}\right). \qquad (7.114)$$

In this limit, the viscoelastic shell relaxes so quickly that the deformation is equivalent to a chamber of radius R_2 in an elastic medium with the specified pressure history.

7.7 Summary and Perspective

Volcano deformation is an important subject in its own right. Fortunately, we are able to utilize and build on results and methods discussed in previous chapters. The Mogi model of a pressurized spherical cavity in a homogeneous half-space, while certainly a highly idealized representation of a magma chamber, has undoubtedly been the most commonly employed volcanic source model. Even this problem can be solved analytically only in an approximate sense, accurate when the chamber radius is small compared to its depth. The standard Mogi solution, corresponding to an isotropic center of dilatation, is simply derived from the Green's functions presented in chapter 3. This approach also yields an equivalent moment tensor representation of the source. The amplitude of the surface deformation scales with the change in magma chamber volume. From deformation data alone, it is not possible to determine the absolute chamber volume or the change in magma pressure.

Eshelby's cut-and-weld procedure provides a means for considering ellipsoidal magma chambers. For an ellipsoidal inclusion in a full-space, the strains within the inclusion are found to be spatially uniform. For a given chamber aspect ratio, an appropriate choice of the transformation strains guarantees a uniform and isotropic state of stress, equivalent to a uniform fluid pressure. The displacements in the surrounding elastic medium can be computed by a convolution of the tractions associated with the transformation strain with the elastic Green's tensors. This is equivalent to the displacements generated by a uniform distribution of centers of dilatation and double forces within the ellipsoidal chamber. In order to approximate the displacements in a half-space, we employ the half-space Green's tensors. For a prolate ellipsoid, Yang et al. (1988) found that the Eshelby solution can be reduced to an appropriate distribution of centers of dilatation and double forces along the line connecting the foci of the ellipsoid. The displacements due to an elongated magma conduit is obtained as a limiting case of the Eshelby solution.

The predicted surface displacements from these models demonstrate that it is generally difficult to discriminate between different source shapes based solely on vertical displacement data. Analysis of the ratio of vertical to horizontal displacements is generally far better at resolving source shape, which may also trade off with depth, and therefore source strength.

Previously discussed dislocation (chapter 3) and crack (chapter 4) solutions are well suited for analysis of dikes and sills. Vertical and near-vertical dikes generate distinctive surface deformation patterns, so it is generally possible to discriminate dikes from other possible volcanic sources. For a dike or sill filled with pressurized but nonflowing magma, the shear tractions vanish, and the magma pressure gradients are "magma-static." This motivates the application of crack models with specified traction boundary conditions, as opposed to dislocation models with specified displacement discontinuities. The boundary element method provides computationally efficient solutions for pressurized cracks in elastic half-spaces.

Typical volcano deformation models treat the earth as a homogeneous elastic half-space. The influence of heterogeneous elastic properties can be included using methods described in chapter 5. Topographic effects, which can be significant on stratovolcanoes, are analyzed in chapter 8. It is certainly a strong idealization to model magma chambers as entirely fluid and the surroundings as linearly elastic right up to the magma chamber walls. In some instances, magmatic fluids may permeate the surrounding wall rocks; porous media effects are described in chapter 10 that could be brought to bear on this process. Heat will certainly be transferred to the surroundings, leading to possible thermoelastic strains, a subject also touched on in chapter 10. The heating of wall rocks adjacent to a magma chamber may result in stress relaxation within the surroundings. This chapter closed with a discussion of the simple case of a spherical magma chamber surrounded by a concentric viscoelastic shell, exploiting methods described in chapter 6. In a more realistic description, the rheology of the surrounding wall rocks would be a function of the local temperature.

Deformation measurements alone often cannot distinguish magmatic intrusion from hydrothermal phenomenon. Repeated gravity determinations combined with deformation measurements can, in principle, constrain the density of intruded fluids and help discriminate between competing processes. This is discussed in more detail in chapter 9.

7.8 Problems

1. Derive equation (7.3), the equilibrium equation for spherical symmetry.

2. Show that the volume of the uplift for a Mogi source is given by equation (7.25). Explain why the uplift decreases with increasing Poisson's ratio.

3. Derive the solution for the displacement at the free surface due to a center of dilatation (7.14) from the half-space Green's functions (3.93) and (3.94).

4. Show that the Eshelby method can be used to generate the Mogi solution. First, show from the integrals (7.52) that for a sphere of radius a

$$I_1 = I_2 = I_3 = 4\pi/3,$$
$$I_{11} = I_{22} = I_{33} = I_{12} = I_{23} = I_{13} = 4\pi/5a^2. \tag{7.115}$$

From this, show that the shape tensors for a sphere are

$$S_{1111} = S_{2222} = S_{3333} = \frac{7 - 5v}{15(1 - v)},$$

$$S_{1122} = S_{2211} = S_{1133} = S_{2233} = S_{3311} = S_{3322} = \frac{5v - 1}{15(1 - v)}. \tag{7.116}$$

Last, solve for the surface displacements, using the Mindlin Green's functions and assuming that the radius of the chamber is small compared to its depth, so that the Green's functions can be taken out of the integral.

5. Use the Eshelby theory to derive the displacements due to a penny-shaped pressurized crack. Note that for $a_1 = a_2 \gg a_3$, Mura (1987) gives

$$S_{1111} = S_{2222} = \frac{13 - 8v}{32(1 - v)} \pi \frac{a_3}{a_1},$$

$$S_{3333} = 1 - \frac{1 - 2v}{(1 - v)} \frac{\pi}{4} \frac{a_3}{a_1},$$

$$S_{1122} = S_{2211} = \frac{8v - 1}{32(1 - v)} \pi \frac{a_3}{a_1},$$

$$S_{1133} = S_{2233} = -\frac{1 - 2v}{8(1 - v)} \pi \frac{a_3}{a_1},$$

$$S_{3311} = S_{3322} = \frac{v}{1 - v} \left(1 - \frac{4v + 1}{8v} \pi \frac{a_3}{a_1} \right). \tag{7.117}$$

Use these results to show that in the limit $a_3/a_1 \to 0$,

$$\epsilon_{33}^T = \frac{2(1 - v) p}{\mu \pi} \frac{a_1}{a_3},$$

and

$$\epsilon_{11}^T = \frac{(1 - 2v)(1 - v) p}{4\mu(1 + v)}.$$

Show that in the limit $a_3/a_1 \to 0$ that $\epsilon_{33}^c = \epsilon_{33}^T$. Last, combine with equations (7.81) and (7.82) to derive (7.80).

6. Derive equation (7.44). Eshelby redefines the angles l_i such that they point from the observation point $\mathbf{x}$ to the integration point ξ. Because the q_{ijk} are odd functions of $\mathbf{l}$, this results in a sign change:

$$u_i^c(\mathbf{x}) = -\frac{\epsilon_{kj}^T}{8\pi(1 - v)} \int_V \frac{q_{ijk}(\mathbf{l})}{r^2} \mathrm{d}V. \tag{7.118}$$

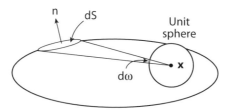

Figure 7.40. A conical volume element extending from the field point **x** to the surface of the ellipsoid. The area element on the ellipsoid is dS, whereas the area subtended by the cone on the surface of a unit sphere is $d\omega$.

First show using the divergence theorem that equation (7.118) can be transformed to a surface integral:

$$u_i^c(\mathbf{x}) = -\frac{\epsilon_{kj}^T}{8\pi(1-\nu)} \int_S q_{ijk}(\mathbf{l}) \frac{r_m n_m}{r^2} dS, \tag{7.119}$$

where S is the surface of the ellipsoid, and **n** is the unit normal to the ellipsoid.

Hint: you may show that $r_m \partial l_i / \partial \xi_m = 0$.

Now consider a unit sphere centered on the observation point **x** (figure 7.40). Show that the area element on the surface of the ellipsoid dS is related to the area element of the unit sphere $d\omega$ by

$$\frac{r_i n_i}{r^2} dS = r\, d\omega. \tag{7.120}$$

Last, combine these results to prove equation (7.44).

7. A unit vertical line force acting on the surface of an elastic half-space produces radial stress $\sigma_{rr} = -(2/\pi r)\cos(\theta)$, where θ is measured relative to the vertical, and $\sigma_{r\theta} = \sigma_{\theta\theta} = 0$ (Timoshenko and Goodier 1970). Use this result and the reciprocal theorem to show that the vertical surface displacement above a vertical dike must be exactly zero. The horizontal displacement there must also vanish by symmetry.

8. This problem develops the expressions needed to compute the surface deformation due to a uniformly pressurized two-dimensional vertical dike in a half-space, using the boundary element method, following the procedure outlined in section 4.6. Recall that a dike can be represented by a sequence of edge dislocations, as in figure 4.4. Here we take the interpretation that a single edge dislocation results from the insertion of a semi-infinite plane of constant thickness, as in figure 3.26 (left) and explored in problem 5 of chapter 3. Discretize the dike plane into evenly spaced elements. The first step is to compute the stress acting on the dike plane due to unit opening in the depth interval $-d_1 < z < -d_2$. From equations (3.77), show that the horizontal stress at depth z is given by

$$\sigma_{11}(z) = \frac{\mu}{\pi(1-\nu)} \left[\frac{\xi}{z^2-\xi^2} + \frac{\xi(\xi-z)}{(z+\xi)^3} \right]_{\xi=d_1}^{\xi=d_2}. \tag{7.121}$$

From this result, compute the matrix that relates the opening in each element to stress change at the midpoint of every element, as in equation (4.61). Setting the cumulative stress change equal to the dike pressure p, invert for the opening in each element.

The next step is to compute the surface displacements given the opening in each element. Use Volterra's formula, as developed in chapter 3, to solve for the displacement on the free surface due to opening in each interval. Show that the horizontal u_1 and vertical u_2 displacements are given by

$$u_1(x_1, x_2 = 0) = -\frac{\Delta u}{\pi} \left[\frac{x_1 \xi}{x_1^2 + \xi^2} + \tan^{-1}\left(\xi/x_1\right) \right]_{\xi=d_1}^{\xi=d_2},$$

$$u_2(x_1, x_2 = 0) = \frac{\Delta u}{\pi} \left[\frac{x_1^2}{x_1^2 + \xi^2} \right]_{\xi=d_1}^{\xi=d_2}. \tag{7.122}$$

Compute the surface displacements given the opening distribution determined earlier.

7.9 References

Agustsson, K., R. Stefansson, A. T. Linde, P. Einarsson, I. S. Sacks, G. B. Gudmundsson, and B. Thorbjarndottir. 2000. Successful prediction and warning of the 2000 eruption of Hekla based on seismicity and strain changes, *EOS Transactions of the American Geophysical Union* **81**(48 Suppl.), Abstract V11B-30.

Amelung, F., S. Jónsson, H. Zebker, and P. Segall. 2000. Widespread uplift and trap door faulting on Galápagos volcanoes. *Nature* **407**(6807), 993–996.

Amoruso, A., and L. Crescenti. 2009. Shape and volume change of pressurized ellipsoidal cavities from deformation and seismic data. *Journal of Geophysical Research* **114**, B02210, doi:10.1029/2008JB005946.

Anderson, E. M. 1936. The dynamics of the formation of cone-sheets, ring-dykes, and caldron subsidence. *Proceedings of the Royal Society of Edinburgh* **56**, 128–157.

Battaglia, M., P. Segall, J. Murray, P. Cervelli, and J. Langbein. 2003a. The mechanics of unrest at Long Valley caldera, California: I. Modelling the geometry of the source using GPS, leveling, and two-color EDM data. *Journal of Volcanology and Geothermal Research* **127**, 195–217.

Battaglia, M., C. Roberts, and P. Segall. 2003b. The mechanics of unrest at Long Valley caldera, California: 2. Constraining the nature of the source using geodetic and micro-gravity data. *Journal of Volcanology and Geothermal Research* **127**, 219–245.

Bonaccorso, A., and P. M. Davis. 1999. Models of ground deformation from vertical volcanic conduits with application to eruptions of Mount St. Helens and Mount Etna. *Journal of Geophysical Research* **104**, 10,531–10,542.

Bonafede, M., M. Dragoni, and F. Quareni. 1986. Displacement and stress fields produced by a center of dilation and by a pressure source in a viscoelastic half-space: application to the study of ground deformation and seismic activity at Campi Flegrei, Italy. *Journal of Geophysical Research* **92**, 14,139–14,150.

Davies, J. H. 2003. Elastic field in a semi-infinite solid due to thermal expansion of a coherently misfitting inclusion. *Journal of Applied Mechanics* **70**, 655–660.

Davis, P. M. 1986. Surface deformation due to inflation of an arbitrarily oriented triaxial ellipsoidal cavity in an elastic half-space, with reference to Kilaeua volcano, Hawaii. *Journal of Geophysical Research* **91**(B7), 7429–7438.

Delaney, P. T., and D. F. McTigue. 1994. Volume of magma accumulation or withdrawal estimated from surface uplift or subsidence, with application to the 1960 collapse of Kilauea volcano. *Bulletin of Volcanology* **56**, 417–424.

Dieterich, J. H., and R. W. Decker. 1975. Finite element modeling of surface deformation associated with volcanism. *Journal of Geophysical Research* **80**, 4094–4102.

Dragoni, M., and C. Magnanensi. 1989. Displacement and stress produced by a pressurized, spherical magma chamber surrounded by a viscoelastic shell. *Physics of the Earth and Planetary Interiors* **56**, 316–328.

Eshelby, J. D. 1957. The determination of the elastic field of an ellipsoidal inclusion and related problems. *Proceedings of the Royal Society* **A241**, 376–396.

Fialko, Y., Y. Khazan, and M. Simons. 2001. Deformation due to a pressurized horizontal circular crack in an elastic half-space, with applications to volcano geodesy. *Geophysical Journal International* **146**, 181–190.

Jackson, M. D., and D. D. Pollard. 1988. The laccolith-stock controversy: new results from the southern Henry Mountains, Utah. *Geological Society of America Bulletin* **100**, 117–139.

Johnson, D. J. 1992. Dynamics of magma storage in the summit reservoir of Kilauea volcano, Hawaii. *Journal of Geophysical Research* **97**, 1807–1820.

Kellog, O. D. 1929. *Potential theory*. Berlin: Springer.

Langbein, J., D. P. Hill, T. N. Parker, and S. K. Wilkinson. 1993. An episode of reinflation of the Long Valley caldera, eastern California: 1989–1991. *Journal of Geophysical Research* **98**, 15,851–15,870.

Larsen, G., K. Grönvold, and S. Thorarinson. 1979. Volcanic eruption through a geothermal borehole at Námafjall, Iceland. *Nature* **278**, 707–710.

Linde, A. T., K. Agústsson, I. S. Sacks, and R. Stefánsson. 1993. Mechanism of the 1991 eruption of Hekla from continuous borehole strain monitoring. *Nature* **365**, 737–740.

Lipman, P. W., J. G. Moore, and D. A. Swanson. 1981. Bulging of the north flank before the May 18 eruption—geodetic data. In P. W. Limpan and D. R. Mullineaux (Eds.), *The 1980 eruptions of Mount St. Helens*, USGS Professional Paper 1250. Washington, DC: U.S. Government Printing Office, pp. 143–155.

McTigue, D. F. 1987. Elastic stress and deformation near a finite spherical magma body: resolution of the point source paradox. *Journal of Geophysical Research* **92**, 12,931–12,940.

Mogi, K. 1958. Relations between the eruptions of various volcanoes and the deformations of the ground surfaces around them. *Bulletin of the Earthquake Research Institute of the University of Tokyo* **36**, 99–134.

Mura, T. 1987. *Micromechanics of defects in solids*. Lancaster: Martinus Nijhoff Publishers.

Murase, T., and A. R. McBirney. 1973. Properties of common igneous rocks and their melts at high temperatures. *Geological Society of America Bulletin* **84**, 3563–3592.

Okada, Y., E. Yamamoto, and T. Ohkubo. 2000. Coswarm and preswarm crustal deformation in the eastern Izu Peninsula, central Japan. *Journal of Geophysical Research* **105**, 681–692.

Owen, S., P. Segall, M. Lisowski, M. Murray, M. Bevis, and J. Foster. 2000. The January 30, 1997, eruptive event on Kilauea volcano, as monitored by continuous GPS. *Geophysical Research Letters* **27**, 2757–2760.

Pollard, D. D., P. T. Delaney, W. A. Duffield, E. T. Endo, and A. T. Okamura. 1983. Surface deformation in volcanic rift zones. *Tectonophysics* **94**, 541–584.

Rivalta, E., and P. Segall. 2008. Magma compressibility and the missing source for some dike intrusions. *Geophysical Research Letters* **35**, L04306, doi:10.1029/2007GL032521.

Routh, D. J. 1892. A treatise on analytical statics 2. London: Cambridge University Press, pp. 196–254.

Rubin, A. M. 1992. Dike-induced faulting and graben subsidence in volcanic rift zones. *Journal of Geophysical Research* **97**, 1839–1858.

Rubin, A. M., and D. D. Pollard. 1988. Dike-induced faulting in rift zones of Iceland and Afar. *Geology* **16**, 413–417.

Sezawa, K. 1931. The plastico-elastic deformation of a semi-infinite solid body due to an internal force. *Bulletin of Earthquake Research Institute University of Tokyo* **9**, 398–406.

Timoshenko, S. P., and J. N. Goodier. 1970. *Theory of elasticity*. New York: McGraw-Hill, pp. 97–104.

Vasco, D. W., L. R. Johnson, and N. E. Goldstein. 1988. Using surface deformation and strain observations to determine deformation at depth, with an application to Long Valley caldera, California. *Journal of Geophysical Research* **93**, 3232–3242.

Walsh, J. B., and R. W. Decker. 1971. Surface deformation associated with volcanism. *Journal of Geophysical Research* **76**(14), 3291–3302.

Yang, X.-M., P. M. Davis, and J. H. Dieterich. 1988. Deformation from inflation of a dipping finite prolate spheroid in an elastic half-space as a model of volcanic stressing. *Journal of Geophysical Research* **93**, 4249–4257.

Yamakawa, N. 1955. On the strain produced in a semi-infinite elastic solid by an interior source of stress. *Zisin (Journal of the Seismologica Society of Japan)* **8**, 84–98.

Yun, S., P. Segall, and H. Zebker. 2006. Constraints on magma chamber geometry at Sierra Negra volcano, Galápagos Islands, based on InSAR observations. *Journal of Volcanology and Geothermal Research* **150**, 232–243.

8

Topography and Earth Curvature

Up to this point, we have treated the earth as a half-space, ignoring both earth curvature and topography. The earth is, of course, an oblate spheroid with mountains, valleys, and volcanic edifices. Except in the most extreme cases, however, the slope of the topography at true scale is modest, and approximate methods accurate for small slope to be presented here are adequate for treating topographic effects. For extreme topography, one would need to resort to boundary element or other numerical procedures.

Semianalytical methods also exist for dislocation sources in a spherical earth (Ben-Menahem et al. 1969; Smylie and Mansinha 1971; Pollitz 1996; Sun et al. 1996). The effects of earth's sphericity, however, are small except at large distances from the deformation source. For all but the very largest earthquakes, the displacements and strains at these distances are extremely small, and the additional mathematical complexity may be unwarranted. We will show that approximate methods can be developed that are adequate for distances from the fault of the order of 1,000 km, which is adequate for most cases of interest.

We will begin by considering the case of irregular surface topography, following Williams and Wadge (2000), who extended the two-dimensional solutions of Ishii and Takagi (1967), Mahrer (1984), and McTigue and Segall (1988) to three dimensions.

The geometry is illustrated in figure 8.1. The surface topography $h(x_1, x_2)$ is measured relative to the plane $x_3 = 0$, usually taken to be the mean elevation in the area of interest. The boundary-value problem to be solved is

$$\frac{\partial \sigma_{ij}}{\partial x_j} = 0, \tag{8.1}$$

with boundary conditions at the free surface and on the fault

$$\sigma_{ij} n_j = 0 \quad \text{on } x_3 = h(x_1, x_2), \tag{8.2}$$

$$u_i^+ - u_i^- = s_i \quad \text{on } \Sigma, \tag{8.3}$$

where n_j is the outward-directed unit normal to the surface $x_3 = h(x_1, x_2)$.

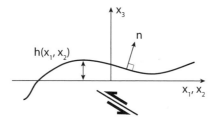

Figure 8.1. Fault beneath topography $h(x_1, x_2)$ with unit normal **n**.

We develop an approximate solution that is accurate for small topographic slopes. In this limit, the unit normal to the free surface has components

$$\mathbf{n} \approx \left[-\frac{\partial h}{\partial x_1}, -\frac{\partial h}{\partial x_2}, 1 \right]^T. \tag{8.4}$$

The topography, $h(x_1, x_2)$, has characteristic height scale H and a horizontal length scale L, such that $H/L \ll 1$. Equations (8.1) can be written in terms of the dimensionless variables $x_i^* = x_i/L$, $h^* = h/H$, and $\sigma_{ij}^* = \sigma_{ij} L/\mu s$. The displacements and stresses can be expanded in powers of the characteristic slope, H/L:

$$u_i^* = u_i^{*(0)} + \frac{H}{L} u_i^{*(1)} + O\left(\frac{H}{L}\right)^2,$$

$$\sigma_{ij}^* = \sigma_{ij}^{*(0)} + \frac{H}{L} \sigma_{ij}^{*(1)} + O\left(\frac{H}{L}\right)^2, \tag{8.5}$$

where O indicates "order of." Neglecting terms of order $(H/L)^2$ and smaller should lead to errors of 10% or less for $H/L \le 0.3$. For small slopes, $H/L \ll 1$, the boundary conditions on the free surface (8.2) are

$$\sigma_{i3}^* - \frac{H}{L} \frac{dh^*}{dx_1^*} \sigma_{i1}^* - \frac{H}{L} \frac{dh^*}{dx_2^*} \sigma_{i2}^* + O\left(\frac{H}{L}\right)^2 = 0 \quad \text{on } x_3 = h(x_1, x_2), \quad i = 1, 2, 3. \tag{8.6}$$

Equation (8.6) accounts for the fact that the free surface is not parallel to the plane $x_3 = 0$. It does not account for the fact that the surface may be located above or below the plane $x_3 = 0$. To do so, we expand the stresses in a Taylor series about $x_3 = 0$:

$$\sigma_{ij}^*(x_3 = h) = \sigma_{ij}^*(x_3 = 0) + \frac{H}{L} h^* \frac{\partial \sigma_{ij}^*}{\partial x_3^*} + O\left(\frac{H}{L}\right)^2, \tag{8.7}$$

which when combined with equation (8.6) yields the approximate free-surface boundary conditions:

$$\sigma_{i3}^* + \frac{H}{L} \left(h^* \frac{\partial \sigma_{i3}^*}{\partial x_3} - \sigma_{1i}^* \frac{\partial h^*}{\partial x_1^*} - \sigma_{2i}^* \frac{\partial h^*}{\partial x_2^*} \right) = 0 \quad \text{on } x_3 = 0, \quad i = 1, 2, 3. \tag{8.8}$$

Now substitute equation (8.5) into the equilibrium equations (8.1) and boundary conditions (8.8), and collect terms of like powers of H/L. Reverting to dimensional variables, the leading order problem, at order $(H/L)^0$, is

$$\frac{\partial \sigma_{ij}^{(0)}}{\partial x_j} = 0,$$

$$\sigma_{j3}^{(0)} = 0 \quad \text{on } x_3 = 0, \quad j = 1, 2, 3,$$

$$u_i^+ - u_i^- = s_i \quad \text{on } \Sigma. \tag{8.9}$$

Notice that this is simply the problem for a fault embedded in a half-space with a plane free surface, the subject of chapter 3.

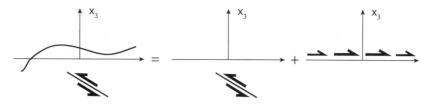

Figure 8.2. Solution to the problem of a fault in an elastic earth with topography can be approximated by a fault in a flat half-space with a correction in the form of distributed surface shear tractions.

The first-order correction due to interaction with the surface topography is given by the governing equations and boundary conditions at order $(H/L)^1$:

$$\frac{\partial \sigma_{ij}^{(1)}}{\partial x_j} = 0,$$

$$\sigma_{j3}^{(1)} = \sigma_{1j}^{(0)} \frac{\partial h}{\partial x_1} + \sigma_{2j}^{(0)} \frac{\partial h}{\partial x_2} - h \frac{\partial \sigma_{j3}^{(0)}}{\partial x_3} \quad \text{on} \quad x_3 = 0, \quad j = 1, 2, 3,$$

$$u_i^+ - u_i^- = 0 \quad \text{on} \quad \Sigma. \tag{8.10}$$

The boundary condition on the plane $x_3 = 0$ for the first-order correction is an applied traction that depends on the known topography and the stress state found solving the system (8.9). Note also that the boundary conditions on the dislocation surface are completely satisfied in the zero-order problem. This is, however, true only if the boundary conditions on the fault are specified in terms of a displacement discontinuity. If the boundary condition is specified in terms of tractions acting on the fault, then these boundary conditions would enter at all values of H/L.

The boundary conditions in equation (8.10) simplify further by use of the equilibrium equations to give

$$\sigma_{j3}^{(1)} = \frac{\partial \left(h \sigma_{1j}^{(0)} \right)}{\partial x_1} + \frac{\partial \left(h \sigma_{2j}^{(0)} \right)}{\partial x_2} \quad \text{on} \quad x_3 = 0, \quad j = 1, 2,$$

$$\sigma_{33}^{(1)} = \frac{\partial \left(h \sigma_{13}^{(0)} \right)}{\partial x_1} + \frac{\partial \left(h \sigma_{23}^{(0)} \right)}{\partial x_2} = 0, \quad \text{on} \quad x_3 = 0, \tag{8.11}$$

where $\sigma_{13}^{(0)}$ and $\sigma_{23}^{(0)}$ vanish everywhere on $x_3 = 0$ by virtue of the boundary conditions at order $(H/L)^0$. The first-order correction is thus equivalent to distributed shear tractions on a half-space (figure 8.2), the form of which depends on the topographic profile, h, and the stresses induced by the dislocation source in a flat half-space. The salient point here is that the first-order correction due to topographic effects is given in terms of quantities that are known after solution of the "zero-order" problem.

There are two remaining steps. The first is to find the displacements $u^{(1)}$ on the plane $x_3 = 0$ by convolving the elastostatic Green's tensors with the traction distributions given by equations (8.11). Employing the notation for the Green's tensors from chapter 3, $u^{(1)}$

becomes

$$
u_i^{(1)}(x_1, x_2, 0) = \int_{-\infty}^{\infty} \int_{-\infty}^{\infty} \left\{ \left[\frac{\partial(h\sigma_{11}^{(0)})}{\partial \xi_1} + \frac{\partial(h\sigma_{12}^{(0)})}{\partial \xi_2} \right] g_i^1(x_1, x_2, 0; \xi_1, \xi_2, 0) \right.
$$
$$
\left. + \left[\frac{\partial(h\sigma_{12}^{(0)})}{\partial \xi_1} + \frac{\partial(h\sigma_{22}^{(0)})}{\partial \xi_2} \right] g_i^2(x_1, x_2, 0; \xi_1, \xi_2, 0) \right\} d\xi_1 d\xi_2. \tag{8.12}
$$

Note that the superscript on the Green's tensors refers to the direction of the point force, whereas the superscript in parentheses on the displacement refers to the order of the term in the power series expansion. From the results in chapter 3, the Mindlin Green's functions reduce to the following expressions when both the source point and the observation point act on the free surface. For a force acting in the 1-direction,

$$
g_1^1 = \frac{1}{2\pi\mu} \left[\frac{(1-\nu)}{r} + \frac{\nu x_1^2}{r^3} \right],
$$
$$
g_2^1 = \frac{\nu}{2\pi\mu} \frac{x_1 x_2}{r^3},
$$
$$
g_3^1 = -\frac{(1-2\nu)}{2\pi\mu} \frac{x_1}{r^2} \qquad r^2 = (x_1 - \xi_1)^2 + (x_2 - \xi_2)^2. \tag{8.13}
$$

Note that the Green's functions for a force acting in the 2-direction are obtained simply by exchanging x_1 and x_2, such that

$$
g_1^2 = \frac{1}{2\pi\mu} \left[\frac{(1-\nu)}{r} + \frac{\nu x_2^2}{r^3} \right],
$$
$$
g_2^2 = \frac{\nu}{2\pi\mu} \frac{x_1 x_2}{r^3},
$$
$$
g_3^2 = -\frac{(1-2\nu)}{2\pi\mu} \frac{x_2}{r^2}. \tag{8.14}
$$

The correction term $u_i^{(1)}(x_1, x_2, 0)$ accounts for the nonhorizontal nature of the free surface but is evaluated on the reference plane $x_3 = 0$, not the actual free surface $x_3 = h$. The second step is thus to compute the displacements at the free surface $x_3 = h$, again keeping the first two terms in a Taylor series expansion:

$$
u_i(x_1, x_2, h) = u_i^{(0)}(x_1, x_2, 0) + u_i^{(1)}(x_1, x_2, 0) + h(x_1, x_2) \frac{\partial u_i^{(0)}}{\partial x_3}(x_1, x_2, 0). \tag{8.15}
$$

The expansion for the displacements (8.15) involves vertical derivatives of the zero-order solution. Because we are primarily interested in deformation on the surface, we often have expressions and/or numerical codes that are restricted to computing the elastic fields at the free surface. It is straightforward, however, to eliminate the vertical derivatives. Consider, for example, the term $\partial u_1^{(0)}/\partial x_3$. Note from Hooke's law that $\sigma_{13} = \mu(\partial u_1^{(0)}\partial x_3 + \partial u_3^{(0)}\partial x_1)$. Since the shear stress vanishes on the plane $x_3 = 0$ in the zero-order problem by equation (8.9), it follows that $\partial u_1^{(0)}/\partial x_3 = -\partial u_3^{(0)}/\partial x_1$. A similar argument leads to $\partial u_3^{(0)}/\partial x_3 = -(\nu/1-\nu)(\partial u_1^{(0)}/\partial x_1 + \partial u_2^{(0)}/\partial x_2)$. In this way, equation (8.15) can be written as

$$u_i(x_1, x_2, h) = u_i^{(0)}(x_1, x_2, 0) + u_i^{(1)}(x_1, x_2, 0) - h(x_1, x_2)\frac{\partial u_3^{(0)}}{\partial x_i}(x_1, x_2, 0), \qquad i = 1, 2,$$

$$u_3(x_1, x_2, h) = u_3^{(0)}(x_1, x_2, 0) + u_3^{(1)}(x_1, x_2, 0) - \frac{\nu}{1-\nu}h(x_1, x_2)\frac{\partial u_i^{(0)}}{\partial x_i}(x_1, x_2, 0). \tag{8.16}$$

8.1 Scaling Considerations

In a parallel analysis, McTigue and Segall (1988) showed that the two-dimensional plane strain analogue of equation (8.12), for vertical displacement, is

$$u_z^{(1)}(x, 0) = -\frac{(1-2\nu)}{4\mu}\int_{-\infty}^{\infty} H(x-\zeta)\,\frac{\partial}{\partial \zeta}\left[h(\zeta)\sigma_{xx}^{(0)}(\zeta, 0)\right]\,\mathrm{d}\zeta, \tag{8.17}$$

where $H(\eta) = -1$ for $\eta < 0$, and $H(\eta) = +1$ for $\eta > 0$. This can be integrated by parts to yield

$$u_z^{(1)}(x, 0) = -\frac{1-2\nu}{2\mu}h(x)\sigma_{xx}^{(0)}(x, 0). \tag{8.18}$$

From this result, it can be shown (problem 1) that

$$u_z(x, h) = u_z^{(0)}(x, 0) - h(x)\epsilon_{xx}^{(0)}(x, 0), \tag{8.19}$$

where $\epsilon_{xx}^{(0)}(x, 0)$ is the surface-parallel normal strain due to the dislocation source in a flat half-space. This simple result reveals two interesting aspects of the topographic correction. First, if the surface-parallel normal strain due to the half-space solution is positive (extension) and topography is above the mean elevation ($h > 0$), then the effect of topography will be to decrease the vertical displacement. In contrast, for compressive surface-parallel strain, the effect of topography is to amplify the uplift in mountains. An important example of the former effect is an inflating magma chamber beneath a volcano. Recall from chapter 7 that an expanding center of dilatation produced surface-parallel extension above the source. For a stratovolcano with elevated topography, we thus expect the topographic correction to be negative—that is, a reduced uplift. The second point is that the topographic effect will be large when the product $h(x)\epsilon_{xx}^{(0)}$ is large—that is, when the topography and the surface-parallel strain are "in phase." This occurs when the spatial scales of the deformation are comparable to the spatial scales of the topography.

More generally, it can be anticipated that the interaction of the source-induced stress field, $\sigma_{ij}^{(0)}$, with the surface topography, h, is strongest when the length scales over which they vary along the surface are similar. Consider one of the equations in (8.11):

$$\sigma_{13}^{(1)} = h\left(\frac{\partial \sigma_{11}^{(0)}}{\partial x_1} + \frac{\partial \sigma_{12}^{(0)}}{\partial x_2}\right) + \sigma_{11}^{(0)}\frac{\partial h}{\partial x_1} + \sigma_{12}^{(0)}\frac{\partial h}{\partial x_2}. \tag{8.20}$$

The stress gradient in the first two terms scales as $\mu s/D^2$, where D is the characteristic dimension of the source. For a point source, this would be the depth. On the other hand, the topographic slope in the last two terms, dh/dy scales like H/L. Therefore, the ratio of the first two terms to the second two terms in equation (8.20) is of order L/D.

For a deep source ($L/D \ll 1$), the second group of terms in equation (8.20) dominates. This simply states that the surface stress, $\sigma_{ij}^{(0)}$, varies slowly compared to the topography, so that the topography interacts with a locally uniform stress. For a shallow source ($L/D \gg 1$), the first two terms in equation (8.20) dominate. In this case, the topography varies slowly compared to the stress, and the stress interacts with a locally horizontal surface. Although the shallow source brings the stress singularity near the surface, the stress falls off so rapidly that it does not interact with the topography.

These dimensional arguments indicate a simple qualitative rule by which one can begin to assess the importance of a topographic correction to modeled surface displacements. The interaction of a source with the irregular surface is significant only when the characteristic dimensions of the source are of the same order as the characteristic horizontal scale of the topography. In addition, the effect is found to be insignificant if the stresses $\sigma_{11}^{(0)}$, $\sigma_{22}^{(0)}$, and $\sigma_{12}^{(0)}$ are not in phase with the topography.

8.2 Implementation Considerations

The integrals (8.12) are convolutions and thus are efficiently performed in the Fourier domain. To compute the influence of actual topography from digital elevation models (DEMs), it is necessary to use numerical fast Fourier transforms (FFTs). In the transform domain, equation (8.12) becomes

$$\mathcal{F}(u_i^{(1)}) = -2\pi i \left\{ \left[k_1 \mathcal{F}(h\sigma_{11}^{(0)}) + k_2 \mathcal{F}(h\sigma_{12}^{(0)}) \right] \mathcal{F}(g_i^1) + \left[k_1 \mathcal{F}(h\sigma_{12}^{(0)}) + k_2 \mathcal{F}(h\sigma_{22}^{(0)}) \right] \mathcal{F}(g_i^2) \right\}. \quad (8.21)$$

Here, we follow Williams and Wadge (2000) and use the transform pairs (A.11). These authors give the transforms of the Green's functions:

$$\mathcal{F}(g_1^1) = \frac{1}{2\pi\mu} \frac{k_1^2 + k_2^2 - \nu k_1^2}{k^3},$$

$$\mathcal{F}(g_2^1) = \frac{1}{2\pi\mu} \frac{-\nu k_1 k_2}{k^3},$$

$$\mathcal{F}(g_3^1) = -\frac{(1-2\nu)}{2\pi\mu} \frac{i k_1}{k^2},$$

$$\mathcal{F}(g_1^2) = \mathcal{F}(g_2^1),$$

$$\mathcal{F}(g_2^2) = \frac{1}{2\pi\mu} \frac{k_1^2 + k_2^2 - \nu k_2^2}{k^3},$$

$$\mathcal{F}(g_3^2) = -\frac{(1-2\nu)}{2\pi\mu} \frac{i k_2}{k^2}, \quad (8.22)$$

where

$$k^2 = k_1^2 + k_2^2. \quad (8.23)$$

8.3 Center of Dilatation beneath a Volcano

Topographic effects can be significant for magma bodies beneath steep-sided volcanoes. Consider, for example, a hypothetical Gaussian volcano, as shown in figure 8.3. The magma body is modeled as a simple center of dilatation directly beneath the peak of the mountain

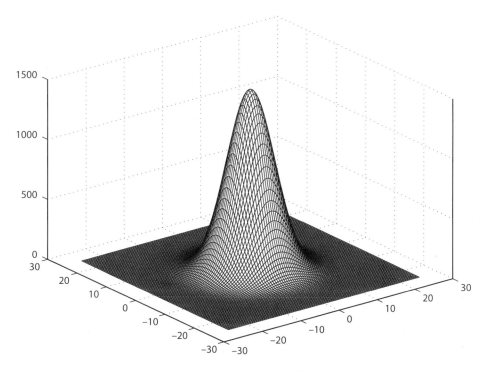

Figure 8.3. A Gaussian volcano, with height $h(r) = h_0 \exp{(-r^2/2\sigma^2)}$ and $\sigma = 5$ km.

at a depth of 5 km. Figure 8.4 illustrates the displacements for different volcano heights and fixed width. We see that relative to the half-space, the uplift is less on a topographic volcano. This should not be a surprise, in that the mountain represents additional elastic material that resists deformation. As the volcano height increases, the uplift begins to flatten out, and with $H/L > 0.1$, for a source at 5 km, localized subsidence occurs at the peak. The characteristic slope at which this feature begins to appear depends on the depth to the pressure source.

Cayol and Cornet (1998) have suggested that topographic effects on deformation have been observed in interferometric SAR data from Mount Etna. Massonnet et al. (1995) first showed that deformation of Mount Etna could be recorded by satellite interferometry. Their result is shown in plate 9. A simple center of dilatation in a homogeneous half-space fits the data reasonably well. There is a slight flattening of the fringe pattern at the top of the volcano that is reminiscent of the topographic effect seen in figure 8.4. Cayol and Cornet (1998) show that correcting for topography introduces the expected flattening. Their study employed a boundary element procedure rather than the perturbation approach described here.

It should also be noted that the depth of the source in the half-space model is uncertain, because it is not clear what elevation to assign to the equivalent half-space. The perturbation analysis suggests that one should take the mean topography as the reference elevation. The reference elevation thus depends on the distribution of measurement sites as well as the topography.

8.4 Earth's Sphericity

Over modest distance scales, we may think of the earth's curvature as a minor perturbation from a flat half-space. Taking the planar free surface to be the tangent to the earth immediately above the source, the effective topography due to the earth's sphericity can be easily calculated from figure 8.5. The radius of the earth is R_e. If r measures the distance on the plane from the

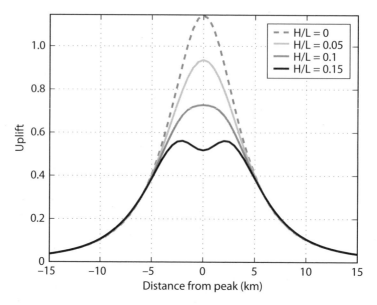

Figure 8.4. Vertical displacement resulting from a center of dilatation beneath the Gaussian volcano of figure 8.3. Results are shown for different ratios of the topographic height-to-length scale of the topography, here taken to be $L = 2\sigma$. Source depth is 5.0 km.

tangent point to an arbitrary observation point, then

$$z = \sqrt{R_e^2 - r^2},$$

(8.24)

where z is defined in figure 8.5. The effective topography $-h$ is

$$-h = R_e - z = R_e - \sqrt{R_e^2 - r^2},$$

$$\approx R_e - R_e \left(1 - \frac{r^2}{2R_e^2}\right),$$

$$h \approx -\frac{r^2}{2R_e},$$

(8.25)

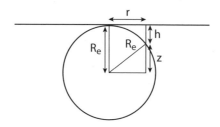

Figure 8.5. Figure for computing the effective topography h due to earth curvature. R_e is the earth's radius; r measures horizontal distance along the tangent plane.

the approximation being valid for $r^2/R_e^2 \ll 1$. Note that the slope due to earth curvature is

$$\frac{dh}{dr} = -\frac{r}{R_e}. \tag{8.26}$$

For the first-order perturbation solution to be accurate, the slope must be reasonably small. However, $dh/dr < 0.1$ even for distances of over 600 km from the surface projection of the deformation source. For distances greater than this, the deformation is so small as to be unmeasurable in all but extreme circumstances.

We must also account for the difference between the Cartesian displacement components in a coordinate system with one axis perpendicular to the tangent plane and the spherical displacement components. If the half-space solution is given in an x, y, z coordinate system, where x is in the local east direction at the point of tangency, y is in the local north, and z is perpendicular to the tangent plane, and the displacements in the spherical system are measured in a radial, tangential, and vertical system connecting the point of tangency and the observation point, then the spherical components u_v, u_r, and u_t are given by

$$u_r = -u_z \sin\theta + (u_x \cos\phi + u_y \sin\phi)\cos\theta,$$

$$u_t = u_y \cos\phi - u_x \sin\phi,$$

$$u_v = u_z \cos\theta + (u_x \cos\phi + u_y \sin\phi)\sin\theta. \tag{8.27}$$

Here, u_r is measured along a great circle connecting the observation point and the tangent point, u_t is orthogonal to this great circle, and u_v is vertical (perpendicular to the sphere). The angle ϕ is the counterclockwise angle between the x direction and the radial direction, and θ is the angular distance between the observation point and the tangent point. If the distance r' is measured along the spheroid, then $\theta = r'/R_e$. Note that r is less than r' by a factor of approximately $1/6(r'/R_e)^2$.

We compare the displacements computed using the first-order perturbation result with calculations for a spherical earth model based on a spherical harmonic basis (Pollitz 1996). The effects of earth curvature are significant only at large distance, and the deformations are significant only at large distance for very deep sources. Thus, we compare the two methods for a dislocation at a depth of 300 km in figure 8.6. The dislocation is effectively a point strike-slip source, with vertical dip and 45-degree strike. The moment is 1.9×10^{19} N-m, $M_W = 6.8$.

The displacements in a spherical earth are slightly greater than those for a half-space (figure 8.6) at distances in excess of approximately 250 km. This is sensible in that for the spherical earth model, there is less elastic material resisting deformation relative to the half-space model. From figure 8.6, we conclude that the perturbation method is accurate to distances of at least 1,000 km. Thus, a corrected half-space result should be accurate for almost all cases of interest. Calculations for crustal earthquakes indicate that corrections for earth curvature are on the order of a few percent of the half-space displacements within a few hundred kilometers of the fault.

8.5 Summary and Perspective

For the study of near-field deformation related to faults and volcanoes, topography and earth curvature are generally second-order effects. There are important exceptions, however. Topographic effects can be important in areas of steep slopes, especially where surface-parallel

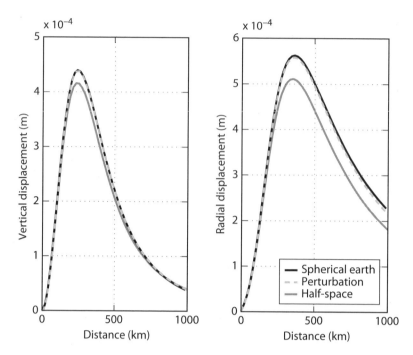

Figure 8.6. Comparison of a spherical model with the corresponding perturbation approximation. Displacements (in meters) due to a point strike-slip fault at a depth of 300 km. The fault strikes 45 degrees, and the displacements are measured along a line trending east. Half-space result is shown for comparsion. Calculations for a spherical earth courtesy of Fred Pollitz.

strains are in phase with the topography. We discuss a perturbation method that adds a first-order correction due to irregular topography to the half-space solution. The first-order correction should be reasonably accurate except in areas of extreme topographic slope. The correction includes two terms: The first is due to vertical gradients in the half-space displacements. The second is equivalent to distributed shear tractions on the surface of a half-space and is most easily computed as a convolution with elastic Green's tensors in the Fourier domain.

A number of studies have examined the differences between spherical earth and half-space models. In some cases, these comparisons have been complicated by assuming different elastic structures (layered versus homogeneous) or different methods for incorporating gravitational effects. Here, we isolate these different effects: elastic layering was examined in chapter 5, and self-gravitation is examined in chapter 9. The aforementioned studies (e.g., Nostro et al. 1999) confirm that earth curvature effects are generally small within a few hundred kilometers of the deformation source. The range of validity of half-space models can be extended to the order of 1,000 km by inclusion of a first-order correction for earth curvature. Clearly, half-space models are inappropriate when analyzing far-field data from the very largest M 9 earthquakes, such as the 1960 Chilean and the 2004 Sumatra–Andaman Islands earthquakes. Breaking the fault into segments and applying the first-order sphericity correction for each segment, however, should lead to reasonable results at intermediate distances. Spherical earth models are required at greater distances. Last, it should be kept in mind that viscoelastic relaxation will act to diffuse deformation away from a fault rupture. For large events, this can lead to detectable signals at even greater distances, increasing the importance of earth curvature.

8.6 Problems

1. Show, using the boundary conditions in the reference half-space configuration, that the plane strain analogue of equation (8.16) can be written as

$$u_z(x, h) = u_z^{(0)}(x, 0) + u_z^{(1)}(x, 0) - \frac{\nu}{2\mu} h(x) \sigma_{xx}^{(0)}(x, 0). \tag{8.28}$$

Using this result and equation (8.18), Hooke's law, and boundary conditions, show that

$$u_z(x, h) = u_z^{(0)}(x, 0) - h(x) \epsilon_{xx}^{(0)}(x, 0),$$

demonstrating equation (8.19).

2. Explore the effect of topography on the surface displacement for a two-dimensional dipping fault employing the result in equation (8.19). Use the expression for the surface-parallel normal strain due to a dipping edge dislocation given in equation (3.72). Choose different topographic profiles, and explore when the topographic effect is significant.

3. Consider a plane strain center of dilatation, equivalent to a line of centers of dilatation extending infinitely in and out of the plane. This provides a simple model of a horizontal volcanic conduit. From the Melan Green's functions in chapter 3, show that the uplift and surface-parallel normal stress for a line of dilatation at $x = 0$, $z = d$ are given by

$$u_z^{(0)}(\xi, 0) = \frac{2(1 - \nu)\Delta V}{\pi d} \frac{1}{1 + \xi^2},$$

$$\sigma_{xx}^{(0)}(\xi, 0) = \frac{4\mu\Delta V}{\pi d^2} \frac{(1 - \xi^2)}{(1 + \xi^2)^2}, \tag{8.29}$$

where $\xi = x/d$, and ΔV is the change in volume per unit length (dimensions of length squared). Next, consider the symmetric topographic profile

$$h(y) = H \frac{1}{1 + \xi^2}. \tag{8.30}$$

Show from these that the first-order correction to the vertical displacement due to topography is

$$u_z = \frac{-2(1 - \nu)\Delta V}{\pi d} \frac{H}{d} \frac{(1 - \xi^2)}{(1 + \xi^2)^3}. \tag{8.31}$$

Plot results for different H/d.

8.7 References

Ben-Menahem, A., S. J. Singh, and F. Solomon. 1969. Static deformation of a spherical earth model by internal dislocations. *Bulletin of the Seismological Society of America* **59**(2), 813–853.

Cayol, V., and F. H. Cornet. 1998. Effects of topography on the interpretation of the deformation field of prominent volcanoes—application to Etna. *Geophysical Research Letters* **25**, 1979–1982.

Ishii, H., and A. Takagi. 1967. Theoretical study on the crustal movements. Part I. The influence of surface topography (two-dimensional SH-torque source). The Science Reports of the Tohoku University, Series 5. *Geophysics* **19**, 77–94.

Mahrer, K. 1984. Approximating surface deformation from a buried strike-slip fault or shear crack in a mildly uneven half-space. *Bulletin of the Seismological Society of America* **74**, 797–803.

Massonnet, D., P. Briole, and A. Arnaud. 1995. Deflation of Mount Etna monitored by spaceborne radar interferometry. *Nature* **375**, 567–570.

McTigue, D. F., and P. Segall. 1988. Displacements and tilts from dip-slip faults and magma chambers beneath irregular surface topography. *Geophysical Research Letters* **15**, 601–604.

Nostro, C., A. Piersanti, A. Antonioli, and G. Spada. 1999. Spherical versus flat models of coseismic and postseismic deformation. *Journal Geophysical Research* **104**, 13,115–13,134.

Pollitz, F. F. 1996. Coseismic deformation from earthquake faulting on a layered spherical earth. *Geophysical Journal International* **125**, 1–14.

Smylie, D. S., and L. Mansinha. 1971. The elasticity theory of dislocations in real earth models and changes in the rotation of the earth. *Geophysical Journal of the Royal Astronomical Society* **23**, 329–354.

Sun, W., S. Okubo, and P. Vanicek. 1996. Global displacements caused by point dislocations in a realistic earth model. *Journal of Geophysical Research* **101**, 8561–8577.

Willams, C. A., and G. Wadge. 2000. An accurate and efficient method for including the effects of topography in three-dimensional elastic models of ground deformation with applications to radar interferometry. *Journal of Geophysical Research* **105**, 8103–8120.

9

Gravitational Effects

It might strike the reader as strange that we have been able to put off a discussion of the effects of gravity for so long. After all, gravity is a dominant force for problems on the scale of the earth. Yet, up to this point, we have ignored all body forces, including gravity, in writing the equilibrium equations. As you will see, the reason we have been able to get away with this is that the preexisting stress state, prior to fault slip or magma chamber inflation, equilibrates the gravitational body forces. Fault slip or magmatic intrusion perturbs this equilibrium state, leading to changes in the elastic fields as well as the gravitational potential. For many spatial and temporal scales of interest, the perturbations in the gravitational forces are small relative to elastic stresses. However, at long spatial scales, gravitational terms cannot be neglected relative to elastic terms in the equilibrium equations. Furthermore, as elastic stresses relax due to viscoelastic processes, gravitational effects become more important and must be taken into account. Last, deformation perturbs the gravitational potential, and therefore the local gravitational acceleration. The change in local gravity can be measured with repeated gravimetry. These measurements can be useful in the analysis of earthquakes and quite important in the study of magmatic processes, where substantial mass movements occur in the subsurface.

We will begin by treating the problem of a radially stratified earth following ideas that date to Love (1911). Our treatment follows the standard approach in normal mode seismology, as described for example in considerably more detail by Dahlen and Tromp (1998).

In the initial state, prior to fault slip or magma chamber inflation, we assume that the density, gravitational potential, and stress are all functions of radius only. The density structure $\rho_0(r)$ and gravitational potential $\phi_0(r)$ are related through Poisson's equation:

$$\nabla^2 \phi_0 = \frac{\partial^2 \phi_0}{\partial r^2} + \frac{2}{r} \frac{\partial \phi_0}{\partial r} = 4\pi G \rho_0, \qquad (9.1)$$

where G is Newton's universal gravitational constant, and r is the radial distance from the earth's center of mass. The gradient of the potential gives the local gravitational acceleration. Assuming radial symmetry,

$$\mathbf{g} = -\nabla \phi_0 = -\frac{\partial \phi_0}{\partial r} \hat{\mathbf{r}} = -g_0 \hat{\mathbf{r}}, \qquad (9.2)$$

where $\hat{\mathbf{r}}$ is a unit vector in the positive radial direction. Combining equations (9.1) and (9.2) yields

$$\frac{1}{r^2} \frac{\partial}{\partial r} \left(r^2 g_0 \right) = 4\pi G \rho_0, \qquad (9.3)$$

which can be integrated to give

$$g_0 = \frac{4\pi G}{r^2} \int_0^r \rho_0(r')r'^2 dr'.$$

(9.4)

Noting that $4\pi r^2 dr$ is the volume of a spherical shell, the acceleration due to gravity in equation (9.4) is equivalent to $g_0 = GM(r)/r^2$, where $M(r)$ is the mass enclosed in radius r. If the limit of integration is taken to be the earth's radius, then $M = M_e$, the mass of the earth.

The equilibrium equations for the initial state are

$$\nabla \cdot \sigma^0 - \rho_0 g_0 \hat{\mathbf{r}} = 0,$$

(9.5)

where σ^0 is the stress in the initial state. Recalling equation (9.2), equations (9.5) become

$$\nabla \cdot \sigma^0 = \rho_0 \nabla \phi_0.$$

(9.6)

We now consider small perturbations in density, gravitational potential, and stress about this reference state

$$\rho = \rho_0 + \delta\rho,$$

$$\phi = \phi_0 + \delta\phi,$$

$$\sigma = \sigma^0 + \delta\sigma,$$

$$\boldsymbol{\xi} = \mathbf{x} + \mathbf{u}.$$

(9.7)

Note that the displacement $\mathbf{u}$ is measured relative to the prestressed reference state, as in equation (9.5), rather than relative to the completely undeformed state. Substituting equation (9.7) into the continuity equation (1.92) and keeping only first-order terms yields

$$\frac{\partial \delta\rho}{\partial t} + \nabla \cdot \left(\rho_0 \frac{\partial u}{\partial t} \right) = 0,$$

(9.8)

since the density in the undeformed state ρ_0 is not time varying. Integrating with respect to time yields

$$\delta\rho = -\nabla \cdot (\rho_0 u) = - \left[\rho_0 (\nabla \cdot u) + u_r \frac{\partial \rho_0}{\partial r} \right].$$

(9.9)

The first term on the right-hand side, the divergence of the displacement field, equals the volumetric strain—a positive volume strain causing a decrease in density. The second term is the change in density due to the radial displacement of preexisting density gradients. If density decreases with radius, then a radial outward displacement causes an increase in density in this Eulerian formulation. (In a Lagrangian formulation, the change in density is $\delta\rho^L = -\rho_0[\nabla \cdot u]$.) Similarly, the perturbed gravitational potential, also in a Eulerian representation, is given by

$$\nabla^2 \delta\phi = 4\pi G \delta\rho.$$

(9.10)

Making use of equations (9.7), the equilibrium equations in the final, perturbed state, including the body force representation of a point dislocation source (as in chapter 3), become

$$\nabla \cdot (\sigma^0 + \delta\sigma^E) - (\rho_0 + \delta\rho)\nabla(\phi_0 + \delta\phi) = M \cdot \nabla\delta(\mathbf{x} - \mathbf{x}_s), \tag{9.11}$$

where the superscript ^{E} has been added as a reminder that this is a Eulerian description of the equilibrium equations, and thus the stress tensor is a Eulerian Cauchy stress. Also, M is the seismic moment tensor, and $\mathbf{x}_s$ is the source location. Ignoring second-order terms and making use of equation (9.6) yields

$$\nabla \cdot \delta\sigma^E - \rho_0\nabla\delta\phi - \delta\rho\nabla\phi_0 = M \cdot \nabla\delta(\mathbf{x} - \mathbf{x}_s). \tag{9.12}$$

To relate the stresses and strains through Hooke's law requires a Lagrangian stress because the elastic properties refer to a parcel of material. Recall from equation (1.89) that for small displacements, the Lagrangian stress can be related to the Eulerian stress via $\delta\sigma = \delta\sigma^E + \mathbf{u} \cdot \nabla\sigma^0$, where we forgo the superscript on the Lagrangian stress, as this will be the stress tensor we use from here on. Thus, equations (9.12) become

$$\nabla \cdot \delta\sigma - \nabla \cdot (\mathbf{u} \cdot \nabla\sigma^0) - \rho_0\nabla\delta\phi - \delta\rho\nabla\phi_0 = M \cdot \nabla\delta(\mathbf{x} - \mathbf{x}_s). \tag{9.13}$$

At this stage of the analysis, it is simplest to assume that the preexisting stress field within the earth is isotropic, as in a purely radial symmetric earth (equation [9.5])—that is, $\sigma_{ij}^0 = -p_0\delta_{ij}$. This is of course not strictly true, since the earth does support deviatoric stresses, but it is perhaps not an unreasonable approximation. Given this approximation, equations (9.13) become

$$\nabla \cdot \delta\sigma + \nabla(\mathbf{u} \cdot \nabla p_0) - \rho_0\nabla\delta\phi - \delta\rho\nabla\phi_0 = M \cdot \nabla\delta(\mathbf{x} - \mathbf{x}_s). \tag{9.14}$$

The second term is often referred to as the *advected prestress*, as it is the preexisting stress carried to the observation point by the deformation. Love (1911) was the first to recognize the importance of this term, noting that the stress at the point (x, y, z) includes "the pressure in the initial state which is displaced to (x, y, z) when the body is strained."

For a hydrostatically prestressed earth, the equilibrium equations in the initial state (9.6) are simply $\nabla p_0 + \rho_0\nabla\phi_0 = 0$. Thus, equations (9.14) become

$$\boxed{\nabla \cdot \delta\sigma - \nabla\rho_0(\mathbf{u} \cdot \nabla\phi_0) - \rho_0\nabla\delta\phi - \delta\rho\nabla\phi_0 = M \cdot \nabla\delta(\mathbf{x} - \mathbf{x}_s).} \tag{9.15}$$

At this point, it would be useful to verify that the signs on the body force terms are physically sensible. The third term in equation (9.15) is sometimes referred to as the *self-gravity*, as it represents an effective force due to the change in gravitational potential caused by deformation. The vertical component of this term can be written as $-\rho_0\delta g\hat{\mathbf{r}}$, illustrating that an increase in the local vertical gravitational acceleration leads to a body force in the negative r direction (downward). Similarly, the fourth term in equation (9.15) can be written as $-\delta\rho g_0\hat{\mathbf{r}}$, indicating that an increase in density also leads to a downward body force.

Making use of equation (9.2), equation (9.15) can be written as

$$\nabla \cdot \delta\sigma - \nabla(g_0\rho_0 u_r) - \rho_0\nabla\delta\phi - \delta\rho g_0\hat{\mathbf{r}} = M \cdot \nabla\delta(\mathbf{x} - \mathbf{x}_s), \tag{9.16}$$

where $\mathbf{u} \cdot \hat{\mathbf{r}} = u_r$. Together with the equation for the perturbing potential (9.10), the equation for the density change (9.9), and Hooke's law, these are 11 equations in the 11 unknowns: $\delta\sigma_{ij}$, u_i, $\delta\rho$, and $\delta\phi$.

9.1 Nondimensional Form of Equilibrium Equations

Before proceeding further, we will nondimensionalize the equilibrium equations following Pollitz (1997) to better understand the magnitude of the various effects. As it happens, we will expand equation (9.16) so that it looks considerably more complicated; however, this will facilitate our understanding of the magnitude of the various terms. In order to keep the notation compact, we will replace $\delta\sigma$ with σ; it should be understood that σ represents the change in the Lagrangian stress from an equilibrated reference state.

The advected prestress, the second term in equation (9.16), can be expanded as

$$-\nabla(u_r \rho_0 g_0) = -u_r \rho_0 \nabla g_0 - g_0 \nabla(u_r \rho_0). \tag{9.17}$$

Also, from equations (9.1) and (9.2), we can write the gradient of the gravitational acceleration as

$$\nabla g_0 = \left(4\pi G\rho_0 - \frac{2}{r} g_0\right) \hat{\mathbf{r}}, \tag{9.18}$$

so that

$$-\nabla(u_r \rho_0 g_0) = -4\pi G\rho_0^2 u_r \hat{\mathbf{r}} + \frac{2u_r \rho_0 g_0}{r}\hat{\mathbf{r}} - g_0 \nabla(u_r \rho_0). \tag{9.19}$$

Thus, making use of equation (9.9), the equilibrium equations (9.16) can be written as

$$\nabla \cdot \sigma - 4\pi G\rho_0^2 u_r \hat{\mathbf{r}} + \frac{2u_r \rho_0 g_0}{r}\hat{\mathbf{r}} - \rho_0 g_0 \nabla u_r - \rho_0 \nabla\delta\phi + \rho_0 g_0 (\nabla \cdot \mathbf{u})\hat{\mathbf{r}} = M \cdot \nabla\delta(\mathbf{x} - \mathbf{x}_s). \tag{9.20}$$

We now proceed to write equations (9.20) in nondimensional form. The displacements are sensibly scaled by the fault slip s, so we define the dimensionless displacement as $\tilde{u} = u/s$. Similarly, a characteristic stress is $\mu s/d$, where d is a characteristic distance, typically a fault dimension or source depth. Distances are also scaled by d, such that $\tilde{\nabla} = d\nabla$. From equations (9.9) and (9.10), we note that $\nabla\delta\phi$ scales with $4\pi G\rho_0 s$, so that in summary, the dimensionless variables are

$$\tilde{u} = \frac{u}{s},$$
$$\tilde{\sigma} = \frac{\sigma d}{\mu s},$$
$$\tilde{\nabla}\delta\tilde{\phi} = \frac{\nabla\delta\phi}{4\pi G\rho_0 s}. \tag{9.21}$$

Substituting equation (9.21) into (9.20) leads to

$$\frac{\mu s}{d^2}\tilde{\nabla} \cdot \tilde{\sigma} - 4\pi G\rho_0^2 s\, \tilde{u}_r \hat{\mathbf{r}} + \frac{2\rho_0 g_0 s}{d}\frac{\tilde{u}_r}{\tilde{r}}\hat{\mathbf{r}} - \frac{\rho_0 g_0 s}{d}\tilde{\nabla}\tilde{u}_r - 4\pi G\rho_0^2 s\tilde{\nabla}\delta\tilde{\phi}$$
$$+ \frac{\rho_0 g_0 s}{d}(\tilde{\nabla} \cdot \tilde{\mathbf{u}})\hat{\mathbf{r}} = \frac{\mu As}{d^4}\tilde{M}\tilde{\nabla}\delta(\mathbf{x} - \mathbf{x}_s), \tag{9.22}$$

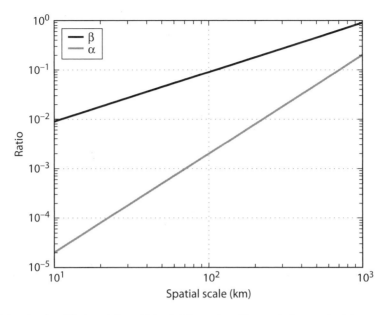

Figure 9.1. Magnitude of little g (β) and big G (α) gravity effects relative to elasticity as a function of spatial scale.

where $A \sim d^2$ is the fault area, and $\tilde{M}$ is the dimensionless moment tensor density. Introducing the dimensionless constants

$$\alpha = \frac{4\pi G \rho_0^2 d^2}{\mu},$$

$$\beta = \frac{\rho_0 g_0 d}{\mu}, \tag{9.23}$$

equations (9.22) become

$$\tilde{\nabla} \cdot \tilde{\sigma} - \alpha \tilde{u}_r \hat{\mathbf{r}} + 2\beta \frac{\tilde{u}_r}{\tilde{r}} \hat{\mathbf{r}} - \beta \tilde{\nabla} \tilde{u}_r - \alpha \tilde{\nabla} \delta \tilde{\phi} + \beta (\tilde{\nabla} \cdot \tilde{\mathbf{u}}) \hat{\mathbf{r}} = \tilde{M} \cdot \tilde{\nabla} \delta(\mathbf{x} - \mathbf{x}_s). \tag{9.24}$$

Choosing values for the parameters $G = 6.7 \times 10^{-11}$ Nm^2 kg^{-2}, $\rho_0 = 2.7 \times 10^3$ kg/m^3, and $\mu = 3 \times 10^{10}$ N/m^2, we find that $\alpha = 2 \times 10^{-7} d^2$ and $\beta = 9 \times 10^{-4} d$, where d is measured in kilometers. Thus, for small length scales d, both the "big G" (α) and "little g" (β) terms are small, and the dislocation source terms are balanced by the elastic stresses; both are order unity (figure 9.1). Note, however, that for a viscoelastic rheology, the shear stresses relax such that at times long compared to the relaxation time of the medium, the dislocation source terms are increasingly balanced by gravitational forces.

For greater distances, the little g terms become more significant; β is of order unity at distances of approximately 1,000 km (figure 9.1). At even greater distances, the big G terms become important. However, for most coseismic applications where we consider displacements tens of km to at most a few hundred km from the fault, big G terms are safely neglected and even the little g effects are small. For the very largest earthquakes, where displacements have been measured thousands of kilometers from the fault, even the big G terms should be included.

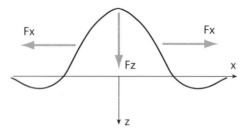

Figure 9.2. Sign of restoring gravitational force for $\rho_0 g_0 \nabla u_z$.

The dimensional analysis shows that it is often reasonable to retain only the elastogravitational terms proportional to β and to write the equilibrium equations as

$$\nabla \cdot \sigma + 2\rho_0 g_0 \frac{u_r}{r} \hat{\mathbf{r}} - \rho_0 g_0 [\nabla u_r - (\nabla \cdot \mathbf{u})\hat{\mathbf{r}}] = m\nabla\delta(\mathbf{x} - \mathbf{x}_s). \tag{9.25}$$

If we restrict attention to the neighborhood of the earth's surface, then $u_r/r \ll \nabla u_r$ and can be neglected. Eliminating this term is equivalent to taking the limit that $r \to \infty$ and replacing the spherical earth model with a half-space. To make this explicit, switch to a Cartesian coordinate system with z downward. This changes the sign on the gradient operator as well as the unit vector, so that equation (9.25) becomes

$$\nabla \cdot \sigma + \rho_0 g_0 \left[\nabla u_z - (\nabla \cdot \mathbf{u})\hat{\mathbf{e}}_z\right] = M \cdot \nabla\delta(\mathbf{x} - \mathbf{x}_s). \tag{9.26}$$

It is worth reiterating the approximations made to derive equation (9.26). Specifically, big G terms proportional to α were neglected. This eliminated self-gravitation, the effect of changes in gravitational potential on equilibrium, as well as one contribution resulting from gradients in g_0. Second, we ignored earth curvature, adopting a half-space approximation, which eliminates the other term due to gradients in g_0.

Before continuing, we examine the sign of the two elastogravitational terms in equation (9.26). The second is proportional to the divergence of the displacement field, or volumetric strain. A positive volumetric strain, or a reduction in density, yields a body force acting in the negative z direction or vertically upward. Conversely, an increase in density results in a force acting in the positive z direction. The first term leads to forces acting in both the horizontal and vertical directions. For example, in figure 9.2, the surface is perturbed upward, resulting in a positive strain $\partial u_z/\partial z$. According to equation (9.26), the force acts in the positive z direction—that is, downward. The sign of the force in the x direction is outward on either side of the uplift. The $\rho_0 g_0 \nabla u_z$ term thus represents a restoring force acting on surfaces that are no longer in hydrostatic equilibrium.

Equation (9.24) also holds for volcanic sources. In this case, we interpret d to be the characteristic dimension of the magma body and s to be the characteristic displacement on the wall of the magma chamber. The equivalent seismic moment, from equation (7.24), is thus given by $\mu \Delta V/d^3$. But the volume change scales with sd^2, so the source term in equation (9.24) is also order unity for volcanic sources. The implication is that, as in the earthquake case, elastogravitational effects are negligible at the short distance scales normally associated with volcano deformation monitoring. With increasing time, however, as shear stresses relax, gravitational effects will become more significant. It is also useful to consider the mass of magma transported from the mantle into a crustal reservoir. The mass addition leads to a term in the equilibrium equations that scales with $\rho_m g_0 d/\mu$, where ρ_m is the magma density. For elastic deformation, this ratio is small (order 10^{-3} for length scales of order 1 km), and the expansion of the magma chamber leads to uplift, as analyzed in chapter 7. At times long compared to the relaxation time of the lower crust or upper mantle, however, the mass loading becomes more important, and the weight of the intruded magma may ultimately lead to subsidence.

9.2 Inclusion in Propagator Matrix Formulation

We can now consider including the effects of gravity, according to equation (9.26), in a plane strain propagator matrix formulation. Recall from chapter 5 that in three dimensions, the problem can be separated into a 4×4 system and a 2×2 system. Since the 2×2 antiplane system contains neither vertical displacements nor vertical body forces, it suffices to focus on the 4×4 plane strain system. From equation (9.26), the stress displacement vector $\mathbf{v}(z)$ again satisfies $d\mathbf{v}(z)/dz = A\mathbf{v}(z)$, with A now given by

$$A = \begin{bmatrix} 0 & k & 1/\mu & 0 \\ \dfrac{-k\lambda}{\lambda + 2\mu} & 0 & 0 & \dfrac{1}{\lambda + 2\mu} \\ \dfrac{4\mu k^2 \left(\lambda + \mu\right)}{\lambda + 2\mu} & k\rho g & 0 & \dfrac{k\lambda}{\lambda + 2\mu} \\ k\rho g & 0 & -k & 0 \end{bmatrix}. \tag{9.27}$$

We proceed as before to find the eigenvalues and eigenvectors of A. There are now four distinct eigenvalues, given by

$$\omega = \pm|k| \left[1 \pm \frac{\rho g}{k\sqrt{\mu(\lambda + 2\mu)}} \right]^{1/2}. \tag{9.28}$$

Recall that without gravity, there were two degenerate eigenvalues $\pm|k|$, so equation (9.28) reduces to the nongravitational elastic solution for $g = 0$. The boundary condition that the displacements decay with increasing depth requires that the real part of two of the eigenvalues be negative. For large wavenumbers (small wavelength), this condition is satisfied. However, for sufficiently small wavenumber, two eigenvalues become purely imaginary, and only one satisfies the condition that $\Re(\omega) < 0$. For a Poisson solid ($\nu = 0.25$), there is only a single decaying solution when $\rho g/\sqrt{3}\mu k > 1$. Choosing parameter values given in the previous section, this occurs for wavelengths of $\sim 1.25 \times 10^4$ km, roughly the diameter of the earth. At these distance scales, big G terms are significant (figure 9.1), and earth curvature certainly can no longer be neglected. Generally, static elastic displacements due to earthquakes are nonnegligible only for distance scales much shorter than this critical wavelength, and this problem is not a practical limitation. This issue does become significant, however, when we consider viscoelastic deformation. For a Maxwell material, the shear modulus $\mu \to 0$ as $t \to \infty$, eventually leading to a loss of regularity for all wavenumbers k. This problem is one with considerable history. Love (1911) considered a homogeneous spherical, elastic, self-gravitating earth and showed that for sufficiently small μ, the earth is gravitationally unstable. We will return to the issue of gravitational stability in the context of viscoelastic models in section 9.4.

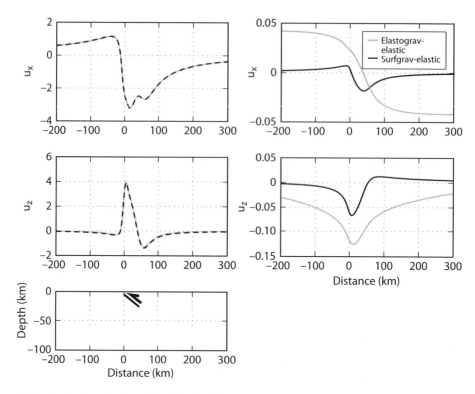

Figure 9.3. Gravitational effects for elastic dislocations. The left side compares three solutions for a 30-degree dipping fault with downdip width of 50 km and 10 m of slip: the solution using the elastogravitational equations (9.27), the approximate "surface gravity" solution, and the elastic solution with no gravity. The differences are generally not visible at this scale. The right side shows the elastogravitational solution and approximate "surface gravity" solutions relative to the nongravitational elastic solution. $\nu = 0.25$.

The eigenvectors of equation (9.27) corresponding to the eigenvalues (9.28) are

$$V = \begin{bmatrix} \mu k^2 - (\lambda + 2\mu)\omega^2 \\ (\lambda + \mu)k\omega - \rho gk \\ -\mu\omega[(\lambda + 2\mu)\omega^2 + k^2\lambda] + \mu k^2 \rho g \\ (\lambda + 2\mu)k\omega(\mu\omega - \rho g) + \mu k^3 \lambda \end{bmatrix} . \tag{9.29}$$

The solution method now follows precisely the approach laid out in chapter 5, with the exception that the solutions to the homogeneous equations make use of the eigenvalues and eigenvectors given by equations (9.28) and (9.29).

Figure 9.3 compares the elastogravitational solution to the nongravitational elastic solution for a crustal thrust fault with 10 meters of slip. The downdip fault width is 50 km (figure 9.3, lower left). Notice that while the effects are small—only a few percent of the maximum displacement, as predicted by the dimensional analysis—they could in principle be detected with modern instrumentation such as GPS or InSAR. For example, with 10 meters of slip, corresponding to 4 meters of uplift, the gravitational effects are of the order of 12 cm or less. It could, however, be quite difficult to distinguish the gravitational effect from other effects such as depth variations in elastic properties, nonuniform slip, and so on. For the largest magnitude 9 earthquakes, however, gravitational effects should be readily detected.

9.3 Surface Gravity Approximation

If we return to equation (9.26) and ignore buoyancy forces ($\rho_0 g_0 \nabla \cdot \mathbf{u} = 0$), the equilibrium equations reduce to

$$\nabla \cdot (\sigma + \rho_0 g_0 u_z I) = M \cdot \nabla \delta(\mathbf{x} - \mathbf{x}_s), \tag{9.30}$$

where I is a 3×3 identity. Notice that by defining a new stress $\sigma^* = \sigma + \rho_0 g_0 u_z I$, the equilibrium equations (9.30) reduce to $\nabla \cdot \sigma^* = M \cdot \nabla \delta(\mathbf{x} - \mathbf{x}_s)$. The corresponding normal boundary condition on σ^* at the free surface is

$$\sigma_{zz}^*(z = 0) = \sigma_{zz}(z = 0) + \rho_0 g_0 u_z(z = 0) = \rho_0 g_0 u_z(z = 0). \tag{9.31}$$

This suggests an approximate solution in which we employ the by now familiar elastic solution but replace the boundary condition at the free surface with one in which the vertical stress is proportional to the vertical displacement $\sigma_{zz}^*(z = 0) = \rho_0 g_0 u_z(z = 0)$. Recall that z is positive downward, so a positive displacement (subsidence) causes a tensile restoring traction at the free surface that acts to pull the free surface up. This approach can be implemented in a convenient way as follows (e.g., Matsu'ura and Sato 1989). Recall from equation (5.120) that the stress-displacement vector at the free surface can be written symbolically as

$$\mathbf{v}(z = 0) = \mathbf{P}(0, H) V \mathbf{c} + \mathbf{P}(0, z_s) \mathbf{f}_s, \tag{9.32}$$

where $\mathbf{P}(0, H)$ and $\mathbf{P}(z, z_s)$ are propagator matrices that carry the solution from the top of the half-space at depth H and the source depth, z_s, respectively. V is a matrix of eigenvectors, $\mathbf{c}$ are coefficients determined by the boundary conditions, and $\mathbf{f}_s$ is the source vector. Premultiplying the right-hand side by a matrix $\mathcal{G}$, defined by

$$\mathcal{G} = \begin{bmatrix} 1 & 0 & 0 & 0 \\ 0 & 1 & 0 & 0 \\ 0 & 0 & 1 & 0 \\ 0 & -\rho g & 0 & 1 \end{bmatrix}, \tag{9.33}$$

gives the two boundary conditions from the last two equations, $\sigma_{12}^* = 0$ and $\sigma_{22}^* - \rho g u_z = 0$. Thus, this surface gravity approximation is implemented, simply by premultiplying the propagators that carry the solution to the free surface by the matrix $\mathcal{G}$.

Figure 9.3 compares the elastogravitational solution to the approximate surface gravity approximation for a 30-degree dipping fault with downdip width of 50 km and a half-space with $\nu = 0.25$. The surface gravity approximation captures about half of the gravitational effect and is somewhat more localized near the fault compared to the full elastogravitational solution based on equation (9.27). As expected, the surface gravity approximation becomes more accurate as the Poisson's ratio approaches 0.5; as the rock becomes incompressible, buoyancy effects diminish.

9.4 Gravitational Effects in Viscoelastic Solutions

We proceed as before with the nongravitational viscoelastic solution as in chapter 6, where we employed the correspondence principle to relate the elastic moduli to appropriate functions in the Laplace domain. From equation (9.28), however, a problem exists for a Maxwell material in the limit as $t \to \infty$, where the resistance to shear strain vanishes. As noted previously, as $\mu \to 0$, there are no longer two independent solutions that decay as a function of depth. Two eigenvalues become imaginary such that, rather than two decaying solutions, one becomes purely oscillatory in z. This is sometimes referred to as the *stability margin*, and eventually happens for all wavenumbers k, so that we cannot appeal to the fact that we are interested only in modest spatial scales, as can be argued in the elastogravitational case.

A related problem is that the eigenvectors in equation (9.29) contain square root functions and thus involve branch cuts. It can be shown (see problem 1) that at the stability margin, a branch cut appears on the positive real axis. The Bromwich integration path for the inverse Laplace transform (appendix A, equation [A.19]) should stay to the right of any singularities. Poles along the positive real axis correspond to *negative* relaxation times; the displacements associated with these poles grow, rather than decay, exponentially with time. Integration around the branch cut on the positive axis results in so-called *growth modes*.

What is going on? The stability problem appears as the resistance to shear relaxes. This suggests that some insight can be gained by considering the equilibrium equations (9.26) in the limit of vanishing shear modulus. In this limit, $\lambda = K$, where K is the bulk modulus. In Cartesian coordinates, balance of forces in the limit $\mu \to 0$ in both the x and z directions reduces to

$$K \frac{\partial (\nabla \cdot \mathbf{u})}{\partial x} + \rho_0 g_0 \frac{\partial u_z}{\partial x} = 0, \tag{9.34a}$$

$$K \frac{\partial (\nabla \cdot \mathbf{u})}{\partial z} + \rho_0 g_0 \frac{\partial u_z}{\partial z} - \rho_0 g_0 \nabla \cdot \mathbf{u} = 0, \tag{9.34b}$$

assuming that horizontal gradients in ρ_0 and g_0 vanish. Differentiating equation (9.34a) with respect to z and equation (9.34b) with respect to x and combining yields

$$\frac{\partial}{\partial x} \left[\rho_0 g_0 \nabla \cdot \mathbf{u} + u_z \frac{\partial}{\partial z} \rho_0 g_0 \right] = 0. \tag{9.35}$$

If, as is common in half-space earth models, the initial density and reference gravity are assumed to be independent of depth, then equilibrium requires that the volumetric strain $\nabla \cdot \mathbf{u}$ be independent of horizontal position. This will generally be inconsistent with the strains generated by a fault in the overlying elastic plate. This incompatibility in equation (9.35) vanishes if the material is incompressible, $\nabla \cdot \mathbf{u} = 0$, or the buoyancy term, $\rho_0 g_0 \nabla \cdot \mathbf{u}$, is neglected in the equations of motion.

Interestingly, the same behavior occurs for a spherical earth model, even retaining the so-called big G terms in the equilibrium equations (Longman 1963), demonstrating that this behavior is not a consequence of the half-space approximation or the neglect of self-gravity terms in the equilibrium equations. We leave it as an exercise (see problem 2) to show that a similar analysis of the fully relaxed equilibrium equations in spherical coordinates yields

$$\left(\frac{\mathrm{d}\rho_0}{\mathrm{d}r} + \frac{\rho_0^2 g_0}{K} \right) (\nabla \cdot \mathbf{u}) = 0. \tag{9.36}$$

Thus, either the material must be incompressible or the initial density gradient must satisfy the *Adams-Williamson equation*:

$$\frac{d\rho_0}{dr} = -\frac{\rho_0^2 g_0}{K}. \tag{9.37}$$

The Adams-Williamson equation (9.37) expresses the density gradient due to self-compression and is simply derived using the chain-rule:

$$\frac{d\rho}{dr} = \frac{d\rho}{dp}\frac{dp}{dr} = \frac{\rho}{K}\frac{dp}{dr} = -\frac{\rho^2 g}{K}. \tag{9.38}$$

The second step follows from the definition of the compressibility (inverse of the bulk modulus); the third from hydrostatic equilibrium.

Evidently the *homogenous*, self-gravitating viscoelastic model is gravitationally unstable in the following sense. If material is pushed downward, it will be compressed and thus will increase in density. If the adjacent deeper rocks are not sufficiently dense, due to the ambient density gradient, there will be a net downward force, and the material sinks unstably. This is, in effect, a *Rayleigh-Taylor instability*, as discussed by Plag and Jüttner (1995).

The preceding argument is made quantitative by considering the *Brunt-Väisälä frequency* (e.g., Dahlen and Tromp 1998, p. 262). If a parcel of material is displaced adiabatically upward by an amount u_r, the density changes by $d\rho = -\rho_0^2 g_0 u_r / K$, as in equation (9.38). Meanwhile, the ambient density at the new depth differs by $d\rho_{amb} = u_r(d\rho_0/dr)$, leading to a buoyancy force $(d\rho_{amb} - d\rho)g_0 = -\rho_0 N^2 u_r$, where N^2 is the square of the Brunt-Väisälä frequency:

$$N^2 = \frac{-g_0}{\rho_0}\left(\frac{d\rho_0}{dr} + \frac{\rho_0^2 g_0}{K}\right). \tag{9.39}$$

Balancing the buoyancy with inertia leads to an equation of motion $\partial^2 u/\partial t^2 + N^2 u_r = 0$, which has solutions of the form $\exp(\pm i Nt)$. Thus, if $N^2 > 0$, a displaced parcel of fluid oscillates with frequency N; however, for $N^2 < 0$, a small displacement grows exponentially in time, implying that the initial density distribution is inherently unstable. For $N = 0$, the displaced parcel is gravitationally stable.

The important point here is that from equation (9.39), a homogeneous medium ($d\rho_0/dr = 0$) is always unstable, so an equilibrium fully relaxed Maxwell limit does not exist. A number of strategies have been employed to overcome this limitation, as discussed in the following sections.

9.4.1 Incompressible Half-Space

There is no stability problem for a homogeneous incompressible half-space, for which $1/K = 0$, and the Brunt-Väisälä frequency is zero (equation [9.39]). It is left as an exercise (problem 3) to show that for an incompressible plane-layered medium, the governing equations are represented by the standard linear form with

$$A = \begin{bmatrix} 0 & k & 1/\mu & 0 \\ -k & 0 & 0 & 0 \\ 4\mu k^2 & \rho g k & 0 & k \\ \rho g k & 0 & -k & 0 \end{bmatrix}, \tag{9.40}$$

and the eigenvalues are given by $\pm|k|$, as was the case for the nongravitational problem. Linearly independent generalized eigenvectors are given by the Jordan decomposition:

$$
\Psi = \begin{bmatrix} \dfrac{1}{-2k\mu \pm \rho g} & \dfrac{\mp 1}{k(2k\mu \mp \rho g)} \\[2mm] \dfrac{1}{\pm 2k\mu - \rho g} & 0 \\[2mm] \dfrac{\mp 2k\mu}{2k\mu \mp \rho g} & \dfrac{-2\mu}{k(2k\mu \mp \rho g)} \\[2mm] 1 & 0 \end{bmatrix}, \tag{9.41}
$$

where the top symbol applies to the $+k$ solution and vice versa. The limit of these solutions as $\mu \to 0$ leads to the fully relaxed Maxwell viscoelastic limit. The displacement stress vector can be constructed from any linearly independent set of eigenvectors. It is thus possible to show that in the limit $\mu \to 0$, equations (9.41) can be written as

$$
\Psi = \begin{bmatrix} 0 & 1 \\ -1 & 0 \\ 0 & 0 \\ \rho g & 0 \end{bmatrix}. \tag{9.42}
$$

As expected, the shear stress vanishes in both eigenvectors. The first expresses hydrostatic equilibrium $\sigma_{zz} = -\rho g u_z$. The second shows that in the fully relaxed limit, the horizontal displacements are decoupled from the vertical displacement and stress.

Following the usual procedure and making use of the correspondence principle leads to generalized solutions in the Laplace transform domain. Recalling that we can scale each of the eigenvectors by arbitrary constants:

$$
V = \begin{bmatrix} -\mathrm{sgn}(k)(s+\beta) & \dfrac{(s+\beta)}{k} \\[2mm] -(s+\beta) & 0 \\[2mm] 2k\mu s & -2\mathrm{sgn}(k)\mu s \\[2mm] \mathrm{sgn}(k)2k\mu s + \rho g(s+\beta) & 0 \end{bmatrix}. \tag{9.43}
$$

It is now possible to follow the procedure of chapter 6 and analytically invert the Laplace transforms. For an incompressible half-space, the determinant is quadratic in s, and there are two relaxation times for each wavenumber. The relaxation spectrum is shown in figure 9.4. Recall that for a Poisson layer and half-space ($\nu = 0.25$), neglecting gravitational effects, there were three relaxation modes. Without gravity, both incompressible relaxation times become unbounded at small wavenumber. With gravity, however, one of the modes reaches a maximum relaxation time at $kH \cong 1/2$ and decreases for smaller wavenumbers. The effect of gravity is thus to accelerate the decay of the long wavelength components of the displacement field.

9.4.2 No-Buoyancy Approximation

The incompressible half-space is well behaved; however, it is restrictive to assume that deeper materials are incompressible, $\nu = 0.5$. A variant that has been (not commonly) used in surface loading calculations (e.g., post-glacial rebound, which are in fact the same as the problems

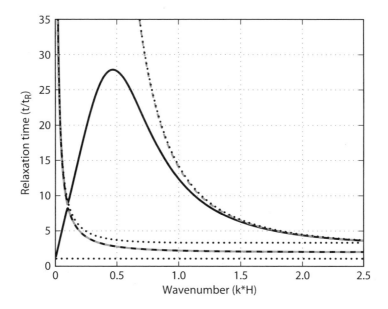

Figure 9.4. Relaxation spectrum for a fault in an elastic gravitating layer overlying a gravitating incompressible viscoelastic half-space (solid line). For comparison, results are shown for an incompressible half-space without gravity (dashed line) and for the nongravitational model in chapter 6 with $\nu = 0.25$ (dotted line).

considered here except for the fact that the loads are applied at the earth's surface rather than in the earth's interior) is to ignore the buoyancy term ($\rho_0 g_0 \nabla \cdot u$) in the equation of motion for the half-space, while retaining general compressible elastic constants. It should be recalled, however, that dimensional analysis suggests that the buoyancy term is comparable to the $\rho_0 g_0 \nabla u_z$ term (both proportional to β), so excluding one at the expense of the other is difficult to justify.

Klemann et al. (2003) show that for a half-space, neglecting buoyancy in the equilibrium equations leads to eigenvalues $\pm k$, $-k[(1 + \varepsilon^2)^{1/2} + \varepsilon]$, and $k[(1 + \varepsilon^2)^{1/2} - \varepsilon]$, where $\varepsilon = \rho g / 2k(\lambda + 2\mu)$. These eigenvalues are well behaved in the limit that $\mu \to 0$. The square roots introduce branch cuts, which necessitate numerical integration of the Bromwich integral; however, in this case, the branch cuts are restricted to the negative half-plane. An even simpler approach is to employ the surface gravity approximation, in which the half-space eigenvalues do not explicitly include gravity. This approximation ignores the buoyancy terms in the equation of motion for both the layer and the half-space. We also expect this solution to be well-behaved in the fully relaxed limit.

9.4.3 Wang Approach

Wang (2005; note erratum in 2007) suggested an approach for including buoyancy effects. He includes the term $-u_r \rho_0 dg_0/dr$ in the equilibrium equations, which is generally excluded in half-space calculations based on dimensional arguments. He further assumes the product $\rho_0 g_0$ to be constant, such that $dg_0/dr = -(g_0/\rho_0)d\rho/dr$, and that the Adams-Williamson density gradient holds. Combining the various expressions yields an additional body force, in the equilibrium equations, to be added to equation (9.26):

$$-\frac{\rho_0^2 g_0^2 u_r}{K} \hat{e}_r. \tag{9.44}$$

It is worth pointing out that from the discussion preceding equation (9.39), the term (9.44) is exactly equal in magnitude and opposite in sign to the buoyancy force in the limit $d\rho_0/dr = 0$. Thus, we expect that the fully relaxed limit including the term (9.44) will be identical to the solution ignoring buoyancy forces, and this is confirmed by numerical calculations. This is of course a consequence of the density gradient satisfying the Adams-Williamson relation.

Including equation (9.44) in the governing equations yields eigenvalues

$$\omega = \pm \left[k^2 + \frac{2}{3} \left(k_g^2 \pm k_g \sqrt{k_g^2 - k^2} \right) \right]^{1/2}, \tag{9.45}$$

where the gravitational wavenumber is given by $k_g = \rho g /[4K(\lambda + 2\mu)/3]^{1/2}$. Notice that k_g remains finite in the limit that the shear modulus vanishes. For $k > k_g$, the eigenvalues (9.45) are complex; however, the real part is nonzero for all k, so in contrast to the standard gravitational case (9.28), there are decaying solutions with increasing depth. The eigenvectors corresponding to the two decaying solutions are complex conjugates. Because the boundary conditions at $z = 0$ are real, this implies that the unknown coefficients are also complex. Last, it should be noted that the self-gravity terms, which have been ignored in Wang's analysis, are of the same order of magnitude as the ∇g term he includes.

A final approach that has been advocated with spherical earth models is to return to the nominal full-gravity equations (9.26), but to exclude the growth modes, corresponding to poles in the positive half-plane, from the Bromwich integral. This is not possible for half-space models because (as seen in problem 1) one branch cut extends from the origin to positive values along the real axis. Thus, any Bromwich contour that includes the pole at the origin but attempts to exclude the growth modes would cross the branch cut.

9.4.4 Comparison of Different Viscoelastic Models

Figures 9.5, 9.6, and 9.7 compare gravitational viscoelastic solutions for a 30-degree dipping thrust fault that cuts completely through a 50-km-thick elastic layer. The fault width in the downdip direction is thus 100 km. Only the postseismic component of the vertical displacement fields are shown. For the no-buoyancy and Wang (2005) approaches, branch cuts must be introduced. For these calculations, a numerical Laplace inversion algorithm due to Hollenbeck (1998) based on the method of de Hoog and others (1982) was employed. Because the Laplace inversion was computed for each Fourier mode and the Laplace inversion algorithm expects the function in the time domain to be real, it was necessary to inverse Laplace transform the real and the imaginary parts of the Fourier transformed displacements separately.

From figure 9.5, it is apparent that at $t = t_R$, where t_R is the Maxwell relaxation time, the surface gravity and no-buoyancy approximations are very similar—both exhibit subsidence relative to the nongravitational solution. This indicates that buoyancy within the elastic layer is not significant. The Wang approach exhibits enhanced subsidence at long wavelengths relative to the previous approximations. The incompressible case is substantially different from the other approximations, in large part because the different elastic structure results in a different coseismic stress distribution, and therefore different time-dependent flow at depth.

After five Maxwell relaxation times, $t/t_R = 5$ (figure 9.6), the viscoelastic displacements have increased, and gravitational effects become more important. At this stage, there is practically no difference between the surface gravity and no-buoyancy approximations. These approximations account for roughly half of the gravitational contribution seen in the Wang approximation. The difference between the Wang and surface gravity/no-buoyancy approximations amounts to roughly 2% of the slip, which for an extremely large earthquake

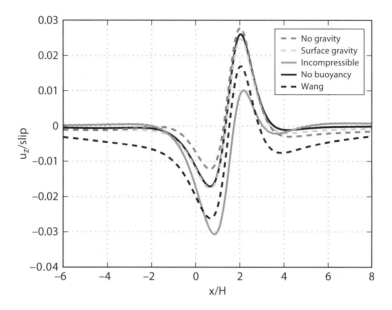

Figure 9.5. Effect of gravity in a viscoelastic solution comparing different approximations. The vertical viscoelastic displacement, excluding the elastic coseismic contribution, is shown per meter of slip for a 30-degree dipping thrust fault that cuts completely through a 50-km-thick elastic layer. $t/t_R = 1$. Distances are normalized by the elastic layer thickness. Five solutions are shown. No gravity: Excludes all gravitational effects. Surface gravity: Excludes buoyancy in both the layer and the half-space and applies the gravitational restoring force at the free surface (section 9.3). Incompressible: Assumes an incompressible half-space. No buoyancy: Ignores $\rho_0 g_0 \nabla \cdot u$ in the equilibrium equations for the half-space (section 9.4.2). Wang: Designates the approach of Wang (2005), as discussed in section 9.4.3.

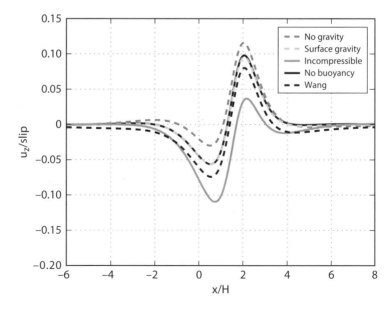

Figure 9.6. Effect of gravity on viscoelastic solutions at $t/t_R = 5$, comparing different approximations. Notation is the same as in figure 9.5.

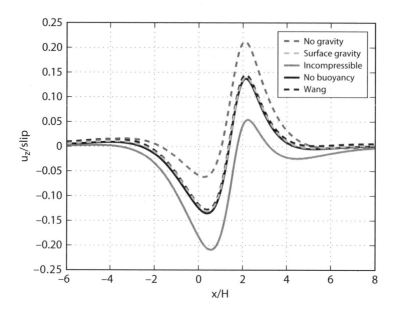

Figure 9.7. Effect of gravity on viscoelastic solutions at $t/t_R = 20$, comparing different approximations. Notation is the same as in figure 9.5.

with 10 to 20 meters of slip is well within the range of what can be measured geodetically. Last, after 20 relaxation times (figure 9.7), there is practically no difference between the surface gravity, no-buoyancy, and Wang approximations, showing as expected that these solutions all converge to the same fully relaxed limit. The incompressible solution appears to be quite different; however, it is important to note that the coseismic elastic displacements are also different. We show next that the cumulative fully relaxed response is similar for all four solutions.

9.4.5 Relaxed Viscoelastic Response

The response over long periods of time is of considerable interest, particularly when considering cycles of repeating earthquakes, as in chapter 12. Figure 9.8 compares the behavior at long times for the incompressible half-space and the surface gravity approximation, including both the elastic and viscoelastic response. At short times, the two solutions are different because the elastic properties of the underlying half-space are different; in the incompressible case, $\nu = 0.5$, whereas in the surface gravity approximation, $\nu = 0.25$. The elastic layer is a Poisson solid ($\nu = 0.25$) in both cases. As time progresses, both the hanging wall and the foot wall subside, and the difference between the two solutions diminishes so that in the fully relaxed limit, they are identical.

Gravitational restoring forces inhibit long-wavelength vertical motions such that vertical displacements are restricted to the region near the fault as stresses relax. Contrast this with the viscoelastic solution without gravity (see figure 6.25), where as $t \to \infty$, the entire foot wall subsided and the entire hanging wall uplifted in a rigid body offset. The horizontal displacements (not shown) are not surprisingly less influenced by gravity; the fully relaxed response is a step function with motion inward toward the fault.

Figure 9.9 illustrates the fully relaxed response for an earthquake that cuts the full elastic layer, comparing the surface gravity and Wang (2005) approximations. In this limit, the Wang and surface gravity results are essentially indistinguishable. The incompressible and surface

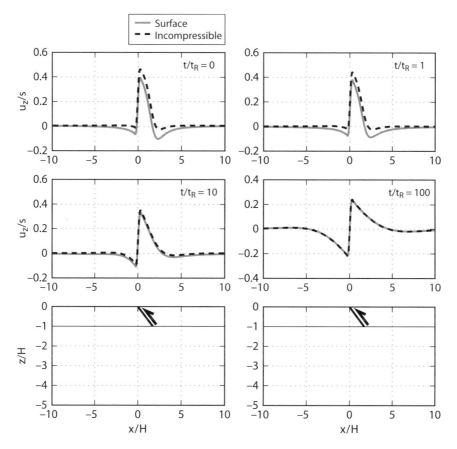

Figure 9.8. Effect of gravity on the vertical displacements for four different dimensionless times. Two solutions are shown: one for an incompressible half-space overlaid by a Poisson elastic layer, the second a surface gravity approximation for a Poisson elastic layer overlying a Poisson half-space. The difference in elastic properties causes the short time response to differ. With increasing time, the effects of gravity become increasingly important, and the differences between the two solutions diminish.

gravity models also agree in the fully relaxed state. That the three models yield the same fully relaxed response is expected, since the Wang approach explicitly negates the effects of buoyancy in the relaxed limit.

9.5 Changes in Gravity Induced by Deformation

Temporal changes in gravity provide additional information that is not available from surface displacement or strain data. Gravity changes are sensitive to changes in subsurface mass distribution as well as deformation. There has been a resurgence of interest in using gravity measurements in volcano and tectonic studies. Part of the reason for this interest relates to advances in instrumentation. Careful use of spring-type gravimeters can measure g to a precision of 5 to 10 μGals, where one μGal $= 10^{-8}$ m/s^2. Newer automated gravimeters may provide higher sensitivity. Absolute gravity meters, which measure the acceleration of a falling test mass can achieve accuracies of a few μGals.

The first-order effect of vertical deformation is to move an observing point relative to the earth's center of mass, leading to the so-called *free air* change in gravity. From equation (9.4),

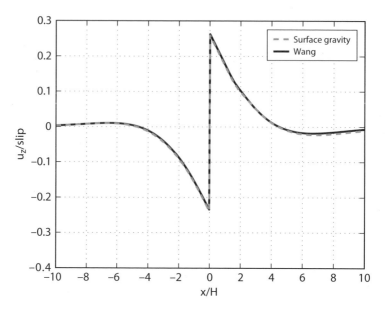

Figure 9.9. Effect of gravity on the vertical displacements at very long nondimensional time, $t/t_R = 10^2$. Results are illustrated for the surface gravity and Wang (2005) approximations.

the theoretical free-air gradient is given by

$$\frac{\partial g_0}{\partial r} = -\frac{2GM_e}{r^3} = -\frac{2g_0}{r}. \tag{9.46}$$

Given $GM_e \simeq 3.99 \times 10^{14}$ m^3/s^2 and $r \simeq 6.36 \times 10^6$ m for the earth's radius, the theoretical free-air gradient is $\simeq -309\ \mu$Gal/m. The actual free-air gradient measured by moving a gravimeter up a ladder, say, can depart from the theoretical value due to local topographic features. While some researchers advocate correcting for the free-air effect using a measured gradient, if the source of deformation is reasonably deep seated such that the vertical displacements vary slowly over distance scales of many kilometers, then local topographic features will displace vertically with the observation point. This suggests that in some circcumstances, it may be more appropriate to use the theoretical value of the free-air gradient than a measured one.

Of more interest are the changes in gravitational potential and acceleration due to deformation. Combining equations (9.10) and (9.9) yields

$$\nabla^2 \delta\phi + 4\pi G \nabla \cdot (\rho_0 u) = 0. \tag{9.47}$$

Equation (9.47) shows that the change in gravitational potential depends on two effects, changes in density due to volumetric strain and displacement of preexisting density gradients:

$$\nabla \cdot \delta g = 4\pi G[\rho_0 (\nabla \cdot u) + u \cdot \nabla \rho_0]. \tag{9.48}$$

Integrating equation (9.47) over a small rectangular volume that encompasses a layer boundary:

$$\int_V \nabla \cdot (\nabla \delta\phi + 4\pi G \rho_0 u)\, dV = 0. \tag{9.49}$$

Making use of the divergence theorem:

$$\int_S (\nabla \delta\phi + 4\pi G\rho_0 u) \cdot \mathbf{n} \, dS = 0. \tag{9.50}$$

Equation (9.50) holds for all choices of surface S so that the integrand must vanish and $(\nabla \delta\phi + 4\pi G\rho_0 u) \cdot \mathbf{n}$ must be continuous across boundaries. In particular, for horizontal layers,

$$\frac{\partial \delta\phi}{\partial z} + 4\pi G\rho_0 u_z \tag{9.51}$$

must be continuous (e.g., Aki and Richards 1980), where $\partial \delta\phi/\partial z = -\delta g_z$ is the vertical component of the change in local gravity. The perturbing potential $\delta\phi$ is also continuous across the boundary.

Above the earth's surface, $\delta\phi$ satisfies Laplace's equation, $\nabla^2 \delta\phi = 0$. Fourier transforming with respect to horizontal coordinates yields

$$\frac{d^2 \overline{\delta\phi}}{dz^2} - (k_x^2 + k_y^2)\overline{\delta\phi} = 0, \tag{9.52}$$

where k_x and k_y are wavenumbers in the two horizontal directions, and the overbar indicates a Fourier transformed quantity. Solutions to this equation are of the form $\overline{\delta\phi} = Ae^{kz}$, where $k = (k_x^2 + k_y^2)^{1/2}$, and A is a constant. We retain only the positive exponent, since z is negative above the free surface. Thus, $d\overline{\delta\phi}/dz = k\overline{\delta\phi}$ for $z \leq 0$. This suggests that we adopt a variable

$$\bar{\Psi} \equiv \frac{d\overline{\delta\phi}}{dz} + 4\pi G\rho_0 \bar{u}_z - k\overline{\delta\phi}, \tag{9.53}$$

which is not only continuous, since $\delta\phi$ is continuous, but also vanishes at the free surface, $z = 0$, since $\rho_0 = 0$ for $z = 0^-$.

Fourier transforming (9.47) with respect to horizontal coordinates, assuming no lateral gradients in reference density, and combining with (9.53) leads to

$$\frac{\partial \bar{\Psi}}{\partial z} = -k\bar{\Psi} + 4\pi G\rho_0 (k\bar{u}_z - ik_x \bar{u}_x - ik_y \bar{u}_y). \tag{9.54}$$

Combining this result with equation (9.53) leads to a system of equations for $\bar{\Psi}$ and $\overline{\delta\phi}$:

$$\frac{\partial}{\partial z}\begin{bmatrix} \overline{\delta\phi} \\ \bar{\Psi} \end{bmatrix} = \begin{bmatrix} k & 1 \\ 0 & -k \end{bmatrix}\begin{bmatrix} \overline{\delta\phi} \\ \bar{\Psi} \end{bmatrix} - 4\pi G\rho_0 \begin{bmatrix} \bar{u}_z \\ i(k_x \bar{u}_x + k_y \bar{u}_y) - k\bar{u}_z \end{bmatrix}. \tag{9.55}$$

This system of equations is of the form $d\mathbf{v}/dz = A\mathbf{v} + \mathbf{f}$, which we have treated many times, beginning in chapter 5. The boundary condition is $\Psi(z = 0^+) = 0$. The propagator matrix for this system is

$$\mathbf{P}(z, z_0) = \begin{bmatrix} e^{(z-z_0)k} & k^{-1}\sinh[(z-z_0)k] \\ 0 & e^{-(z-z_0)k} \end{bmatrix}. \tag{9.56}$$

Having previously solved for the displacements, such that the source terms in equations (9.55) are known, the system of equations can be integrated to give the perturbing gravitational potential and its vertical derivative.

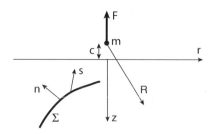

Figure 9.10. Definition sketch for Walsh and Rice analysis of change in gravity due to slip s across dislocation surface Σ. A test mass m is held at height c above the earth's surface by force F.

Here, we have assumed that the self-gravity terms, depending on the perturbing potential, have been excluded from equilibrium equations. To include these terms, it would be necessary to combine the potential equations with the stress-displacement and equilibrium equations, yielding a coupled 6×6 system. However, as shown in section 9.1, the coupling terms, proportional to G, are small for most quasi-static problems of interest in earthquake and volcano studies, so this is generally unnecessary. It is thus appropriate to solve for the displacements ignoring self-gravity effects and then to introduce the displacements into the inhomogeneous term in equation (9.55).

In the following, we will examine an alternative method for computing the change in gravity due to motion across dislocation surfaces in a homogeneous elastic half-space.

Walsh and Rice (1979) derived an elegant expression for computing the change in gravitational acceleration due to relative displacement s across a dislocation surface Σ. They consider a point mass m located at a height c above the earth's surface (figure 9.10). A vertical force $F = mg$ must be applied in order to hold the mass in position. Our objective is to determine δF due to displacements in the earth associated with slip on faults, or dilation of magma chambers, from which we can compute $\delta g = \delta F / m$, the change in the vertical component of gravity.

Displacing the point mass vertically changes the potential energy of the mass by $\delta E = F \delta c$. Gravitational attraction of the mass gives rise to stresses σ_{ij} in the earth. Slip on the fault redistributes mass and thus changes the potential energy of the earth–mass system. Both processes are assumed to occur reversibly. Let the unit normal to the surface Σ be designated n_j. Displacement across the dislocation changes the potential energy by

$$\delta E = \int_\Sigma \left(\sigma_{ij} + \frac{1}{2} \tilde{\sigma}_{ij} \right) s_i n_j d\Sigma \tag{9.57}$$

(equation [3.114]), where s is the relative displacement acting across the dislocation surface Σ, and $\tilde{\sigma}_{ij}$ is the stress change due to displacement on Σ at fixed c. To connect with equation (3.114), note that the stress change is $\tilde{\sigma}_{ij} = \sigma_{ij}^f - \sigma_{ij}^0$.

Now consider two cases in which the mass is displaced and the fault slips. In one, first displace the point mass, perturbing the stress from σ_{ij} to $\sigma_{ij} + (\partial \sigma_{ij} / \partial c) \delta c$, and then displace the surface Σ by s_i in the perturbed stress field. In the second case, reverse the order, and displace the surface Σ by s_i first, changing the force to $F + \delta F$, and then displace the mass subjected to the perturbed force. Since both processes are assumed reversible, the net change in potential energy must be the same. In the first case, displacing the mass first, the change in energy is

$$\delta E = F \delta c + \int_\Sigma \left[\sigma_{ij} + \frac{1}{2} \tilde{\sigma}_{ij} + \left(\frac{\partial \sigma_{ij}}{\partial c} \right) \delta c \right] s_i n_j \, d\Sigma. \tag{9.58}$$

In the second case, displacing the fault first, the change in energy is

$$\delta E = (F + \delta F)\delta c + \int_\Sigma \left(\sigma_{ij} + \frac{1}{2}\tilde{\sigma}_{ij} \right) s_i n_j \, d\Sigma. \tag{9.59}$$

Equating (9.58) and (9.59) leads to

$$\delta F = \int_\Sigma \left(\frac{\partial \sigma_{ij}}{\partial c} \right) s_i n_j \, d\Sigma. \tag{9.60}$$

Given that $\delta F = m\delta g$,

$$\boxed{\delta g = \int_\Sigma S_{ij} s_i n_j \, d\Sigma,} \tag{9.61}$$

where the tensor S_{ij} is defined by

$$S_{ij} = \frac{1}{m} \frac{\partial \sigma_{ij}}{\partial c}. \tag{9.62}$$

The S_{ij} are computed from the stresses induced in the half-space due to the gravitational attraction of the point mass. The gravitational force at each point in the body is $d\mathbf{f} = (G\rho m \, dV/R^2)\mathbf{e}_R$, where ρ is the density of the half-space, dV is a volume increment, and $\mathbf{e}_R$ is a unit vector pointing from the volume element to the point mass. From the definition of the elastic Green's functions (3.7), the displacements due to the distributed body forces are thus

$$u_k(\mathbf{x}) = Gm \int \frac{\rho(\boldsymbol{\xi})}{R^2} g_k^n(\mathbf{x}, \boldsymbol{\xi})\mathbf{e}_n(\boldsymbol{\xi}) \, dV(\boldsymbol{\xi}), \tag{9.63}$$

where $g_k^n(\mathbf{x}, \boldsymbol{\xi})$ are the half-space Green's tensors (3.93) and (3.94), and the components of the unit vector are $e_r = r/R$ and $e_z = (z+c)/R$, where $R^2 = r^2 + z^2 = x^2 + y^2 + z^2$. The stresses are then determined by Hooke's law, $\sigma_{ij} = C_{ijkl}u_{k,l}$. Walsh and Rice (1979) used a different approach to find the stresses, from which they derive

$$S_{rr} = \frac{G\rho z}{R^3} \left(1 - \frac{3r^2}{R^2} \right),$$

$$S_{zz} = \frac{G\rho z}{R^3} \left(1 - \frac{3z^2}{R^2} \right),$$

$$S_{\theta\theta} = \frac{G\rho z}{R^3},$$

$$S_{rz} = -\frac{3G\rho r z^2}{R^5}. \tag{9.64}$$

Note that $S_{kk} = 0$. A surprising implication of this result is that for any isotropic (volumetric) source for which we can write $s_i n_j = s\delta_{ij}$, the gravity change is identically zero. We thus anticipate no change in gravity due to deformation alone—that is, apart from addition or subtraction of fluid mass—from dilation or contraction of a spherical magma chamber.

Walsh and Rice (1979) also consider the gravity change resulting from a dipping two-dimensional fault. Take the x direction to be the horizontal perpendicular to fault strike and y to be the direction parallel to strike. Slip is restricted to the x–z plane. Take the fault normal to be $n_j = \delta_{nj}$ and the slip to be $s_i = s\delta_{is}$. Thus, equation (9.61) becomes

$$\delta g = \int_\Sigma s S_{sn} \, d\Sigma, \tag{9.65}$$

where S_{sn} is the tensor rotated into downdip and fault-normal coordinates. In Cartesian coordinates, the S_{ij} are

$$S_{xx} = \frac{G\rho z}{R^3}\left(1 - \frac{3x^2}{R^2}\right),$$
$$S_{zz} = \frac{G\rho z}{R^3}\left(1 - \frac{3z^2}{R^2}\right),$$
$$S_{xz} = -\frac{3G\rho\, xz^2}{R^5}. \tag{9.66}$$

S_{sn} is then found by rotating the S_{ij} into a fault-parallel and perpendicular coordinate system, leading to

$$\delta g = \frac{3G\rho}{2}\int_\Sigma \frac{z}{R^5}\left[2xz\cos 2\delta - (x^2 - z^2)\sin 2\delta\right] s(\eta) \, d\Sigma, \tag{9.67}$$

where η is the coordinate in the downdip direction, y is the coordinate in the along-strike direction, and $d\Sigma = d\eta dy$ is the area element on the fault. Carrying out the integration in the y direction from $-\infty$ to ∞ yields

$$\delta g = 2G\rho \int \frac{z}{(x^2 + z^2)}\left[2xz\cos 2\delta - (x^2 - z^2)^2\sin 2\delta\right] s(\eta) \, d\eta. \tag{9.68}$$

This result can be compared with the vertical displacements from Volterra's formula for a dipping two-dimensional fault (see equation [3.57]):

$$u_z = \frac{1}{\pi}\int \frac{z}{(x^2 + z^2)}\left[2xz\cos 2\delta - (x^2 - z^2)^2\sin 2\delta\right] s(\eta) \, d\eta. \tag{9.69}$$

Thus,

$$g_z = 2\pi G\rho u_z. \tag{9.70}$$

The gravity change is thus equivalent to a Bouger correction associated with moving rock from regions of subsidence to areas of uplift. For reasonable densities, this amounts to a change of roughly $-200\ \mu$Gal per meter of elevation change (including the free-air effect). Walsh and Rice (1979) cite data from a number of earthquakes that range from $-197\ \mu$Gal/m (the 1964 Alaska, earthquake) to $-215\ \mu$Gal/m (the 1971 San Fernando, California, earthquake).

Okubo (1991, 1992) also computed the gravity change due to dislocations in a homogeneous elastic half-space by directly integrating the volumetric strains and surface displacements. The integral solution of Poisson's equation (9.47) is

$$
\delta\phi(\mathbf{x}) = G \int_V \frac{\nabla \cdot (\rho_0 \mathbf{u})}{||\mathbf{x} - \boldsymbol{\xi}||} \, dV_\xi,
$$

$$
= G\rho_0 \int_V \frac{\nabla \cdot \mathbf{u}}{||\mathbf{x} - \boldsymbol{\xi}||} \, dV_\xi + G \int_V \frac{\mathbf{u} \cdot \nabla\rho_0}{||\mathbf{x} - \boldsymbol{\xi}||} \, dV_\xi. \tag{9.71}
$$

The first integral gives the change due to dilatation, while the second gives the change due to displacement of preexisting density gradients. For a half-space with spatially uniform density ρ_0, the second integral is nonzero only on the free surface; for $\rho_0(\boldsymbol{\xi}) = \rho_0 H(\xi_3)$, with ξ_3 positive down, $\nabla\rho_0 = \rho_0[0, 0, \delta(\xi_3)]^T$. From this, equation (9.71) reduces to

$$
\delta\phi(\mathbf{x}) = G\rho_0 \left[\int_V \frac{\nabla \cdot \mathbf{u}(\boldsymbol{\xi})}{||\mathbf{x} - \boldsymbol{\xi}||} \, dV_\xi + \int_{\xi_3=0} \frac{u_3(\xi_1, \xi_2)}{||\mathbf{x} - \boldsymbol{\xi}||} \, d\xi \right]. \tag{9.72}
$$

Displacements and strains can, of course, be computed from any of the various forms of Volterra's formula (e.g., equation [3.15]). The integrations in equation (9.72) are given by Okubo (1991) for point dislocation sources and for finite rectangular dislocations by Okubo (1992). The change in geoid height ΔH is found directly from the change in gravitational potential; the geoid being the equipotential surface most closely corresponding to mean sea level:

$$
\Delta H = -\frac{\delta\phi}{g_0}, \tag{9.73}
$$

where here g_0 is the standard gravity at the earth's surface, and ΔH is positive upward.

Only recently it has become possible to measure gravity changes accompanying the very largest earthquakes from space. Han et al. (2006) report a ± 15 μGal change in gravity accompanying the 2004 M 9.2 Sumatra–Andaman Islands earthquake using the Gravity Recovery and Climate Experiment (GRACE) satellites. Because the measurement is made from space, there is no free-air correction. Plate 10 shows the observed and predicted gravity field. The predicted gravity change was computed by numerically integrating the volumetric strain and the displacement of the sea floor and the density contrast at the Moho. Because the measurements were made from space, they offer the potential for learning about suboceanic earthquakes without the costly deployment of sea-floor instrumentation, albeit currently only for the very largest earthquakes.

9.5.1 Gravity Changes and Volcano Deformation

One subject that continues to receive considerable attention is gravity change caused by volcanic inflation. Deformation measurements alone, no matter how accurate, cannot reveal the cause of the uplift. As you saw in chapter 7, surface deformation data can often be interpreted in terms of a volume change at depth within the earth. These data, however, cannot determine the cause of the volume change. Possible sources of volume change include pressurization of hydrothermal reservoirs, thermal expansion, and intrusion of magma. From the point of view of predicting volcanic hazards, it is imperative to distinguish magmatic from nonmagmatic sources of inflation and to determine as much as possible about the composition and degree of volatile saturation of any intruded magmas.

Gravity changes, combined with deformation measurements, can help distinguish between competing processes. Crudely speaking, deformation data provide constraints on ΔV, while

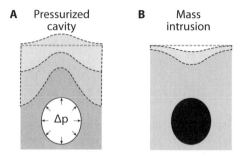

Figure 9.11. Intrusion of magma into a preexisting magma chamber. A: First the chamber is pressurized with fixed fluid mass. B: New magma is added, holding the pressure constant.

gravity-change data constrain the mass of any intruded fluids ΔM. From this, it is possible to estimate the density of intruded fluids. Because aqueous fluids have densities a factor of 2.5 to 3.0 less than magmas, it should be possible to discriminate between hydrothermal and magmatic processes by combining repeated gravity and deformation observations.

It is useful to separate deformation effects on gravity from the effect of fluid migration. Consider a preexisting melt-filled magma chamber (figure 9.11). Both gravity change and deformation measurements are insensitive to the preexisting state of the reservoir. Separate the intrusion of new magma into two steps. In the first step (figure 9.11A), the reservoir pressure increases without the addition of new magma. In the second step, new magma is added to the chamber, maintaining the melt pressure constant (figure 9.11B). The second step causes a change in gravitational acceleration related to the net mass of intruded magma, but only a small deformation (section 9.1). The change in the vertical component of gravity due to the increased mass is computable directly from Newton's law:

$$\delta g_z(\mathbf{x}) = G \int \frac{\rho(\xi)\mathbf{e}_z}{r^2} \, dV(\xi), \qquad (9.74)$$

where $\mathbf{e}_z$ is the z component of the unit vector, as in equation (9.63). Thus, if the deformation effects can be accounted for, gravity measurements can reveal useful information about the nature of any intruded fluids.

There are significant simplifications that occur for spherically symmetric volume sources, particularly the point center of dilatation. The Walsh and Rice (1979) results demonstrate that for a purely isotropic source in a homogeneous elastic half-space, the deformation effect on gravity is identically zero. Thus, if one is able to correct for the free-air effect by directly measuring vertical displacement and also remove any nonmagmatic effects, then the only remaining gravity signal should be due to mass changes. It is important to emphasize that this simplification is not generally true for nonspherically symmetric sources.

Recall that the surface displacements due to an expanding point source are directed radially from the source and are given by equation (7.14), which we repeat here:

$$u_z^{\Delta V} = \frac{(1-v)\Delta V}{\pi} \frac{d}{(r^2 + d^2)^{3/2}}, \qquad (9.75)$$

where $r^2 = x^2 + y^2$. These displacements do not include the effects of mass loading, as in figure 9.11B. The displacements at the free surface due to mass loading can be estimated from the Mindlin Green's functions (3.93), assuming that the radius of the chamber is small compared to its depth:

$$u_z^{\Delta M} = -\frac{\rho g \Delta V}{4\pi\mu}\left[\frac{2(1-\nu)}{R} + \frac{d^2}{R^3}\right], \tag{9.76}$$

where $R^2 = r^2 + d^2$. Evaluating the ratio of the inflation-induced deformation to the mass-loading contribution for $\nu = 1/4$, $\mu = 10^{10}$ Pa, a density of 3×10^3 kg/m^3, and a depth of 3 km yields $u_z^{\Delta V}/u_z^{\Delta M}$ at $r = 0$ of more than a factor of 10^2, which justifies neglect of the mass-loading-induced displacements in most cases.

The gravitational attraction due to a point mass ΔM is also radial, which means that the spatial dependence is the same as in equation (9.75),

$$\Delta g_z - \gamma u_z = G\Delta M \frac{d}{(r^2 + d^2)^{3/2}}, \tag{9.77}$$

where γ is the free-air gradient. Thus, the ratio of the estimated mass change to the estimated volume change, accounting for the free-air effect but ignoring mass loading, yields an estimate of the density of the intruded material

$$\rho = \frac{(1-\nu)}{\pi G}\left(\frac{\Delta g_z}{u_z} - \gamma\right). \tag{9.78}$$

This simple result (9.78) can be utilized to interpret the *gravity gradient* $\Delta g_z/u_z$ in terms of the intruded fluid density. A few words of caution are, however, warranted. The formula (9.78) holds *only* for a spherical point source. In this special case, the spatial dependence of the free-surface displacement and the gravity change are the same. For other shapes, this is not the case. Battaglia et al. (2003a; 2003b) found that for even slightly elliptical magma chambers (aspect ratio of 2 to 1), density estimates based on equation (9.78) can be biased by as much as 40%. In chapter 7, we found that if the true source is a prolate ellipsoid, assuming a spherical source biases the source to greater depth. The greater the depth, the greater the mass needed to explain the measured gravity signal. Thus, the mass change is biased to large values. One might suspect that there would be a similar bias in the volume change. However, the spherical source is more efficient at generating vertical deformation than the prolate ellipsoid, so there is relatively small bias in the volume change. The net effect is to bias the density to large values.

Second, if the magma chamber is not spherical, there are nonzero deformation effects that must be accounted for. As we have seen, both volumetric strains and displacement of layers with differing density contribute to changes in gravity. At this point in time, these effects have not yet been adequately explored for nonspherical magma chambers. The conclusion is that one should be cautious about interpreting gravity data using equation (9.78).

Last, as discussed in chapter 7, one needs to account for the compressibility of the magma in determining the ratio of mass-to-volume change (Johnson 1992). From equation (7.32), we have

$$\rho_m = \frac{\delta m}{\delta V}\left(1 + \frac{\beta_m}{\beta_c}\right)^{-1}, \tag{9.79}$$

where β_m and β_c are the magma and chamber compressibilities, respectively. If the melt is saturated with respect to volatile phases, the ratio β_m/β_c can become significantly greater than 1 (see chapter 7).

Last, there are practical issues, beyond model selection, when interpreting repeated gravity measurements that should be considered. It is important to propagate errors in the uplift and gravity measurements into the estimated density. In order to limit the error in the free-air correction to under 5 μGal, it is necessary to know the uplift at the gravity stations to 1.6 cm. While this is certainly within current measurement technology, in some volcanic environments, it is nontrivial to limit errors in height change to this accuracy at all gravity stations.

Another major limiting factor in the interpretation of gravity measurements is the contaminating effect of changes in groundwater level. It is not difficult to estimate the gravity effect of a change in water table within an infinitely extensive flat, unconfined aquifer. The well-known gravity change due to a slab of thickness T is $\Delta g = 2\pi G \rho T$, where ρ is the density. For a porous medium, we replace T with the porosity ϕ times the change in water table height Δh, leading to

$$\frac{\Delta g_z}{\Delta h} = 42\, \phi \quad \mu\text{Gal/m}. \tag{9.80}$$

For porous sediments, gravity can change by as much as 20 μGal per meter of water-level change. It can be quite challenging to correct for this effect in volcanic environments, where water table monitoring is often nonexistent. In many cases of interest, uncertainties in water table effects are larger than the intrinsic measurement errors and thus dominate the error budget in determining changes in gravity.

9.5.2 An Example from Long Valley Caldera, California

Battaglia et al. (2003b) reported gravity change measurements at Long Valley caldera in California. The data are summarized in plate 11. The gravity changes are negative over the resurgent dome (plate 11A), primarily reflecting the free-air effect; as stations moved farther from the earth's center of mass, the gravitational acceleration decreased. A combination of leveling and GPS measurements document the uplift during the time period the gravity measurements were collected. These data show uplift of as much as 450 mm on the resurgent dome (plate 11B). Water-level measurements were made in a number of monitoring wells around Long Valley, allowing for an approximate correction due to the effects of variations in groundwater mass. The gravity data, corrected for the free-air effect and the effects of water table variations, are shown in plate 11c. There remain significant increases in residual gravity on the resurgent dome of as much as 50 $\pm$ 10 μGal. A few sites in the southwest portion of the caldera showed large increases in residual gravity; however, these are not significant given the uncertainties.

As discussed briefly in chapter 7, Battaglia et al. (2003a) found that the deformation data are reasonably well fit with a prolate ellipsoidal magma chamber with an aspect ratio of 0.5 at a depth of 6 km. Assuming this geometry, Battaglia et al. (2003b) estimate a density of 1180 to 2330 kg/m^3 for the intruding fluid. Failure to account for the ellipsoidal shape of the source results in a 40% bias (to larger values) in the density estimate. The calculations of Battaglia et al. (2003b) did not account for changes in gravity induced by deformation, as in equation (9.61), and therefore cannot be considered as final.

9.6 Summary and Perspective

Gravitational effects may be the least well understood aspect of earthquake and volcano deformation modeling and remain the subject of ongoing research. We began by examining small perturbations from an initial, radially symmetric, self-equilibrated reference state. This leads to terms in the momentum balance due to changes in density, gravitational potential, and advection of the initial stress. Dimensional analysis indicates that except at the very longest spatial scales, of order 1,000 km, self-gravity—the effect of changes in gravitational potential on equilibrium—as well as gradients in the reference gravitational acceleration can be neglected. Half-space models, even accounting for a first-order correction due to earth curvature as in chapter 8, become inaccurate at longer distance scales; spherical earth models are required to analyze deformations at these distances.

Elastic deformations are only slightly influenced by gravity except at spatial scales greater than roughly several hundred kilometers. Thus, except for very large earthquakes, gravitational contributions to coseismic deformation are small. The surface gravity approximation, which ignores buoyancy associated with volumetric strain, appears to recover roughly half of the full gravitational effect. The loss of regularity that arises at extremely long wavelengths when including buoyancy is generally of no consequence, because half-space models break down at these distance scales.

Gravitational effects are considerably more important for viscoelastic deformations, particularly at times following the earthquake much greater than material relaxation times. A compressible viscoelastic half-space with uniform initial density is gravitationally unstable. In contrast, an initial density profile that follows the Adams-Williamson self-compression gradient is stable. Including the effects of gradients in gravitational acceleration, assuming that $\rho_0 g_0$ is independent of depth and that $d\rho_0/dz$ follows the Adams-Williamson equation, appears to be a reasonable means for including compressibility in viscoelastic calcultions. We noted that this set of assumptions for a homogeneous half-space is equivalent to adding a restoring force that precisely negates the induced buoyancy force. Buoyancy effects diminish with increasing time after the earthquake such that the surface gravity approximation becomes asymptotically equivalent to the more complete theory. For very long times after an earthquake, which will be of interest in discussion of earthquake cycle models (chapter 12), the surface gravity and incompressible approximations are equivalent and consistent with isostatic equilibrium.

Computation of gravitationally consistent viscoelastic deformation on the very largest spatial scales requires spherical earth models. These are beyond our scope; however, a number of recent studies (Piersanti et al. 1995; Pollitz 1997; and Tanaka et al. 2006, 2007) have analyzed this problem. The differences in approach stem from differences in handling self-gravity effects, whether or not compressibility is included, and in the numerical procedures for inverting the Laplace transforms.

Measurements of the change in local gravitational acceleration or change in geoid height provide information that is not accessible from other geodetic or strain observations. Satellite measurements of gravity change provide constraints on long wavelength deformation in oceanic environments, where data are otherwise limited to sea-floor geodetic or strain installations. Gravity change in volcanic environments can provide important constraints on the nature of intruding fluids that cannot be obtained from measurements of deformation alone. The Walsh-Rice procedure provides a simple integral representation for change in gravity resulting from arbitrary dislocation sources in a homogeneous elastic half-space. Their result demonstrates that there is no change in gravity due to deformation for a spherically symmetric source. For this geometry, the gravity change results only from the free-air effect and redistribution of fluid mass. For more general magma chamber geometries, one has to correct for the effect of changes in density due to volumetric strain as well as displacement of preexisting density gradients in order to estimate subsurface fluid mass redistributions.

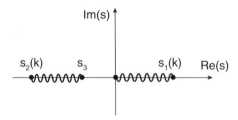

Figure 9.12. Branch cuts and branch points for the elastogravitational half-space, including buoyancy (problem 1).

9.7 Problems

1. This problem explores the effect of the branch cuts due to the square root functions in the viscoelastic gravitational problem. The function $\sqrt{z}$, where z is complex, has a branch point at $z = 0$ and a branch cut that is commonly taken along the negative real axis. The eigenvalues of interest in equation (9.28) are $-|k|(1 \pm \epsilon)^{1/2}$, where $\epsilon = \rho g / k\{\bar{\mu}(s)[\bar{\lambda}(s)+2\bar{\mu}(s)]\}^{1/2}$; there are branch cuts due to (1) the square-root $\{\bar{\mu}(s)[\bar{\lambda}(s) + 2\bar{\mu}(s)]\}^{1/2}$ and (2) the square root $(1 - \epsilon)^{1/2}$. In the first case, when the argument of the square root is negative, ϵ is imaginary, and the eigenvalues are complex conjugates. Because the matrix A is real, the eigenvectors are also conjugates. Thus, as discussed by Klemann et al. (2003, appendix B), the solution vector can be found as the real part of the appropriate matrix products.

   In the second case, $[1 - \epsilon(s)]^{1/2}$, the branch point is at $\epsilon(s) = 1$, and the branch cut corresponds to all values of s such that $\epsilon > 1$.

   (a) Show that the branch points in the s plane are given by the two roots s_1, s_2:

$$s = \frac{1}{\tau_R} \left[\frac{\mu \Gamma}{2K} \left(1 \pm \sqrt{1 + \frac{16}{3} \Gamma^{-1}} \right) - 1 \right]^{-1}, \tag{9.81}$$

   where s_1 corresponds to the positive sign and s_2 to the negative sign, and

$$\Gamma = \left(\frac{kK}{\rho g} \right)^2, \tag{9.82}$$

   where k is the wavenumber, and K is the bulk modulus.

   (b) Next show that the branch cuts, corresponding to all values $\epsilon(s) > 1$, span the real axis over the interval (see figure 9.12)

$$0 \le s \le s_1(k),$$

$$s_2(k) \le s \le s_3 = -\frac{1}{\tau_R} \frac{K}{K + \frac{4}{3}\mu}. \tag{9.83}$$

   (c) Show that in the limit as $k \to \infty$, the two branch cuts shrink to the points 0 and $-(1/\tau_R) \left[K/(K + \frac{4}{3}\mu) \right]$.

2. This problem develops a proof of equation (9.36). Start by taking the fluid limit of the equilibrium equations (9.15) in spherical polar coordinates (see equation [1.99]). Note that the r component of the gradient operator is $\partial/\partial r$, while the θ component is $1/r\partial/\partial\theta$. Show

that equilibrium in the r direction, without dislocation sources, can be written as

$$\frac{\partial}{\partial r}\left[\lambda\Delta - \rho_0(\delta\phi + g_0 u_r)\right] + \rho_0 g_0 \Delta + (\delta\phi + g_0 u_r)\frac{\partial\rho_0}{\partial r} = 0, \tag{9.84}$$

where $\Delta = \nabla\cdot u$. Assuming that neither ρ_0 nor g_0 vary laterally in the earth, then equilibrium in the θ direction can be written as

$$\frac{\partial}{\partial\theta}\left[\lambda\Delta - \rho_0(\delta\phi + g_0 u_r)\right] = 0. \tag{9.85}$$

Use these results to show that

$$\left(\frac{d\rho_0}{dr} + \frac{\rho_0^2 g_0}{K}\right)\frac{\partial\Delta}{\partial\theta} = 0. \tag{9.86}$$

Excluding the case where the volumetric strain Δ is a function of r only, which would require globally extensive strains from a localized source, leads directly to equation (9.36).

3. Show that in the limit of an incompressible medium, $1/\lambda \to 0$, the governing equations can be written in the form $d\mathbf{v}(z)/dz = A\mathbf{v}(z)$, with A given by equation (9.40).

4. In a linear viscoelastic material, the stress can be computed as a convolution of the deviatoric strain rate with a shear relaxation function $\Psi(t)$ (e.g., Malvern 1969; Flugge 1975). For a Maxwell material,

$$\Psi(t) = \mu e^{-t/\tau_R}, \tag{9.87}$$

where μ is the elastic shear modulus, and $\tau_R = \eta/\mu$ is the Maxwell relaxation time.

We can use this to estimate a time when the stability margin is met and the regularity condition is no longer satisfied (Klemann et al. 2003). First, from equation (9.28), show that the critical shear modulus μ_c is

$$\mu_c = \frac{3}{8}K\left[\sqrt{1 + \frac{16}{3}\left(\frac{\rho_0 g_0}{kK}\right)^2} - 1\right]. \tag{9.88}$$

From this result, derive a wavenumber-dependent critical time as

$$t_c = -\tau_R \ln\left\{\frac{3}{8}\frac{K}{\mu}\left[\sqrt{1 + \frac{16}{3}\left(\frac{\rho_0 g_0}{kK}\right)^2} - 1\right]\right\}. \tag{9.89}$$

Plot t_c/τ_R as a function of $\ln(1/k)$.

9.8 References

Aki, K., and P. G. Richards. 1980. *Quantitative seismology: theory and methods*. New York: W. H. Freeman.

Battaglia, M., P. Segall, J. Murray, P. Cervelli, and J. Langbein. 2003a. The mechanics of unrest at Long Valley caldera, California: 1. Modelling the geometry of the source using GPS, leveling, and two-color EDM data. *Journal of Volcanology and Geothermal Research* **127**, 195–217.

Battaglia, M., C. Roberts, and P. Segall. 2003b. The mechanics of unrest at Long Valley caldera, California: 2. Constraining the nature of the source using geodetic and micro-gravity data. *Journal of Volcanology and Geothermal Research* **127**, 219–245.

Dahlen, F. A., and J. Tromp. 1998. *Theoretical global seismology*. Princeton, NJ: Princeton University Press.

de Hoog, F. R., J. H. Knight, and A. N. Stokes. 1982. An improved method for numerical inversion of Laplace transforms. *Society for Industrial and Applied Mathematics Journal on Scientific and Statistical Computing* **3**, 357–366.

Flugge, W. 1975. *Viscoelasticity*, 2nd ed. New York: Springer-Verlag.

Han, S.-C., C. K. Shum, M. Bevis, C. Ji, and C.-Y. Kuo. 2006. Crustal dilatation observed by GRACE after the 2004 Sumatra-Andaman earthquake. *Science* **313**, 658–662.

Hollenbeck, K. J. 1998. INVLAP.M: A MATLAB function for numerical inversion of Laplace transforms by the de Hoog algorithm. http://www.isva.dtu.dk/staff/karl/invlap.htm.

Johnson, D. J. 1992. Dynamics of magma storage in the summit reservoir of Kilauea volcano, Hawaii. *Journal of Geophysical Research* **97**, 1807–1820.

Klemann, V., P. Wu, and D. Wolf. 2003. Compressible viscoelasticity: stability of solutions for homogeneous plane-earth models. *Geophysical Journal International* **153**, 569–585.

Longman, I. M. 1963. A Green's function for determining the deformation of the earth under surface mass loads. 2. Computations and numerical results. *Journal of Geophysical Research* **68**, 485–496.

Love, A. E. H. 1991. *Some problems in geodynamics*. Cambridge, UK: Cambridge Press.

Malvern, L. E. 1969. *Introduction to the mechanics of continuous media*. Englewood Cliffs, NJ: Prentice Hall, pp. 306–327.

Matsu'ura, M. and T. Sato. 1989. A dislocation model of the earthquake cycle at convergent plate boundaries. *Geophysical Journal* **96**, 23–32.

Okubo, S. 1991. Potential and gravity changes raised by point dislocations. *Geophysical Journal International* **105**, 573–586.

Okubo, S. 1992. Gravity and potential changes due to shear and tensile faults in a half-space. *Journal of Geophysical Research* **97**, 7137–7144.

Piersanti, A., G. Spada, R. Sabadini, and M. Bonafede. 1995. Global postseismic deformation. *Geophysical Journal International* **120**, 544–566.

Plag, H.-P., and H.-U. Jüttner. 1995. Rayleigh-Taylor instabilities of a self-gravitating earth. *Journal of Geodynamics* **20**, 267–288.

Pollitz, F. F. 1997. Gravitational viscoelastic postseismic relaxation on a layered spherical earth. *Journal of Geophysical Research* **102**, 17,921–17,941.

Tanaka, Y., J. Okuno, and S. Okubo. 2006. A new method for the computaion of global viscoelastic post-seismic deformation in a realistic earth model (I)—vertical displacement and gravity variation. *Geophysics Journal International* **164**, 273–289.

Tanaka, Y., J. Okuno, and S. Okubo. 2007. A new method for the computaion of global viscoelastic post-seismic deformation in a realistic earth model (II)—horizontal displacement. *Geophysics Journal International* **170**, 1031–1052.

Walsh, J. B., and J. R. Rice. 1979. Local changes in gravity resulting from deformation. *Journal of Geophysical Research* **84**, 165–170.

Wang, R. 2005. The dislocation theory: a consistent way of including the gravity effect in (visco)elastic plane-earth models. *Geophysical Journal International* **161**, 191–196.

———. 2007. Erratum to "The dislocation theory: a consistent way of including the gravity effect in (visco)elastic plane-earth models." *Geophysical Journal International* **170**, 857.

Plate 1. Waterfall created by a fault scarp near the northern end of the Chi-Chi, Taiwan, rupture. Photo courtesy of J. Mori.

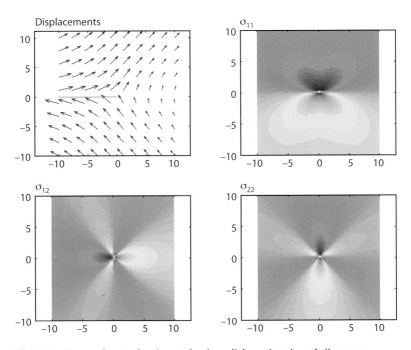

Plate 2. Stress due to horizontal edge dislocation in a full-space.

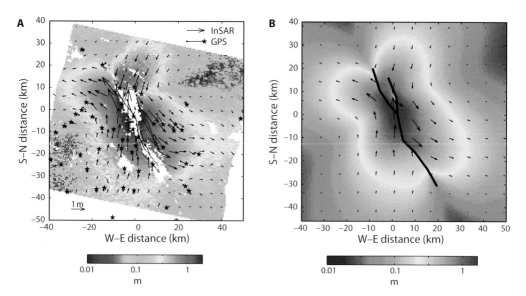

Plate 3. A: Horizontal displacements during the 1999 Hector Mine, California, earthquake derived from a combination of GPS and InSAR. The color scale indicates the magnitude of the horizontal displacement from InSAR. The regularly spaced short vectors indicate displacements derived from InSAR, while the irregularly spaced vectors with stars are GPS-derived displacements. After Fialko et al. (2001). B: Model-predicted displacements from an elastic dislocation model that follows the fault trace. The slip distribution is shown later in plate 4. After Jónsson et al. (2002).

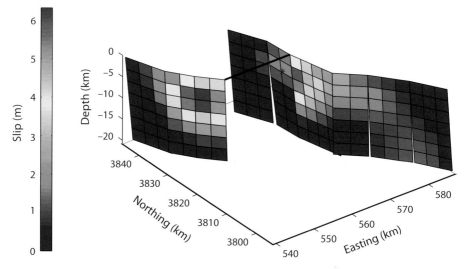

Plate 4. Distributed slip model for the Hector Mine earthquake using rectangular dislocations as a basis. From Jónsson et al. (2002).

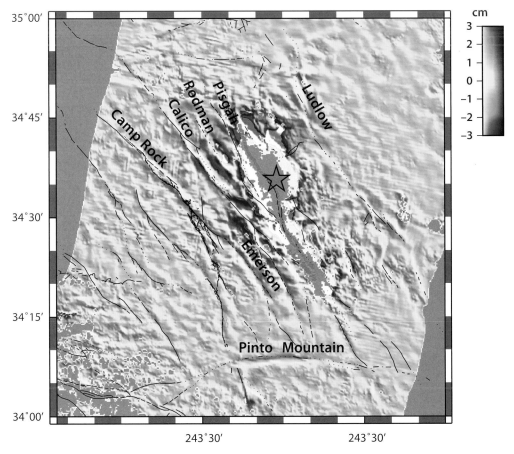

Plate 5. High-pass filtered interferogram of the 1999 Hector Mine earthquake. Rupture zone of the quake is shown in red. Notice the phase variations across the Pinto Mountain, Calico, and Rodman faults. After Fialko et al. (2002).

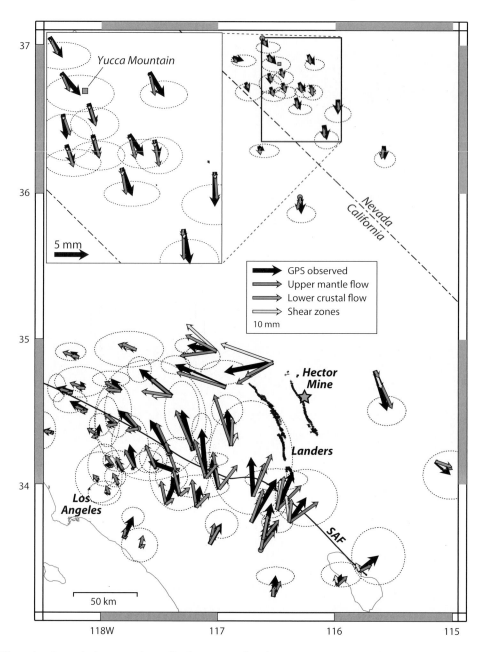

Plate 6. Cumulative transient displacement for the seven-year period following the 1999 Hector Mine earthquake in southern California. The nearby Landers earthquake occurred in 1992. The displacements were determined by fitting a linear plus logarithmic function to the GPS position time series. Observed horizontal displacements are shown with 95% confidence ellipses. The enlargement illustrates displacements in the Yucca Mountain region of Nevada. Also shown are predicted displacements for three models: distributed flow in the lower crust, with a viscosity of 1.2×10^{18} Pa-s in the depth range 20–28 km; distributed flow in the upper mantle, with a viscosity in excess of 10^{21} Pa-s in the lower crust, decreasing to a constant 10^{19} Pa-s below 40 km; and model shear zones that allow complete relaxation of the coseismic stress change on narrow shear zones that cut both the lower crust and the mantle beneath the Landers and Hector Mine ruptures. From Freed et al. (2007).

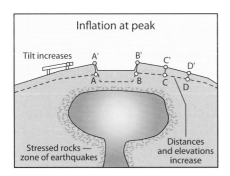 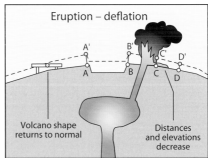

Plate 7. Model of inflation–deflation cycles. During the inflationary phase, magma accumulates in a shallow chamber. Eruption, or injection of a dike into an adjacent rift zone, leads to deflation. Adapted from the Hawaiian Volcano Observatory website, http://hvo.wr.usgs.gov/.

A **B**

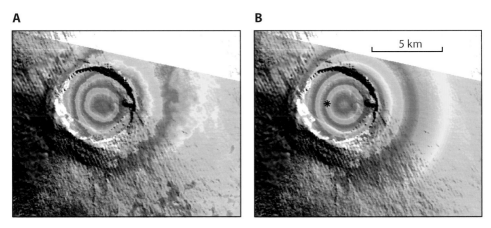

Plate 8. SAR interferogram of Darwin volcano in the Galápagos. A: Data from 1992 to 1998. B: Predicted interferogram based on a center of dilation located at a depth of 3 km beneath the star. Each color cycle represents 5 cm of displacement in the line of sight between the satellite and the ground. Because the line of sight is not vertical, the pattern is not purely radially symmetric. After Amelung and others (2000).

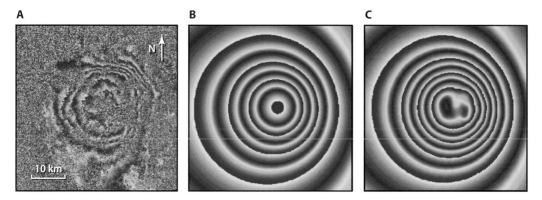

Plate 9. A: Interferogram of Mount Etna between October 1992 and October 1993. Each color cycle represents 28 mm of displacement in the line of sight between the satellite and the ground. The sign of the signal is deflation; the maximum line-of-sight displacement is roughly 130 ± 10 mm. B: The predicted displacement field for the best fitting center of dilatation in a homogeneous half-space. C: The predicted displacement field for the best fitting source accounting for topography. From Cayol and Cornet (1998).

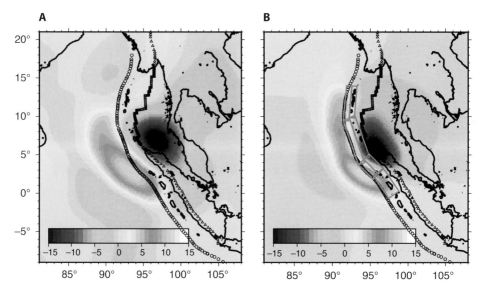

Plate 10. Gravity change (μGal) associated with the 2004 Sumatra–Andaman Islands earthquake as detected by the GRACE satellite mission. A: Observations of average difference between the 2005–2003 and 2005–2004 gravity fields, filtered to suppress signals far from the epicenter. B: Predicted gravity change from a seismically derived dislocation model (gray rectangles). The model prediction is spatially filtered to the same scales retained in the observations. Line of circles marks location of trench. After Han et al. (2006).

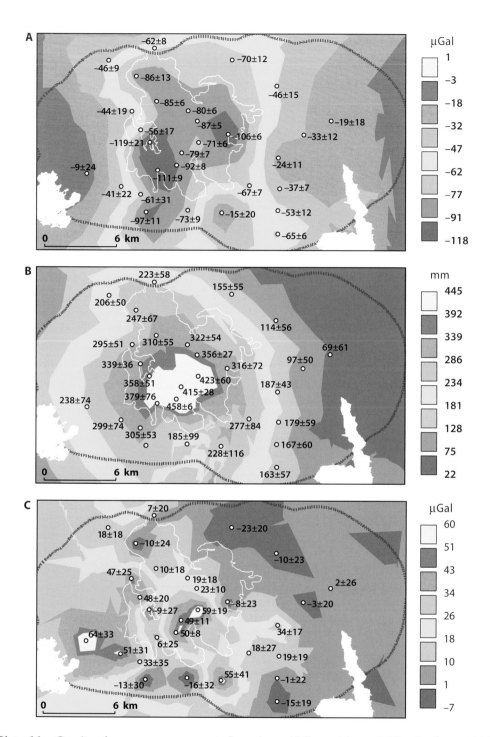

Plate 11. Gravity change measurements from Long Valley caldera, California, from 1982 to 1999. The caldera outline is shown as a purple dotted line. The outline of the resurgent dome is shown in white. A: Gravity change measurements, corrected for tides and loop closure. Numbers indicate mean values and one standard deviation errors. B: Vertical displacements during the same time interval. C: Residual gravity changes, including corrections for the free-air effect and water table variations. After Battaglia et al. (2003b).

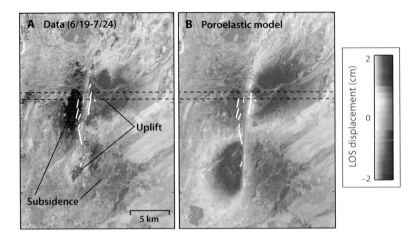

Plate 12. Left: Unwrapped interferogram showing the line-of-sight (LOS) displacement following the 17 June 2000 earthquake in the South Iceland Seismic Zone. The image shows the displacement between 19 June and 24 July. Regions of uplift and subsidence are marked. The red line approximates the fault trace, while white lines mark observed surface ruptures; slip was right-lateral. Effects of a second earthquake during the data interval, which are generally small in this region, have been removed using a three-dimesional dislocation calculation. Right: Simulated interferogram showing the predicted LOS displacement assuming a full relaxation of induced pore pressures, with $v_u = 0.31$ and $v = 0.27$, during the 35-day interval. Dashed lines mark the profile shown later in figure 10.17. After Jónsson et al. (2003).

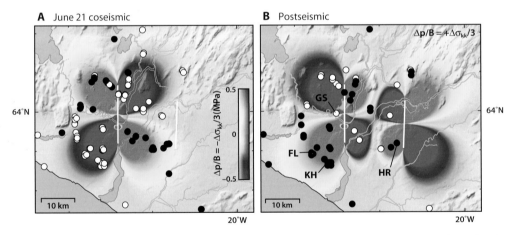

Plate 13. Water-level changes accompanying and following the June 2000 earthquakes in the South Iceland Seismic Zone. Black circles represent wells in which the water level increased during the indicated time interval. White circles represent wells in which the water level decreased during the indicated time interval. Notice that water levels increased in the compressional regions and decreased in the extensional regions. For the postseismic period, the negative of the coseismic pressure changes associated with the two earthquakes is shown, as would be appropriate if the crust fully drained during that interval. Stress change is computed at a depth of 0.5 km. Two-letter codes indicate wells shown in figure 10.14. After Jónsson et al. (2003).

10

Poroelastic Effects

Up to this point, we have considered the earth's crust to be a solid nonporous elastic, or viscoelastic, medium. In actuality, the crust is porous and, except for the very shallow near-surface region, at least partially liquid saturated. That rock is a multiphase composite consisting of solid and liquid filled pores adds a significant richness to its mechanical behavior. The liquid phase flows under gradients in pore-fluid pressure. Changes in fluid pressure induce strains, *and* conversely, changes in stress or strain induce changes in pore pressure.

To see this at an intuitive level, consider a fluid-saturated sponge. Rapidly squeezing the sponge compacts the pore space and causes the pressure of the water in the pores to rise. The converse is that decreasing the water content by allowing the sponge to dry causes the sponge to shrink. This duality has important geophysical implications, and adds significant mathematical complexity to the solution of boundary value problems in poroelastic media. Because fluid pressure responds to strain, measuring pore-pressure changes in wells can be used to infer strain changes in the earth. In effect, water wells in confined aquifers can be employed as strainmeters (figure 10.1). That changes in fluid pressure cause deformation is apparent in many places where large volumes of fluid have been extracted from subsurface reservoirs. Decreases in pore pressure cause subsidence, in extreme cases reaching magnitudes of up to 10 meters and horizontal strains of as much as a few percent (Yerkes and Castle 1970), and in some cases, triggering earthquakes (figure 10.2). Variations in fluid pressure also lead to more subtle deformations that in some cases can be misinterpreted as tectonic or volcanic in origin.

As a way of introducing some basic terminology, consider two idealized experiments. First, imagine an experiment in which a cube of fluid-saturated rock is subject to an isotropic stress. The sample is jacketed with an impermeable membrane that prevents fluid from leaving (or entering) the sample (figure 10.3). This is referred to as an *undrained test*, since drainage of the fluid is restricted by the impermeable membrane. The deformation can be effectively undrained, even without the impermeable membrane, if the sample is deformed very rapidly. If the deformation is imposed rapidly, and one considers only short times before fluid flow has had a chance to occur, the deformation is effectively undrained.

The other end member is a fully *drained test*. In this case, the sample is deformed so slowly that the pore pressures always maintain equilibrium with an external reservoir at constant pressure (figure 10.3). Deformation at long times after the application of a load approaches a fully drained state. This is an excellent analogy with viscoelasticity, where at short times, a viscoelastic medium exhibits an *unrelaxed* elastic response, whereas at long times, it exhibits a fully *relaxed* response. The undrained poroelastic response corresponds to the unrelaxed viscoelastic response, while the drained poroelastic response corresponds to the fully relaxed viscoelastic response.

As with viscoelasticity, the solution to either a completely undrained or a fully drained poroelastic problem is given by an equivalent elastic solution. Indeed, all of the analyses we have treated so far in this book have assumed that pore pressures everywhere remain constant—that is, our solutions have implicitly been fully drained. In this limit, the problem reduces to an elasticity problem with appropriate—that is, drained—elastic constants. Undrained poroelastic problems are also equivalent to conventional elastic problems;

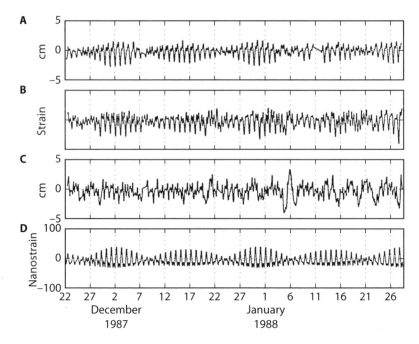

Figure 10.1. Strains induce pore pressures that can be monitored in water wells. A: Water level in a well near Parkfield, California, along the San Andreas fault. B: The volumetric strain recorded by a borehole dilatometer roughly 20 km from the water well. C: The observed barometric pressure. D: Theoretical volumetric strain tide. Notice that the water level responds clearly to the tidal strains. After Roeloffs (1995).

however, in this case the elastic constants are modified by the fluid-filled pores. Because pore fluids have finite compressibility, the undrained bulk modulus of rock is stiffer than its drained counterpart. For isotropic rocks, the shear modulus is not affected by changes in pore pressure.

Consider an idealized experiment in which a cube of porous rock is subjected to a step change in confining stress, $-\Delta\sigma_{kk}/3$ (figure 10.4). The applied compression causes the rock to contract instantaneously by an amount $-\Delta\sigma_{kk}/3K_u$, where K_u is the *undrained bulk modulus*.

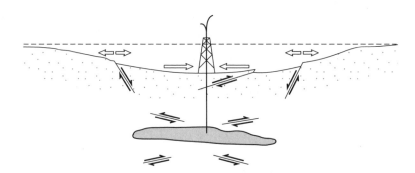

Figure 10.2. Schematic illustration of deformation associated with decreasing reservoir pressure in a hydrocarbon reservoir. Open arrows indicate the sense of horizontal strain. After Segall (1989).

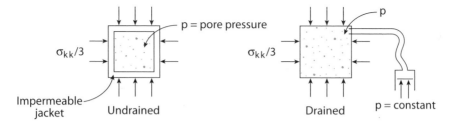

Figure 10.3. Left: A jacketed sample undergoing undrained deformation. Right: A sample hydraulically connected to an external reservoir that is kept at constant pressure. The latter is experiencing fully drained deformation.

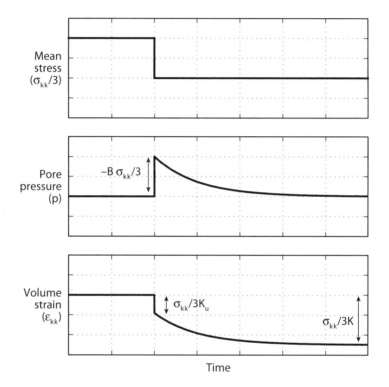

Figure 10.4. Results of an idealized experiment on an unjacketed sample. At some time, the sample is suddenly compressed; the mean normal stress decreases by an amount $\Delta\sigma_{kk}/3$. The pore pressure instantaneously increases by $p = -B\,\Delta\sigma_{kk}/3$ and with time relaxes back to the initial value as pore fluid flows out of the sample. The sample instantaneously compacts by an amount proportional to K_u but then continues to contract as fluid drains out of the sample. After Roeloffs (1995).

Simultaneously, the pore pressure increases by an amount

$$p = -B\Delta\sigma_{kk}/3. \qquad (10.1)$$

Equation (10.1) introduces a new constant B known as *Skempton's pore pressure coefficient*. It is defined as the ratio of induced pore pressure to the mean normal stress for undrained loading. As time proceeds, fluid flows out of the sample, and the pore pressure will return to its initial value. In the fully drained state, $p = 0$. As fluid flows out, the rock becomes more compliant, and the volumetric strain gradually approaches $-\Delta\sigma_{kk}/3K$, where K is the ordinary, or *drained bulk modulus*.

Our next step is to generalize these concepts to appropriate tensor constitutive laws analogous to the generalized Hooke's law for elasticity. The theory was originally introduced by M. A. Biot (1941). Useful reviews include Rice and Cleary (1976), Roeloffs (1995), and Wang (2000).

10.1 Constitutive Laws

There are two ways of developing constitutive laws. One starts strictly from macroscopic continuum mechanics principles, the other from a micromechanical perspective. Both are useful; we will start here with the continuum perspective.

10.1.1 Macroscopic Description

We consider small changes in stress and pore-fluid pressure from an initially equilibrated state and thus restrict attention to linearized constitutive equations, analogous to Hooke's law. Biot (1941) first suggested that the strains are linear functions of the stresses and pore-fluid pressure. In the most general anisotropic case, we can write this as

$$\epsilon_{ij} = S_{ijkl}\sigma_{kl} + A_{ij}\,p, \tag{10.2}$$

where S_{ijkl} is the compliance tensor, the inverse of the C_{ijkl} in the generalized Hooke's law (1.100). Equation (10.2) introduces a second-rank tensor A_{ij} that relates strain and pore pressure. It should be noted that the pore pressure p and stress are properly defined as changes from an equilibrated state for which $\epsilon_{ij} = 0$. It proves awkward to always write δp; however, the reader should remember that $p = 0$ implies that the pore pressure remains constant, not that the pore pressure vanishes. In most cases, we will consider only isotropic materials, in which the first term on the right-hand side of equation (10.2) reduces to the usual isotropic form of Hooke's law. The most general isotropic form of the second term is $A_{ij} = constant \times \delta_{ij}$. Biot took this constant to be $1/H$. Here, for reasons that will become clear momentarily, we write this constant as $(1 - 2\nu)\alpha/2\mu(1 + \nu)$, so that

$$2\mu\epsilon_{ij} = \sigma_{ij} - \frac{\nu}{1+\nu}\sigma_{kk}\delta_{ij} + \frac{(1 - 2\nu)\alpha}{1+\nu}p\delta_{ij}. \tag{10.3}$$

We leave it as an exercise to show that with this form of the constitutive equations, equation (10.3) reduces to the standard form of Hooke's law for an *effective stress*, where the effective stress is defined to be $\sigma'_{ij} \equiv \sigma_{ij} + \alpha p\delta_{ij}$. (Note that we continue to use the extensional stress positive notation, which explains the plus sign in the definition of the effective stress.) For drained deformation $p=0$, equation (10.3) reduces to the isotropic form of Hooke's law. Thus, the elastic constants in equation (10.3) are drained constants. For isotropic materials (10.3), the shear strains are independent of changes in pore-fluid pressure. However, for a general anisotropic material, as in equation (10.2), changes in pore-fluid pressure can cause shear strains.

A complete constitutive description requires that we also account for the amount of fluid stored within a representative elementary volume of rock. Biot (1941) originally employed the *change in fluid content per unit volume of solid*; however, here we follow Rice and Cleary (1976) and employ the *mass of pore fluid per unit solid volume*, which we write as $m = \rho v$, where ρ is the fluid density, and v is the volume of the pore fluid per unit volume of the solid, the latter measured in the *reference state*. We further assume that the pores are fully saturated with fluid. (Note that the porosity is the volume of the pore space per unit volume of the solid in the *current* state.) Linearizing for small changes in m, we have

$$\Delta m = \rho_0 \Delta v + v_0 \Delta \rho,$$
$$= \rho_0 (\Delta v + v_0 \beta p), \tag{10.4}$$

where the fluid compressibility is defined as $\beta = (1/\rho)(\partial \rho / \partial p)$. The change in fluid volume is in general a function of the state variables σ_{ij} and p. Thus, in the general anisotropic case, we write

$$\Delta v = A'_{ij} \sigma_{ij} + \Omega p, \tag{10.5}$$

where Ω is a newly introduced scalar. We now show (following Biot [1941]) that the existence of an internal energy density guarantees that $A'_{ij} = A_{ij}$. Specifically, define $U(\epsilon_{ij}, \Delta v)$ to be the internal energy per unit volume, where that volume is measured in the reference state. (Note that the energy per unit mass is U/ρ_{r0}, where ρ_{r0} is the rock density, solid plus pore fluid, in the reference state.) Thus,

$$dU = \sigma_{ij}\, d\epsilon_{ij} + p\, dv. \tag{10.6}$$

Thermal contributions to the internal energy are not included here, but see section 10.3. Define a free energy $\Psi(\sigma_{ij}, p)$ as

$$\Psi = U - \sigma_{ij}\epsilon_{ij} - p\Delta v, \tag{10.7}$$

such that $d\Psi$ is given by

$$d\Psi = -\epsilon_{ij} d\sigma_{ij} - \Delta v\, dp,$$
$$= \left(\frac{\partial \Psi}{\partial \sigma_{ij}}\right)_p d\sigma_{ij} + \left(\frac{\partial \Psi}{\partial p}\right)_{\sigma_{ij}} dp. \tag{10.8}$$

Comparing, we see directly that

$$\epsilon_{ij} = -\left(\frac{\partial \Psi}{\partial \sigma_{ij}}\right)_p,$$
$$\Delta v = -\left(\frac{\partial \Psi}{\partial p}\right)_{\sigma_{ij}}. \tag{10.9}$$

Because $d\Psi$ is a total differential, the mixed partial derivatives must be equal:

$$\frac{\partial^2 \Psi}{\partial p \partial \sigma_{ij}} = \frac{\partial^2 \Psi}{\partial \sigma_{ij} \partial p}, \tag{10.10}$$

which leads to the conclusion that

$$\frac{\partial \epsilon_{ij}}{\partial p} = \frac{\partial \Delta v}{\partial \sigma_{ij}}, \tag{10.11}$$

and hence from equations (10.2) and (10.5), that $A'_{ij} = A_{ij}$. The isotropic form of (10.5) is thus

$$\Delta v = \frac{(1 - 2v)\alpha}{2\mu(1 + v)} \sigma_{kk} + \Omega p, \tag{10.12}$$

so that equation (10.4) becomes

$$\Delta m = \rho_0 \left[\frac{(1 - 2v)\alpha}{2\mu(1 + v)} \sigma_{kk} + (v_0 \beta + \Omega) p \right]. \tag{10.13}$$

For an undrained test, as in figure 10.3 (left), $\Delta m = 0$. Since Δm is the change in fluid mass per unit volume of rock, the latter *measured in the reference state*, the volumetric strain of the rock does not change Δm. We now use the definition of Skempton's coefficient (10.1) for undrained deformation to write equation (10.13) as

$$\Delta m = \frac{(1 - 2v)\alpha \rho_0}{2\mu(1 + v)} \left[\sigma_{kk} + \frac{3}{B} p \right]. \tag{10.14}$$

For isotropic poroelastic media, there are four material constants. In the form of equations (10.3) and (10.14), they are the shear modulus μ and drained Poisson's ratio v of classical elasticity, Skempton's coefficient B, and the Biot pore pressure coefficient α. For a variety of reasons, many other choices of material constants are used in the literature. Compare isotropic linear elasticity with isotropic linear poroelasticity. In the elastic case, there are only two independent constants, but at various times, it is convenient to use any two of the following: μ, λ, v, E, K. In the case of poroelasticity, there are four independent constants, giving rise to a veritable alphabet soup of material parameters. Kumpel (1991) and Wang (2000) provide more detail on the relationships between the various constitutive parameters.

One widely used parameter is the undrained Poisson's ratio v_u. We leave it as an exercise (problem 2) to show that α can be written in terms of v_u as

$$\alpha = \frac{3(v_u - v)}{B(1 + v_u)(1 - 2v)}. \tag{10.15}$$

Using equation (10.15) to replace α in (10.3) and (10.14) yields the widely cited Rice and Cleary (1976) form of the poroelastic constitutive equations:

$$2\mu\epsilon_{ij} = \sigma_{ij} - \frac{v}{1 + v} \sigma_{kk} \delta_{ij} + \frac{3(v_u - v)}{B(1 + v)(1 + v_u)} p\delta_{ij},$$

$$\Delta m = \frac{3\rho_0(v_u - v)}{2\mu B(1 + v)(1 + v_u)} \left(\sigma_{kk} + \frac{3}{B} p \right). \tag{10.16}$$

The solid volume strain can be related to changes in mean stress and pore pressure by taking the contraction of equation (10.3):

$$\epsilon_{kk} = \frac{\sigma_{kk}}{3K} + \frac{\alpha p}{K}.$$ (10.17)

Equation (10.17) reveals that if rock is free from constraints ($\sigma_{kk} = 0$), it will contract by an amount $\alpha p/K$ when subject to a decrease in pore pressure ($p < 0$). On the other hand, if the rock is perfectly constrained ($\epsilon_{kk} = 0$), it will be driven into tension $\sigma_{kk} = -3\alpha p$ by a reduction in pore pressure. An alternative form for the voumetric strain in terms of undrained parameters and fluid mass alterations can be found in problem 3.

10.1.2 Micromechanical Description

We now turn to a micromechanical description that relates the macroscopic constitutive parameters to the properties of the rock, its solid grains, and its pores. To do so, imagine a porous medium subjected to an external confining pressure and internal fluid pressure such that it is possible to experimentally alter the confining pressure by an amount $p_c = -\sigma_{kk}/3$ and the internal fluid pressure p_f, where we have in this section only added the subscript f on the fluid pressure. Following Berryman and Pride (2002; see also Zimmerman 1991), define the differential pressure as

$$p_d = p_c - p_f.$$ (10.18)

Denote the volume of the solid (including pores) V and the volume of the pore space V_ϕ. For small changes in pressure, then

$$-\frac{\delta V}{V} = -\epsilon_{kk} = \frac{p_d}{K} + \frac{p_f}{K_s},$$ (10.19)

where K is the drained bulk modulus, as seen by setting $p_f = 0$. For vanishing differential pressure, $p_d = 0$, $-\epsilon_{kk} = p_f/K_s$. Since the external and internal pressures are equal, the resulting volumetric strain is that of the solid components only, not the pores. This demonstrates that K_s is the bulk modulus of the solid grains making up the rock. For a pure quartz sandstone, for example, K refers to the sandstone, while K_s refers to the quartz grains.

We also introduce a similar relation for the change in pore volume:

$$-\frac{\delta V_\phi}{V_\phi} = \frac{p_d}{K_p} + \frac{p_f}{K_\phi},$$ (10.20)

where K_p and K_ϕ are two new moduli. Last, the volume change of the fluid phase is proportional to the fluid compressibility:

$$-\frac{\delta V_f}{V_f} = \frac{p_f}{K_f} = \beta p_f.$$ (10.21)

Note that equation (10.17) can be written as $\epsilon_{kk} = -p_c/K + \alpha p_f/K$. Comparing this to equation (10.19) yields

$$\boxed{\alpha = 1 - K/K_s}$$ (10.22)

(Nur and Byerlee 1971). Figure 10.5 shows that an effective pressure $p_c - \alpha p_f$, where α is given by equation (10.22) does a good job of explaining experimental data. In the limit that the drained

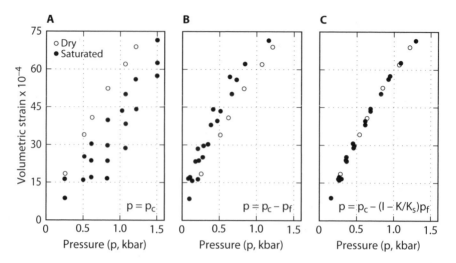

Figure 10.5. Laboratory measurements of volumetric strain as a function of different measures of pressure. The rock is Weber sandstone, with porosity of 6%. Open circles are dry, with no internal pore pressure. Solid circles are fluid saturated, with pore pressure p_f. In A, the pressure is confining pressure, p_c. In B, it is the Terzaghi effective stress $p_c - p_f$, whereas in C, it is the effective stress $p_c - \alpha p_f$. α is computed from equation (10.22), with K_s the bulk modulus of quartz and K the slope of the open circles in A. After Nur and Byerlee (1971).

bulk modulus of the rock is small compared to the bulk modulus of the solid constituents, as would be the case for unconsolidated or loosely consolidated materials, $K/K_s \to 0$ and $\alpha \to 1$, leading to the so-called *Terzaghi effective stress*, $p_c - p_f$.

To determine B in terms of the properties of the constituents, note that equation (10.4) can be written, in the current notation, as

$$\Delta m / \rho_0 = \delta V_\phi + \phi p_f / K_f, \tag{10.23}$$

where $\phi = v_0$ is the pore-volume fraction in the reference state. However, from equation (10.20), we have

$$\delta V_\phi = -\phi \frac{p_c}{K_p} + \phi \left(\frac{1}{K_p} - \frac{1}{K_\phi} \right) p_f. \tag{10.24}$$

Combining this with equation (10.23) and noting from equation (10.1) that for undrained deformation $p_f = B p_c$, we find that

$$B = \frac{1}{K_p} \left(\frac{1}{1/K_p + 1/K_f - 1/K_\phi} \right). \tag{10.25}$$

The dependence of δV_ϕ on confining pressure is, from equation (10.12), $-\alpha / K$. By comparison to equation (10.24), we see that

$$\frac{\phi}{K_p} = \frac{\alpha}{K}, \tag{10.26}$$

which can be used to eliminate K_p, leading to

$$B = \frac{1/K - 1/K_s}{1/K - 1/K_s + \phi \left(1/K_f - 1/K_\phi\right)} \qquad (10.27)$$

(e.g., Rice and Cleary 1976). For soils and unconsolidated rocks, $K_s \gg K$. In addition, for a relatively incompressible fluid such as water $\phi/K_f \ll 1/K$ and $\phi/K_\phi \ll 1/K$. In these limits, B tends toward unity. For hard rocks including sandstones, marbles, and granites, B ranges from 0.5 to 0.88 (Wang 2000, table C1).

In summary, it is possible to relate the two independent macroscopic constitutive parameters, either α and B, or ν_u and B, to the intrinsic moduli K, K_s, K_p, and K_ϕ.

10.2 Field Equations

The quasi-static equilibrium equations are (equation [1.97]),

$$\frac{\partial \sigma_{ij}}{\partial x_j} + f_i = 0. \qquad (10.28)$$

Stress compatibility conditions are established by introducing the constitututive equations (10.3) into the strain compatibility relations (1.83). Making use of the equilibrium equations, this yields equations analogous to the Beltrami-Michell equations in elasticity:

$$\sigma_{ij,kk} + \frac{1}{1+\nu}\sigma_{kk,ij} - \frac{\nu}{1+\nu}\sigma_{kk,mn}\delta_{ij} + \frac{(1-2\nu)\alpha}{1+\nu}(p_{,kk}\delta_{ij} + p_{,ij}) = -(f_{i,j} + f_{j,i}). \qquad (10.29)$$

Taking the contraction of equation (10.29) yields

$$\nabla^2 \sigma_{kk} + \frac{2\alpha(1-2\nu)}{(1-\nu)}\nabla^2 p = -\left(\frac{1+\nu}{1-\nu}\right)f_{k,k}, \qquad (10.30)$$

which will soon prove useful.

Conservation of fluid mass requires

$$\frac{\partial q_i}{\partial x_i} + \frac{\partial m}{\partial t} = 0, \qquad (10.31)$$

where q_i is fluid mass flux. Darcy's law relating fluid mass flux to pore pressure gradients is

$$q_i = -\rho_0 \frac{\kappa_{ij}}{\eta}\left(\frac{\partial p}{\partial x_j} - \rho_0 g \delta_{3j}\right), \qquad (10.32)$$

where κ_{ij} is the permeability tensor, with dimensions of length squared, and η is fluid viscosity. It is assumed here that gravity acts in the 3-direction. Combining equations (10.32) and (10.31) yields

$$\frac{\rho_0}{\eta}\frac{\partial}{\partial x_i}\left[\kappa_{ij}\left(\frac{\partial p}{\partial x_j} - \rho_0 g \delta_{3j}\right)\right] = \frac{\partial m}{\partial t}. \qquad (10.33)$$

Before continuing further, it is worthwhile to consider the limiting case of fluid flow through a nondeforming, rigid rock mass. With $\Delta v = 0$, $dm/dt = \rho_0\phi\beta dp/dt$, where the pore volume fraction in the reference state is identified with porosity, ϕ. Combined with equation (10.33), this yields

$$\frac{1}{\phi\eta\beta}\frac{\partial}{\partial x_i}\left[\kappa_{ij}\left(\frac{\partial p}{\partial x_j} - \rho_0 g\delta_{3j}\right)\right] = \frac{\partial p}{\partial t}. \tag{10.34}$$

Equation (10.34) is widely used in modeling fluid flow in situations where the deformation of the rock is safely neglected. For homogeneous and isotropic permeability, equation (10.34) reduces to a homogeneous diffusion equation with hydraulic diffusivity given by

$$c = \frac{\kappa}{\phi\eta\beta}, \tag{10.35}$$

where κ is the scalar permeability. We now return to the more general poroelastic case. In what follows, we assume that the permeability is isotropic and homogeneous and ignore the hydrostatic term $\rho_0 g\delta_{3j}$. Thus, equation (10.33) reduces to

$$\frac{\rho_0\kappa}{\eta}\nabla^2 p = \frac{\partial m}{\partial t}. \tag{10.36}$$

Taking the Laplacian of equation (10.14) and combining with the compatibility equation (10.30) yields an expression relating pore pressure and fluid mass alterations:

$$\nabla^2 m = \frac{(1-2\nu)\alpha\rho_0}{2\mu(1+\nu)}\left[\frac{3(1-\nu)-2\alpha B(1-2\nu)}{B(1-\nu)}\nabla^2 p - \frac{1+\nu}{1-\nu}f_{k,k}\right]. \tag{10.37}$$

Using this expression to eliminate p from equation (10.36) leads to a diffusion equation in fluid mass content:

$$c\nabla^2 m = \frac{\partial m}{\partial t} - \Gamma f_{k,k} \tag{10.38}$$

(Biot 1941). Note that the hydraulic diffusivity, which is given by

$$c = \frac{\kappa}{\eta}\frac{2\mu(1-\nu)}{(1-2\nu)}\left[\frac{B(1+\nu)}{3\alpha(1-\nu)-2B\alpha^2(1-2\nu)}\right], \tag{10.39}$$

is more complex than in the case of a rigid rock skeleton but is still proportional to the permeability and inversely proportional to the fluid viscosity (Rice and Cleary 1976). The divergence of the body forces appears as a source term in the diffusion equation, where

$$\Gamma = \frac{\kappa\rho}{\eta}\left[\frac{B(1+\nu)}{3(1-\nu)-2B\alpha(1-2\nu)}\right]. \tag{10.40}$$

As noted by Rice and Cleary (1976), equation (10.14) can be introduced into (10.38), yielding a diffusion equation in a linear combination of pore pressure and mean normal stress:

$$c\nabla^2\left(\sigma_{kk} + \frac{3}{B}p\right) = \frac{\partial}{\partial t}\left(\sigma_{kk} + \frac{3}{B}p\right),$$

(10.41)

where here we have omitted the body force term.

Before moving on, it is worth considering where we stand in terms of variables and governing equations. There are a total of 20 field variables σ_{ij}, ϵ_{ij}, u_i, p, Δm, and q_i, all functions of position. Fortunately, there are an equal number of equations: 3 equilibrium equations (10.28), 1 equation for conservation of fluid mass (10.31), 6 strain-displacement relations (1.4), 7 constitutive equations for the solid (10.3) and fluid (10.14) phase, and 3 equations representing Darcy's law (10.32).

The equilibrium equations can be cast in terms of displacements and pore pressure. The kinematic relations between strain and displacement, together with equation (10.3), allow us to write the stress in terms of the pore pressure and displacements as

$$\sigma_{ij} = \mu\left(\frac{\partial u_i}{\partial x_j} + \frac{\partial u_j}{\partial x_i}\right) + \frac{2\mu v}{1-2v}\frac{\partial u_k}{\partial x_k}\delta_{ij} - \alpha p\delta_{ij}.$$

(10.42)

Substituting this expression into the equilibrium equations (10.28) yields

$$\mu\nabla^2 u_i + \frac{\mu}{(1-2v)}\frac{\partial^2 u_j}{\partial x_i \partial x_j} - \alpha\frac{\partial p}{\partial x_i} + f_i = 0.$$

(10.43)

Equation (10.43) illustrates that gradients in pore pressure enter the equilibrium equations like body forces. If we somehow knew the pore pressure distribution, we could solve for the deformation field by inserting the pore-fluid pressure gradients into equation (10.43). The recognition that the pore pressure gradients appear in the equilibrium equations as body forces suggests a Green's function approach, in which we solve for the displacements resulting from an equivalent body force distribution given by $f_i = -\alpha\partial p/\partial x_i$ (see section 10.8). However, it should be emphasized that in general, the pore-fluid pressure is unknown and must be solved for together with the solid deformations.

Using equation (10.17) to replace the mean normal stress with the volumetric strain in equation (10.41) yields

$$c\nabla^2(u_{k,k} - \omega p) = \frac{\partial}{\partial t}(u_{k,k} - \omega p),$$

(10.44)

where $\omega = (\alpha B - 1)/KB$. It can be shown (see problem 4) that equation (10.44) can be written in the form of a diffusion equation in which the divergence of the displacements acts as a source term:

$$S_\alpha \frac{\partial p}{\partial t} - \frac{\kappa}{\eta} \nabla^2 p = -\alpha \frac{\partial}{\partial t} (\nabla \cdot \mathbf{u}), \tag{10.45}$$

where the storage coefficient S_α is given by equation (10.117).

Equations (10.43) and (10.44) or (10.45) represent four equations in the unknowns u_i and p. The critical point here is that *the equilibrium and flow equations are fully coupled.* Physically, this arises because, as noted at the beginning of this chapter, deformation induces pore pressure changes and conversely gradients in pore pressure induce deformation. Mathematically, this arises because $\partial p/\partial x_i$ appears in the equilibrium equations (10.43), while $\nabla \cdot \mathbf{u}$ appears in the pore pressure diffusion equation. There are some special situations in which the governing equations uncouple, greatly facilitating solution. Unfortunately, this is not generally the case.

10.3 Analogy to Thermoelasticity

As anyone who has put a non-ovensafe dish on the stove knows, temperature changes also induce deformation. Assuming small changes in stress and temperature, the strains can be written as linear functions of these variables. Thus, the constitutive equations for linear thermoelastic medium are

$$2\mu\epsilon_{ij} = \sigma_{ij} - \frac{\nu}{1+\nu}\sigma_{kk}\delta_{ij} + 2\mu\Lambda_{ij}(T - T_0). \tag{10.46}$$

In the isotropic case, $\Lambda_{ij} = \Lambda\delta_{ij}$, where Λ is the coefficient of linear thermal expansion. Compare these equations with (10.3). In poroelasticity, it is necessary to keep track of the pore-fluid content per unit volume of rock. In thermoelasticity, the analogous quantity is entropy per unit volume (where again that volume is measured in the reference state), which can be taken as a function of stress and temperature $s(\sigma_{ij}, T)$. For small changes, ds can be approximated as a linear function of the stress and temperature, paralleling equation (10.5):

$$T ds = T\Upsilon_{ij} d\sigma_{ij} + \rho_{r0}c_p dT, \tag{10.47}$$

where $\Upsilon_{ij} = (\partial s/\partial \sigma_{ij})_T$, and $c_p = T/\rho_{r0}(\partial s/\partial T)_{\sigma_{ij}}$ is the specific heat capacity at constant stress. The change in internal energy is

$$dU = \sigma_{ij} d\epsilon_{ij} + T ds \tag{10.48}$$

(e.g., Malvern 1969). Defining a free energy as $U - \sigma_{ij}\epsilon_{ij} - sT$, and following a parallel argument to that in section 10.1.1, one can show that in the isotropic case equation (10.47) reduces to

$$T ds = \Lambda T d\sigma_{kk} + \rho_{r0}c_p dT. \tag{10.49}$$

The first law of thermodynamics in the limit of small strains is

$$\frac{\partial U}{\partial t} = \sigma_{ij}\frac{\partial \epsilon_{ij}}{\partial t} - \nabla \cdot \mathbf{q}_h \tag{10.50}$$

(e.g, Boley and Weiner 1960; Malvern 1969), where $\mathbf{q}_h$ is the heat flux. Comparing equation (10.50) to the time derivative of (10.48) leads to

$$\nabla \cdot \mathbf{q}_h + T\frac{\partial s}{\partial t} = 0. \tag{10.51}$$

Fourier's law for heat conduction, analogous to Darcy's law, is

$$\mathbf{q}_h = -k_t \nabla T, \tag{10.52}$$

where k_t is the thermal conductivity. Combining equations (10.52) and (10.49) with (10.51) yields the heat conduction equation:

$$\frac{\partial}{\partial x_i}\left(k_{ij}\frac{\partial T}{\partial x_j}\right) = \rho_{r0}c_p\frac{\partial T}{\partial t} + T\Lambda\frac{\partial \sigma_{kk}}{\partial t}. \tag{10.53}$$

The primary difference between thermoelasticity and poroelasticity is that with thermoelasticity, the coupling term, the second term on the right-hand side of equation (10.53), is generally negligible. The ratio of second to third terms in equation (10.53) is $\rho_{r0}c_p\Delta T / T\Lambda\Delta\sigma_{kk}$. Typical values of parameters are: $\rho_{r0}c_p\sim 3$ MPa/°C and $\Lambda\sim 10^{-5}$/°C. Thus, for $\Delta T/T\sim 0.1$ and $\Delta\sigma_{kk} = 10$ MPa, the ratio of the second to third terms is 3×10^3. In plain terms, deforming rock elastically does not cause a significant increase in the rock's temperature. On the other hand, compressing rock does significantly increase the pore pressure. Because the thermoelastic coupling term is normally safely neglected, one can generally first solve for the temperature distribution and then solve for the stresses and displacement given that temperature field. As in equations (10.43), the gradient in temperature appears in the equilibrium equations like a body force.

More generally, one should consider the effects of temperature change on the pore-fluid phase. While coupled thermoporoelasticity is beyond our present scope, the interested reader may consult Palciauskas and Domenico (1982) and McTigue (1986) and references therein. Obvious applications of a more complete theory include volcanic and geothermal settings.

10.4 One-Dimensional Deformation

There are not many poroelastic solutions available in the literature for faults or magmatic sources. Those that are known are typically quite complex. To get a flavor for the behavior of poroelastic systems, we consider the classic problem of a uniform load suddenly introduced on the earth's surface. Deformation occurs only in the vertical z direction. In this case, the diffusion equation uncouples from the deformation equations.

In two-dimensional plane strain ($\epsilon_{yy} = 0$), the equilibrium equations, without body forces, are

$$\frac{\partial \sigma_{xx}}{\partial x} + \frac{\partial \sigma_{xz}}{\partial z} = 0, \tag{10.54}$$

$$\frac{\partial \sigma_{xz}}{\partial x} + \frac{\partial \sigma_{zz}}{\partial z} = 0. \tag{10.55}$$

The stress normal to the plane is determined by the plane strain condition, $\epsilon_{yy} = 0$, such that

$$\sigma_{yy} = v(\sigma_{xx} + \sigma_{zz}) - \frac{3(v_u - v)}{B(1 + v_u)}p. \tag{10.56}$$

Also for plane strain, the diffusion equation (10.41) reduces to

$$\left(c\nabla^2 - \frac{\partial}{\partial t}\right)\left[\sigma_{xx} + \sigma_{zz} + \frac{3}{B(1+\nu_u)}p\right] = 0. \tag{10.57}$$

For one-dimensional deformation in which $\partial/\partial x = 0$, the equilibrium equations require that the shear stress and vertical normal stress are independent of depth z. Thus, since the shear stress vanishes at the free surface, it must vanish everywhere. The vertical stress is constant and equal to the load applied on the free surface.

For deformation in the z direction only, $\epsilon_{xx} = 0$, so that from equation (10.56), the inplane horizontal normal stress is given by

$$\sigma_{xx} = \frac{\nu}{(1-\nu)}\sigma_{zz} - \frac{3(\nu_u - \nu)}{B(1+\nu_u)(1-\nu)}p. \tag{10.58}$$

Taking advantage of equation (10.58) and the equilibrium equations for one-dimensional deformation, equation (10.57) reduces to

$$\frac{3c(1-\nu_u)}{B(1+\nu_u)}\frac{\partial^2 p}{\partial z^2} = \frac{\partial}{\partial t}\left[\frac{3(1-\nu_u)}{B(1+\nu_u)}p + \sigma_{zz}\right]. \tag{10.59}$$

10.4.1 Step Load on the Free Surface

Consider a homogeneous and isotropic poroelastic half-space. At time $t = 0$, a uniform load F is suddenly introduced at $z = 0$ and held constant:

$$\sigma_{zz}(z = 0, t) = -F H(t), \tag{10.60}$$

where $H(t)$ is the Heavyside function. We consider the surface $z = 0$ to be permeable, so the pore pressure at the free surface does not change—that is,

$$p(z = 0, t) = 0. \tag{10.61}$$

Taking the Laplace transform (A.13) of equation (10.59) yields

$$\frac{d^2\tilde{p}}{dz^2} - \frac{s}{c}\tilde{p} = -F\frac{\gamma}{c}, \tag{10.62}$$

where s is the transform variable, $\tilde{p}$ is the transformed pressure, and

$$\gamma = \frac{B(1+\nu_u)}{3(1-\nu_u)}. \tag{10.63}$$

The solution to equation (10.62) that fits the boundary condition at the free surface and vanishes at $z \to \infty$ is

$$\tilde{p} = \frac{F\gamma}{s}\left[1 - \exp\left(-\sqrt{\frac{s}{c}}z\right)\right]. \tag{10.64}$$

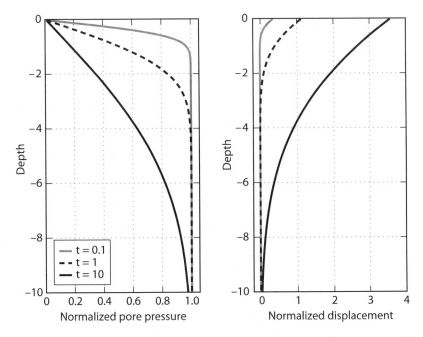

Figure 10.6. Solution for one-dimensional deformation with a unit vertical load imposed at the free surface at $t = 0$. Left: Pore pressure as a function of depth for $c = 1$. Pore pressure is normalized by the undrained response $B(1 + v_u)/3(1 - v_u)$. Right: Vertical displacement as a function of depth for $c = 1$. Displacement is normalized by $(v_u - v)/\mu(1 - v)(1 - v_u)$.

Inverting the Laplace transform (e.g., Mathews and Walker 1970, pp. 240–242) gives

$$p(z, t) = F \frac{B(1 + v_u)}{3(1 - v_u)} \operatorname{erf}\left(\frac{z}{\sqrt{4ct}}\right). \tag{10.65}$$

(For an alternative derivation, see problem 6.) The pore pressure is shown as a function of depth in figure 10.6. The application of the load suddenly increases the pore pressure everywhere by $F B(1 + v_u)/3(1 - v_u)$—that is, the undrained response. With time, fluid flows to the free surface and the pressure decline diffuses into the earth.

The displacements can be computed by first solving for the vertical strain:

$$\epsilon_{zz} = \frac{1 - 2v}{2\mu(1 - v)}\sigma_{zz} + \frac{3(v_u - v)}{2\mu B(1 + v_u)(1 - v)}p,$$

$$= -F\left[\frac{1 - 2v}{2\mu(1 - v)} - \frac{(v_u - v)}{2\mu(1 - v)(1 - v_u)}\operatorname{erf}\left(\frac{z}{\sqrt{4ct}}\right)\right]. \tag{10.66}$$

The strain includes an instantaneous elastic part and a time-dependent part. The elastic strain is uniform and integrates to an infinite displacement. We are primarily interested in the time-dependent response. Subtracting the instantaneous strain and integrating from infinite depth to depth z, we find

$$u_z = -\frac{(v_u - v)}{2\mu(1 - v)(1 - v_u)}F\int_{\infty}^{z}\operatorname{erfc}\left(\frac{z'}{\sqrt{4ct}}\right)dz', \tag{10.67}$$

where erfc$(\cdot) = 1 - $ erf$(\cdot)$ is the complementary error function. A change of variables and $\zeta = z'/(4ct)^{1/2}$ yields

$$u_z = \frac{(\nu_u - \nu)\sqrt{ct}}{\mu(1 - \nu)(1 - \nu_u)} F \int_{z/\sqrt{4ct}}^{\infty} \text{erfc}(\zeta) \, d\zeta. \tag{10.68}$$

The integral of the complementary error function is written $i$erfc, where

$$i\text{erfc}(z) = \frac{1}{\sqrt{\pi}} e^{-z^2} - z \, \text{erfc}(z) \tag{10.69}$$

(Abramowitz and Stegun 1972). From this, we derive

$$u_z(z, t) = F \frac{(\nu_u - \nu)\sqrt{ct}}{\mu(1 - \nu)(1 - \nu_u)} \left[\frac{1}{\sqrt{\pi}} \exp\left(-\frac{z^2}{4ct}\right) - \frac{z}{\sqrt{4ct}} \text{erfc}\left(\frac{z}{\sqrt{4ct}}\right) \right]. \tag{10.70}$$

Note that at the free surface, this reduces to

$$u_z(z = 0, t) = F \frac{(\nu_u - \nu)}{\mu(1 - \nu)(1 - \nu_u)} \sqrt{\frac{ct}{\pi}}, \tag{10.71}$$

so that the displacement increases with $\sqrt{t}$. The displacement as a function of depth and time is shown in figure 10.6.

10.4.2 Time-Varying Fluid Load on the Free Surface

Roeloffs (1988) considered the case of a periodically varying fluid pressure load on the free surface. This could be taken to be a reservoir with seasonal variations in water level. In this case, the boundary condition is given in terms of $p(z = 0, t)$. The vertical stress at the free surface is given by $\sigma_{zz}(z = 0, t) = -p(z = 0, t)$. We thus return to equation (10.59), and look for solutions that are harmonic in time:

$$\sigma_{zz} = -p_0 e^{i\omega t},$$
$$p(z, t) = \tilde{p}(z) e^{i\omega t}. \tag{10.72}$$

Equation (10.59) reduces to

$$\frac{d^2}{dz^2} \frac{\tilde{p}}{p_0} - \frac{i\omega}{c} \frac{\tilde{p}}{p_0} = -\frac{i\omega\gamma}{c}. \tag{10.73}$$

The solution to equation (10.73) satisfying the boundary condition is

$$\frac{\tilde{p}}{p_0} = (1 - \gamma) \exp\left[-(1 + i)\sqrt{\frac{\omega}{2c}} z\right] + \gamma. \tag{10.74}$$

The behavior of equation (10.74) is best understood in terms of the magnitude and phase of the pore pressure change, as shown in figure 10.7. For the uncoupled case $\gamma = 0$, the magnitude of the pore pressure change decreases monotonically with depth. However, for larger values of γ, the magnitude of the pore pressure reaches a minimum. In all cases, the asymptotic limit as $z \to \infty$ is γ, as is easily seen from equation (10.74). The peak pore pressure at depth lags the surface load over the normalized depth range $0 < (\omega/2c)^{1/2} z < \pi$, but leads the surface pore pressure change over the depth range $\pi < (\omega/2c)^{1/2} z < 2\pi$.

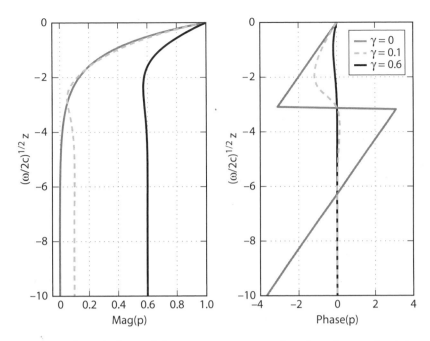

Figure 10.7. Magnitude and phase of the pore pressure as a function of normalized depth for an oscillatory pore pressure applied at the free surface assuming one-dimensional deformation. The magnitude of the pore pressure is normalized to the maximum surface pressure, where the phase is relative to the phase of the applied load. The parameter γ is defined by equation (10.63).

10.5 Dislocations in Two Dimensions

As has been emphasized, poroelastic solutions in even two dimensions are considerably more difficult than their elastic counterparts. The exception of course is antiplane strain, which we employed much earlier as a model for an infinitely long strike-slip fault. In this case, fault slip induces no change in mean normal stress, so there is no induced pore pressure change and no poroelastic effect.

Rice and Cleary (1976) used complex potential methods to derive the plane strain solution for a two-dimensional edge dislocation in a poroelastic medium. (This problem has also been explored using displacement potentials, similar to the method used in section 10.6.) Rice and Cleary (1976) found that

$$\sigma_{\theta\theta} = \frac{\mu \Delta u (\nu_u - \nu)}{2\pi (1 - \nu_u)(1 - \nu)} \frac{\sin(\theta)}{r} \left[2e^{-r^2/4ct} - \frac{1 - \nu}{\nu_u - \nu} - \frac{4ct}{r^2} \left(1 - e^{-r^2/4ct} \right) \right],$$

$$\sigma_{r\theta} = \frac{\mu \Delta u (\nu_u - \nu)}{2\pi (1 - \nu_u)(1 - \nu)} \frac{\cos(\theta)}{r} \left[\frac{1 - \nu}{\nu_u - \nu} - \frac{4ct}{r^2} \left(1 - e^{-r^2/4ct} \right) \right],$$

$$\sigma_{rr} = \frac{\mu \Delta u (\nu_u - \nu)}{2\pi (1 - \nu_u)(1 - \nu)} \frac{\sin(\theta)}{r} \left[\frac{4ct}{r^2} \left(1 - e^{-r^2/4ct} \right) - \frac{1 - \nu}{\nu_u - \nu} \right],$$

$$p = \frac{\mu \Delta u B (1 + \nu_u)}{3\pi (1 - \nu_u)} \frac{\sin(\theta)}{r} \left[1 - e^{-r^2/4ct} \right], \tag{10.75}$$

where Δu is the fault slip, and r, θ are a polar coordinate system centered at the dislocation end (figure 10.8). Superposition of two edge dislocations of opposite sign approximates a fault in a

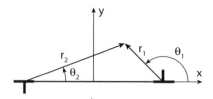

Figure 10.8. Definition sketch for a pair of dislocations in a two-dimensional poroelastic medium. A polar coordinate system is erected at each dislocation.

full space, with depth much greater than the fault length. It is apparent from these expressions (10.75) that the total stress is time varying, not simply the effective stress.

The strains are readily computed from the constitutive equations:

$$\epsilon_{\theta\theta} = \frac{-\Delta u(v_u - v)}{4\pi(1 - v_u)(1 - v)} \frac{\sin(\theta)}{r} \left[\frac{(1 - v)(1 - 2v_u)}{v_u - v} + \frac{4ct}{r^2} \left(1 - e^{-r^2/4ct} \right) \right],$$

$$\epsilon_{rr} = \frac{-\Delta u(v_u - v)}{4\pi(1 - v_u)(1 - v)} \frac{\sin(\theta)}{r} \left[\frac{(1 - v)(1 - 2v_u)}{v_u - v} + 2e^{-r^2/4ct} - \frac{4ct}{r^2} \left(1 - e^{-r^2/4ct} \right) \right],$$

$$\epsilon_{r\theta} = \frac{\Delta u(v_u - v)}{4\pi(1 - v_u)(1 - v)} \frac{\cos(\theta)}{r} \left[\frac{1 - v}{v_u - v} - \frac{4ct}{r^2} \left(1 - e^{-r^2/4ct} \right) \right]. \tag{10.76}$$

The displacements can be determined following the procedure laid out in problem 9 of chapter 3. After considerable computation, this yields

$$u_r = \frac{-b(v_u - v)\sin(\theta)}{4\pi(1 - v_u)(1 - v)} \left[\frac{(1 - v)(1 - 2v_u)}{v_u - v} \ln(r) + \frac{2ct}{r^2} \left(1 - e^{-r^2/4ct} \right) - \frac{1}{2} E_1 \left(r^2/4ct \right) \right],$$

$$+ \frac{b}{2\pi} \theta \cos\theta$$

$$u_\theta = \frac{b(v_u - v)\cos(\theta)}{4\pi(1 - v_u)(1 - v)} \left\{ \frac{(1 - v)(1 - 2v_u)}{v_u - v} [1 - \ln(r)] + \frac{2ct}{r^2} \left(1 - e^{-r^2/4ct} \right) + \frac{1}{2} E_1 \left(r^2/4ct \right) \right\}$$

$$- \frac{b}{2\pi} (\theta \sin\theta + \cos\theta), \tag{10.77}$$

where the exponential integral $E_1(z)$ is defined as follows:

$$E_1(z) = \int_z^\infty \frac{e^{-t}}{t} dt. \tag{10.78}$$

The reader should verify that these displacements yield the strains in equation (10.76), as well as the proper drained and undrained limits. Note that the derivative $dE_1(z)/dz = -e^{-z}/z$. Limiting values are given by $E_1(z \to \infty) \to 0$, and $E_1(z \to 0) \to -\gamma' - \ln z$, where γ' is Euler's constant $\gamma' = 0.57721\ \ldots$. The fully drained limit, $t \to \infty$, yields terms (such as $\ln[4ct]$) that are not in the elastic solution; however, these terms do not contribute to the strain.

The displacements are plotted in figure 10.9 for one particular location near a model fault. Notice that both components of displacement initially equal the undrained response and asymptotically approach the drained state, as they must. However, the behavior, at

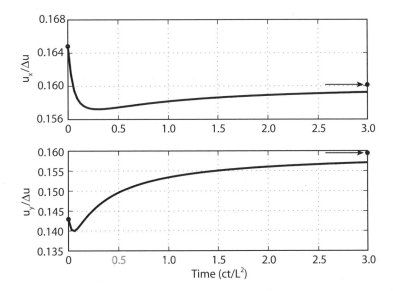

Figure 10.9. Time-dependent displacements for an edge dislocation with unit slip from equation (10.77). The geometry is shown in figure 10.8. The fault is centered at the origin and has length of 2.0. Displacement is shown for point at $x = 2.5$, $y = 1$. The circles denote the undrained and drained displacements. $\nu_u = 0.36$, $\nu = 0.2$.

this position, is not monotonic. The fault-parallel component u_x displacement decreases as one would expect, given the difference between the drained and undrained Poisson's ratio; however, it reaches a minimum and approaches the infinite time asymptote from below. The fault-normal component u_y is perhaps even more surprising. Even though the drained value is greater than the undrained value, the displacement first decreases before increasing. While neither component of displacement is monotonic at this particular location, it is at other locations for this model.

These results indicate that it is not safe to assume that the deformation varies monotonically from the undrained to the drained states in more complex situations where analytical solutions do not exist. For thrust and normal faults sufficiently long along-strike that they can be represented with edge dislocations, it is possible to show that the undrained and drained displacements on the free surface are identical (see problem 10). However, this cannot be taken to ensure that there are no transient deformations.

Results (10.75), (10.76), and (10.77) are for a permeable fault plane; note that the pore pressure in equation (10.75) vanishes on the fault plane. The fault-normal pore pressure gradient, however, is nonzero, which by Darcy's law implies fluid flow across the fault. Rudnicki (1987) analyzed the corresponding problem of a dislocation suddenly introduced on an impermeable fault. Rudnicki and Roeloffs (1990) give the full stress and pore pressure distributions for plane strain dislocations moving at a steady rate on both permeable and impermeable faults.

10.6 Inflating Magma Chamber in a Poroelastic Half-Plane

There are very few analytical poroelastic solutions, particularly with a free surface, which is unfortunate because data are generally collected at or near the earth's surface. There is one class of problems that is tractable with a free surface: those involving a spherically symmetric

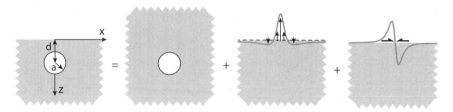

Figure 10.10. The solution for an expanding cylindrical source can be constructed as the sum of three solutions: the full-space solution for an expanding source plus distributed normal and shear loads. Because the expanding magma chamber in a full-space is a pure shear source, only the second and third contributions are time dependent.

volume source. As we found in chapter 7, it is possible to construct the solution for a spherical magma chamber in a half-space starting with the solution for a full-space and then applying tractions on the plane $z = 0$ that negate the induced shear and normal stresses (figure 10.10). The stress field due to the spherical pressure source in a full-space is a pure shear; the mean normal stress is everywhere zero. This special characteristic allows us to construct a solution for a spherical magma chamber in a poroelastic half-space.

We start with the solution for an expanding magma chamber in a full-space. The mean stress and therefore the induced pore pressure is zero; the poroelastic problem at this stage is no different from a drained elastic solution. We then compute the tractions acting on the putative free surface, $z = 0$. Since there are no induced pore pressures, these stresses are time invariant. We can thus add shear and normal tractions on a poroelastic half-space that exactly cancel the imposed stresses. The problem thus reduces to solving for the displacements caused by appropriately distributed shear and normal loads on a poroelastic half-space. As in chapter 7, this solution matches the boundary conditions on the free surface exactly but those on the boundaries of the magma chamber only approximately. The solution should be accurate as long as the depth of the source is substantially greater than its radius.

In this analysis, there is no transfer of magmatic fluids into the surrounding crust. We simply consider pore-fluid flow induced by deformation. This is one limiting case in which all magmatic fluids reside within the magma chamber. An alternative model would involve leakage of magmatic fluids into the crust.

For simplicity here, we actually treat the two-dimensional plane strain analogue of the Mogi problem. The source is an infinitely long cylinder extending in the y direction. The radius of the cavity is a, and its depth from the free surface is d. The magma chamber expands by an amount ΔV per unit length in the y direction. Adopting a coordinate system with z downward, the boundary conditions are

$$\sigma_{zz}(z = 0, t \geq 0) = 0,$$

$$\sigma_{xz}(z = 0, t \geq 0) = 0,$$

$$p(z = 0, t \geq 0) = 0. \tag{10.79}$$

Following the procedure described in chapter 7, we construct the solution for an expanding cylindrical source in a full-space. The displacements decay as $1/r$, where r is the distance from the source. On the plane $z = 0$, the vertical component of the displacement is

$$u_z(z = 0) = -\frac{\Delta V}{2\pi} \frac{d}{d^2 + x^2}. \tag{10.80}$$

The normal N and shear T tractions acting on the plane $z = 0$ resulting from expansion of the chamber in a full-space are

$$N = \frac{\mu \Delta V}{\pi} \frac{x^2 - d^2}{(d^2 + x^2)^2},$$

$$T = \frac{2\mu \Delta V}{\pi} \frac{xd}{(d^2 + x^2)^2}. \tag{10.81}$$

There are a variety of methods for solving plane poroelasticity problems using both stress and displacement potentials (Wang 2000). We find it convenient here to employ displacement potentials as introduced by McNamee and Gibson (1960a; 1960b) and discussed further by Wang (2000). As shown by Wang (2000), these methods all derive from a generalization of the Boussinesq-Papkovitch potentials in classical elasticity theory introduced by Biot (1956). Define displacement functions, E and S, such that

$$u_x = -\frac{\partial E}{\partial x} + z\frac{\partial S}{\partial x},$$

$$u_z = -\frac{\partial E}{\partial z} + z\frac{\partial S}{\partial z} - (3 - 4v_u)S,$$

$$p = -\frac{2\mu(1 - v)}{(1 - 2v)\alpha}\left[\nabla^2 E - \frac{2(v_u - v)}{1 - v}\frac{\partial S}{\partial z}\right]. \tag{10.82}$$

Introducing these expressions into the plane strain Navier form of the equilibrium equations (10.43) and the plane strain diffusion equation (10.57) leads to uncoupled equations in the two potentials:

$$\nabla^2 S = 0,$$

$$c\nabla^4 E = \frac{\partial}{\partial t}\nabla^2 E, \tag{10.83}$$

where ∇^2 here indicates the two-dimensional Laplacian operator, and ∇^4 the biharmonic operator. The general solution to the biharmonic diffusion equation can be written in terms of the sum of two potentials $E = E_1 + E_2$ such that

$$c\nabla^2 E_1 = \frac{\partial E_1}{\partial t},$$

$$\nabla^2 E_2 = 0,$$

$$\nabla^2 S = 0. \tag{10.84}$$

Introducing equation (10.82) into the constitutive equations (10.42) leads to expressions for the stresses in terms of the potentials

$$\sigma_{xx} = 2\mu \left[\frac{\partial^2 E}{\partial z^2} + z\frac{\partial^2 S}{\partial x^2} - 2\nu_u \frac{\partial S}{\partial z} \right],$$

$$\sigma_{zz} = 2\mu \left[\frac{\partial^2 E}{\partial x^2} + z\frac{\partial^2 S}{\partial z^2} - 2(1-\nu_u)\frac{\partial S}{\partial z} \right],$$

$$\sigma_{xz} = 2\mu \left[-\frac{\partial^2 E}{\partial x\partial z} + z\frac{\partial^2 S}{\partial x\partial z} - (1-2\nu_u)\frac{\partial S}{\partial x} \right] \tag{10.85}$$

(e.g., Wang 2000).

In order to solve the differential equations (10.84), Fourier transform in the x direction and Laplace transform in time (see appendix A), reducing the partial differential equations to ordinary differential equations in the depth variable z:

$$\left(\frac{\partial^2}{\partial z^2} - k^2 - \frac{s}{c} \right) E_1 = 0,$$

$$\left(\frac{\partial^2}{\partial z^2} - k^2 \right) E_2 = 0,$$

$$\left(\frac{\partial^2}{\partial z^2} - k^2 \right) S = 0, \tag{10.86}$$

where k is the horizontal wavenumber, and s is the Laplace transform variable. The solutions to equations (10.86) that decay with depth z are

$$E_1 = A_1 e^{-nz},$$

$$E_2 = A_2 e^{-|k|z},$$

$$S = A_3 e^{-|k|z}, \tag{10.87}$$

where

$$n = \sqrt{k^2 + \frac{s}{c}}. \tag{10.88}$$

We also Fourier transform the boundary conditions so that the the normal load is $\bar{N}(k)$ and the shear load is $\bar{T}(k)$. Assuming an instantaneous expansion of the magma chamber, the boundary conditions for the normal loading problem in the Fourier-Laplace domain are

$$\bar{\sigma}_{zz}(z=0) = -\frac{\bar{N}(k)}{s},$$

$$\bar{\sigma}_{xz}(z=0) = 0,$$

$$\bar{p} = 0. \tag{10.89}$$

The three constants A_1, A_2, A_3 are determined by the three boundary conditions (10.89). Solving the resulting system of equations leads to

$$A_1 = -\frac{c\gamma_2 \bar{N}(k)}{\mu s D},$$

$$A_3 = \frac{\bar{N}(k)}{2\mu|k|D},$$

$$A_2 = -\frac{n}{|k|}A_1 + \frac{1-2\nu_u}{|k|}A_3, \tag{10.90}$$

where

$$\gamma_2 = \frac{\nu_u - \nu}{1 - \nu},$$

$$D = 2c\gamma_2(n|k| - k^2) - s. \tag{10.91}$$

The vertical displacement on the free surface $z = 0$ becomes

$$u_z(z = 0) = nA_1 + |k|A_2 - (3 - 4\nu_u)A_3, \tag{10.92}$$

which leads directly to

$$u_z^N(z = 0) = -2(1 - \nu_u)A_3,$$

$$= -\frac{(1 - \nu_u)\bar{N}(k)}{\mu|k|D}, \tag{10.93}$$

where the superscript indicates that these are the displacements due to the normal load.

Boundary conditions for the shear loading problem are

$$\bar{\sigma}_{zz}(z = 0) = 0,$$

$$\bar{\sigma}_{xz}(z = 0) = -\frac{\bar{T}(k)}{s},$$

$$\bar{p} = 0. \tag{10.94}$$

Solving for the coefficients leads to

$$A_1 = \frac{i\,\mathrm{sgn}(k)c\gamma_2 \bar{T}(k)}{\mu s D},$$

$$A_3 = -\frac{i\bar{T}(k)}{2\mu k D},$$

$$A_2 = -A_1 + \frac{2(1 - \nu_u)}{|k|}A_3. \tag{10.95}$$

Introducing the coefficients into equation (10.92), the vertical displacements at the free surface due to the distributed shear load are

$$u_z^T(z=0) = \frac{i\bar{T}(k)}{2\mu ks\,D}[s(1-2v_u) + 2c\gamma_2(|k|n - k^2)]. \qquad (10.96)$$

The Laplace transforms in equations (10.93) and (10.96) can be inverted analytically. The trick, following Wang (2000), is to remove the $(k^2 + s/c)^{1/2}$ term from the denominator by completing the square. In detail, multiplying the numerator and denominator by a factor of $-[s + 2c\gamma_2(n|k| + k^2)]$ leads to expressions with

$$s[s + 4c\gamma_2 k^2(1-\gamma_2)], \qquad (10.97)$$

in the denominator. The following inverse transforms are required:

$$\mathcal{L}^{-1}\left(\frac{\sqrt{s+a}}{s+b}\right) = \frac{e^{-at}}{\sqrt{\pi t}} + \sqrt{a-b}\,e^{-bt}\mathrm{erf}\sqrt{(a-b)t} \qquad (10.98)$$

(e.g., Wang 2000, table 7.1), and

$$\mathcal{L}^{-1}\left[\frac{\sqrt{s+a}}{s(s+b)}\right] = \int_0^t \left[\frac{e^{-at}}{\sqrt{\pi t}} + \sqrt{a-b}\,e^{-bt}\mathrm{erf}\sqrt{(a-b)t}\right]dt,$$

$$= \frac{1}{\sqrt{a}}\mathrm{erf}\sqrt{at} + \frac{1}{b\sqrt{a}}(a-b)\mathrm{erf}\sqrt{at}$$

$$- \frac{\sqrt{a-b}}{b}e^{-bt}\mathrm{erf}\sqrt{(a-b)t}, \qquad (10.99)$$

the latter accomplished via application of the convolution theorem for Laplace transforms (A.18).

The resulting displacement field in the Fourier domain is given by

$$2\mu k^2 u_z(z=0,t) = |k|(1-v)\bar{N} + ikv\bar{T} + (1-v)\left(|k|\bar{N} - ik\bar{T}\right)\left\{\mathrm{erf}(\sqrt{ck^2t})\right.$$

$$\left. + \sqrt{1-\beta}e^{-\beta ck^2t}\mathrm{erfc}[\sqrt{c(1-\beta)k^2t}]\right\}, \qquad (10.100)$$

where

$$\sqrt{1-\beta} = 1 - 2\gamma_2. \qquad (10.101)$$

The limiting value corresponding to undrained deformation is simply

$$2\mu k^2 u_z(z=0,t=0) = 2|k|(1-v_u)\bar{N} - (1-2v_u)ikv\bar{T}, \qquad (10.102)$$

and the corresponding drained value as $t \to \infty$ is equivalent to equation (10.102), with v replacing v_u. These expressions can be inverted by taking the Fourier transforms of (10.81) and inverting (10.102). For the drained case,

$$u_z(z = 0, t \to \infty) = -\frac{(3 - 4\nu)\Delta V}{2\pi}\frac{d}{d^2 + x^2}. \tag{10.103}$$

Adding these displacements due to the surface loads to the full-space term (10.80) yields the total displacement:

$$u_z(z = 0, t \to \infty) = -\frac{2(1 - \nu)\Delta V}{\pi}\frac{d}{d^2 + x^2}, \tag{10.104}$$

which agrees with the displacements computed using Green's function methods (see chapter 8, problem 3, and Segall 1985, appendix C). The corresponding undrained response is again given by replacing the drained with undrained Poisson's ratio.

Last, the vertical velocity is easily computed from the displacements (10.100):

$$2\mu v_z(z = 0, t) = \beta(1 - \nu)\left[\bar{N} - i\,\mathrm{sgn}(k)\bar{T}\right]\left\{\sqrt{\frac{c}{\pi t}}e^{-ck^2t}\right.$$
$$\left. -c|k|\sqrt{1 - \beta}e^{-\beta ck^2t}\mathrm{erfc}[\sqrt{c(1 - \beta)k^2t}]\right\}. \tag{10.105}$$

Displacements and velocities found by numerically inverting the Fourier transforms (10.100) and (10.105) are shown in figure 10.11. The displacements appear to be rather unremarkable, smoothly varying from the undrained to drained values. However, the vertical velocities tell a more interesting story. At early times, the central region above the magma chamber rises. The flanking regions, however, actually subside relative to the undrained displacements. The subsidence of the flanking regions reverses at later times, causing the net postintrusive displacement to be everywhere upward.

This behavior results from transient pore-fluid flow induced by the deformation. The region above the magma chamber dilates and thus experiences a decrease in pore-fluid pressure, whereas the pore pressure increases in the flanking regions. As pore fluid flows from the flanking region toward the center of uplift, it causes a short-term subsidence of the flanks. At later times, the pore fluid flows in from greater distance, eventually repressurizing the shallow crust, leading to uplift. Needless to say, actual intrusions are not instantaneous. An extension of this analysis could account for a finite period of inflation. The results here (10.105) should be accurate as long as the duration of the source inflation is relatively short compared to the characteristic time for fluid diffusion.

10.7 Cumulative Poroelastic Deformation in Three Dimensions

While it is difficult to compute solutions for dislocations in three-dimensional poroelastic half-spaces using purely analytical techniques, it is straightforward to determine the cumulative postseismic response as the crust relaxes from the completely undrained state, immediately after the earthquake, to the fully drained, or infinite time, response. One simply computes the displacement fields for a dislocation in a homogeneous elastic half-space using both undrained and drained Poisson's ratios and takes the difference, $u_{poro} = u(\nu_u) - u(\nu)$.

The cumulative poroelastic response is shown for a 70-km-long strike-slip fault that ruptures from the surface to a depth of 15 km in figure 10.12. Notice that the maximum displacements are on the order of a few percent of the fault slip and are concentrated near

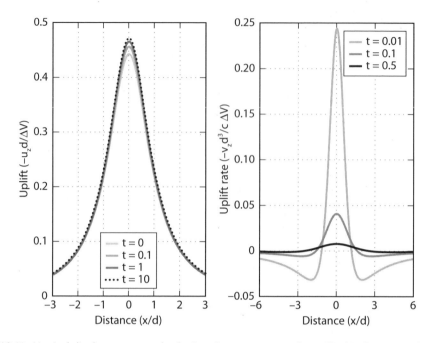

Figure 10.11. Vertical displacements and velocites due to an expanding cylindrical magma chamber in a poroelastic half-plane. Times are normalized by the characteristic diffusion time, d^2/c, where c is the hydraulic diffusivity, and d is the source depth. The volume change per unit along-strike length is ΔV. For these calculations, $\nu_u = 0.3$, $\nu = 0.25$.

the ends of the fault. For an infinitely long fault, the solution is antiplane strain (chapter 2), for which there are no changes in mean-normal stress and therefore no poroelastic pressure change. Thus, it is not surprising that the poroelastic response is concentrated near the ends of the rupture.

The cumulative response for a thrust fault is shown in figure 10.13. In this case, the maximum displacements are on the order of 15% of the coseismic slip. As in the strike-slip case, the poroelastic displacements are concentrated near the fault ends. For an infinitely long dip-slip fault, the drained and undrained surface displacements are identical (see problem 10), so again it is not surprising that the porelastic deformations are concentrated near the ends of the fault.

Transient poroelastic deformation has been observed following several crustal earthquakes. Peltzer et al. (1996) observed postseismic rebound in dilational steps along the 1992 Landers ruptures, using satellite radar interferometery. They suggested that these stepovers dilated and subsided during the earthquake. Postseismic pore-fluid flow into these areas of low pore pressure lead to uplift, as observed in the interferograms.

Jónsson and others (2003) observed broader scale transient deformation following two M 6 earthquakes in south Iceland in June of 2000, also using satellite interferometry. Plate 12 (left) shows the observed displacements in the line of sight (LOS) between the spacecraft and the ground from 19 June (two days after the first earthquake) to 24 July, 37 days after the earthquake. Due to the steep incidence angle, the LOS displacements are most sensitive to vertical displacement. The data display a quadrantal pattern of displacement, with LOS shortening (uplift) in two quadrants and LOS extension (subsidence) in the remaining two quadrants, although the uplift in the southwest quadrant is not as well developed as it is in the northeast quadrant. Slip was right-lateral, so the northeast and southwest quadrants

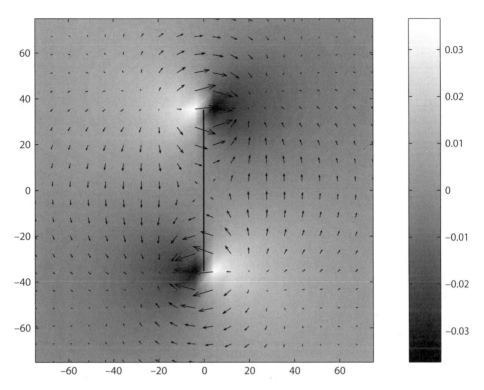

Figure 10.12. Cumulative poroelastic response for a 70-km-long strike-slip fault that extends to 15 km depth with one meter of slip. Horizontal displacements are shown with vectors (in meters), and vertical displacements by gray scale (in meters). The maximum horizontal displacement is 0.03 m. Distance scale is in kilometers. Calculation assumes $v = 0.2$ and $v_u = 0.33$.

dilated during the earthquake. The two dilational quadrants were observed to subside during the earthquake in coseismic interferograms, as expected. The dilational quadrants would be anticipated to rebound as pore pressure increased following the earthquake. This is what is observed in the interferograms in plate 12. The remaining two quadrants compressed coseismically and subsided following the earthquake, as pore fluid flowed out of the regions of high pore pressure.

A simple calculation showing the predicted range change as the crust relaxed from the undrained to the fully drained state is shown in plate 12 (right). The similarity between the observed and predicted images suggests that poroelastic effects caused the postseismic deformation. It also suggests that the characteristic fluid diffusion time was shorter than the 35-day repeat time of the InSAR satellite orbit. This is supported by additional observations that show little if any deformation in the following months, suggesting that any remaining poroelastic effects were at or below the noise level of the InSAR observations (see also figure 10.14).

South Iceland is an active geothermal area, and there are numerous wells in the region in which water level is monitored. Jónsson et al. (2003) showed that the coseismic and postseismic water-level changes in these wells were generally consistent with the predicted spatial pattern of pore pressure change associated with the two earthquakes (plate 13), adding additional support to the poroelastic interpretation of the InSAR data. Not only are the spatial patterns of deformation and water-level change consistent with a poroelastic process, but the temporal evolution of the changes also agree. This is illustrated in figure 10.14, which shows

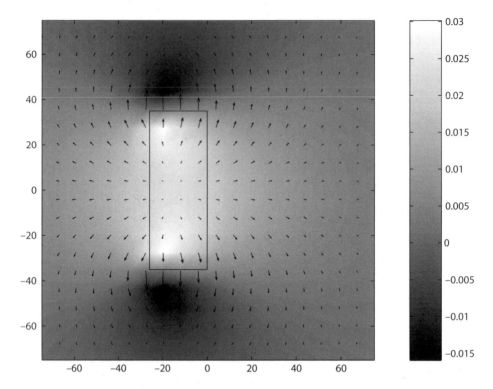

Figure 10.13. Cumulative poroelastic response for a 70-km-long east dipping thrust fault with one meter of slip. Surface projection of the dislocation is shown. Horizontal displacements are shown with vectors (in meters), and vertical displacements by gray scale (in meters). The maximum horizontal displacement is 0.04 m. Distance scale is in kilometers. Calculation assumes $\nu = 0.2$ and $\nu_u = 0.33$.

the temporal evolution of the deformation as recorded by satellite interferometry (right scale) and the water-level histories in selected wells. Notice that the deformation was essentially complete approximately two weeks after the earthquake and that the water levels equilibrated on similar timescales. Taken together, the spatial and temporal patterns in the deformation and water wells are in good agreement and strongly support a poroelastic response following these earthquakes.

10.8 Specified Pore Pressure Change

In general, the deformation and pore pressure fields are coupled through the equilibrium equations for the solid and the diffusion equation governing pore-fluid flow. In many cases of interest, it is not possible to solve the fully coupled equations analytically. If, on the other hand, one has independent constraints on the pore pressure change distribution, then it is possible to solve directly for the deformation fields. Uncoupling the pore pressure and deformation fields in this manner makes the poroelastic problem completely analogous to (uncoupled) thermoelasticity and permits us to take advantage of well-known methods in this field.

We first show that the poroelastic displacements are generated by a distribution of centers of dilatation with magnitude proportional to $\alpha p(\mathbf{x})$. As noted earlier, $-\alpha \partial p / \partial x_i$ enters the equilibrium equations (10.43) in the same form as body forces f_i. We make use of the

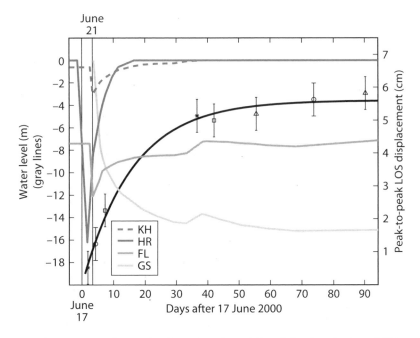

Figure 10.14. Deformation and water level as a function of time following the June 2000 earthquakes in south Iceland. Water-level variations are shown on the left axis for four wells designated KH, HR, FL, and GS. Well locations are shown in plate 13. Peak line-of-sight (LOS) range change determined from satellite radar interferograms is shown with error bars (scale on the right axis). The times of the two earthquakes are denoted by vertical lines. After Jónsson et al. (2003).

elastostatic Green's tensors, $g_i^k(\mathbf{x}, \boldsymbol{\zeta})$, which relate forces $f_k(\boldsymbol{\zeta})$ to displacements $u_i(\mathbf{x})$, via

$$\mu u_i(\mathbf{x}) = \int_V f_k(\boldsymbol{\zeta}) g_i^k(\mathbf{x}, \boldsymbol{\zeta}) \, dV_\zeta. \tag{10.106}$$

The equivalent body force distribution is given by the gradients in pore pressure $-\alpha \partial p / \partial x_k$, such that

$$\mu u_i(\mathbf{x}) = -\alpha \int_V \frac{\partial p}{\partial \zeta_k} g_i^k(\mathbf{x}, \boldsymbol{\zeta}) \, dV_\zeta. \tag{10.107}$$

Integrating by parts, and assuming that we are considering localized pore pressure disturbances, such that $p(|\boldsymbol{\zeta}| \to \infty) = 0$, yields

$$\mu u_i(\mathbf{x}) = \alpha \int_V p(\boldsymbol{\zeta}) \frac{\partial g_i^k(\mathbf{x}, \boldsymbol{\zeta})}{\partial \zeta_k} \, dV_\zeta. \tag{10.108}$$

Note that $\partial g_i^k(\mathbf{x}, \boldsymbol{\zeta}) / \partial \zeta_k$ is the displacement due to a sum of force couples acting along three orthogonal axes—that is, a center of dilatation (chapter 7). Equation (10.108) shows that the displacement field can be constructed by summing the displacements due to centers of dilatation with magnitude determined by the local change in pore pressure scaled by α.

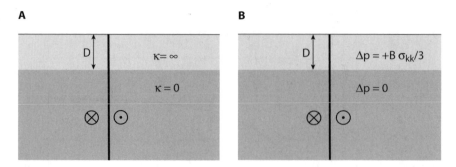

Figure 10.15. A: Fault in a poroelastic medium with a very permeable layer overlying an impermeable half-space. B: During the postseismic period following the earthquake, the layer experiences a pore pressure recovery $\Delta p = B\sigma_{kk}/3$ that is opposite in sign to the coseismic pressure change, where σ_{kk} is the coseismic stress change. There is no postseismic pore pressure change in the impermeable lower region.

This is physically reasonable, as equation (10.17) shows that the local volumetric strain is proportional to $\alpha p(\mathbf{x})$.

Define the displacement Green's function for a center of dilatation as

$$g_i^{CD}(\mathbf{x},\,\zeta) \equiv \frac{\partial g_i^k(\mathbf{x},\,\zeta)}{\partial \zeta_k}. \tag{10.109}$$

It follows then that

$$u_i(\mathbf{x},\,t) = \frac{\alpha}{\mu} \int_V p(\zeta,\,t) g_i^{CD}(\mathbf{x},\,\zeta)\, \mathrm{d}V_\zeta, \tag{10.110}$$

and

$$\sigma_{ij}(\mathbf{x},\,t) = \alpha \left[\int_V p(\zeta,\,t) G_{ij}(\mathbf{x},\,\zeta)\, \mathrm{d}V_\zeta - p(\mathbf{x},\,t)\delta_{ij} \right], \tag{10.111}$$

where the stress Green's functions are obtained from the displacements by

$$G_{ij} = \frac{\partial g_i^{CD}}{\partial x_j} + \frac{\partial g_j^{CD}}{\partial x_i} + \frac{2\nu}{1 - 2\nu} \frac{\partial g_k^{CD}}{\partial x_k}\delta_{ij}. \tag{10.112}$$

One example of an application of this approach is to determine how the surface deformation changes if the near-surface materials are more permeable than the rocks at greater depth. This is of interest in the south Iceland example discussed in the previous section, where the permeability is fracture dominated and expected to diminish with depth in the crust. We can get a feeling for the effect of depth-dependent permeability by considering the extreme case of an infinitely permeable layer of thickness D overlying an impermeable half-space (figure 10.15). The earthquake induces an instantaneous pore pressure change $\Delta p = -B\sigma_{kk}/3$, where $\sigma_{kk}/3$ is the coseismic change in mean stress. Because flow is inhibited at depth, there is no further change in pore pressure in the lower, impermeable medium. In the upper layer, however, the induced pore pressure very rapidly dissipates. Thus, the postseismic displacements are those generated by drainage of the coseismic pore pressure change in the shallow layer—that is, the postseismic pressure change is $\Delta p = +B\sigma_{kk}/3$ for $z < D$.

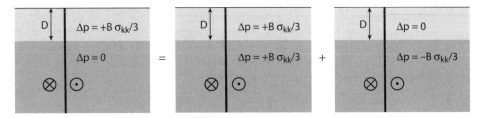

Figure 10.16. Drainage of the near-surface layer can be represented as the sum of two solutions. In the first, the entire half-space drains. This solution can be computed by elementary methods. The second solution involves repressurization of the deeper medium. This calculation is numerically stable, as the integrand never becomes singular when the observation points are restricted to the free surface.

Numerical evaluation of the integral (10.108) can be difficult, as the kernels are singular when the observation and source points are equal—that is, when $\mathbf{x} = \boldsymbol{\zeta}$. A trick that proves useful in this case is to break the problem in figure 10.15 into two parts using the superposition principle, as shown in figure 10.16. Here the problem is divided into one part in which the entire half-space drains and one part in which only the deeper medium is repressurized. The displacements due to the first problem are simply obtained by differencing the elastic displacement fields obtained with drained and undrained Poisson's ratios in a half-space. Integrals to compute the second problem on the right-hand side of figure 10.16 are nonsingular, as the field and source points are never colocated, given that we compute the deformation only on the free surface.

Setting the pressure change in equation (10.108) equal to the coseismic pressure change, as in figure 10.16, and making use of equation (10.15) yields

$$u_i(\mathbf{x}) = \frac{(v_u - v)}{(1 + v_u)(1 - 2v)} \int_V \sigma_{kk}(\boldsymbol{\zeta}) \frac{\partial g_i^k(\mathbf{x}, \boldsymbol{\zeta})}{\partial \zeta_k} \, dV_\zeta. \tag{10.113}$$

The Green's functions in a full-space (7.20) are given by

$$\frac{\partial g_i^k(\mathbf{x}, \boldsymbol{\zeta})}{\partial \zeta_k} = \frac{1}{8\pi\mu} \left(\frac{1 - 2v}{1 - v} \right) \frac{x_i - \zeta_i}{R^3}. \tag{10.114}$$

As discussed in chapter 7, the free-surface correction terms for a center of dilation are equivalent to multiplying the full-space expressions by a factor of $4(1 - v)$, so the displacements are finally given by

$$u_i(\mathbf{x}) = \frac{(v_u - v)}{2\pi(1 + v_u)} \int_V \sigma_{kk}(\boldsymbol{\zeta}) \frac{x_i - \zeta_i}{R^3} \, dV_\zeta, \tag{10.115}$$

where the integral is taken over the domain $z > D$. The simulated range change is shown in figure 10.17A for different thicknesses of the permeable layer, D. Not surprisingly, the greater the depth of drainage, the larger the displacement. However, the amplitude of the signal also depends on the difference between the drained and undrained Poisson's ratios. As these parameters are not well known for crustal materials, it is difficult to predict a priori the magnitude of the poroelastic effect.

Scaling the different plots to the same maximum range change is shown in figure 10.17B. This shows that without accurate knowledge of the poroelastic constants, it is not possible to determine the depth of drainage from the surface displacements. One might have expected

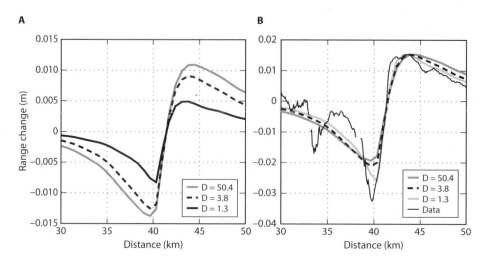

Figure 10.17. A: Calculation of the postseismic poroelastic range change for the June 2000 earthquakes in the South Iceland Seismic Zone, for three different depths of drainage, D. In all cases, the upper layer is assumed to drain rapidly, whereas the lower medium remains undrained. Left: Unscaled. B: Scaled so that the maximum amplitudes are equal. Also shown is the observed range change along a transect across the fault, from data between dashed lines in plate 12. After Jónsson et al. (2003).

that the spatial scale of the deformation would be larger when the drainage is deeper. However, the spatial scale of the postseismic poroelastic deformation is set by the scale of the coseismic pore pressure change. This in turn is controlled by the dimensions of the rupture. Even if only a shallow layer drains, this drainage will be spatially extensive and difficult to distinguish from drainage of a deeper layer. It is important to emphasize that this calculation cannot reveal the time dependence of the deformation. Nevertheless, it can provide useful insights into the effect of an inhomogenous permeability distribution.

10.9 Summary and Perspective

Poroelastic effects have only recently been recognized as contributing to observable crustal deformation. On the other hand, poroelastic changes have long been of interest in fault mechanics studies, as changes in pore-fluid pressure directly influence frictional resistance on faults through the effective stress principle. The theory of poroelasticity is based on linearized constitutive laws that relate changes in pore pressure to strain and account for changes in fluid mass stored within a representative volume of rock. For an isotropic medium, this requires two new material constants in addition to the two constants that characterize classical elastic materials. Consideration of idealized experiments in which either the external confining pressure or the internal pore pressure is varied yields relationships between these poroelastic parameters and the compressibility of the bulk rock, pore fluid, solid grains, and pores.

Pore pressure gradients appear in the equilibrium equations analogous to body forces. Combining the constitutive relations with conservation of pore-fluid mass and Darcy's law leads to a diffusion equation that is fully coupled to the solid equilibrium equations. This coupling significantly complicates solution of boundary value problems, particularly for faults and magma bodies in a half-space. A limited number of exact solutions have been found that provide insights into the often counterintuitive behavior of the coupled poroelastic problem. At the time of this writing, however, exact solutions have not been found for dislocation sources in a poroelastic half-space.

Comparisons to data have often relied on the fact that the undrained (instantaneous) and fully drained (infinite time) responses are equivalent to classical elasticity problems with undrained and drained Poisson's ratios, respectively. Caution is required, however, because there is no guarantee that the displacement or strain fields vary monotonically between the undrained and drained limits. In the admittedly limited case where the pore pressure distribution is known a priori, the displacement field can be computed as a convolution of the pressure field with Green's functions equivalent to centers of dilatation. This is sensible in that each parcel of material in the earth expands or contracts by an amount proportional to the pore pressure change within that element.

10.10 Problems

1. Show that equation (10.3) reduces to the standard form of Hooke's law for an effective stress, σ'_{ij}, where the effective stress is defined to be $\sigma'_{ij} \equiv \sigma_{ij} + \alpha p \delta_{ij}$.

2. Prove equation (10.15)—namely, that

$$\alpha = \frac{3(\nu_u - \nu)}{B(1 + \nu_u)(1 - 2\nu)}.$$

3. Show that the constitutive equations (10.16) can be recast as

$$2\mu\epsilon_{ij} = \sigma_{ij} - \frac{\nu_u}{1 + \nu_u}\sigma_{kk}\delta_{ij} + \frac{2\mu B \Delta m}{3\rho_0}\delta_{ij}.$$

From this, show that the volumetric strain can be written in terms of fluid mass alterations and undrained parameters as

$$\epsilon_{kk} = \frac{\sigma_{kk}}{3K_u} + \frac{B\Delta m}{\rho_0},$$

where the undrained bulk modulus is given by $K_u = 2\mu(1 + \nu_u)/3(1 - 2\nu_u)$. This shows that Skempton's coefficient also has the interpretation as the ratio of volumetric strain ϵ_{kk} to change in fluid content $\Delta m/\rho_0$ when the mean normal stress vanishes.

4. Show that the diffusion equation (10.44) can be cast in the following form:

$$S_\alpha \frac{\partial p}{\partial t} - \frac{\kappa}{\eta}\nabla^2 p = -\alpha\frac{\partial}{\partial t}(\nabla \cdot \mathbf{u}), \tag{10.116}$$

where the storage coefficient is defined as

$$S_\alpha = \frac{3\alpha(1 - 2\nu)(1 - \alpha B)}{2\mu B(1 + \nu)}. \tag{10.117}$$

This form shows that the divergence of the displacements acts as a source term in the pore pressure diffusion equation.

5. Show that S_α as defined in problem 4 can be written as

$$S_\alpha = \frac{\alpha}{K_s} + \phi\left(\frac{1}{K_f} - \frac{1}{K_\phi}\right). \tag{10.118}$$

6. Solve equation (10.59) for a constant normal load on the earth's surface using a combination of Laplace and Fourier transforms to reduce the partial differential equation to an algebraic equation. Note that in order to extend the domain of z to $-\infty$ to ∞ to make use of the Fourier transform, and to match the boundary condition for the vertical stress on the plane $z = 0$, we need to take $\sigma_{zz}(z = 0, t) = -2F\,H(t)[H(z) - 1]$, as in appendix B. Following the procedures there leads to the result given in equation (10.65).

7. Prove that the undrained pore pressure change for one-dimensional deformation is

$$p = -\sigma_{zz}\frac{B(1 + v_u)}{3(1 - v_u)},$$

so equation (10.65) has the correct limiting response.

8. Derive equation (10.74) from the governing differential equation and boundary conditions.

9. Show that the stresses due to an edge dislocation (10.75) have the appropriate drained and undrained limits.

10. Show that for an infinitely long dip-slip fault (plane strain), the free-surface displacements at time $t = 0$ are identical to those in the limit $t \to \infty$.

10.11 References

Abramowitz, M., and I. A. Stegun. 1972. *Handbook of mathematical functions*. New York: Dover Publications. pp. 295–329.

Berryman, J. G., and S. R. Pride. 2002. Models for computing geomechanical constants of double-porosity materials from constituents' properties. *Journal of Geophysical Research* **107**(3), 10.1029/2000JB000108.

Biot, M. 1941. General theory of three-dimensional consolidation. *Journal of Applied Physics* **12**, 155–164.

———. 1956. General solutions of the equations of elasticity and consolidation for a porous material. *Journal of Applied Mechanics* **78**, 91–96.

Boley, B. A., and J. H. Weiner. 1960. *Theory of thermal stresses*. New York: John Wiley and Sons.

Jónnson, S., P. Segall, R. Pedersen, and G. Björnsson. 2003. Post-earthquake ground movements correlated to pore-pressure transients. *Nature* **4242**, 179–183.

Kumpel, H. J. 1991. Poroelasticity: parameters reviewed. *Geophysical Journal International* **105**, 783–799.

Malvern, L. E. 1969. *Introduction to the mechanics of continuous media*. Englewood Cliffs, NJ: Prentice Hall.

Mathews, J., and R. L. Walker. 1970. *Mathematical methods of physics*, 2nd ed. Menlo Park, CA: Benjamin/Cummings.

McNamee, J., and R. E. Gibson. 1960a. Displacement functions and linear transforms applied to diffusion through porous elastic media. *Quarterly Journal of Mechanics and Applied Math* **13**, 98–111.

———. 1960b. Plane strain and axially symmetric problems of the consolidation of a semi-infinite clay stratum. *Quarterly Journal of Mechanics and Applied Math* **13**, 210–227.

McTigue, D. F. 1986. Thermoelastic response of a fluid-saturated porous rock. *Journal of Geophysical Research* **91**, 9533–9542.

Nur, A., and J. Byerlee. 1971. An exact effective stress law for elastic deformation of rock with fluids. *Journal of Geophysical Research* **76**, 6414–6419.

Palciauskas, V. V., and P. A. Domenico. 1982. Characterization of drained and undrained response of thermally loaded repository rocks. *Water Resources Research* **19**, 281–290.

Peltzer, G., P. Rosen, F. Rogez, and K. Hudnut. 1996. Postseismic rebound in fault step-overs caused by pore fluid flow. *Science* **273**(5279), 1202–1204.

Rice, J. R., and M. P. Cleary. 1976. Some basic stress diffusion solutions for fluid-saturated elastic porous media with compressible constituents. *Reviews of Geophysics and Space Physics* **14**, 227–241.

Roeloffs, E. 1988. Fault stability changes induced beneath a reservoir with cyclic variations in water level. *Journal of Geophysical Research* **93**(3), 2107–2124.

———. 1995. Poroelastic techniques in the study of earthquake-related hydrologic phenomena. *Advances in Geophysics* **37**, 135–195.

Rudnicki, J. W. 1987. Plane strain dislocations in linear elastic diffusive solids. *Journal of Applied Mechanics* **109**, 545–552.

Rudnicki, J. W., and E. Roeloffs. 1990. Plane strain shear dislocations moving steadily in linear elastic diffusive solids. *Journal of Applied Mechanics* **112**, 32–39.

Segall, P. 1985. Stress and subsidence resulting from subsurface fluid withdrawal in the epicentral region of the 1983 Coalinga earthquake. *Journal of Geophysical Research* **90**, 6801–6816.

———. 1989. Earthquakes triggered by fluid extraction. *Geology* **17**, 942–946.

Wang, H. F. 2000. Theory of linear poroelasticity. Princeton, NJ: Princeton University Press.

Yerkes, R. F., and R. O. Castle. 1970. Surface deformation associated with oil and gas field operations in the United States. In *Land Subsidence*, UNESCO Publication 89, vol. 1. Paris: International Association of the Science of Hydrology, pp. 55–66.

Zimmerman, R. W. 1991. *Compressibility of sandstone*. Amsterdam: Elsevier.

11

Fault Friction

We began our study with dislocation models of faulting, in which the slip on a fault is prescribed. In chapter 4, we explored models in which the shear stress in the slipping zone is specified, leading to mixed stress-displacement boundary conditions. In neither of these classes of models does the slip on the fault arise naturally due to the physical properties of the fault zone and its interactions with its elastic surroundings. In order to develop such models, we must consider the constitutive properties of faults—the topic of the present chapter. This also leads to a discussion of fault stability and earthquake nucleation. Earthquakes are not simply slip on faults, but fast, inertially limited slip that causes elastic waves to be radiated into the surrounding rock. The study of earthquake nucleation is in part motivated by the need to understand the physics of how earthquakes initiate, but also by the hope that the nucleation process may lead to detectable strain signals. At the time of this writing, even very sensitive and well-situated instruments have failed to detect convincing strain precursors. While this does not diminish the importance of studying the physical processes involved, it does dampen the hope that such signals will provide useful warnings of impending earthquakes. Knowledge of fault friction is also key to understanding creeping faults, postseismic slip, and recently discovered transient slip phenomena.

While fault zones in nature have finite thickness, they are often modeled mathematically as a surface, most often planar, bounding two elastic regions. This may seem a severe oversimplification, but it is in part motivated by the understanding that deformation within gouge zones tends to *localize* to a very narrow zone under conditions that permit unstable fault slip. There may be conditions when deformation tends to spread or delocalize; however, we will not consider such issues in the necessarily abbreviated treatment here.

In the simplest case, Amonton-Coulomb friction states that the resistive stress along a frictional surface is given by the product of the normal stress acting across that surface σ and a coefficient of friction f (we avoid the notation μ to avoid confusion with the elastic Lame parameter):

$$\tau = f\sigma. \tag{11.1}$$

It is instructive to consider an analogue model of a fault system consisting of a block in frictional contact with a flat surface connected to an elastic spring (figure 11.1). The spring has stiffness k and is driven at constant load point velocity, v_{plate}. While clearly oversimplified, this *single degree of freedom spring-slider system* does embody a few key elements present in crustal faulting: a frictional contact surface modeling the fault, a spring representing the elasticity of the earth's crust (necessary to store elastic strain energy that is ultimately released in fast slip), and a constant load point velocity, mimicking the constant plate motion that supplies energy to plate boundary faults in the earth.

The driving force acting on the block is given by the difference between the *load point displacement*, $v_{\mathrm{plate}}t$, and the displacement of the block δ multiplied by the spring stiffness

$$\tau_{drive} = k(v_{\mathrm{plate}}t - \delta), \tag{11.2}$$

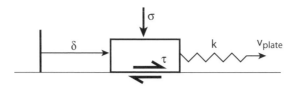

Figure 11.1. Single degree of freedom spring-slider system. A block is pressed against a flat interface with normal stress σ. Frictional resistance to slip is τ. A spring with stiffness k is attached to the block, the far end of which displaces at constant rate v_{plate}. The cumulative displacement of the block is δ.

where k is taken to be the spring stiffness per unit of frictional contact area, with dimensions of stress per unit displacement. Ignoring inertial effects—that is, assuming that all slip is quasi-static—equilibrium requires that the driving stress given by equation (11.2) equal the frictional resisting stress given by equation (11.1):

$$k(v_{plate}t - \delta) = f\sigma. \tag{11.3}$$

The introductory physics explanation for unstable *stick-slip* motion is that the static friction coefficient, which applies to surfaces in stationary contact, exceeds the dynamic friction coefficient, which holds during sustained sliding. Once the static frictional resistance is overcome, the stress exceeds that required to continue slip, resulting in an instability. In nature, however, transitions are not instantaneous but occur over some finite slip or timescale.

11.1 Slip-Weakening Friction

A more realistic interpretation of the transition from static to dynamic friction is so-called *slip-weakening friction* in which the friction coefficient weakens as a function of slip $f(\delta)$. As long as equation (11.3) is satisfied, the forces balance, and the block is in static equilibrium. Now consider what happens if the block is perturbed slightly by a small displacement $d\delta$. A forward displacement decreases the driving stress (the spring shortens). At the same time, the frictional resistance decreases by an amount $(\partial f/\partial\delta)d\delta$, where $(\partial f/\partial\delta) < 0$. The forces will be unbalanced in the forward direction, favoring further slip, if the frictional resistance drops faster than the driving force

$$\text{unstable}: \quad \sigma\frac{\partial f}{\partial\delta} < -k < 0, \tag{11.4}$$

since $\partial\tau_{drive}/\partial\delta = -k$. On the other hand, the system is stable if the frictional resistance decreases more slowly than the elastic system unloads:

$$\text{stable}: \quad -k < \sigma\frac{\partial f}{\partial\delta}. \tag{11.5}$$

Notice that the stability of the spring-slider system depends on (1) the elastic stiffness k, (2) the normal stress σ, and (3) the rate of frictional weakening $\partial f/\partial\delta$. Thus, stability is not a property of the frictional behavior alone, but an interaction between the frictional interface and the elastic loading system. Specifically, increasing the spring stiffness k can stabilize the system, whereas increasing the normal stress tends to destabilize the system. This phenomenon is well known in laboratory rock mechanics, where stiff testing machines are required to prevent rock samples from failing catastrophically (e.g., Jaeger and Cook 1976). Figure 11.2 shows that the higher the normal stress, the stiffer the machine must be to prevent instability.

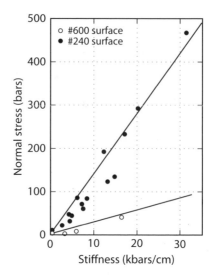

Figure 11.2. Normal stress at which there was a transition from stable sliding to stick-slip behavior as a function of machine stiffness. The different symbols refer to different surface roughness. After Dieterich (1978).

The behavior of the slip-weakening system can be understood with the graphical construction shown in figure 11.3. The heavy curve shows the stress-displacement behavior of the fault surface. The strength initially increases but eventually reaches a peak strength beyond which the strength decreases with increasing slip. The straight lines with slope $-k$ represent the elastic unloading line for the system; the stress decreases along this line with increasing slip δ for fixed load point displacement. From equation (11.3), the stress vanishes when $v_{\text{plate}}t = \delta$,

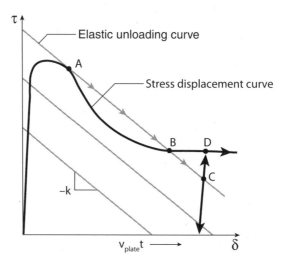

Figure 11.3. Instability in a single degree of freedom elastic system with slip weakening friction. Solid line represents the strength curve for the fault interface $\tau(\delta)$. Straight lines with slope $-k$ are the elastic unloading lines. The intersection of the two represents mechanical equilibrium; the applied stress equals the frictional resistance. With increasing time t, the elastic line moves to the right and stress increases. See text for discussion. After Rice (1983).

so that the unloading line intersects the horizontal axis at the load point displacement $v_{\text{plate}}t$. The intersection of the unloading curve with the strength-displacement curve represents the equilibrium configuration of the system; at this point the driving stress exactly balances the resisting stress.

As the load point displacement increases, the elastic line moves to the right, and the stress increases until the unloading line is tangent to the strength-displacement curve. At this point, marked A, the system is in unstable equilibrium, and $\sigma \partial f / \partial \delta = -k$. A small increase in slip causes the resisting stress to drop faster than the driving stress. The next potential equilibrium point is marked B, indicating that the fault displacement must at least jump from A to B. However, inertia may lead to a dynamic overshoot, with the system continuing to a point D. Static equilibrium must be satisfied as soon as the block stops sliding, so the frictional strength must suddenly decrease to C. Further loading causes the system to move toward point D, where it rejoins the stress displacement curve. The maximum overshoot occurs when there is no seismic radiation and all of the elastic energy released is converted to work against friction.

The slip-weakening model has several positive attributes. First, it provides a physically plausible transition from "static" to "dynamic" friction. Note that an instantaneous transition from static to dynamic friction implies that the system is always unstable, which is inconsistent with laboratory results that indicate that increasing the stiffness of the loading system can stabilize slip. Second, the model predicts that stability depends on normal stress, as observed in the lab (figure 11.2).

On the other hand, the simple slip-weakening model has significant defects. First, there is no mechanism for restrengthening or healing. As indicated in figure 11.3, the system would be expected to slide stably following a single dynamic slip event. Early earthquake models based on slip-weakening friction required ad hoc restrengthening, whereby the system was arbitrarily reset to the origin of the strength-slip curve some time following the earthquake. Second, the lack of rate dependence in the model means that the tangent point (point A in figure 11.3) is unstable to even infinitesimal perturbations in stress. Thus, this model taken literally predicts that small stress variations due, for example, to solid earth tides would trigger earthquakes. Yet careful studies have found only a very weak tendency for earthquakes to correlate with tidal stress cycles. This is strong evidence for some rate sensitivity in the earthquake system, although it need not necessarily be in the fault constitutive properties.

11.2 Velocity-Weakening Friction

It might seem reasonable to postulate that fault strength degrades with slip rate, as opposed to slip. However, as we shall see, direct velocity-weakening friction leads to mechanical inconsistencies. Consider that the fault strength is a decreasing function of slip speed alone, $\tau(v)$. At equilibrium, the frictional resistance exactly balances the applied stress, given by equation (11.2). Perturbing this solution about the equilibrium value, we have

$$\left. \frac{\partial \tau}{\partial v} \right|_{v_0} \Delta v = -k\Delta\delta,$$

$$\left. \frac{\partial \tau}{\partial v} \right|_{v_0} \frac{\mathrm{d}\Delta v}{\mathrm{d}t} = -k\Delta v, \tag{11.6}$$

where the subscript v_0 indicates that the derivative is evaluated at the initial velocity. Equation (11.6) is a first-order differential equation with solution

$$\Delta v = \exp\left(-kt / \left. \frac{\partial \tau}{\partial v} \right|_{v_0} \right). \tag{11.7}$$

Thus, for velocity weakening, $\partial \tau / \partial v|_{v_0} < 0$, the velocity perturbations grow exponentially, regardless of the stiffness k. A full nonlinear analysis also shows that direct velocity weakening is *unconditionally unstable*. This is not only inconsistent with laboratory tests as noted earlier, but also leads to nonphysical behavior when applied to faults in elastic continua. As described in chapter 4, in a Fourier description of fault slip, the effective fault stiffness is a function of wavelength, with stiffness increasing with decreasing wavelength. Since velocity-weakening friction is unstable for all stiffnesses, it is therefore unstable for all wavelengths. Thus, an infinitesimal patch of fault could become unstable independent of the rest of the fault. Such unphysical behavior cannot occur with slip weakening friction because a small fault patch is very stiff and will thus respond stably to small perturbations in stress or slip.

It is worthwhile to review field and laboratory observations and consider their implications for required fault frictional properties:

1. Faults in nature exhibit both unstable stick-slip cycles and stable creep. The best known example of stable creep is the central creeping zone of the San Andreas fault. In addition, geodetic and seismic data point to stable deformation at depth below the seismically active zone (typically greater than 15 km on the San Andreas fault). This is often attributed to distributed plastic flow associated with a temperature dependent brittle-ductile transition, although as we shall see, such transitions are also expected for deformation localized to a narrow fault zone.

2. Most faults exhibit rapid earthquake slip followed by decades to millennia of locking. There must be some process for restrengthening the fault following rapid slip.

3. Laboratory experiments (figure 11.2) show that the tendency for instability is enhanced by decreasing the system stiffness or increasing the normal stress.

4. The weak correlation between tidal stressing and earthquake nucleation requires some rate dependence in the system. Direct velocity weakening is ruled out because faults with velocity-weakening friction are unconditionally unstable.

In the 1970s and early 1980s, laboratory friction experiments led to the development of *rate and state (history)–dependent* constitutive laws that when applied to model faults in elastic continua exhibit the required behavior just described. These friction laws are the subject of the next section.

11.3 Rate and State Friction

We will review very briefly here some of the observations that led to the development of rate- and state-dependent friction laws. In the first set of experiments, rock surfaces were placed in contact under fixed normal stress. The stress required to initiate slip was measured as a function of the time the surfaces were held in contact (Dieterich 1972). These experiments showed that the nominal coefficient of static friction increased roughly with the logarithm of the time of static contact (figure 11.4).

The second set of experiments pioneered by Dieterich (1978; 1979), referred to as *velocity stepping tests*, probe the velocity dependence of friction. In these experiments, a sliding surface is held at constant normal stress. A servo-control system simulates a very stiff testing machine in order to inhibit instability. The surfaces are slipped for some time at a fixed load point velocity v_1 until the resistive shear stress achieves a stable value. The load point velocity is then rapidly changed to v_2 (see figure 11.5).

The stress response to step changes in velocity is shown schematically in figure 11.5. Actual laboratory data are shown in figure 11.6. These experiments illustrate a number of important features that motivated the development of specific constitutive laws by Dieterich (1979) and Ruina (1983).

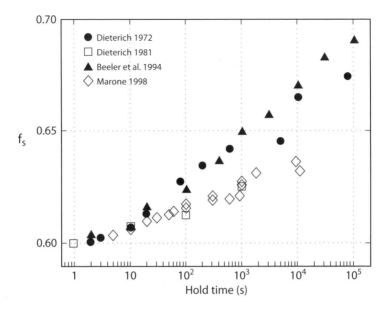

Figure 11.4. Variation in static friction coefficient f_s with hold time. Initially bare rock surfaces (solid symbols) and fault gouge (open symbols). The data have been offset to a friction coefficient of 0.6 at 1 second so that they represent relative changes in friction. After Marone (1998).

1. For slip at a constant rate, there appears to be a unique *steady-state* frictional resistance that depends on slip speed, $\tau_{ss}(v)$—that is, the frictional resistance asymptotically approaches $\tau_{ss}(v)$ regardless of past slip history. In addition to these changes, there may be slow increases in friction with cumulative offset. While not necessarily well understood, this is often attributed to the buildup of gouge on the sliding surfaces.

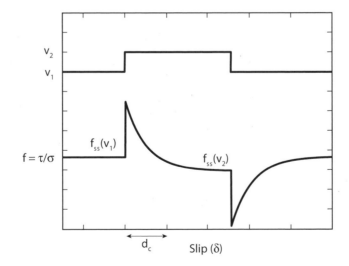

Figure 11.5. Schematic illustration of a velocity stepping test. The load point velocity (upper curve) is suddenly changed from v_1 to v_2. The frictional resistance shows an instantaneous increase followed by a decay to a lower value. The final value may be less than the initial friction, termed *steady-state velocity weakening*, as in this example, or greater than the initial friction, termed *steady-state velocity strengthening*.

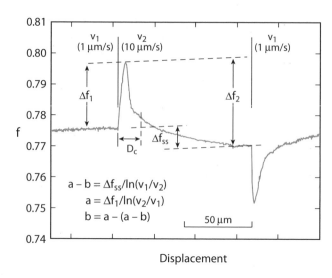

Figure 11.6. Laboratory results for a velocity stepping test, showing transient and steady-state changes in friction. The rapid response to the velocity step Δf_1 gives the constitutive parameter a. The subsequent evolution of the friction coefficient with slip Δf_2 yields the parameter b, so the steady-state change Δf_{ss} is given by $a - b$. This sample exhibits steady-state velocity weakening. Experiment at 150 mpa normal stress. After Kilgore et al. (1993).

2. Following a step increase in velocity, the friction instantaneously increases (figure 11.5). In contrast, an instantaneous decrease in friction follows a step decrease in slip rate. This behavior is referred to as *direct velocity strengthening*, $\partial \tau / \partial v > 0$. The experiments show that the instantaneous change in friction varies roughly with the logarithm of the change in slip speed—that is, $\tau = \tau_0 + a \log(v_2/v_1)$.

3. Following the abrupt increase in friction for a velocity increase, the friction decreases with increasing displacement. Similarly, the friction gradually increases following a rapid drop due to a step decrease in the sliding velocity. The gradual decay to a new steady state occurs over one or more characteristic displacement scales d_c that depend on surface roughness or the presence and properties of fault gouge but do not depend on slip speed. The final approach to the new steady-state friction is exponential in slip.

4. While the direct dependence of τ on v is always strengthening (item 2), the steady-state frictional dependence on slip speed $\partial \tau_{ss}/\partial v$ can be either positive or negative. That is the final value of friction following a step increase in speed can be either less than the initial friction (as in figure 11.5) or more.

Empirical rate and state friction laws postulate that the friction coefficient depends on the current slip rate v and one or more *state variables* that characterize the state of the sliding surfaces:

$$\tau = \sigma f(v, \text{state}) \tag{11.8}$$

(Ruina 1983). The vector of state variables can be written as $[\theta_1, \theta_2, \theta_3, \ldots]$, which in the limit of a single state variable reduces to the scalar θ. Dieterich associates the state variable with the average time of asperity contacts. For surfaces sliding at constant rate, the average contact time is proportional to the average asperity size and inversely proportional to slip speed. More generally, state variables can be any parameter that characterizes the state of the surface, such as grain size or porosity of fault gouge. Implicit in this description is the understanding that state applies to a point, or small area, on the fault and that changes in state at one point do not directly influence state elsewhere.

Equation (11.8) is not complete; one must specify how state varies with slip or time. For a single state variable θ, these *state evolution laws* are of the general form (Ruina 1983)

$$\frac{d\theta}{dt} = g(\sigma, v, \theta). \tag{11.9}$$

The most commonly used form of the constitutive law with a single state variable (11.8) is

$$\tau = \sigma \left(f_0 + a \log \frac{v}{v_0} + b \log \frac{\theta}{\theta_0} \right). \tag{11.10}$$

Here, f_0 is a nominal coefficient of friction that typically takes on values near 0.6 for many rock types under normal crustal conditions. v_0 and θ_0 are normalizing constants such that $f = f_0$ when $v = v_0$ and $\theta = \theta_0$. The normalization constant θ_0 is arbitrary, it will prove to be convenient to choose $\theta_0 = d_c / v_0$. Notice that for fixed θ, the friction varies directly with logarithm of slip speed, as suggested by laboratory experiments. Assuming, as is always observed experimentally, that $a > 0$, equation (11.10) exhibits direct velocity strengthening $\partial \tau / \partial v > 0$. (Note that in some formulations, including that by Ruina (1983), the term $b \log \theta / \theta_0$ was written as Θ. In this case, the constant b appears in the state evolution law. The use of Θ changes the appearance of the constitutive equations but not their meaning).

While equation (11.10) can be viewed solely as an empirical fit to laboratory data, there is some basis for the particular form of the equation from a micromechanical understanding of frictional processes. If the real area of asperity contacts is denoted A_c, whereas the nominal surface area is A, then the normal stress acting at contacts is $\sigma_c = \sigma A / A_c$, since the net force must be the same. The normal stress at contacts can be quite large if A_c is a small fraction of A. Similarly, the shear stress acting on contacts is $\tau_c = \tau A / A_c$. Taking the ratio of these expressions:

$$\frac{\tau_c}{\sigma_c} = \frac{\tau}{\sigma}. \tag{11.11}$$

Similarly, equation (11.11) can be written as $\tau = \tau_c A_c / A$. Equation (11.11) explains why the macroscopic friction coefficient $f = \tau / \sigma$ is independent of normal stress (Bowden and Tabor 1964). It also helps to rationalize why friction coefficients do not vary greatly for different rock types—weak materials with low shear strength τ_c tend to have low indentation strengths σ_c and hence larger contact areas for a given applied normal stress (e.g., Nakatani 2001).

The shear strength of the asperity contacts is believed to be controlled by thermally activated creep processes (e.g., Stesky 1978; Chester 1994). The shear strength of most materials is a positive function of strain rate, which for constant thickness of the shearing layer will be proportional to slip speed. Thus, it is reasonable to assume $\tau_c(v)$, where $\partial \tau_c / \partial v \geq 0$. As discussed later in this section, the real area of contact depends on the state of the surface, which can be described by the state variable θ such that

$$\tau = \tau_c(v) \frac{A_c(\theta)}{A}. \tag{11.12}$$

The thermally activated creep process can be understood following Rice et al. (2001) and Nakatani (2001), who built on earlier work of Heslot et al. (1994), and others. Following the notation of Rice et al. (2001), for a single activation process $v = v_0 \exp(-E/k_B T)$, where E is an activation energy, k_B is the Boltzman constant and T is temperature. Here, v_0 is the frequency of thermally activated jumps multiplied by a characteristic atomic-scale displacement in each completed jump. The activation energy is taken to be of the form $E = E_1 - \tau_c \Omega$, where $\tau_c \Omega$ is the work done by the shear stress during the thermally activated process, and Ω is an activation volume (e.g., Poirier 1985). Thus,

$$v = v_0 \exp\left[\frac{-(E_1 - \tau_c \Omega)}{k_B T}\right]. \tag{11.13}$$

Combining equations (11.11) and (11.13) leads to

$$\frac{\tau}{\sigma} = \frac{E_1}{\sigma_c \Omega} + \frac{k_B T}{\sigma_c \Omega} \log \frac{v}{v_0}. \tag{11.14}$$

Comparison with equation (11.10) demonstrates that the empirical parameter a is associated with

$$a = \frac{k_B T}{\sigma_c \Omega}. \tag{11.15}$$

Rice et al. (2001) and Nakatani (2001) use laboratory estimates of a and σ_c to determine an activation volume of order a few atomic volumes. Rice et al. (2001) estimate an activation energy E_1 of order 100 kJ/mol. Nakatani (2001) obtains an average estimate of 200 kJ/mol over the temperature range of 25°C to 800°C but discusses evidence for a switch in mechanisms at 400°C.

We now turn to the state evolution equation (11.9). For a single state variable and constant normal stress, two widely used evolution equations are

$$\frac{d\theta}{dt} = 1 - \frac{\theta v}{d_c}, \tag{11.16}$$

and

$$\frac{d\theta}{dt} = -\frac{v\theta}{d_c} \ln\left(\frac{v\theta}{d_c}\right). \tag{11.17}$$

The first equation (11.16) is referred to as the *aging law* in that it predicts that state increases linearly with time for surfaces in stationary contact, consistent with Dieterich's view that state represents time of asperity contact. When combined with equation (11.10), this predicts that the frictional strength increases with logarithm of time as observed experimentally (figure 11.4). The second law (11.17), often referred to as the *slip law*, predicts no change in state when the slip speed vanishes, as $lim_{x \to 0} x \log x = 0$. Ruina (1983) showed that the slip law, when coupled to an elastic system under applied shear stress also predicts behavior consistent with an increase in friction during apparent static contact. This occurs because even though the load point is fixed, the surface creeps to relax the shear stress, and the state evolves during that slip.

There has been a great deal of discussion about the micromechanical interpretation of the state variable and its evolution. Two possible mechanisms are most often cited for giving rise to the increase in friction with time of apparent contact (e.g., Dieterich 1978). One is for plastic creep of asperity contacts, which tends to increase the real area of contact and thus overall shear resistance of the surface. The second possibility is that the area of contact remains constant but that the degree of adhesion increases, due possibly to the diffusion of adsorbed surface contaminants away from the contact points.

Dieterich and Kilgore (1994) conducted an ingenious set of experiments for directly observing frictional contacts under both normal and shear loads. They illuminate contact surfaces between transparent materials, including glass and lucite plastic, with monochromatic light. Where the surfaces are in contact, the light passes through so that the contacts appear bright. Where the surfaces are not contacting, the light is scattered at the interfaces, so these areas appear dark. With this methodology, the authors are able to show that the actual area of contact scales linearly with normal stress $A_c = \sigma A / \sigma_c$, and increases with log time for surfaces in stationary contact at constant normal load. They also were able to measure the change in contact area with velocity during velocity stepping tests. Contact area was observed to scale inversely with slip speed, consistent with $\theta = d_c / v$, under steady sliding. The mean contact diameter, assuming circular contacts, was found to be approximately equal to d_c determined independently from the variation of friction following velocity steps. Last, Dieterich and Kilgore (1994) were able to quantitatively reproduce the observed recovery in friction using estimates of the change in contact area from their microscopic images.

We can combine Dieterich and Kilgore's (1994) observation that contact area increases with logarithm of time with equation (11.14) as follows. From the experimental data, take $A_c = A_c^0 + A_1 \log \theta / \theta_0$, where A_c^0 gives the contact area at $\theta = \theta_0$. Thus,

$$\frac{E_1}{\sigma_c \Omega} = \frac{E_1 A_c^0}{\sigma \Omega A} + \frac{E_1 A_1}{\sigma \Omega A} \log \frac{\theta}{\theta_0}. \tag{11.18}$$

This suggests that $f_0 = E_1 A_c^0 / \sigma \Omega A$ and that the constitutive parameter b is given by $f_0 A_1 / A_c^0$. Data for glass at 10 MPa (figure 11.7) suggests that $A_1 / A_c^0 \sim 0.03$, which yields $b \sim 0.02$ for f_0 of 0.6, which is of the same order of magnitude as experimentally derived estimates. Note that by the same arguments, the coefficient a in equation (11.14) should increase with contact time; however, this can be viewed as a second-order effect.

Rate and state friction laws have proved extremely useful in understanding slip behavior. This is true, even though a full micromechanical understanding of the processes that are responsible for the macroscopic behavior remains uncertain. (The same is true for linear elastic behavior; the theory was found to be useful well before the underlying atomic scale physics that gives rise to the macroscopic behavior was understood.) Improved knowledge of the underlying processes would, however, allow more confident extrapolations from laboratory scale to natural faults.

The data in figure 11.7 are for slide-hold-slide experiments. Small amounts of slip can occur during the nominal hold, leading to changes in state for the slip law. Dieterich and Kilgore (1994) reported that experiments under true stationary contact, without applied shear stress, yielded similar behavior. If slip during nominally stationary contact gives rise to state evolution, then frictional restrengthening during slide-hold-slide tests should depend on machine stiffness. Beeler et al. (1994) varied machine stiffness and found that the rate of strengthening with log time was independent of stiffness, inconsistent with the slip law prediction.

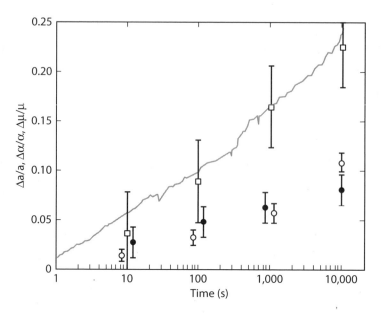

Figure 11.7. Change in contact area and friction during slide-hold-slide tests (from Dieterich and Kilgore 1994). During these experiments, the surface first slides at constant slip speed. The load point is then held at fixed displacement during the "hold," and then sliding is resumed. The continuous line and open circles represent normalized change in the contact area. The continuous line gives the maximum change in contact area, while the open circles give the change in area at the peak change in friction. The solid circles indicate the relative change in the friction coefficient. The open squares indicate changes in area during microindentation experiments. Experiments on glass at 10 MPa normal stress.

The peak strength data from the slide-hold-slide tests in Beeler et al. (1994) and Dieterich and Kilgore (1994) support the aging law. On the other hand, the slip law may be more consistent with velocity stepping experiments. For the slip law, the normalized strength decays with the same distance scale regardless of the magnitude of the jump in slip speed (figure 11.8B). The slip law is also antisymmetric with respect to positive and negative velocity steps. On the other hand, for the aging law, the apparent weakening distance increases with the magnitude of the positive velocity step but exhibits only small strength changes for step decreases in velocity (figure 11.8A). Ruina (1980, figure 10) showed two order of magnitude step increases and decreases that are nearly antisymmetric, supporting the slip law. Bayart et al. (2006) showed similar results with yet larger velocity steps. Nakatani (2001) also concludes that the slip law is more consistent with experimental observations. It is fair to say that at the time of this writing, considerable uncertainty exists as to both the relevant micromechanics and even the most appropriate state evolution equations for modeling fault slip. Indeed, it is not clear that a single-state evolution law is appropriate for all conditions of interest.

The discussion so far has been limited to conditions of constant normal stress; however, normal stress can be expected to vary during slip due to a number of processes. First, faults are never precisely planar, and departures from planarity necessarily lead to changes in normal stress with slip. Second, slip on faults that bound rocks of differing elastic properties changes the fault-normal stress even for perfectly planar faults. Last, changes in pore-fluid pressure lead to changes in the effective stress borne by solid contacts.

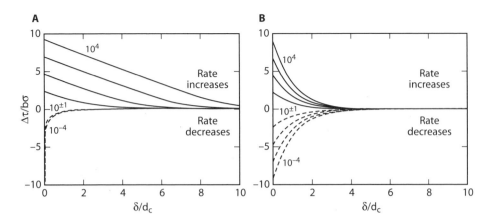

Figure 11.8. Normalized stress as a function of normalized displacement for step changes in velocity. Numbers indicated magnitude of velocity jump. A: Aging law. B: Slip law. Step increases are solid lines; decreases are dashed lines. After Rubin and Ampuero (2008).

General considerations suggest that for variable normal stress, one could consider modifying equation (11.9) to

$$\frac{d\theta}{dt} = g_1(v, \theta) + \frac{d\sigma}{dt} g_2(\sigma, v, \theta) \tag{11.19}$$

(Rice et al. 2001), where g_1 could be either equation (11.16) or (11.17). Linker and Dieterich (1992) conducted experiments with step changes in normal stress and concluded that an appropriate form for g_2 is $-\alpha\theta/b\sigma$, where α is a newly introduced nondimensional parameter. For rapid changes in normal stress, such that θ does not evolve due to slip (constant δ), it follows that $\theta/\theta_0 = (\sigma/\sigma_0)^{-\alpha/b}$, where σ_0 and θ_0 are the normal stress and state prior to the imposed change in normal stress. For constant velocity sliding, the rate of change of friction is

$$\dot{\tau} = \sigma b \frac{\dot{\theta}}{\theta} + f\dot{\sigma}, \tag{11.20}$$

where the overdot indicates a time derivative. For rapid changes in normal stress, so that state does not evolve due to slip, Linker and Dieterich (1992), predict $\dot{\theta} = -(\alpha\theta/b\sigma)\dot{\sigma}$, so that

$$\dot{\tau} = (f - \alpha)\dot{\sigma}. \tag{11.21}$$

Thus, $f - \alpha$ acts as an effective friction coefficient for rapid changes in normal stress at fixed sliding velocity.

The concept of steady-state friction that depends only on normal stress and slip speed is a central concept in rate and state friction theory. For the single degree of freedom spring-slider (illustrated in figure 11.1), the steady-state slip speed is the load point speed, v^∞. For both evolution equations (11.16) and (11.17), the steady-state value of the state variable θ is $\theta_{ss} = d_c/v^\infty$. The steady-state frictional resistance is found by substituting v_{ss} and θ_{ss} into

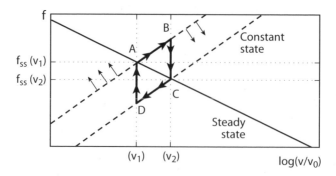

Figure 11.9. Phase plane f versus $\log(v/v_0)$. The solid line indicates the steady-state friction line with slope $a - b$, drawn for steady-state velocity weakening. Lines of constant state are dashed; arrows point in the direction of state evolution. Lines connecting the points marked A, B, C, and D correspond to the evolution of the system during a velocity stepping test, as illustrated in figure 11.5.

equation (11.10), such that the steady-state friction is

$$\tau_{ss} = \sigma \left[f_0 + (a - b) \log \left(\frac{v_{ss}}{v_0} \right) \right]. \tag{11.22}$$

Thus, $(a - b)$ measures the slope of the steady-state frictional dependence on log velocity. If $a > b$, the slope is positive, and friction is *steady-state velocity strengthening*. Alternatively, if $a < b$, the slope is negative, and the friction is *steady-state velocity weakening*.

It often proves helpful to visualize the behavior of the spring-slider system with rate and state friction in terms of a phase space with axes of f and $\log(v/v_0)$, as introduced by Ruina (1983). First, note that the steady-state behavior (11.22) is simply a line with slope $a - b$ in this space. Rapid changes, which occur at constant state, are also straight lines, with slope a. A velocity stepping test as illustrated in figure 11.5 can be understood by the trajectory in phase space illustrated in figure 11.9. The experiment starts at steady state with velocity v_1. An instantaneous jump in load point velocity to v_2 causes the friction to instantaneously increase by $a \log(v_2/v_1)$, to point B. The velocity is then held fixed as the friction decays to a new steady-state value appropriate to the new slip speed $f_{ss}(v_2)$, point C. A step decrease in velocity back to v_1 causes an instantaneous decrease in friction (point D) and then a gradual increase back to the steady-state friction appropriate for that slip speed, $f_{ss}(v_1)$, point A.

11.3.1 Linearized Stability Analysis

In this section, we determine whether the spring-slider system at steady state, with the slider moving at speed v^∞, is stable to small perturbations in either stress or slip speed. The analysis, following Ruina (1983), which for simplicity is restricted to constant normal stress and neglects inertia, consists of two steps. First, we linearize the governing equations about the steady-state solution, which yields a system of ordinary differential equations in terms of the perturbations in slip speed, stress, and state. We then seek solutions to these equations of the form $\exp(st)$. If the real part of s is positive, then perturbations grow exponentially. Alternatively, if the real part of s is negative, then perturbations are damped with time and the system is stable.

At equilibrium, the slider is moving at v^∞, and the frictional resistance balances the spring force:

$$\sigma \left(f_0 + a \log \frac{v^\infty}{v_0} + b \log \frac{v_0 \theta_{ss}}{d_c} \right) = k(v^\infty t - \delta). \tag{11.23}$$

Perturbing equation (11.23) around steady state, $v = v^\infty + \Delta v$, $\theta = \theta_{ss} + \Delta \theta$, equivalent to keeping the first two terms in a Taylor series expansion about steady state, we find

$$\sigma \left(f_0 + a \log \frac{v^\infty}{v_0} + a \frac{\partial}{\partial v} \log \frac{v}{v_0} \bigg|_{ss} \Delta v + b \log \frac{v_0 \theta_{ss}}{d_c} + b \frac{\partial}{\partial \theta} \log \frac{v_0 \theta}{d_c} \bigg|_{ss} \Delta \theta \right) = k(v^\infty t - \delta - \Delta \delta),$$

$$\sigma \left(a \frac{\partial}{\partial v} \log \frac{v}{v_0} \bigg|_{ss} \Delta v + b \frac{\partial}{\partial \theta} \log \frac{v_0 \theta}{d_c} \bigg|_{ss} \Delta \theta \right) = -k \Delta \delta,$$

$$\sigma \left(\frac{a}{v^\infty} \Delta v + \frac{b v^\infty}{d_c} \Delta \theta \right) = -k \Delta \delta. \tag{11.24}$$

Perturbing either of the evolution equations (11.16) or (11.17) about steady state yields the same linearized expression:

$$\frac{\partial \Delta \theta}{\partial t} = \Delta \dot\theta = \frac{\partial \dot\theta}{\partial v} \bigg|_{ss} \Delta v + \frac{\partial \dot\theta}{\partial \theta} \bigg|_{ss} \Delta \theta,$$

$$= -\frac{1}{v^\infty} \Delta v - \frac{v^\infty}{d_c} \Delta \theta. \tag{11.25}$$

Differentiating the last expression in (11.24) with respect to time and combining with (11.25):

$$\sigma \frac{a}{v^\infty} \Delta \dot v = -\sigma \frac{b v^\infty}{d_c} \Delta \dot\theta - k \Delta v,$$

$$\Delta \dot\theta = -\frac{1}{v^\infty} \Delta v - \frac{v^\infty}{d_c} \Delta \theta. \tag{11.26}$$

Equations (11.26) represent two linear differential equations in two unknowns Δv and $\Delta \theta$. We now seek solutions of the form

$$\Delta v = V e^{st},$$
$$\Delta \theta = \Theta e^{st}. \tag{11.27}$$

Substituting the preceding into equation (11.26) and combining leads to a single equation that is quadratic in s:

$$\sigma a s^2 + \left[\sigma(a-b) \frac{v^\infty}{d_c} + k v^\infty \right] s + k v^\infty \frac{v^\infty}{d_c} = 0. \tag{11.28}$$

The behavior of the system is determined by the real part of s. First note that in the limit of infinite spring stiffness, $k \to \infty$, both roots of equation (11.28) are real and negative (see problem 1 and figure 11.10). This implies that the system is stable for sufficiently large spring stiffness, which is consistent with both laboratory behavior and simple slip-weakening friction. Second, notice that $s=0$ is not a solution to equation (11.28) for $k > 0$. Thus, when

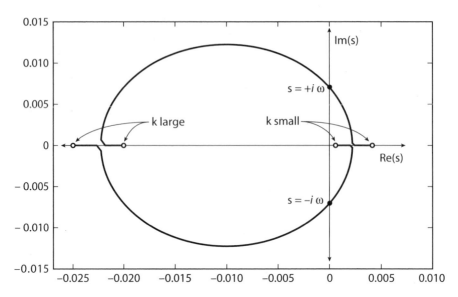

Figure 11.10. An s-plane plot showing the behavior of the roots as a function of spring stiffness k. In the limit that k is very large, both roots are real and negative. The roots cross the imaginary axis as a pair at $\pm i\omega$ at the critical spring stiffness. As k decreases from the critical stiffness, the roots eventually become real and positive. The figure is drawn for $a/(b - a) = 2$.

the real part of s changes sign, the roots must cross the imaginary axis as a pair, which we label $s = \pm i\omega$, as in figure 11.10. Substituting this into equation (11.28) and noting that both real and imaginary parts of the resulting expression must be satisfied yields a pair of algebraic equations:

$$-\sigma a\omega^2 + kv^\infty \frac{v^\infty}{d_c} = 0,$$

$$\sigma(a - b)\frac{v^\infty}{d_c} + kv^\infty = 0. \tag{11.29}$$

The second equation yields the critical spring stiffness:

$$k_{crit} = \frac{\sigma(b - a)}{d_c}. \tag{11.30}$$

This important result, due to Ruina (1983), shows that if the spring stiffness is greater than k_{crit}, the system will respond stably to small perturbations. If the spring stiffness is less than the critical value, small perturbations will grow in amplitude, the slip-speed eventually becoming infinite. As in the case of simple slip-weakening friction, increasing normal stress causes the system to become more unstable. Increasing normal stress increases the critical stiffness, making it increasingly likely that $k < k_{crit}$. The result (11.30) shows that the tendency for instability also depends on the difference $b - a$, which determines the slope of the steady-state weakening line. Since spring stiffness is a positive quantity, the system is stable to small perturbations if $a > b$ for all stiffnesses. Note with reference to equation (11.22) that

$\partial \tau_{ss}/\partial \log v = a - b$, so it is possible to write the critical stiffness as

$$k_{crit} = -\frac{\sigma}{d_c}\frac{\partial \tau_{ss}}{\partial \log v}. \tag{11.31}$$

This latter form makes it particularly clear that the system is linearly stable if the friction is steady-state velocity strengthening, $\partial \tau_{ss}/\partial \log v > 0$. For steady-state velocity weakening, $\partial \tau_{ss}/\partial \log v < 0$, the system is potentially linearly unstable, if $k < k_{crit}$.

While the second equation in (11.29) gives the critical spring stiffness, the first gives ω, which is the frequency of oscillation at neutral stability $k = k_{crit}$. Specifically,

$$\omega = \sqrt{\frac{b-a}{a}\frac{v^{\infty}}{d_c}}. \tag{11.32}$$

Numerical results are shown in figure 11.11 for a spring-slider system slightly perturbed from steady state. We note that when the stiffness equals the critical value, the solution is purely oscillatory and the phase plane trajectories are stable orbits. When the stiffness exceeds the critical value, the oscillations decay with time and the phase trajectories spiral into the steady-state fixed point. In contrast, when the stiffness is slightly less than the critical stiffness, the oscillations grow with time, and eventually become inertially limited. As expected, the numerical results confirm the predictions of the linearized stability analysis.

The stability analysis highlights the importance of the velocity dependence of steady-state friction, parameterized by $a - b$. Laboratory experiments indicate that $\partial \tau_{ss}/\partial \log v$ varies with temperature. Indeed, equation (11.15) indicates that the direct effect should scale linearly with temperature, and it is also sensible that creep of contacting asperities, if that is indeed what controls state, will be temperature dependent. Blanpied et al. (1995) conducted experiments on simulated granite fault gouge over a range of temperatures at an effective normal stress of 400 MPa and fluid pressure of 100 MPa. Their results show a fair amount of scatter but generally that $\partial \tau_{ss}/\partial \log v$ is negative at temperatures between roughly 100 and 300°C (figure 11.12). At higher temperatures between 350°C and 600°C, friction is steady-state velocity strengthening, with $a - b$ with increasing temperature. At temperatures below 100°C, the behavior is weakly velocity strengthening. Other lab experiments, however, show steady-state velocity weakening of bare surfaces at room temperature, although $\partial \tau_{ss}/\partial \log v$ tends to increase (become positive) with increasing gouge thickness (Marone 1998).

According to equation (11.31), faults are unstable only when $\partial \tau_{ss}/\partial \log v < 0$, suggesting that earthquakes can nucleate only at depths corresponding to temperatures between 100 and 300°C. For a thermal gradient of 25°C/km, this corresponds to a depth range of roughly 3 to 13 km, which agrees quite well with the depth range of earthquake nucleation along the San Andreas fault system.

11.4 Implications for Earthquake Nucleation

The stability results from the previous section have important implications for earthquake nucleation. We found that for unstable slip to occur, the stiffness of the slipping patch must be less than a critical value. It is reasonable to idealize the slipping zone as a circular crack of radius r and use the elastic solution from chapter 4 to compute the effective stiffness of the crack. Computing the effective stiffness $\Delta \tau/2u$ using the maximum displacement at the center

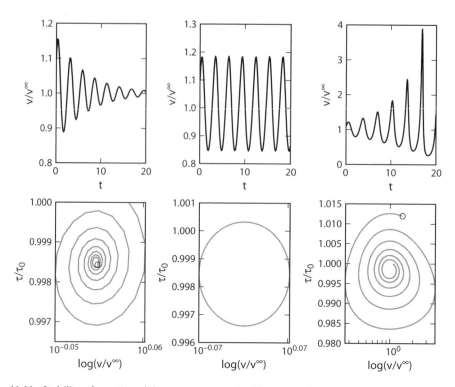

Figure 11.11. Stability of a spring-slider system perturbed from steady state. Top: Velocity time history. Bottom: Phase trajectories in $\tau - \log v$ space. The left column is for $k/k_c = 1.2$. The middle column is for $k/k_c = 1$. The right column is for $k/k_c = 0.85$. The open circle represents the final point. Computations were done with the aging law.

of the crack from equation (4.74) and assuming $v = 0.25$:

$$k = \frac{7\pi\mu}{24r}. \tag{11.33}$$

Combining this with the critical stiffness from equation (11.30), we find that the minimum crack radius for unstable slip is

$$r_{crit} = \frac{7\pi\mu d_c}{24(\sigma - p)(b - a)}, \tag{11.34}$$

Where $\sigma - p$ is the effective fault-nomal stress. The slipping zone must grow to at least this size for slip to become unstable. For a depth of 7 km, the effective stress is on the order of 100 MPa, assuming hydrostatic pore pressure. Laboratory data (figure 11.12) suggest that $b - a \sim 0.005$ or less. There has historically been a great deal of uncertainty about the appropriate value of d_c to use in modeling earthquakes. Laboratory data are generally characterized by values in the range of 1 to 100 μm; however, d_c is known to scale with surface roughness and gouge particle size. Since natural faults may be rougher than laboratory analogues, many workers have suggested that d_c is considerably larger in nature than in the laboratory. This has remained a subject of debate; some geologic observations have suggested that the active deformation zones are not only very narrow but relatively smooth on the scale of earthquake slip distances. For the sake of argument, here we use $d_c = 10$ μm as

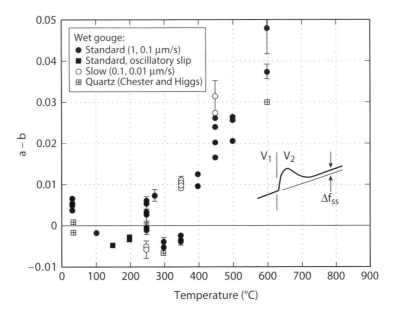

Figure 11.12. Velocity dependence of steady-state friction, $a - b$, of granite surfaces as a function of temperature, T °C, after Blanpied et al. (1995). Points labeled Standard used velocity steps from 0.1 to 1 micron/second, whereas those labeled Slow used velocity steps from 0.01 to 0.1 micron/second.

representative of lab values. From equation (11.34) and assuming a shear modulus of 3×10^4 MPa, this yields a critical crack dimension of 0.5 meters. For d_c of 100 μm, the critical radius is 5 meters. The presence of microearthquakes in some mines is not inconsistent with these rupture dimensions and hence laboratory estimates of d_c.

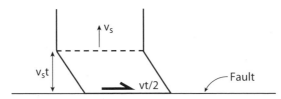

Figure 11.13. Diagram to illustrate the radiation damping stress. The fault slips at rate v, so at time t, the upper surface has displaced by an amount $vt/2$. An s wave has propagated a distance $v_s t$ from the surface.

The full nonlinear friction equations combined with spring-slider elasticity can be solved using standard algorithms for coupled differential equations. Ignoring inertial effects leads to situations in which the slip speed increases without bound. Rather than explicitly include inertia as $m \mathrm{d}v/\mathrm{d}t$, we use the so-called *radiation damping approximation*, in which the equation of motion becomes

$$k(v^{\infty} t - \delta) - \sigma f(v, \theta) = \eta v, \tag{11.35}$$

where $\eta=\mu/2v_s$, μ is the shear modulus, and v_s is the shear wave velocity (Rice 1993). The radiation damping stress can be simply understood with reference to figure 11.13, which illustrates one side of a fault. The fault slip rate is v, so that at time t, the upper surface has displaced by an amount $vt/2$. A plane s wave radiated from the fault has traveled a distance $v_s t$ from the surface in the same time interval. Thus, the medium adjacent to the fault has strained by $\gamma = (vt/2)/(v_s t) = v/2v_s$, and the radiation stress is $\mu v/2v_s$. While $m dv/dt$ is correct for a lumped mass, relative to faulting in a continuum, it overpredicts dynamic overshoot. Radiation damping, on the other hand, predicts no dynamic overshoot.

It is convenient to express the system as a series of coupled first-order differential equations. Differentiating equation (11.35) with respect to time when combined with (11.16) or (11.17) and the time derivative of (11.10) yields the following system:

$$\dot{v} = \left(\frac{\eta}{\sigma} + \frac{a}{v}\right)^{-1} \left(\frac{\dot{\tau}}{\sigma} - \frac{b\dot{\theta}}{\theta}\right), \tag{11.36}$$

$$\dot{\theta} = 1 - \frac{\theta v}{d_c} \quad \text{or} \quad -\frac{v\theta}{d_c} \ln\left(\frac{v\theta}{d_c}\right). \tag{11.37}$$

$$\dot{\tau} = k(v^\infty - v). \tag{11.38}$$

Numerical simulation of the equations governing the spring-slider system (11.36), (11.37), and (11.38) exhibit many of the features of natural earthquake cycles. In calculations here, the governing equations were first normalized: slip velocity by v^∞, time and state by d_c/v^∞, and shear stress by σ. Choosing $a/b < 1$ and a spring stiffness less than k_{crit} results in behavior as illustrated in figure 11.14. Orbits in the τ/σ versus $\log(v)$ phase space show stress accumulation at very low slip rates, followed by a nucleation phase in which the slider accelerates at nearly constant stress. Stress drop at high slip speed is followed at first by rapid deceleration along the steady-state line and then post-slip deceleration at nearly constant stress. These simple equations thus exhibit a natural transition from essentially locked to rapid slip, as well as recovery of strength following rapid slip.

Returning to the nucleation phase, we can obtain an approximate analytical solution for earthquake nucleation with aging law friction following Dieterich (1992). During the relatively short nucleation phase, it is reasonable to ignore the stressing due to plate motion; therefore, v^∞ can be neglected in equation (11.38). It is also reasonable to ignore radiation damping until very late in the nucleation phase, so η can be ignored in equation (11.36). Dieterich (1992) employed the aging law and further assumed that $\theta v/d_c \ll 1$ in equation (11.37)— this may be regarded as a "no-healing limit" since the healing term is ignored in the state evolution equation. The logic for this approximation is that during nucleation, the slip speed far exceeds the steady-state value (equal to the plate velocity) and that state, dependent on the time since the last slip event, will also be well above the steady-state value. However, as the fault accelerates, θ decreases ($\dot{\theta} < 0$), so the no-healing limit may be violated during nucleation (Rubin and Ampuero 2005). Assuming that healing can be ignored, equations (11.36), (11.37), and (11.38) reduce to

$$\frac{dv}{dt} = \frac{v^2}{a} \left(\frac{b}{d_c} - \frac{k}{\sigma}\right). \tag{11.39}$$

Notice that the system is unstable if the term in brackets is positive, since the acceleration increases with v^2. On the other hand, if the bracketed term is negative, the system is stable.

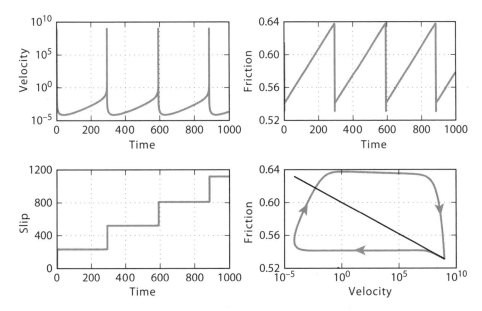

Figure 11.14. Numerical simulation of a spring-slider system with rate- and state-dependent friction exhibits stick-slip cycles when $a/b < 1$ and $k/k_{crit} < 1$. Upper left: v/v^∞ versus tv^∞/d_c, showing long periods of extremely slow slip rate punctuated by regularly spaced rapid slip events. Upper right: τ/σ versus tv^∞/d_c, showing slow stress accumulation followed by abrupt stress drops during fast slip. Lower left: δ/d_c versus tv^∞/d_c. Lower right: Phase plane plot. The line indicates the steady-state frictional response. Calculation for $a/b = 0.75$ and $k/k_{crit} = 0.1$.

This leads to a critical stiffness given by

$$k_{crit} = \frac{\sigma b}{d_c}, \tag{11.40}$$

which is of the same form as equation (11.30) with $b - a$ replaced by b. The critical stiffness in equation (11.30) was derived by considering small perturbations from steady sliding. On the other hand, equation (11.40) is based on the assumption that the system is far from steady state.

Equation (11.39) can be integrated, with initial conditions $v(t = 0) \equiv v_0$, to yield

$$\frac{v}{v_0} = \left(1 - \frac{Hv_0 t}{a}\right)^{-1}, \tag{11.41}$$

where H is the term in brackets in equation (11.39), with units of 1/length. Because we have ignored inertial effects, equation (11.41) predicts that the slip speed becomes infinite at time $t = t_{inst} = a/Hv_0$. Integrating (11.41) gives the fault slip as a function of time:

$$s(t) = -\frac{a}{H} \log\left(1 - \frac{Hv_0 t}{a}\right) = -\frac{a}{H} \log\left(1 - \frac{t}{t_{inst}}\right). \tag{11.42}$$

In the no-healing limit, θ evolves exponentially with slip $\theta = \theta_0 \exp(-s/d_c)$. From this and equations (11.41) and (11.42), it can be shown that the no-healing approximation, $v\theta/d_c \gg 1$, remains valid only if $k < \sigma(b - a)/d_c$ (Rubin and Ampuero 2005).

Important insights into earthquake nucleation with rate and state friction have been gained by numerical simulations, mostly with one-dimensional faults and two-dimensional elasticity. For antiplane strain, the equation of motion becomes

$$\tau^{\infty}(x) - \frac{\mu}{2\pi} \int_{-\infty}^{\infty} \frac{\partial s/\partial \xi}{x - \xi} d\xi - \sigma f(v, \theta) = \eta v. \tag{11.43}$$

The first two terms on the left side represent the stress acting to drive slip, while $\sigma f(v, \theta)$ is the frictional resistance to slip, σ being interpreted as the effective normal stress. τ^{∞} is the remotely applied stress acting on the fault, whereas the stress change due to slip s is given by the Hilbert transform of the slip gradient, equation (4.64) in chapter 4. The difference between the stress and the frictional resistance is balanced by radiation damping ηv. Because rate and state friction postulates that the fault is always slipping, albeit sometimes at vanishingly small rates, we can take the integral in equation (11.43) to extend over the entire fault and avoid the mixed boundary conditions that characterize traditional crack models. As described in chapter 4, the stress interaction term in the Fourier domain is $-(\mu/2)|n|\mathcal{F}(s)$, where $\mathcal{F}(s)$ is the Fourier transform of the slip, and n is the wavenumber. (We reserve k for stiffness in this chapter.) Differentiating equation (11.43) with time yields

$$\left(\frac{\eta}{\sigma} + \frac{a}{v}\right) \dot{v} = \frac{\dot{\tau}^{\infty}}{\sigma} - \frac{\mu}{2\sigma}|n|v - \frac{b\dot{\theta}}{\theta}, \tag{11.44}$$

where it is implicit that the spatially variable slip speed v and state θ are Fourier transformed quantities. Equation (11.44) together with an evolution law, specified $\dot{\tau}^{\infty}$, and appropriate initial conditions describes the evolution of the system.

Figure 11.15 shows a simulation in which the frictional properties are uniform with $a/b = 0.3$, but the initial velocity and therefore stress are spatially heterogeneous. The computation is initiated well above steady state. Accelerating slip is found to concentrate in a zone of fixed length, even though longer wavelengths are present in the initial velocity distribution that are more compliant than the nucleation length (Dieterich 1992). We can gain some insight into this behavior by ignoring radiation damping and remote stressing in equation (11.44). In the no-healing limit, this reduces to

$$\frac{dv}{dt} = \frac{v^2}{a} \left(\frac{b}{d_c} - \frac{\mu|n|}{2\sigma}\right), \tag{11.45}$$

which suggests a critical wavenumber $n_{crit} = 2\sigma b/\mu d_c$. For velocity fluctuations at smaller wavenumber, the perturbation accelerates proportional to velocity squared. Indeed, filtering the heterogeneous initial velocity distribution for wavenumbers smaller than the critical value and picking the maximum initial velocity accurately predicts the location of the eventual nucleation zone, as long as the no-healing limit is satisfied. Given that $n = 2\pi/\lambda$ and that the wavelength λ is twice the length of the nucleation zone, this suggests that the half-length of the growing slip zone is $l \cong \pi\mu d_c/4b\sigma$. Rubin and Ampuero (2005) show that the fixed length solution with finite stress (zero stress intensity factor) at the edge of the nucleation zone has a half-length of $1.377\mu d_c/b\sigma$, not far from the simple estimate of $\pi\mu d_c/4b\sigma$. We refer to $1.377\mu d_c/b\sigma$ as the *Dieterich length*, since it was first noted by Dieterich (1992). The Dieterich

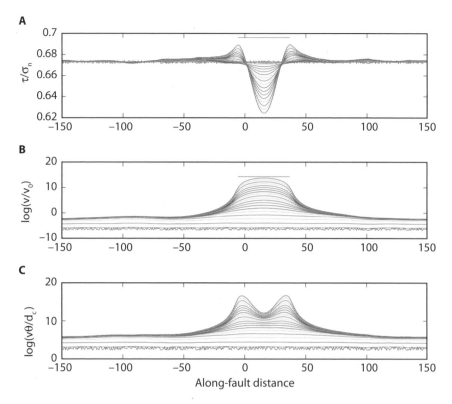

Figure 11.15. Simulation of earthquake nucleation in a two-dimensional elastic full-space. Each curve represents a snapshot in time. The horizontal axis is along-fault dimension (meters). A: Shear stress normalized by effective normal stress. B: Slip speed $\log(v/v_0)$. C: $\log(v\theta/d_c)$. Computations are for $d_c = 1$ mm, effective normal stress of 100 MPa, $a = 0.003$, $b = 0.01$, and the aging law. The initial velocity distribution includes a small component of white noise. Bar indicates a length of $2 \times 1.377 \mu d_c/b\sigma$.

length is smaller than the critical length for slip to nucleate, which corresponds to the critical stiffness (11.30):

$$L_c = \frac{\pi \mu d_c}{4(b-a)\sigma}, \tag{11.46}$$

by a factor of $b/(b-a)$. The assumption that healing can be neglected throughout nucleation, however, is valid only for $a/b < 0.38$ (Rubin and Ampuero 2005). For larger a/b, the state decreases sufficiently rapidly that $v\theta/d_c$ is driven toward a constant value, near unity, that depends on a/b.

For $a/b > 0.38$, the behavior is qualitatively different (figure 11.16), with nucleation occurring as a crack that expands quasi-statically to a limiting length, at which point, the slip becomes dynamic. Rubin and Ampuero (2005) have shown that it is possible to understand this behavior using fracture mechanics concepts. This approach is motivated by numerical results, which show that the stress within the interior of the slipping zone is nearly uniform (figure 11.16). Specifically, they balance an effective fracture energy at the crack front G_c with an estimate of the crack extension force G.

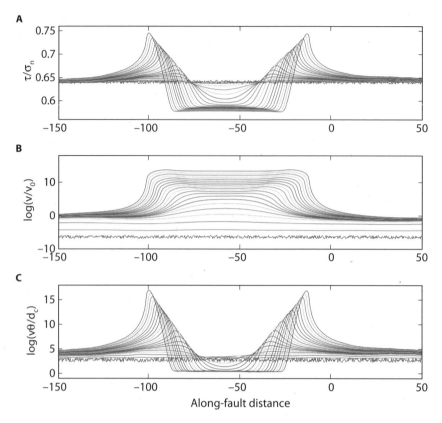

Figure 11.16. Simulation of earthquake nucleation in a two-dimensional elastic full-space. Each curve representes a snapshot in time. The horizontal axis is along-fault dimension (meters). A: Shear stress normalized by effective normal stress. B: Slip speed $\log(v/v_0)$. C: $\log(v\theta/d_c)$. Computations are for $d_c = 1$ mm, $a = 0.008$, $b = 0.01$, with the aging law.

To a reasonable approximation, the slip speed in the interior of the crack, v, is uniform. In addition, the interior of the crack is nearly steady state (figure 11.16C), such that the stress there is $\tau_{in} \sim \sigma[f_0 + (a - b) \log(v/v_0)]$. In front of the crack, the slip speed increases rapidly at nearly constant state from the background value to high value at the crack tip. If the state variable in front of the crack is given by θ_i, then the peak stress is approximately $\tau_p \sim \sigma[f_0 + a \log(v/v_0) + b \log(v_0\theta_i/d_c)]$. The stress drop from the peak at the crack tip to the stress inside the crack (figure 11.17) is

$$\Delta\tau^p \equiv \tau_p - \tau_{in} \sim \sigma b \log(v\theta_i/d_c). \tag{11.47}$$

The fracture energy G_c is given by the area under the slip-weakening curve, $G_c = \Delta\tau^p s_c/2$, where s_c is the slip distance over which the stress drops from τ_p to τ_{in}. Healing can be neglected in the weakening zone behind the crack tip; in the no-healing limit, the state evolution equation (11.16) integrates directly to $\theta = \theta_0 \exp(-s/d_c)$. Substituting into the friction law (11.10) yields

$$\tau = \sigma \left(f' + a \log \frac{v}{v_0} - b \frac{s}{d_c} \right). \tag{11.48}$$

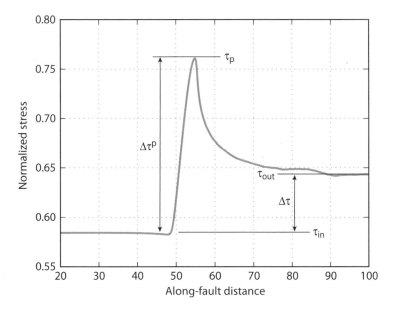

Figure 11.17. Stress distribution in the neighborhood of a growing crack. All stresses are normalized by the fault-normal stress. τ_{out} is the stress far from the crack, τ_{in} the stress in the interior of the slipping crack, and τ_p is the peak stress at the crack tip. The stress drop is $\Delta\tau$, and the peak-to-residual stress drop is $\Delta\tau^p$.

This shows that the friction law exhibits linear slip weakening in the no-healing limit. Thus, at constant slip speed, $d\tau/ds = -\sigma b/d_c$. From this, we deduce that $s_c \sim d_c \Delta\tau^p/b\sigma$, so that

$$G_c = \frac{d_c(\Delta\tau^p)^2}{2b\sigma} = \frac{b\sigma d_c}{2}\left[\log\left(\frac{v\theta_i}{d_c}\right)\right]^2. \tag{11.49}$$

The energy release rate is given by $G = (\pi/2)(\Delta\tau)^2 L/\mu$ (chapter 4), where L is crack half-length, and $\Delta\tau$ is the stress drop (figure 11.17). If the slip speed outside the crack is v_{out}, then the stress in front of the crack can be approximated by $\tau_{out} \sim \sigma[f_0 + (a-b)\log(d_c/\theta_i v_0) + a\log(v_{out}\theta_i/d_c)]$, so that the stress drop is

$$\Delta\tau \equiv \tau_{out} - \tau_{in} \sim \sigma[(b-a)\log(v\theta_i/d_c) + a\log(v_{out}\theta_i/d_c)]. \tag{11.50}$$

Thus, the energy release rate is

$$G = \frac{\pi L(b-a)^2\sigma^2}{2\mu}\left[\log\left(\frac{v\theta_i}{d_c}\right) + \frac{a}{b-a}\log\left(\frac{v_{out}\theta_i}{d_c}\right)\right]^2. \tag{11.51}$$

The crack grows stably as long as the rate of energy release balances the fracture energy. Equating G from equation (11.51) with G_c from (11.49) yields the equilibrium crack half-length:

$$L = \frac{b\mu d_c}{\pi(b-a)^2\sigma}\left[\frac{\log(v\theta_i/d_c)}{\log(v\theta_i/d_c) + a\log(v_{out}\theta_i/d_c)/(b-a)}\right]^2. \tag{11.52}$$

If the fault in front of the crack is above steady state such that $v_{out}\theta_i/d_c > 1$, the term in brackets is less than one. As the slip speed inside the crack increases, v eventually dominates v_{out}, such

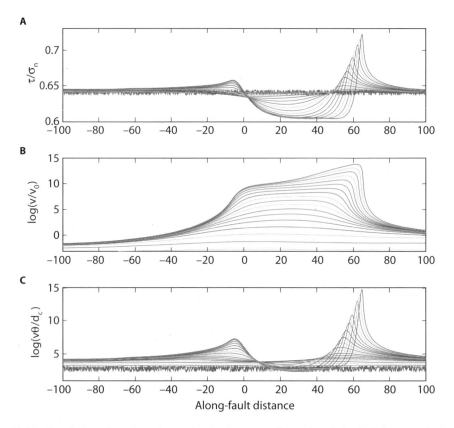

Figure 11.18. Simulation of earthquake nucleation in a two-dimensional elastic full-space. Each curve represents a snapshot in time. The horizontal axis is along-fault dimension (meters). A: Shear stress normalized by effective normal stress. B: Slip speed $\log(v/v_0)$. C: $\log(v\theta/d_c)$. Computations are done for $d_c = 1\ mm$, $a = 0.008$, $b = 0.01$, and the slip law.

that the term in brackets approaches one from below, leading to a maximum equilibrium half-length $L_\infty = b\mu d_c/\pi(b-a)^2\sigma$. L_∞ is a factor of $[b/(b-a)]^2$ greater than the Dieterich length, which could be a substantial difference if $b-a$ is small, making detection of a nucleation phase under these circumstances considerably more likely.

The Dieterich length and the asymptotic length L_∞ apply only for the aging law, whereas L_c, the critical length for nucleation (11.46), applies to both commonly used evolution laws, as they have the same linearized form about steady state. The behavior for the slip law, however, is yet different (figure 11.18). Simulations with the slip law show a growing unidirectional slip pulse as opposed to cracklike growth.

We now return to the question of the potential for detecting premonitory strain and take equation (11.42) to give the slip as a function of time. A reasonable value for H can be found by setting the stiffness to the critical value given by equation (11.30), which yields $s(t) = d_c \log(1 - t/t_{inst})$. The predicted strain will of course depend on the relative geometry of the strainmeter and the nucleation patch. For simplicity, we assume strike slip on a vertical fault directly below a strainmeter capable of measuring shear strain. From results presented in chapter 3, it is possible to show that for this geometry, the shear strain is given by

$$\gamma = \frac{sA}{8\pi d^3},\qquad\qquad (11.53)$$

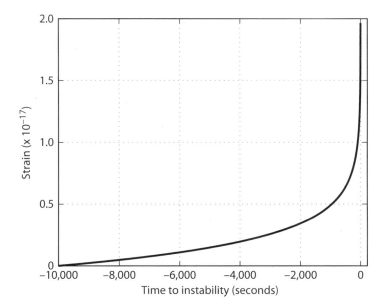

Figure 11.19. Predicted shear strain history for a nucleation patch 7 km below the strainmeter. Parameters are chosen to be plausible based on laboratory measurements.

where A is the fault area, computed from equation (11.34), and d is the depth to the nucleation patch. Combining, we have

$$\gamma = \frac{d_c}{8d^3} \left[\frac{7\pi\mu d_c}{24(\sigma - p)(b - a)} \right]^2 \log\left(1 - \frac{t}{t_{inst}}\right). \qquad (11.54)$$

To make an order of magnitude estimate of the nucleation strain, we choose the following parameter values: a shear modulus of 3×10^4 MPa and an effective normal stress of 17 MPa/km depth, which yields roughly 120 MPa for a depth of 7 km, a typical hypocentral depth on the San Andreas fault. We take $b - a \sim 0.005$ and d_c of 10 μm. The characteristic nucleation time starting from the plate velocity of 10^{-9} m/s is $d_c/v_0 = 10^4$ seconds, or a few hours. The predicted strain from the simple model is shown in figure 11.19. The maximum strain change at 1 second before the instability is of the order of a few parts in 10^{-17} or 10^{-8} nanostrain, well below the current level of detectability. The decay of strain with distance cubed effectively prohibits surface detection of deep nucleation with current instrumentation. At the time of this writing, a number of experiments are under way that attempt to place strain and seismic instruments in the near field of the nucleation source. Figure 11.20 shows the strain between $t = d_c/v_0$ and one second for a range of depths and d_c consistent with laboratory data. We predict that precursory strains can be measured only by placing sensors extremely close to the nucleation zone.

11.5 Nonlinear Stability Analysis

Nonlinear stability analyses of the single degree of freedom spring-slider system have yielded significant insights (Rice and Gu 1983; Ranjith and Rice 1999). Unlike the linearized stability analysis, which predicts the same behavior for either of the commonly used state evolution laws, the nonlinear analysis predicts different behavior for the two laws.

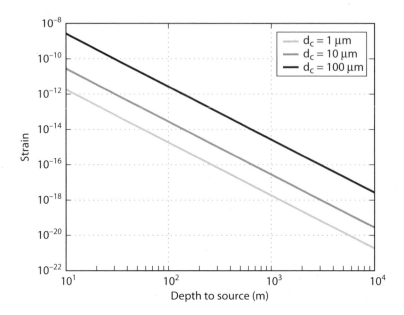

Figure 11.20. Strain accumulating between 10^3 s and 1 s prior to instability as a function of depth of the nucleation patch for different values of d_c.

Begin by considering the case of zero load point velocity. If processes evolve sufficiently rapidly, it is reasonable to ignore the slow stressing due to plate motion and set the load point speed to zero. With $v_0 = 0$, there is no steady solution, and the slider slip speed evolves either to zero or to infinity. For steady-state velocity weakening ($a < b$) and $k < k_{crit}$, the phase plane separates into two fields; see figure 11.21A. The boundary between the two domains is a straight line parallel to, but above, the steady-state line. For initial conditions above the stability boundary, the slider accelerates to infinite slip speed. For initial conditions below the stability boundary, the slider may accelerate for some time before reaching a maximum velocity and slowing down, ultimately tending toward zero slip speed. The offset between the steady-state line and the stability boundary depends on the evolution law. For the slip law, equation (11.17), the offset is

$$\Delta\tau = \frac{bd_ck}{b-a} = \frac{b\sigma k}{k_{crit}} \tag{11.55}$$

(Rice and Gu 1983). A similar behavior is observed with the aging law, equation (11.16) as illustrated in Figure 11.22. However, in this case, the offset between the stability boundary and the steady-state line is given by

$$\Delta\tau = b\sigma \, \log\left(\frac{k_{crit}}{k_{crit}-k}\right) \tag{11.56}$$

(Ranjith and Rice 1999). For $a > b$, the system responds stably and the slider speed eventually evolves to zero. This is illustrated for the slip law in figure 11.21B. Note that for initial stresses far above the steady-state stress, the slider first accelerates before reaching a maximum slip speed and then slowing down. For initial stresses sufficiently below steady-state stress, the slider decelerates monotonically. There is again a boundary separating the two different behaviors,

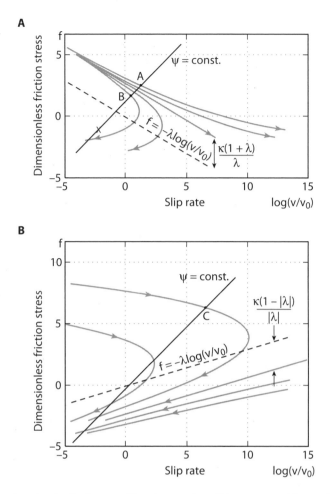

Figure 11.21. Behavior of a single degree of freedom spring-slider system with zero load point motion $v_0 = 0$ and the slip evolution law, equation (11.17), from Rice and Gu (1983). Dashed line is steady-state friction. A: Velocity weakening friction, $(b - a)/a = 0.6$. B: Velocity strengthening friction, $(b - a)/a = -0.3$. All variables are nondimensionalized; f is nondimensional friction stress, $(f - f_0)/a$, Ψ is the state variable, $\lambda = (b - a)/a$, and $\kappa = kd_c/a\sigma$. $\kappa = 1$ for these plots. See text for description.

which is parallel to but offset below the steady-state line by an amount

$$\Delta\tau = \frac{kd_c b}{a - b}. \tag{11.57}$$

These results can be extremely useful in understanding the response of the system to rapid changes in stress, due for example to neighboring earthquakes. Recall that rapid changes in stress occur at constant state, which with the nondimensionalization in figure 11.21 appear as straight lines with slope of 1. For a velocity-weakening fault, a stress step that brings the system to the point labeled A in figure 11.21A will cause the fault to accelerate to instability—a triggered earthquake or aftershock. In contrast, a fault experiencing a smaller stress step so that it moves to the point labeled B will experience an accelerating but ultimately stable creep response. Similarly, for velocity-strengthening friction, a point experiencing a stress step to C will exhibit a triggered creep response that may model afterslip following an earthquake.

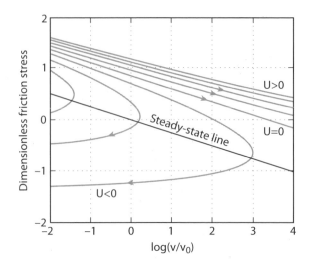

Figure 11.22. Behavior of a single degree of freedom spring-slider system with zero load point motion $v_0 = 0$ and the aging evolution law, equation (11.16), from Ranjith and Rice (1999). $U = 0$ marks the stability boundary; see text.

The system behavior including nonzero load point velocity is somewhat more complex. A nonlinear analysis with the aging law, equation (11.16), was studied by Ranjith and Rice (1999). The behavior depends on the magnitude of the stiffness k relative to the critical value k_{crit}. With nonzero load point velocity, v_0, and $k = k_{crit}$, perturbations from steady motion at v_0 induce periodic oscillations. For small perturbations, the oscillations are sinusoidal, but for larger perturbations, they are markedly nonharmonic, with fairly rapid stress drops and slow stress accumulations. For $k > k_{crit}$, the motion is stable, with the slider velocity tending toward v_0. On the other hand, for $k < k_{crit}$, the system is unstable in the sense that the slider tends to infinite slip speed if inertial effects are neglected.

For the slip evolution law, equation (11.17), the behavior is similar except that instabilities can exist even when $k > k_{crit}$ if the perturbations are sufficiently large. A sample phase plane plot for $a < b$ from Rice and Gu (1983) is shown in figure 11.23. Note that the stability boundary encloses the steady-state solution such that small perturbations from steady state are stable, as expected from the linearized analysis. As was the case with zero load point velocity, a rapid increase in stress can move the system out of the stable region. Interestingly, forcing the system to low slip speeds at nearly constant stress could result in moving the system onto an unstable trajectory when nonzero load point motion is included.

11.6 Afterslip

Transient slip has been observed or inferred following numerous earthquakes. This phenomenon is generally referred to as *afterslip* or *postseismic slip*. Afterslip was first observed following the 1966 Parkfield earthquake on the San Andreas fault (Smith and Wyss 1968) and is inferred to have resulted from slip on the shallow fault in response to coseismic slip at greater depths (figure 11.24). Afterslip below the seismogenic zone is also to be expected where, due to temperature-dependent frictional properties, the fault transitions from steady-state velocity weakening to steady-state velocity strengthening behavior. As discussed in chapter 6, for strike-slip faults, it can be difficult to distinguish deep afterslip from distributed viscoelastic deformation. Last, there are examples of afterslip inferred to have taken place along strike of

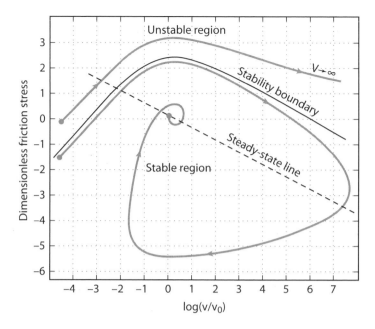

Figure 11.23. Stability boundary and trajectories for a system with imposed load point motion and the slip evolution law, equation (11.17), after Rice and Gu (1983). This particular example is for nondimensional stiffness of 1, which is twice the critical stiffness.

the coseismic rupture at seismogenic depths, perhaps suggesting lateral variations in frictional properties.

The coseismic rupture must decelerate from seismic slip rates following the earthquake. During fast coseismic rupture, the fault slips many times d_c so that θ is driven toward steady state appropriate to dynamic rupture speeds, $\theta_{dyn} \to d_c/v_{dyn}$, where dynamic rupture speeds, v_{dyn}, are of order 1 m/s. Numerical simulations employing the radiation damping approximation show that the fault tends to decelerate at approximately constant stress (figure 11.14). This is not the case with calculations that include inertia rather than radiation damping (Rice and Tse 1986), where large dynamic overshoot is observed. Assuming deceleration at nearly constant stress, and ignoring radiation damping, equation (11.36) simplifies to

$$a\frac{\dot{v}}{v} + b\frac{\dot{\theta}}{\theta} = 0. \tag{11.58}$$

As the fault decelerates, the friction drops far below steady state such that $v\theta/d_c \ll 1$. For the aging law, $\dot{\theta} \simeq 1$, so that state increases linearly with time, $\theta = \theta_{dyn} + t$, where $\theta_{dyn} = d_c/v_{dyn}$. Making this substitution in equation (11.58) and solving for slip speed yields

$$v = v_{dyn}(1 + v_{dyn}t/d_c)^{-b/a} \simeq v_{dyn}(v_{dyn}t/d_c)^{-b/a}. \tag{11.59}$$

The second form in equation (11.59) is valid for $t \gg d_c/v_{dyn}$. For d_c of 10 to 100 μm, and $v_{dyn} \sim 1$ m/s, d_c/v_{dyn} is on the order of 0.01 to 0.1 ms, so that this approximation is generally valid. To determine whether decelerating slip could be detected with geodetic or

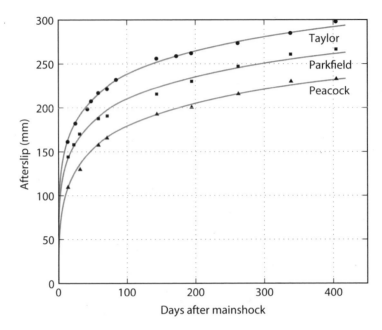

Figure 11.24. Afterslip following the 1966 Parkfield earthquake as measured by creepmeters. Data from Smith and Wyss (1968). Data are compared to the simple prediction from equation (11.70). After Marone and others (1991).

strain instrumentation, it is instructive to ask how long it takes for the slip speed to drop to the plate velocity. Rearranging equation (11.59) yields

$$t_{plate} = \frac{d_c}{v_{dyn}} \left(\frac{v_{dyn}}{v_{plate}} \right)^{a/b}. \tag{11.60}$$

Choosing generic parameter values with a/b in the range of 0.5 to 0.95, and assuming $v_{dyn}/v_{plate} \sim 10^9$, yields t_{plate} ranging from 0.3 to 3×10^4 seconds. The displacement during the deceleration phase is obtained by integrating equation (11.59). Neglecting displacement that accumulates at times $t < \theta_{dyn}$—that is setting $\delta = 0$ at $t = \theta_{dyn}$—yields

$$\delta = \frac{d_c}{\left(\frac{b}{a} - 1 \right)} \left[1 - \left(\frac{t v_{dyn}}{d_c} \right)^{1 - \frac{b}{a}} \right]. \tag{11.61}$$

The maximum displacement is $a d_c / (b - a)$, which is of the order of d_c unless the fault is very nearly velocity neutral. Whether this slip can be detected will depend on the magnitude of d_c and whether the friction is close to velocity neutral or strongly weakening.

The loading term due to plate motion becomes important as the slip speed decreases below the plate velocity. In this case, equation (11.36) simplifies to

$$a \frac{\dot{v}}{v} + b \frac{\dot{\theta}}{\theta} = \frac{k v_{plate}}{\sigma}. \tag{11.62}$$

Rubin and Ampuero (2005) define the end of the afterslip period to be the time at which the deceleration vanishes, $\dot{v} = 0$. This is easily seen from equation (11.62) to be

$$t_{post} = \frac{b\sigma}{kv_{plate}} - \frac{d_c}{v_{dyn}} \simeq \frac{b\sigma}{kv_{plate}} = \frac{d_c}{v_{plate}} \frac{k_c}{k}, \tag{11.63}$$

where we have used k_c from equation (11.40). Since the stiffness scales inversely with rupture dimension, it is reasonable to associate k_c/k with the ratio of the size of the final rupture to that of the nucleation zone. Given laboratory estimates of d_c, d_c/v_{plate} is of the order of 10^4 to 10^5 seconds, so that t_{post} could easily range from months to years. However, most of the transient displacement accumulates during fast slip shortly after the earthquake.

Most attention in the literature has focused on afterslip outside of the coseismic rupture zone. With reference to the nonlinear stability analyses, transient slip can be triggered with either velocity-strengthening or velocity-weakening friction. The conditions for stable transients with velocity-weakening friction, however, are limited. For example, in figure 11.21A, point B experiences a transient increase in slip speed. If the stress change is slightly larger to point A in figure 11.21A, the slip accelerates to inertial rates. On the other hand, if the stress change is slightly smaller, no transient is induced. For this reason, and because laboratory data suggest a transition to stable friction below, and perhaps above, the seismogenic zone, most attention has focused on afterslip in velocity-strengthening regions surrounding the coseismic rupture zone.

If the aseismically slipping zone expands with time, the process cannot be accurately modeled with a single degree of freedom model with constant stiffness. Nevertheless, it is hoped that some insight can be gained from such an analysis. Assume that a velocity-strengthening region outside the rupture is initially at steady state, at slip rate v^∞, prior to the imposition of a rapid change in stress $\Delta\tau$ due to the earthquake. For an instantaneous (constant state) change in stress, the slip rate immediately increases to

$$v = v^\infty \exp\left[\Delta\tau/a\sigma\right]. \tag{11.64}$$

Numerical simulations (figure 11.25) show that for stiff systems, $kd_c/\sigma(a - b)$ of order one or more, the slider decelerates following the instantaneous step increase. On the other hand, if the system is sufficiently compliant, relative to $\sigma(a - b)/d_c$, the slider accelerates at nearly constant stress to steady state, after which it decelerates along the steady-state line (figure 11.25). In this case, the maximum slip speed can be estimated as the velocity at which the perturbed stress reaches steady state:

$$v_{max} = v^\infty \exp\left[\Delta\tau/(a - b)\sigma\right]. \tag{11.65}$$

If $a - b$ within the velocity-strengthening regions is comparable in magnitude to $b - a$ within weakening regions where earthquakes nucleate, then as long as the afterslip zones are large in comparison to nucleation dimensions, the condition $kd_c/\sigma(a - b) \ll 1$ is likely to be met.

For sufficiently low stiffness, the system tends to decelerate along the steady-state line (figure 11.25). The governing differential equation for steady-state friction is

$$\sigma(a - b)\frac{\dot{v}}{v} = k(v^\infty - v). \tag{11.66}$$

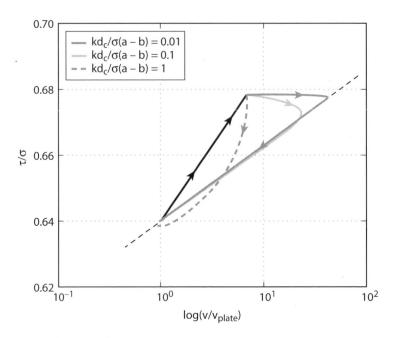

Figure 11.25. Spring-slider simulation of afterslip induced by step change in shear stress. Sudden increase in stress causes an increase in slip speed at constant state. The following response depends on the system stiffness. For sufficiently compliant stiffness, the slider first accelerates at nearly constant stress and then decelerates along the steady-state line (dashed). $d_c = 1$ mm, $a = 0.02$, $b = 0.01$.

Equation (11.66) is separable and integrates directly, noting that

$$\int \frac{\mathrm{d}v}{v(v^\infty - v)} = \frac{1}{v^\infty} \ln\left(\frac{v}{v - v^\infty}\right). \tag{11.67}$$

The initial conditions for the deceleration phase are given by $v(t = 0) = v_{max}$, where v_{max} is given by equation (11.65). Thus,

$$v(t) = \frac{v^\infty}{1 - Ce^{-t/t_c}},$$

$$C = \frac{v_{max} - v^\infty}{v_{max}},$$

$$t_c = \frac{\sigma(a - b)}{kv^\infty}. \tag{11.68}$$

Note that the characteristic decay time depends both on frictional properties as well as on the system stiffness. Larger slipping zones are more compliant, and thus will experience longer transients, all other things being equal. Integrating equation (11.68) yields displacement

$$\delta(t) = \frac{\sigma(a - b)}{k} \ln\left(\frac{e^{t/t_c} - C}{1 - C}\right) = \frac{\sigma(a - b)}{k} \ln\left[\frac{v_{max}}{v^\infty}\left(e^{t/t_c} - 1\right) + 1\right]. \tag{11.69}$$

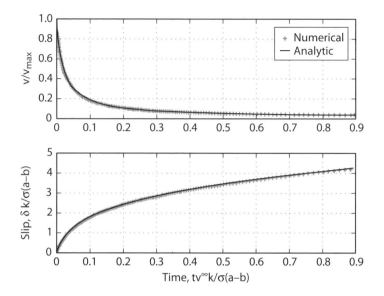

Figure 11.26. Velocity (top) and slip (bottom) for transient afterslip corresponding to the lowest stiffness simulation in figure 11.25. Numerical results are compared with analytical approximations for slip speed (11.68) and slip (11.69).

The approximations for slip rate (11.68) and slip (11.69) are in excellent agreement with numerical simulations as long as the system is sufficiently compliant, $kd_c/\sigma(a - b) \ll 1$ (figure 11.26). For stiffer systems, the trajectories do not follow the steady-state line, and the approximations break down.

At short times $t \ll t_c$, the contribution due to plate loading is small. Ignoring v^∞ in equation (11.66) leads to displacements

$$\tilde{\delta}(t) = \frac{\sigma(a - b)}{k} \ln \left[\frac{kv_{max}}{\sigma(a - b)}t + 1 \right] \tag{11.70}$$

(Scholz 1990). While it is tempting to add steady-state slip $v^\infty t$ to the transient displacements $\tilde{\delta}(t)$, this is not correct in that it fails to account for the frictional response to the stressing associated with steady load point motion. Nevertheless, Marone et al. (1991) were able to fit $\tilde{\delta}(t) + v^\infty t$ to creep data following the 1966 Parkfield earthquake (figure 11.24). While equation (11.69) and $\tilde{\delta}(t) + v^\infty t$ are not equivalent, they are identical for $t \ll t_c$ (see problem 4).

Equation (11.70) has led to the conclusion that frictional afterslip, and any associated deformation, is logarithmic in time; however, the spring-slider model is unlikely to properly represent slip on faults in nature. For a fault in an elastic medium, the size of the aseismically slipping region is not known a priori and will generally change with time. This suggests that caution should be used when applying results for the single degree of freedom spring-slider system to faults in a continuum. One example of a continuum calculation in which unstable slip was nucleated within a velocity weakening region and allowed to propagate into a velocity strengthening region is shown in figure 11.27. As the rupture propagates into the velocity-strengthening region, the maximum slip speed decreases, as does the strength of the stress concentration at the rupture front. The interior of the dynamic rupture decelerates rapidly during the postseismic phase.

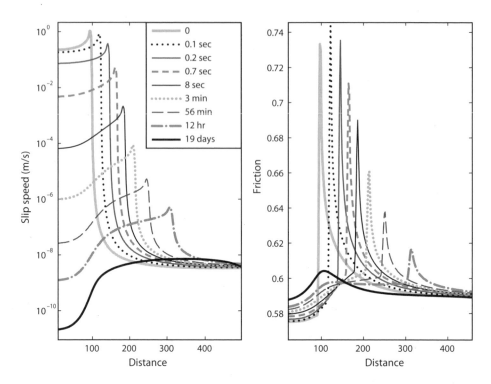

Figure 11.27. Continuum simulation of a fast rupture (assuming the radiation damping approximation to elastodynamics) propagating into a velocity-strengthening region. Interior ($x < 150$) is velocity weakening; exterior ($x > 150$) is velocity strengthening. Left: Slip speed at selected times. Right: Ratio of shear to normal stress. Model is symmetric; only the right half is illustrated. $a = 8 \times 10^{-3}$ everywhere, whereas b varies from 0.01 in the center of the fault to 6×10^{-3} at roughly 250 m from the fault center, so that $a - b$ varies from -2×10^{-3} to 2×10^{-3}, and a/b is 0.8 in the interior. d_c is 1 mm; $\sigma = 100$ MPa, $\mu = 11.5$ GPa.

Figure 11.28 (top) compares the velocity history of the center of the velocity-weakening region with the prediction from equation (11.59). In general, the simple analytical expression predicts the behavior quite well. For the chosen parameters, the slip rate drops to the plate velocity in $\sim 10^4$ seconds. Figure 11.28 (bottom) compares the afterslip history within the rupture zone with that in the velocity-strengthening region. Postseismic slip within the rupture zone accumulates very rapidly, with most of the slip occurring within the first second, in what could be considered to be part of the dynamic rupture. The slip in the velocity-strengthening region is roughly linear in log time, for times greater than approximately 10^5 seconds; however, as can be seen from figure 11.27, the rapidly slipping zone expands so that it is not possible to define a single stiffness appropriate to the afterslip region as required for equations (11.69) or (11.70).

11.7 Transient Slip Events

The advent of large, continuously recording GPS networks led to the recognition of transient deformation events, quite independent of the postseismic transients that follow large earthquakes. Aseismic transients have been observed in subduction zones (Hirose et al. 1999; Dragert et al. 2001), but are not limited to these environments (e.g., Cervelli et al. 2002).

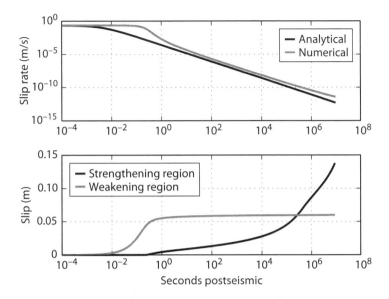

Figure 11.28. Slip histories for the simulation shown in figure 11.27. Top: Slip rate at the fault center compared with the prediction from equation (11.59). Bottom: Average slip within rate weakening region (gray line) and rate strengthening region (dark line).

The subduction transients have been interpreted to result from transient slip events at depths below the locked mega-thrust zone. In some cases, the transients clearly propagate along strike of the subduction zones. Durations last from days to fractions of a decade; equivalent moments range up to M 7.

In some cases, the transient slip is accompanied by nonvolcanic tremor (Rogers and Dragert 2003; Obara et al. 2004). Recent work has shown that, in southwest Japan, the tremor is caused by shear slip on the plate interface (Shelley et al. 2007). While the frictional properties of the fault interface are undoubtedly important in understanding transient slip, the mechanics of this fascinating process are only beginning to be revealed.

11.8 Summary and Perspective

Consideration of friction allows for a fully self-consistent and deterministic description of fault behavior. For this reason, numerous studies are moving from kinematic models to those based on frictional constitutive laws. Some form of rate- and state-dependent friction not only explains considerable laboratory data, but in combination with continuum elasticity also rationalizes observed fault phenomenon, including both stable and unstable sliding, restrengthening following rapid slip, and afterslip both within and adjacent to dynamic ruptures. No other competing theory allows for such quantitative predictions.

An accepted microstructural understanding of laboratory observations remains elusive, which inhibits extrapolation of laboratory data to field conditions. In addition, there are few laboratory determinations of the important temperature, and hence depth dependence, of frictional parameters. The direct velocity dependence of friction can be considered to be reasonably well understood. The empirical evidence for time-dependent strengthening of surfaces in nominal stationary contact is also well established. However, the appropriate state evolution law, or indeed even whether a single evolution law is appropriate for all circumstances, remains unresolved. Experiments show that rapid changes in normal stress result in both rapid and transient changes in frictional resistance, which has been interpreted to indicate that state evolves with normal stress as well as time.

Linearized stability analysis of a single-degree of freedom spring-slider system shows that the system is always stable to small perturbations for steady-state velocity strengthening friction. An unstable response results for steady-state velocity weakening friction if the spring stiffness is less than a critical value. This leads to the concept of a critical rupture nucleation dimension, which for laboratory-derived constitutive parameters and expected effective normal stresses is of the order of a few meters. Numerical computations with the aging form of the state evolution equation on a one-dimensional fault (two-dimensional elasticity) show that with mildly velocity-weakening friction, the nucleation zone expands quasi-statically for some distance. Similar calculations with the slip law predict unidirectional slip pulses. Although there is some variability in behavior, these results suggest that detection of preinstability strain transients from remote measurements will be extremely difficult. Specially designed experiments to place instruments very near to expected earthquake sources are ongoing at the time of this writing. Successful detection of the aseismic nucleation phase would be extremely important for constraining and validating models of earthquake nucleation, irrespective of whether this ultimately leads to short-term predictions of impending earthquakes.

Efforts to understand post-earthquake deformation have been significantly impacted by modern understanding of fault friction. Within the earthquake source region, slip decelerates, at nearly constant stress, to plate rates in a very short time (order seconds to less than one day), accumulating slip of order d_c. Postseismic slip outside the coseismic slip zone on velocity strengthening sections of the fault may be much larger. A single degree of freedom spring-slider that is rapidly loaded by an external stress change first speeds up and then decelerates. For a sufficiently compliant system, the deceleration is at steady-state friction. This leads to postseismic transient slip that increases logarithmically in time, consistent with some observations. However, numerical calculations employing the radiation damping approximation to elastodynamics find an expanding afterslip zone, which cannot be characterized by a constant stiffness as in a simple spring-slider model. These results indicate that a spatially averaged slip does not increase linearly with logarithm of time. Methods for retrieving estimates of frictional parameters from postseismic observations remain an area of active research.

The discussion here has been limited to isothermal and drained conditions. This should not be taken to imply that thermal and/or pore pressure effects are insignificant. Dilatancy acts to reduce pore pressures within faults and may thus stabilize slip against frictional weakening. As slip rates increase, frictional heating may activate a number of weakening mechanisms (Rice 2006), including in the extreme case melting, which have not been considered here.

The strong nonlinearity inherent in the rate-state formulation complicates analytical treatment, and much recent work has been numerical. This is particularly true when considering spatially variable frictional properties. Laboratory experiments demonstrate that rate- and state-dependent frictional parameters vary with temperature, which implies a significant depth variation in the earth. The next chapter reviews earthquake cycle models that include depth-dependent frictional properties, consistent with laboratory measurements.

11.9 Problems

1. Show that the spring-slider system with rate-state friction is linearly stable in the limit of infinite spring stiffness. Specifically, show that the roots of equation (11.28) in the limit $k \to \infty$ are $s = -kv^{\infty}/a\sigma$ and $s = -v^{\infty}/d_c$. Since both roots are real and negative, the system is always stable in this limit.

2. Show that for the spring-slider system, if one includes inertial effects, the critical spring stiffness is given by

$$k_{crit} = \frac{\sigma(b-a)}{d_c}\left(1 + \frac{mv^2}{\sigma a d_c}\right),$$
(11.71)

where m is the mass per unit contact area of the slider, and v is slip speed.

3. Note the behavior of equation (11.10) is singular at $v = 0$. Rice et al. (2001) suggest a physical basis for regularizing the behavior in this limit. They point out that for low applied stress, one needs to consider backward as well as forward atomic-scale jumps, which are equally likely in the limit of no applied stress. Replace equation (11.13) with the following:

$$v = v_0 \left\{ \exp\left[\frac{-(E_1 - \tau_c\Omega)}{k_B T}\right] - \exp\left[\frac{-(E_1 + \tau_c\Omega)}{k_B T}\right] \right\}.$$
(11.72)

Note the signs on the $\tau_c\Omega$ terms; for backward jumps the stress acts in the opposite direction to the slip and therefore does negative work. Following the same procedure as for the forward-jump-only case described in the text, and assuming that the associations (11.15) and (11.18) apply, derive a regularized form of equation (11.10) as

$$\tau = a\sigma \operatorname{arcsinh}\left\{ \frac{v}{2v_0} \exp\left[\frac{f_0 + b\log(\theta/\theta_0)}{a}\right] \right\}.$$
(11.73)

Since $\operatorname{arcsinh}(x) = \log[x + (x^2 + 1)]^{1/2}$, show that equation (11.73) reduces to (11.10) in the limit that the argument is large compared to unity, but vanishes when v goes to zero from positive values.

4. Derive equation (11.70) and show that it is equivalent to equation (11.69) at short times. Show that the slip rate associated with equation (11.70) at $t = 0$ is v_{max}, so adding a steady velocity v_∞ violates the initial condition $v(t = 0) = v_{max}$.

11.10 References

Bayart, E., A. M. Rubin, and C. Marone. 2006. Evolution of fault friction following large velocity jumps, *EOS Transactions of the American Geophysical Union* **87**(52, Suppl.), Abstract S31A-0180 2006 Fall Meeting.

Beeler, N. M., T. E. Tullis, and J. D. Weeks. 1994. The roles of time and displacement in the evolution effect in rock friction. *Geophysical Research Letters* **21**(18), 1987–1990.

Blanpied, M. L., D. A. Lockner, and J. D. Byerlee. 1995. Frictional slip of granite at hydrothermal conditions. *Journal of Geophysical Research* **100**, 13,045–13,064.

Bowden, F. P., and D. Tabor. 1964. *The friction and lubrication of solids, part 2*. New York: Oxford Press.

Cervelli, P., P. Segall, K. Johnson, M. Lisowski, and A. Miklius. 2002. Sudden aseismic fault slip on the south flank of Kilauea volcano. *Nature* **415**, 1014–1018.

Chester, F. 1994. Effects of temperature on friction: constitutive equations and experiments with quartz gouge. *Journal of Geophysical Research* **99**, 7247–7261.

Dieterich, J. H. 1972. Time-dependent friction of rocks. *Journal of Geophysical Research* **77**, 3690–3697.

———. 1978. Time-dependent friction and the mechanics of stick-slip. *Pure and Applied Geophysics* **116**, 790–806.

———. 1979. Modeling of rock friction. 1. Experimental results and constitutive equations. *Journal of Geophysical Research* **84**, 2161–2168.

———. 1992. Earthquake nucleation on faults with rate and state dependent strength. *Tectonophysics* **211**, 115–134.

Dieterich, J. H., and B. D. Kilgore. 1994. Direct observation of frictional contacts: new insights for state-dependent properties. *Pure and Applied Geophysics* **143**, 283–302.

Dragert, H., K. Wang, and T. S. James. 2001. A silent slip event on the deeper Cascadia subduction interface. *Science* **292**, 1525–1528.

Heslot, F., T. Baumberger, B. Perrin, B. Caroli, and C. Caroli. 1994. Creep, stick-slip, and dry friction dynamics: experiments and a heuristic model. *Physical Review E* **49**, 4973–4988.

Hirose, H., K. Hirahara, F. Kimata, N. Fujii, and S. Miyazaki. 1999. A slow thrust slip event following the two 1996 Hyuganada earthquakes beneath the Bungo Channel, southwest Japan. *Geophysical Research Letters* **26**, 3237–3240.

Jaeger, J. C., and Cook, N. G. W. 1976. *Fundamentals of rock mechanics*, 2nd ed. London: Chapman and Hall.

Kilgore, B. D., M. L. Blanpied, and J. H. Dieterich. 1993. Velocity-dependent friction parameters of granite over a wide range of conditions. *Geophysical Research Letters* **20**, 903–905.

Linker, M. F., and J. H. Dieterich. 1992. Effects of variable normal stress on rock friction—observations and constitutive equations. *Journal of Geophysical Research* **97**, 4923–4940.

Marone, C. 1998. Laboratory-derived friction laws and their application to seismic faulting. *Annual Reviews of Earth and Planetary Science* **26**, 643–696.

Marone, C. J., C. H. Scholz, and R. Bilham. 1991. On the mechanics of earthquake afterslip. *Journal of Geophysical Research* **96**, 8441–8452.

Nakatani, M. 2001. Conceptual and physical clarification of rate and state friction: frictional sliding as a thermally activated rheology. *Journal of Geophysical Research* **106**, 13,347–13,380.

Obara, K., H. Hirose, F. Yamamizu, and K. Kasahara. 2004. Episodic slow slip events accompanied by non-volcanic tremors in southwest Japan subduction zone. *Geophysical Research Letters* **31**, L23602, doi:10.1029/2004GL020848.

Poirier, J.-P. 1985. *Creep of crystals*. New York: Cambridge University Press.

Ranjith, K, and J. R. Rice. 1999. Stability of quasi-static slip in a single degree of freedom elastic system with rate and state dependent friction. *Journal of Mechanics and Physics of Solids* **47**, 1207–1281.

Rice, J. R. 1983. Constitutive relations for fault slip and earthquake instabilities. *Pure and Applied Geophysics* **121**, 443–475.

———. 1993. Spatio-temporal complexity of slip on a fault. *Journal of Geophysical Research* **98**, 9885–9907.

———. 2006. Heating and weakening of faults during earthquake slip. *Journal of Geophysical Research* **111**, B05311, doi:10.1029/2005JB004006.

Rice, J. R. and J.-C. Gu. 1983. Earthquake aftereffects and triggered seismic phenomena. *Pure and Applied Geophysics* **121**(2), 187–219.

Rice, J. R., N. Lapusta, and K. Ranjith. 2001. Rate and state dependent friction and the stability of sliding between elastically deformable solids. *Journal of Mechanical and Physical Solids* **49**, 1865–1898.

Rice, J. R., and S. T. Tse. 1986. Dynamic motion of a single-degree of freedom system following a rate and state dependent friction law. *Journal of Geophysical Research* **91**, 521–530.

Rogers, G., and H. Dragert. 2003. Episodic tremor and slip on the Cascadia subduction zone: the chatter of silent slip. *Science* **300**(5627), 1942–1943, doi: 10.1126/science.1084783.

Rubin, A. M., and J.-P. Ampuero. 2005. Earthquake nucleation on (aging) rate- and-state faults. *Journal of Geophysical Research* **110**, B11312, doi:10.1029/2005JB003686.

———. 2008. Earthquake nucleation on rate-and-state fault—aging and slip laws. *Journal of Geophysical Research* **113**, B01302, doi:10.1029/2007JB005082.

Ruina, A. L. 1980. Friction laws and instabilities: a quasistatic analysis of some dry frictional behavior. PhD Thesis, Brown University.

———. 1983. Slip instability and state variable friction laws. *Journal of Geophysical Research* **88**, 10,359–10,370.

Scholz, C. H. 1990. *The mechanics of earthquake faulting.* Cambridge, UK: Cambridge University Press.

Shelly, D. R., G. C. Beroza, and S. Ide. 2007. Non-volcanic tremor and low-frequency earthquake swarms. *Nature* **446**, 305–307.

Smith, S. W., and M. Wyss. 1968. Displacements on the San Andreas fault subsequent to the 1966 Parkfield earthquake. *Bulletin of the Seismological Society of America* **58**, 1955–1973.

Stesky, R. M. 1978. Mechanisms of high temperature frictional sliding in westerly granite. *Canadian Journal of Earth Sciences* **15**, 361–375.

12

Interseismic Deformation and Plate Boundary Cycle Models

Strain accumulates in the crust adjacent to major faults during the interseismic period between large earthquakes. Measurement of strain accumulation is a first-order predictor of future seismic hazard since elastic strain is ultimately released in earthquakes. Higher rates of strain accumulation should generally be associated with higher fault slip rates and thus either larger or more frequent earthquakes. To make these ideas quantitative requires mathematical models, the predictions of which can be compared to observations. Chapter 2 introduced the elastic screw-dislocation model of interseismic deformation for very long strike-slip faults as well as the backslip concept. We will begin this chapter with a discussion of elastic models of interseismic deformation for strike-slip and dip-slip faults, including those that properly account for far-field motions of tectonic plates. We next build on results from chapters 6 and 9 to construct kinematic plate boundary cycle models that account for viscoelastic behavior of rock below the brittle elastic crust. Concepts of fault friction described in chapter 11 are employed to develop dynamical models of repeating earthquakes where rapid slip events arise naturally from the interaction of the elastic fault system with the fault constitutive law. Efforts to include more realistic constitutive behavior of rocks, including nonlinear rheologies, depth, and temperature-dependent properties, naturally drives modeling studies to numerical approaches. Our focus here is on what can be gleaned from relatively simple analytic and quasi-analytic models, rather than a complete review of the current state of the art in numerical modeling of interseismic deformation.

12.1 Elastic Dislocation Models

Chapter 2 introduced the back-slip concept for an infinitely long strike-slip plate-boundary fault. During the interseismic period, the upper, locked portion of the fault accumulates a *slip deficit*, relative to the far-field plate motion, at a rate $\dot{s} = v_{plate}$. If the average recurrence time for large earthquakes is T, the cumulative slip deficit at the end of the interseismic period is $v_{plate}T$. Coseismic slip of amount $s = v_{plate}T$ on the upper, previously locked fault, leads to a net rigid body offset and complete relaxation of elastic strain in the crust adjacent to the fault (figure 2.10).

The back-slip concept follows from the assumption that *interseismic deformation + coseismic deformation = long-term block motion*. Rearranging, *interseismic deformation = long-term block motion − coseismic deformation*. In other words, interseismic deformation can be modeled as the sum of the long-term rigid body motion and back slip—shallow slip at the long-term rate acting in the opposite direction to the long-term, geologic slip. If the strike-slip fault is planar and attention is restricted to areas far from fault ends, the block motion produces no strain off the fault. Thus, back slip produces the same velocity field as an infinitely deep dislocation slipping at the long-term rate below a shallow locked portion of the fault (as in equation [2.35]).

Of course, faults do not extend to infinite depth and elastic behavior ultimately gives way to inelastic deformation with increasing depth and temperature. Nevertheless, it is

Figure 12.1. Steady slip below a locked seismogenic zone added to coseismic slip on the shallow fault leads to rigid body motion. Note that for dip-slip faults, the rigid body motion includes nonphysical vertical displacements of the earth far from the fault.

possible to consider back slip as an internally consistent abstraction, one that when added to coseismic deformation yields rigid block motion. On a spherical earth, however, the far-field displacements are not properly described by Cartesian motion. Rather, velocities of points distant from plate boundaries are best described as rotation of rigid spherical caps. These motions are quantified mathematically by Euler poles of rotation. Before addressing rigid plate motions, we will consider dip-slip faulting in a half-space geometry, which necessitates consideration of issues not present with strike-slip faults.

12.1.1 Dip-Slip Faults

Consider a dip-slip fault that is locked at shallow depth with steady interseismic slip at long-term rate $\dot{s}$ extending to great depth below (figure 12.1). Addition of shallow coseismic slip with magnitude $\dot{s}T$ leads to rigid body displacement of the entire medium. In particular, the vertical component of rigid body motion is $\dot{s}\sin(\delta)$, where δ is the fault dip. It is clearly unphysical for the entire earth distant from the fault to displace vertically. This complication does not arise for long, planar strike-slip faults in an elastic half-space, where there are no vertical displacements. The surface velocity fields corresponding to the models in figure 12.1 are shown in figure 12.2. Because the buried dislocation (figure 12.1, left) extends to infinite depth, the interseismic vertical displacement is nonzero in the far field. A buried edge dislocation in an elastic half-space, with nonzero dip, is thus not a tenable model for interseismic deformation in dip-slip environments. Fundamentally, one cannot neglect inelastic deformation and gravitational forces when considering deformation over long periods of time.

Savage (1983) introduced a widely used back-slip model for subduction zone environments. This model was intended to describe deformation of the hanging wall, which until the advent of sea-floor geodetic systems, is where essentially all deformation measurements were located. He suggests that one consider the long-term steady-state pattern of displacement separately from the episodic motions associated with the earthquake cycle. Take the convergence rate relative to a fixed site on the hanging wall far from the trench to be v_{plate} (figure 12.3). Over the long term, the slab subducts at the convergence rate with reverse slip occurring on the plate interface. At greater depth, the relative motion may grade into distributed shearing between the plates. Savage assumed that steady long-term plate motion produces no deformation of the hanging wall. This is certainly only a first-order approximation; deformation of the accretionary prism and uplifted marine terraces found along many subduction margins attest to long-term deformation of the hanging wall. As illustrated in the left column of figure 12.3, the long-term velocity field is assumed to be one in which the subducting plate converges toward the overriding plate with relative velocity v_{plate}, and there is no vertical displacement.

Between earthquakes, the locking of the subduction interface may be thought of as applying a back slip over the seismogenic portion of the fault in such as way as to negate the steady-state slip. Because the long-term motion is reverse, back slip in subduction zones is equivalent to steady normal slip on the fault (figure 12.3, middle column). By assumption, the steady-state

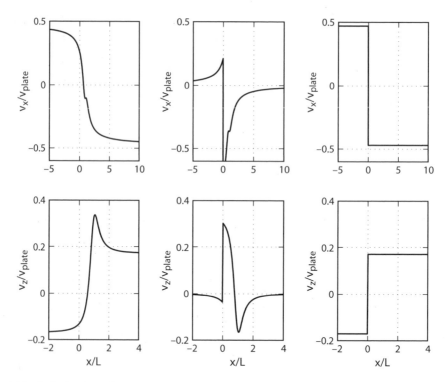

Figure 12.2. Deformation due to a buried edge dislocation, corresponding to the diagrams in figure 12.1. The left column gives the interseismic velocity field, the center column gives the corresponding coseismic displacements, and the right column gives the sum of the interseismic and coseismic displacements. Velocities are normalized by the fault slip rate, while the horizontal distance scale is normalized by the downdip extent of coseismic rupture, L. Note that the horizontal distance scales are different for the horizontal and vertical components.

component causes no deformation of the overriding plate, so interseismic strain accumulation can be completely modeled by the back slip (figure 12.3, right column).

The back-slip model predicts interseismic compression of the overriding plate, with landward (positive) velocities decreasing with distance from the trench (figure 12.3, right). During the interseismic period, the region nearest the trench subsides, as it is pulled down by the subducting plate, whereas the area above the downdip end of the locked zone is uplifted (figure 12.3, right). Depending on the position of the coastline relative to the trench, the coastal area will generally tilt down toward the trench during the interseismic period, and then rebound upward during plate rupturing earthquakes.

In the Savage back-slip interpretation, the slip rate on the subduction interface is equal to the far-field plate convergence rate v_{plate}. (In the case of oblique convergence, motion parallel to the trench is easily included by adding appropriate screw dislocations, as in chapter 2.) In this way, the subducting slab is neither extending nor contracting as it would if the steady-state velocity of the descending slab differed from the plate convergence rate.

Caution is required if applying the Savage back-slip interpretation at either the fault trace or on the foot-wall side of the fault. If the slip rate is set equal to the convergence rate, the back slip introduces a fictitious discontinuity in the interseismic velocity field at the fault trace. For this reason, figure 12.3 shows the deformations only on the hanging-wall side of the fault. Because the interseismic velocities in the buried edge dislocation model (figure 12.1) and the

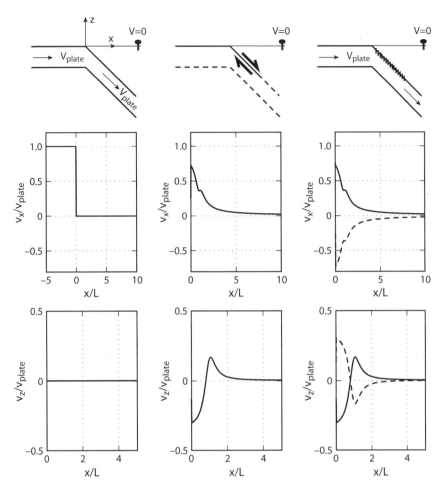

Figure 12.3. Back-slip model in which the strain associated with a locked subduction zone is decomposed into rigid body displacement plus a "virtual" normal slip on the locked interface. v_x and v_z are the horizontal and vertical velocities relative to a fixed site far from the trench on the hanging-wall side of the fault. The left column gives the assumed long-term velocity fields, the middle column gives the contribution due to fault locking, and the right column gives the interseismic velocity field (solid line) and the corresponding coseismic displacement (dashed line). Velocities are normalized by the far-field plate convergence rate, while the horizontal distance scale is normalized by the downdip extent of coseismic rupture. Note that the horizontal distance scales are different for the horizontal and vertical components and that only the hanging-wall side is illustrated in the middle and right columns.

elastic back-slip model (figure 12.3) differ only by rigid body displacements, the strains and tilts are identical.

It is common in subduction zone modeling to include a linear tapering of the slip rate from the plate rate to fully locked through a transition zone. The linear slip-rate profile is motivated more by convenience than physics. A more physically motivated model would be to consider that the transitional region slips at a constant resistive stress. A yet more sophisticated model would employ a constitutive law, such as rate and state friction that transitions from steady-state velocity weakening to steady-state velocity strengthening behavior with depth, to determine the appropriate slip-rate distribution. As a first step, we use methods described in

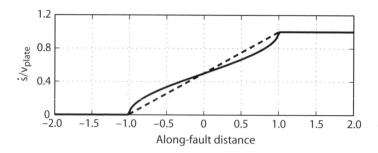

Figure 12.4. Slip-rate transition from a locked fault to a steadily creeping fault, assuming constant resistive stress in the interval $-1 \leq z \leq 1$. Effects of the free surface are not included. The dashed line illustrates a linear transition between the locked and fully creeping zone.

chapter 4 to determine the appropriate slip-rate profile in a two-dimensional elastic full-space. The time derivative of equation (4.7), assuming that the length of the transition zone ($2a$) is time invariant, can be written as

$$\dot{B}(z) = \frac{-2}{\mu\pi} \frac{1}{\sqrt{a^2 - z^2}} \int_{-a}^{a} \frac{\dot{\tau}(\xi)\sqrt{a^2 - \xi^2}}{z - \xi} d\xi + \frac{v^{plate}}{\pi\sqrt{a^2 - z^2}}, \tag{12.1}$$

where the net slip rate at the end of the transition zone ($z = a$) is specified to be v^{plate}. By assumption, the shear stress through the transition zone is constant in time, so the integral vanishes, leaving

$$\dot{B}(z) = \frac{v^{plate}}{\pi\sqrt{a^2 - z^2}}. \tag{12.2}$$

The slip rate then is given by integrating $\dot{B}(z)$, which yields

$$\dot{s}(z) = \int_{-a}^{z} \dot{B}(\xi)d\xi,$$

$$= \frac{v^{plate}}{\pi} \int_{-a}^{z} \frac{d\xi}{\sqrt{a^2 - z^2}},$$

$$= \frac{v^{plate}}{\pi} \sin^{-1}(\xi/a)|_{-a}^{z},$$

$$= \frac{v^{plate}}{\pi} \left[\sin^{-1}\left(\frac{z}{a}\right) + \frac{\pi}{2} \right]. \tag{12.3}$$

The slip-rate profile is illustrated, and compared to a linear transition, in figure 12.4. It is important to emphasize that equation (12.3) applies to a two-dimensional full-space. The appropriate slip-rate distribution, including the effects of the free surface, can easily be determined, however, using boundary element techniques.

12.2 Plate Motions

Plate tectonics holds that the earth's surface is covered with large tectonic plates in relative motion with respect to one another. A theorem attributed to Euler says that small incremental motions of a rigid plate can be completely and uniquely described by an axial rotation. The instantaneous velocity at points on the earth's surface are given by the equation

$$\mathbf{v}(\mathbf{r}) = \mathbf{\Omega} \times \mathbf{r}, \tag{12.4}$$

where $\mathbf{v}(\mathbf{r})$ is the velocity vector as a function of geocentric position $\mathbf{r}$ (i.e., the vector from the center of the earth to the position on the earth's surface), and $\mathbf{\Omega}$ is the angular velocity vector of the plate on which $\mathbf{r}$ lies.

Using the permutation tensor introduced in chapter 1 to put equation (12.4) into indicial form, $v_i = e_{ijk}\Omega_j r_k$, it is straightforward to show that equation (12.4) can be expressed in matrix form as

$$\begin{bmatrix} v_1 \\ v_2 \\ v_3 \end{bmatrix} = \begin{bmatrix} 0 & r_3 & -r_2 \\ -r_3 & 0 & r_1 \\ r_2 & -r_1 & 0 \end{bmatrix} \begin{bmatrix} \Omega_1 \\ \Omega_2 \\ \Omega_3 \end{bmatrix}. \tag{12.5}$$

This can also be written compactly as $\mathbf{v}(\mathbf{r}) = Q(\mathbf{r})\mathbf{\Omega}$, where $Q(\mathbf{r})$ represents the matrix in equation (12.5). It is worth noting that plate tectonics predicts purely horizontal motions. A unit vector at $\mathbf{r}$ in the vertical direction is $\mathbf{r}/||\mathbf{r}||$, thus

$$v_{\mathrm{r}} = \frac{\mathbf{r}}{||\mathbf{r}||} \cdot \mathbf{v} = \frac{\mathbf{r}}{||\mathbf{r}||} \cdot (\mathbf{\Omega} \times \mathbf{r}) = 0, \tag{12.6}$$

since the cross product produces a vector that is orthogonal to $\mathbf{r}$. Measured velocities at a number of stations $\mathbf{r}$ on a given plate can be used to estimate the Euler vector for that plate. The Euler vector has three components, yet each geodetic station provides only two horizontal velocities. The minimum number of stations needed to determine the Euler vector is thus two.

The relative velocity of two points on the same plate located at $\mathbf{r}^A$ and $\mathbf{r}^B$ is

$$\Delta\mathbf{v}^{AB} = \mathbf{\Omega} \times (\mathbf{r}^A - \mathbf{r}^B). \tag{12.7}$$

We are often more interested in the relative motion of the plates than their absolute motion. The relative motion $\mathbf{\Omega}^{AB}$ is given by $\mathbf{\Omega}^{AB} = \mathbf{\Omega}^A - \mathbf{\Omega}^B$, where $\mathbf{\Omega}^A$ and $\mathbf{\Omega}^B$ are the absolute poles of rotation for plates A and B, respectively. The relative motion between two points on different plates is thus

$$\begin{aligned} \mathbf{v}^A - \mathbf{v}^B &= \mathbf{\Omega}^A \times \mathbf{r}^A - \mathbf{\Omega}^B \times \mathbf{r}^B, \\ &= \mathbf{\Omega}^A \times \mathbf{r}^A - \mathbf{\Omega}^B \times \mathbf{r}^B + \mathbf{\Omega}^B \times \mathbf{r}^A - \mathbf{\Omega}^B \times \mathbf{r}^A, \\ &= (\mathbf{\Omega}^A - \mathbf{\Omega}^B) \times \mathbf{r}^A - \mathbf{\Omega}^B \times (\mathbf{r}^A - \mathbf{r}^B). \end{aligned} \tag{12.8}$$

For a point on the plate boundary, $\mathbf{r}^A = \mathbf{r}^B$, so the last term in equation (12.8) vanishes. The relative motion, equivalent to the long-term slip rate on the plate boundary $\dot{\mathbf{s}}^{AB}$, is given simply by

$$\dot{\mathbf{s}}^{AB} \equiv \mathbf{v}^A - \mathbf{v}^B = (\mathbf{\Omega}^A - \mathbf{\Omega}^B) \times \mathbf{r}, \tag{12.9}$$

or in indicial notation

$$\dot{s}_i^{AB} = Q_{ij}(\mathbf{r})\Omega_j^{AB}, \tag{12.10}$$

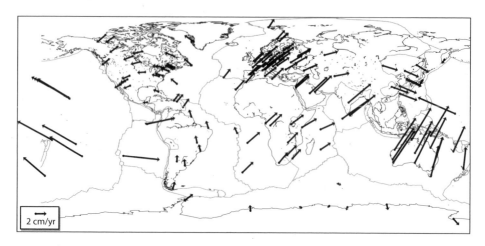

Figure 12.5. Site velocities used in the estimation of the ITRF2005 absolute rotation poles by Altamimi et al. (2007). These authors used horizontal velocities of 152 sites with an error less than 1.5 mm/yr to estimate absolute rotation poles of 15 tectonic plates.

where $\mathbf{\Omega}^{AB} = \mathbf{\Omega}^A - \mathbf{\Omega}^B$ is the relative rotation pole, and $Q_{ij} = \varepsilon_{ijk}r_k$. For points not colocated on the plate boundary, the second term in equation (12.8) takes into account the relative separation between the stations.

Figure 12.5 shows a recent determination of average site velocities at stations removed from plate boundaries that were employed by Altamimi et al. (2007) to estimate the rotation poles of 15 major tectonic plates. The rotation of the major plates, including North America, Eurasia, Pacific, and Australia, are clearly visible in this figure.

Plate tectonics, of course, describes the motion of rigid plates. Most of the time, the boundaries of the plates are locked, and elastic stresses accumulate. At discrete times, the plate boundaries fail, either slipping in earthquakes or opening during rifting events associated with dike intrusion, thereby relaxing the accumulated stresses. In addition to the elastic strain accumulation and release cycle, there may also be permanent inelastic deformations that accumulate at or near plate boundaries.

12.3 Elastic Block Models

One way to model deformation that is both kinematically consistent with rigid plate tectonics far from plate boundaries and at the same time includes the effects of interseismically locked boundary faults is to start with a velocity field described by appropriate Euler poles and then effect fault locking using the back-slip approach. This was apparently first done by Prawirodirdjo et al. (1997) using codes developed by R McCaffrey (2002). A similar method was presented by Meade and Hager (2005) and used to describe the crustal deformation of southern California assuming a series of microplates or blocks. A simplified version was described by Murray and Segall (2001).

The fundamental hypothesis is that the velocity at a point $\mathbf{r}$ on the earth's surface can be represented by

$$\mathbf{v}(\mathbf{r}) = \mathbf{\Omega} \times \mathbf{r} - \mathbf{v}^{bs}(\mathbf{r}), \qquad (12.11)$$

where the first term represents the long-term, plate tectonic contribution and the second the back slip due to interseismic fault locking. Implicit in equation (12.11) is the understanding that **r** is located on the plate with motion described by Euler pole $\mathbf{\Omega}$. From Volterra's formula in chapter 3, we can write the back-slip contribution:

$$v_k^{bs}(\mathbf{r}) = R_{XE} \int_{\Sigma} G_{ik}[R_{SC}(\mathbf{r}), \, \xi] \dot{s}_i(\xi) d\Sigma(\xi), \qquad (12.12)$$

where G are kernels relating fault slip rate to velocity at the earth's surface, and $\dot{s}$ is the slip rate (e.g., equation [3.10]). As in chapter 3, the G in equation (12.12) are related to the elastostatic Green's functions through

$$G_{ik}(\mathbf{x}, \, \xi) = C_{ijpq} n_j \frac{\partial g_k^p(\mathbf{x}, \, \xi)}{\partial \xi_q}. \qquad (12.13)$$

In equation (12.12), the operator R_{SC} maps the position vector from spherical to local Cartesian coordinates, and R_{XE} rotates the displacements from a local x, y, z coordinate system to topocentric east, north, up coordinates. In practical applications, the integral in equation (12.12) may be replaced by a summation over discrete sources. Note furthermore that the Green's functions could include corrections for earth's sphericity (chapter 8) or elastic inhomogeneity (chapter 5). An advantage of this approach is that, assuming the back slip is limited to relatively shallow depths, the deformation due to the back slip will be concentrated near the fault, so earth curvature effects are generally small (chapter 8).

The back-slip rate on the boundary between plates A and B at point ξ on the fault (12.10) is given by

$$\dot{s}_i^{AB}(\xi) = Q_{ij}[R_{CS}(\xi)] \Omega_j^{AB}, \qquad (12.14)$$

where the operator R_{CS} maps the coordinate ξ from Cartesian to spherical coordinates. Equations (12.12) and (12.14) can be combined:

$$v_k^{bs}(\mathbf{r}) = \left\{ R_{XE} \int_{\Sigma} G_{ik}[R_{SC}(\mathbf{r}), \, \xi] Q_{ij}[R_{CS}(\xi)] d\Sigma(\xi) \right\} \Omega_j. \qquad (12.15)$$

This form reveals that the back-slip contribution can be expressed as a linear operator acting on the relative Euler vector. In indicial form, this can be written as $v_k^{bs} = B_{kj} \Omega_j$, where the back-slip tensor B is everything in the braces in equation (12.15). The velocity at point **r** is then given by the rigid plate component plus the sum of contributions from all locked faults (12.11).

Several workers have used elastic block models to describe the kinematics of complex tectonic regions such as southern California. In such cases, there can be considerable ambiguity as to how to define the block boundaries. This may be in part due to distributed inelastic deformation that is not well described by discrete faults. Distributed plastic deformation and the fact that fault structures evolve over time imply that the long-term velocity field cannot be completely described by rigid blocks. An advantage of the block models is that they impose a kinematic consistency (appropriate to rigid blocks), and that the fault slip rates are completely fixed by the relative rotation rates (equation [12.14]). The latter results in fewer parameters to be estimated in geodetic modeling studies than allowing the slip rates to vary arbitrarily as a function of position on the fault.

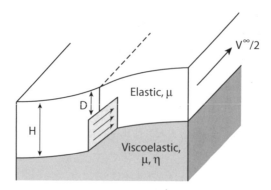

Figure 12.6. Elastic plate of thickness H overlying a viscoelastic half-space. The sides of the elastic plate are driven at fixed velocity $v^\infty/2$. Viscosity of the viscoelastic medium is η. Fault slips by an amount $v^\infty T$ for depths $z \leq D$ every T years. For depths greater than D, the fault creeps at constant rate v^∞.

Vertical strike-slip faults not parallel to small circles about the relevant rotation pole will necessarily involve either contraction or extension across the fault. Meade and Hager (2005) handle this by allowing the dislocations to open or interpenetrate, suggesting that this yields an approximation to localized conjugate thrust or normal faulting. Dip-slip faults also present a challenge for existing methods, even in areas dominated by strike-slip faulting, such as southern California. Further development of the block modeling approach will require kinematically consistent long-term velocity fields to which back slip can be added to model the effects of elastic strain accumulation.

12.4 Viscoelastic Cycle Models

A limitation of purely elastic models is that they fail to account for the flow of rock at depth in the earth over times comparable to the repeat time of large plate boundary earthquakes. Chapter 6 discussed time-dependent deformation following earthquakes due to viscoelastic flow below the seismogenic zone. This section explores earthquake cycle models that include viscoelastic behavior of the lower crust and upper mantle below the brittle-elastic seismogenic zone. These models are kinematic in the sense that the earthquakes are introduced at specified times, rather than arising naturally due to the stress and frictional properties of the fault. In particular, earthquakes with specified slip are introduced periodically with recurrence interval T. This should not be taken as an assertion that earthquakes in nature are truly periodic. Rather, it is a statement that some interval of time is required, depending on the rate of far-field plate motion and the amount of coseismic slip, for the strain relaxed in an earthquake to recover.

12.4.1 Viscoelastic Strike-Slip Earthquake Cycle Models

Chapter 6 analyzed the deformations resulting from a dislocation suddenly introduced into an elastic layer overlying a Maxwell viscoelastic half-space. This analysis included only the transient effect induced by the earthquake and did not incorporate the steady interseismic loading between events. Savage and Prescott (1978) extended the strike-slip viscoelastic model to include interseismic loading. In this model, the elastic layer, of thickness H, overlies a Maxwell viscoelastic half-space (figure 12.6). The fault is locked interseismically to depth D, for times $0 < t < T$, where T is the recurrence interval, but slips steadily at the far-field plate rate v^∞ below D. While this model is idealized—for example, it is more reasonable to consider that

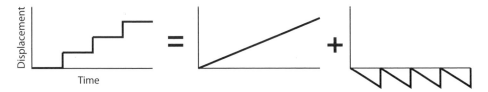

Figure 12.7. Decomposition of a repeated sequence of earthquakes into steady slip plus an elementary earthquake sequence.

the fault below D slips at a rate that depends on the stresses and fault-zone rheology—it does introduce important concepts that can be applied more broadly in viscoelastic earthquake cycle models. Models with stress-driven fault creep are discussed briefly in section 12.4.3.

The Savage-Prescott approach builds on the back-slip concept of previous sections. Specifically, we consider interseismic deformation on a right-lateral fault to be composed of two components. The first component is steady-state right-lateral creep on the fault at the the far-field plate rate v^∞. Following some initial startup (which can be assumed to have occurred a very long time ago), this process induces no strain in the elastic plate. The second component of motion applies only to the locked part of the fault and consists of steady back slip, or left-lateral slip, at rate v^∞ for $0 < t < T$ that exactly cancels the steady creep, punctuated by coseismic slip of amplitude $v^\infty T$. This sawtooth function, referred to by Savage and Prescott (1978) as an elementary earthquake sequence (figure 12.7), can be repeated an arbitrary number of times to build up a sequence of repeating earthquakes.

For the elementary earthquake cycle, the slip history on the fault is given by

$$\Delta u(t) = -v^\infty t \qquad 0 \le t \le T, \qquad 0 \le z \le D,$$
$$\Delta u(t) = 0 \qquad t < 0 \text{ and } t > T, \qquad 0 \le z \le D. \tag{12.16}$$

The Laplace transform (A.13) of this function is

$$\mathcal{L}[\Delta u(t)] = \frac{-v^\infty}{s^2} \left[1 - (sT + 1)e^{-sT} \right] . \tag{12.17}$$

Using the slip function (12.17) with equation (6.34) gives the elementary earthquake cycle in the Laplace domain. One can use the Bromwich inversion formula (A.19) to find the solution in the time domain. Summing an infinite sequence of elementary earthquakes leads to the desired earthquake sequence shown in figure 12.7 (right).

The results of Savage and Prescott (1978) can be derived more simply without making use of the back-slip approach. Equation (6.39) gives the viscoelastic response to a single earthquake with coseismic slip Δu. In order for the fault slip averaged over an earthquake cycle to keep pace with the the far-field motion, $\Delta u = v^\infty T$. Combining with equation (6.39), the transient motion due to a single earthquake at time t_{eq} is thus

$$v(x, t) = \frac{v^\infty T}{\pi t_R} e^{-(t-t_{eq})/t_R} \sum_{n=1}^{\infty} \frac{1}{(n-1)!} \left(\frac{t - t_{eq}}{t_R} \right)^{n-1} F_n(x, D, H), \tag{12.18}$$

where $F_n(x, D, H)$ is defined in equation (6.32) and gives the spatial dependence on plate thickness H and depth of earthquake slip D. t_R is the Maxwell relaxation time given by $t_R = 2\eta/\mu$, where η is viscosity and μ is shear modulus. Taking the time of the most recent earthquake to be $t_{eq} = 0$, then the cumulative effect of N regularly spaced plate-rupturing earthquakes at times $t_{eq} = -kT,\ k = 0, 1, \ldots, N$ is

$$
v(x,\, t) = \frac{v^{\infty} T}{\pi\, t_R} e^{-t/t_R} \sum_{n=1}^{\infty} \frac{F_n(x,\, D,\, H)}{(n-1)!} \sum_{k=0}^{N} e^{-kT/t_R} \left(\frac{t + kT}{t_R} \right)^{n-1}. \tag{12.19}
$$

In equation (12.19), the sum over k represents the effects of past earthquakes, the contributions from which decrease with increasing k. The sum over n is the contribution of the different modes in the solution. We saw in the discussion of a single earthquake in chapter 6 that the first modes dominate near the fault and at short times after an earthquake.

Equation (12.19) can be written more compactly as

$$
v(x,\, t) = \frac{1}{\pi} \sum_{n=1}^{\infty} \mathcal{T}_n(t/t_R,\, T/t_R)\, F_n(x,\, D,\, H), \tag{12.20}
$$

where

$$
\mathcal{T}_n(t/t_R,\, T/t_R) = \frac{v^{\infty} T}{t_R} \frac{e^{-t/t_R}}{(n-1)!} \sum_{k=0}^{N} e^{-kT/t_R} \left(\frac{t + kT}{t_R} \right)^{n-1}. \tag{12.21}
$$

For the $n = 1$ term,

$$
v_1(x,\, t) = \frac{v^{\infty} T}{\pi\, t_R} e^{-t/t_R} F_1(x,\, D,\, H) \sum_{k=0}^{N} e^{-kT/t_R}. \tag{12.22}
$$

The sum over k can be evaluated by multiplying both sides of equation (12.22) by $(e^{-T/t_R} - 1)$ and noticing that all but the first and last terms cancel, so that

$$
(e^{T/t_R} - 1) \sum_{k=0}^{N} e^{-kT/t_R} = e^{T/t_R} - e^{-NT/t_R}. \tag{12.23}
$$

For N large, the sum is thus

$$
\sum_{k=0}^{\infty} e^{-kT/t_R} = \frac{e^{T/t_R}}{e^{T/t_R} - 1}. \tag{12.24}
$$

The term $\mathcal{T}_1(t/t_R,\, T/t_R)$ thus becomes

$$
\mathcal{T}_1(t/t_R,\, T/t_R) = \frac{v^{\infty} T}{t_R} e^{-t/t_R} \left(\frac{e^{T/t_R}}{e^{T/t_R} - 1} \right), \tag{12.25}
$$

and is exponentially decaying in time. Similarly, the $n = 2$ term becomes (problem 2)

$$
\mathcal{T}_2(t/t_R,\, T/t_R) = \frac{v^{\infty} T}{t_R} e^{-t/t_R} \left(\frac{e^{T/t_R}}{e^{T/t_R} - 1} \right) \left[\frac{t}{t_R} + \frac{T}{t_R} \left(\frac{1}{e^{T/t_R} - 1} \right) \right]. \tag{12.26}
$$

Each mode has an exponential character, although higher order modes become more nearly steady in time. The precise behavior depends on the ratio of the recurrence time to

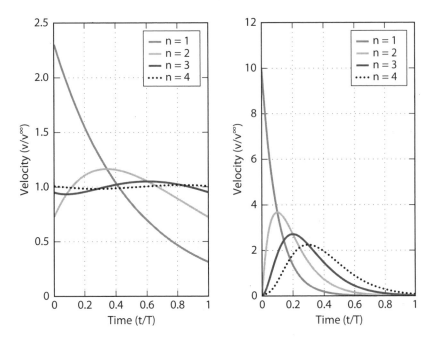

Figure 12.8. Time dependence of the first four modes $\mathcal{T}_n(t/t_R, T/t_R)$, for $T/t_R = 2$ (left) and $T/t_R = 10$ (right). The rate for each mode is normalized by v^∞. Note that with increasing order n, the functions become more nearly steady at rate v^∞.

the Maxwell relaxation time. The first four temporal modes for $T/t_R = 2$ and $T/t_R = 10$ are shown in figure 12.8. Notice that as n increases, the temporal dependence becomes more nearly steady state throughout the earthquake cycle, at rate v^∞. Larger n correspond to slip at a greater depth intervals through $F_n(x, D, H)$ (see figure 6.11). Thus, at sufficiently great depth, the equivalent elastic half-space solution is simply steady sliding at v^∞. For small T/t_R, this depth is approached very quickly, as can be seen in figure 12.8. For larger T/t_R, one must go to greater depth for the behavior to approach steady slip. The average slip rate in each depth interval is equal to v^∞ (see problem 3).

Figure 12.9 shows the maximum difference from steady slip as a function of order n for different values of T/t_R. Remarkably, for $T/t_R \sim 1$, we need to include only one or two terms in the expansion. Everything below that can be approximated by steady slip on a vertical fault in an elastic half-space. For $T/t_R \sim 10$, we need roughly 20 terms in the expansion to obtain an accurate result.

Figure 12.10 shows the velocities predicted by the Savage-Prescott model when the earthquakes break the entire elastic layer ($D = H$), for different time intervals and two different values of T/t_R. Notice that for $T/t_R = 1$, the velocity profile changes only modestly during the earthquake cycle. For $x/H \geq 5$, the velocity is always close to half the plate rate. For $T/t_R = 5$, corresponding to a shorter Maxwell relaxation time, there are more pronounced transients during the earthquake cycle. Early in the cycle, the velocities in the range $1 \leq x/H \leq 5$ exceed the long-term slip rate, whereas late in the cycle, the velocities are less than the long-term slip rate. For $T/t_R = 5$, it appears that the velocity curves are asymptotic to a value that is different from half the plate velocity. In fact, all the curves do approach $v^\infty/2$, although this occurs a substantial distance from the fault.

We next treat the case in which coseismic slip extends to depth $D < H$ and the fault creeps at constant rate $\dot{s}$ for $D < z < H$. Take an infinite sequence of repeating slip events, as in

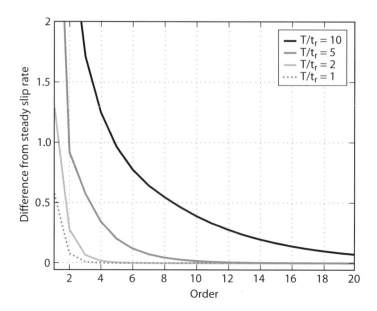

Figure 12.9. Maximum difference between the equivalent slip rate and v^∞, normalized by v^∞, as a function of order n for different values of T/t_R. For recurrence intervals comparable to the Maxwell relaxation time, one need only include a few terms in the expansion. Higher order (deeper) contributions can be approximated by steady slip in an elastic half-space. For recurrence times on the order of 10 Maxwell relaxation times, one must include on the order of 20 terms.

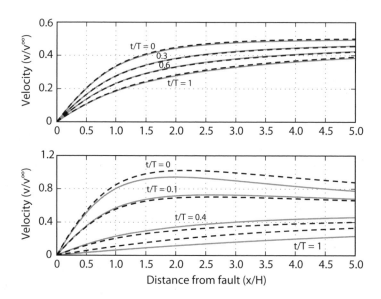

Figure 12.10. Interseismic velocity distribution as a function of distance from a vertical strike-slip fault according to the Savage and Prescott model for $T/t_R = 1$ (top) and $T/t_R = 5$ (bottom). $D = H$. The solid lines include enough terms to approximate the full solution. The dashed lines show the solution with only one term. Notice that this approximation is good for the $T/t_R = 1$ case but not accurate for $T/t_R = 5$. Note also that the velocity scales are different for the two different values of T/t_R.

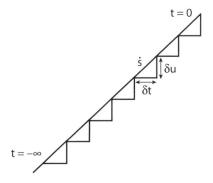

Figure 12.11. A sequence of infinitesimal slip events stretching from $t = -\infty$ to $t = 0$.

equation (12.19), in the limit that the recurrence interval and coseismic slip become vanishingly small, $\delta u = \dot{s}\delta\tau$ (figure 12.11). In an elastic earth, this would result in a velocity field $\dot{s}\hat{u}^{\text{elastic}}(x)$, where $\hat{u}^{\text{elastic}}(x)$ is the elastic displacement field due to unit fault slip uniform over the appropriate depth interval. With reference to equation (2.36), the elastic solution for creep at a constant rate from depth D to the bottom of the lithosphere, depth H is given by

$$\dot{s}\hat{u}^{\text{elastic}}(x) = \frac{\dot{s}}{\pi}\left[\tan^{-1}\left(\frac{x}{D}\right) - \tan^{-1}\left(\frac{x}{H}\right)\right]. \tag{12.27}$$

To this, we must add the contribution from viscoelastic flow, consisting of an infinite sequence of events, extending from $t_{eq} = -\infty$ to $t_{eq} = 0$ evaluated at $t = 0$. Making use of equation (12.18), replacing $v^{\infty}T$ with $\dot{s}dt_{eq}$, and integrating over time:

$$v(x, t = 0) = \frac{\dot{s}}{\pi}\sum_{n=1}^{\infty}\frac{F_n(x, D, H)}{(n-1)!}\int_{-\infty}^{0}e^{t_{eq}/t_R}\left(\frac{-t_{eq}}{t_R}\right)^{n-1}\frac{dt_{eq}}{t_R}. \tag{12.28}$$

Here, $F_n(x, D, H)$ denotes surface deformation due to slip from the bottom of the coseismic slip zone D to the bottom of the elastic plate, depth H. Note that equation (6.39), from which this derives, includes only the transient viscoelastic response, hence the necessity of including the elastic response separately. Change variables $\zeta = -t_{eq}/t_R$ and reverse the order of integration such that

$$v(x, t = 0) = \frac{\dot{s}}{\pi}\sum_{n=1}^{\infty}\frac{F_n(x, D, H)}{(n-1)!}\int_{0}^{\infty}e^{-\zeta}\zeta^{n-1}d\zeta. \tag{12.29}$$

The integral is the Gamma function, which for integer values of n becomes

$$\int_{0}^{\infty}e^{-\zeta}\zeta^{n-1}d\zeta = \Gamma(n) = (n-1)! \tag{12.30}$$

(Abromowitz and Stegun 1972, equation 6.1.6). Thus, the surface velocity due to a creeping fault is, combining equation (12.29) with the elastic term (12.27),

$$v_c(x) = \dot{s} \left[\frac{1}{\pi} \sum_{n=1}^{\infty} F_n(x, D, H) + \hat{u}^{\text{elastic}}(x) \right],$$

$$= \frac{\dot{s}}{\pi} \left[\sum_{n=1}^{\infty} F_n(x, D, H) + \tan^{-1}\left(\frac{x}{D}\right) - \tan^{-1}\left(\frac{x}{H}\right) \right], \quad (12.31)$$

where the subscript $_c$ denotes the surface velocity due to fault creep. Note that the creep solution is steady, so that the displacements are simply $v_c t$.

It is perhaps surprising to note that the surface displacement due to steady fault creep is precisely the fully relaxed response to a *single* slip event over the same depth interval. To see this, compare $v_c t$ from equation (12.31) with equation (6.37) in the limit $t \to \infty$, accounting for the fact that the latter was for slip from the surface to depth D, rather than from D to H, as in this section. The slip per event is $\Delta u = \dot{s} t$. The correspondence between deformation due to steady sliding and the fully relaxed response to a single slip event also occurs for dip-slip faulting, as will be shown later in this chapter.

12.4.2 Comparison to Data from San Andreas Fault

We can compare the prediction of the simple elastic layer over Maxwell half-space cycle model with the Southern California Earthquake Center (SCEC) velocity data from the Carrizo Plain introduced in chapter 2. The average earthquake recurrence time (T) on this section of the San Andreas from paleoseismic studies is roughly 220 years (the 1995 Working Group on California Earthquake Probabilities estimated a recurrence interval of 206 years [WGCEP 1995]), with the last great event being the 1857 Fort Tejon earthquake. We thus fix $t = 144$ years and $t/T = 0.65$. The remaining parameters that can be estimated from the data are the elastic layer thickness (H), relaxation time (t_R), and the average slip rate. Figure 12.12A shows the misfit to the model as a function of elastic layer thickness and relaxation time. The minimum misfit is for $H = 14.3$ km, and $t_R = 60$ years, corresponding to a viscosity of $\eta = \mu t_R/2 = 1.9 \times 10^{19}$ Pa-s, assuming a shear modulus of 2×10^{10} Pa. A comparison of the observations with model predictions is shown in figure 12.13. While the fit to the data is good, the minimum in the misfit space is broad, and a range of models fit the data nearly as well. For example, the predicted velocity field for a 22-km-thick elastic layer overlying a viscoelastic half-space with a relaxation time of 180 years ($\eta = 5.7 \times 10^{19}$ Pa-s) is indistinguishable from that of the optimal model. The trade-off between relaxation time and elastic layer thickness can be understood by considering how the observed velocity field, which is broadly distributed, can arise in this model. One extreme is with a thick elastic layer, which causes the postseismic flow to be deep and therefore results in a broadly distributed deformation field. For a thinner elastic layer, the deformation is broadly distributed when T/t_R is greater than one, and the observation time t is at least modestly late in the earthquake cycle (see figure 12.10). Thus, for a fixed recurrence interval, the data require a shorter relaxation time, hence the trade-off between decreasing elastic layer thickness and relaxation time.

The best fitting slip rate is $\sim$44.2 mm/yr, which is considerably higher than geologic estimates of 33.9 $\pm$ 2.9 mm/yr for the past 3,700 years determined by Sieh and Jahns (1984). The corresponding slip per event is 9.7 meters. As can be seen from figure 12.12B, the slip rate is insensitive to H, and t_R along the minimum in misfit. One possible explanation for the discrepancy between the model slip rate and geologic estimates is the effect of the Hosgri fault. Including 3 mm/yr on the Hosgri fault (modeled simply with a buried screw dislocation) decreases the estimated slip rate on the San Andreas to 39.1 mm/yr; 5 mm/yr on the Hosgri

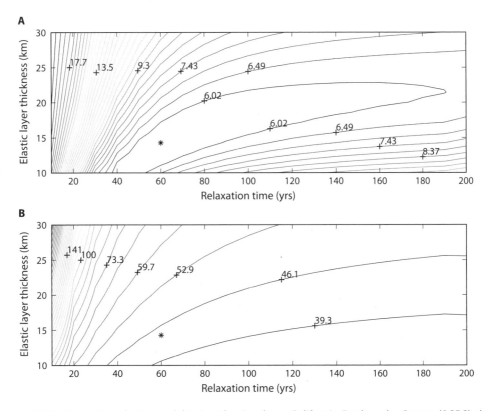

Figure 12.12. Fit to viscoelastic model using the Southern California Earthquake Center (SCEC) data from the Carrizo Plain (figure 2.16). A: Misfit contoured as a function of elastic layer thickness H and Maxwell relaxation time t_R. Best fitting model is shown with an asterisk. B: Estimated long-term slip rate for the different values of H and t_R.

drops the rate on the San Andreas to 36.6 mm/yr, which is indistinguishable from the geologic estimate. Note that in either case, the inferred relaxation time exceeds 60 to 70 years. Another possibility is that the estimated recurrence interval is inaccurate.

The comparison between observations and predictions (figure 12.13) is reasonable, considering the assumptions made, but not particularly better than the simple screw-dislocation model introduced in chapter 2. Qualitatively, both elastic and viscoelastic models fit the data equally well. On the other hand, we have only utilized data taken at a single epoch in time well after the last great earthquake. Observations made at different times in the earthquake cycle should provide additional constraints on the model parameters. In particular, we reconsider the fit to the post-1906 strain rate data from northern California first discussed in chapter 6.

To model the post-1906 data, we assume a 220-year recurrence interval for the northern San Andreas fault and a 25 mm/yr slip-rate (Schwartz et al. 1998; WGCEP 1999). The transient decay in strain rate in the early part of the twentieth century is insensitive to these assumptions. Again we solve for the best fitting elastic layer thickness and viscoelastic relaxation time. Results are shown in figure 12.14. The data again favor a rather thin elastic layer. The striking observation is that the post-1906 data favor a relaxation time of 25 to 30 years, far shorter than inferred from the Carrizo Plain data. The inferred relaxation time, corresponding to a viscosity of 8×10^{18} Pa-s, is not strongly sensitive to the elastic layer thickness. Comparing figures 12.14 and 12.12, we conclude that there is no single viscosity that is consistent with both the

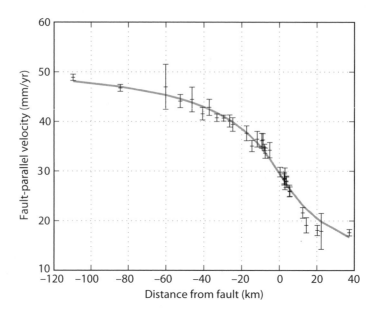

Figure 12.13. Fit to the Southern California Earthquake Center (SCEC) data from the Carrizo Plain assuming a simple elastic layer over a Maxwell viscoelastic half-space. The model prediction corresponds to the best fitting parameters: Elastic layer thickness of 14.2 km, Maxwell relaxation time of 60 years, and a slip rate of 44.2 mm/yr. No slip on the Hosgri fault is included.

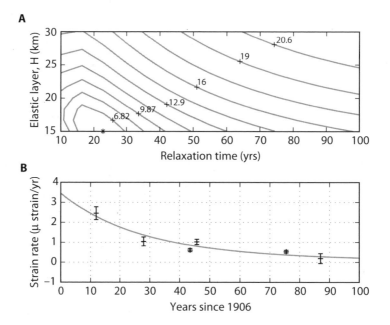

Figure 12.14. Fit to post-1906 strain-rate data assuming a simple elastic layer overlying a viscoelastic half-space. A: Misfit to the strain rate data as a function of elastic layer thickness and viscoelastic relaxation time. Asterisk indicates best fitting parameters. B: Comparison of model predicted and observed strain rate. Data from Kenner and Segall (2000).

post-1906 strain data and the interseismic Carrizo Plain data. The post-1906 data require a viscosity at least a factor of two less than the interseismic Carrizo Plain data. Actually, the difference is likely to be substantially greater than that, since the lower relaxation times in figure 12.12A correspond with unrealistically thin elastic layers, given that earthquakes such as the 1989 Loma Prieta earthquake ruptured to depths of ~16 km.

These results point to significant shortcomings in the elastic layer over Maxwell half-space model. The discrepancy between fits to the contemporary Carrizo Plain velocity distribution and post-1906 strain rate data can be explained by a number of model limitations. One possibility is that the rheology of the relaxing lower crust or upper mantle is stress dependent, such that the effective viscosity decreases under the high stresses acting immediately after the earthquake but increases as the stresses relax. Another possibility, considered in the following section, is that the fault below the seismogenic zone creeps in response to the stress change imposed by the earthquake (Johnson and Segall 2004). Before considering models with stress driven afterslip, we briefly examine the effects of multiple viscoelastic layers with differing relaxation times.

Johnson et al. (2007) considered an extension of the Savage-Prescott model including multiple relaxing layers representing the lower crust, mantle lithosphere, and upper mantle. The corresponding elastic problem was analyzed using propagator methods outlined in chapter 5 (see problem 6). Representative results are shown in figure 12.15. The first example has a low-viscosity lower crust overlying a more viscous mantle. At short times, this system behaves like a low-viscosity channel as analyzed in chapter 6; at longer times, the mantle layers relax. As was the case with the homogeneous Maxwell half-space, the surface velocities exceed the plate velocity early in the cycle but lag the plate velocity late in the cycle. The behavior is quite different when the mantle viscosity is less than that of the lower crust and uppermost mantle (figure 12.15B). In this case, the deformation is concentrated near the fault by the relatively high viscosity crust and uppermost mantle. (Maxwell relaxation times for these layers are comparable to the earthquake recurrence interval.) The velocity profile at distances greater than 50 km from the fault are rather flat, but at rates below the long-term slip rate. A similar result occurs with an intermediate-viscosity lower crust (figure 12.15C). These results demonstrate that caution must be used when inferring slip rates from the apparent asymptote of a velocity profile. In these cases, the apparent asymptotes substantially underestimate the actual long-term slip rate.

Johnson et al. (2007) were able to find a layered viscoelastic model that is consistent with postseismic relaxation following the 1992 Landers earthquake, triangulation measurements spanning the interval 1932 to 1977, GPS measurements of the contemporary velocity field, and paleoseismic data along the San Andreas fault. Their results imply that the effective viscosity of the upper mantle is a factor of 100 to 1000 less than the uppermost mantle, and less than or equal to that of the lower crust.

12.4.3 Viscoelastic Models with Stress-Driven Deep-Fault Creep

Several methods have been developed to include both the effects of distributed viscoelastic flow beneath the elastic crust and stress-driven creep on discrete faults beneath the seismogenic zone. Two such models are briefly described here, without going into details of the derivations.

Li and Rice (1987) employed a generalization of the Elsasser coupling model introduced in chapter 6, in which a long strike-slip fault slips periodically, with repeat time T, at depths $-L \leq z \leq 0$ (figure 12.16). Basal shear acts on the base of the elastic layer, which has thickness $H \geq L$. Below the seismogenic layer, at depths $-H \leq z \leq -L$, the fault creeps at constant shear stress. Stress equilibrium in the elastic plate is given by equation (6.3), where σ is interpreted

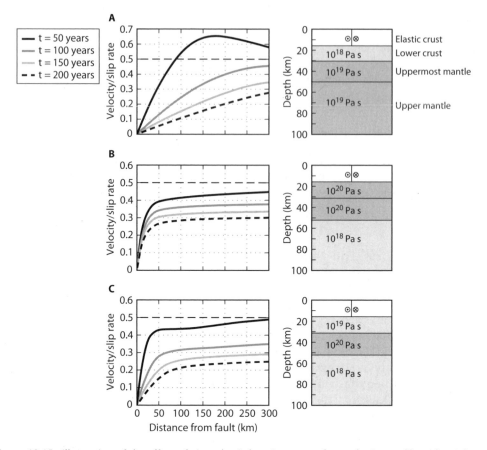

Figure 12.15. Illustration of the effect of viscoelastic layering on surface velocity profiles, after Johnson et al. (2007). Curves show normalized velocity at different times following the last earthquake. Earthquake cycle time is 200 years. A: Low-viscosity lower crust. B: Low-viscosity upper mantle. C: Low-viscosity upper mantle and intermediate-viscosity lower crust.

as the fault-parallel shear stress. As in the Elsasser model, the thickness-averaged stress and strain are proportional (equation [6.4]). The basal shear drag, assuming the asthenospheric layer obeys a Maxwell viscoelastic rheology (6.18), is

$$\frac{b}{\mu}\frac{\partial \tau(y, t)}{\partial t} + \frac{h}{\eta}\tau = \frac{\partial u(y, t)}{\partial t} - V_0(y), \qquad (12.32)$$

where y is the fault-perpendicular coordinate.

Li and Rice (1987) introduce b as an effective elastic coupling thickness, chosen to be $(\pi/4)^2 H$ so that the short-term elastic response is consistent with the corresponding fully elastic solution. $V_0(y)$ specifies the velocity distribution at the base of the asthenosphere, assumed independent of time, that loads the system. The basal velocity matches the plate velocity in the far field; $V_0(y \rightarrow \pm\infty) = \pm v_{plate}/2$. Interestingly, it turns out that the surface deformation is insensitive to the detailed form of the distribution $V_0(y)$. This arises because the time-dependent surface deformation is controlled by asthenospheric flow driven by slip on the fault, which concentrates stress at the base of the fault.

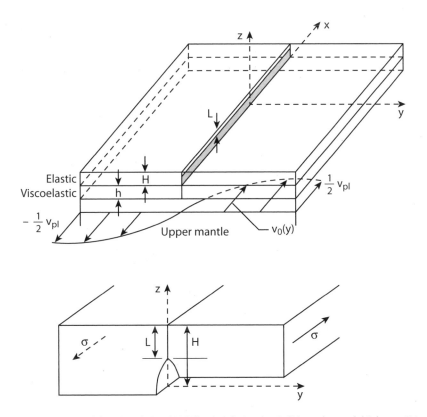

Figure 12.16. Geometry of the Li and Rice (1987) model. An elastic lithosphere of thickness H is coupled through an asthenosphere of thickness h to the upper mantle. The velocity distribution of the upper mantle flow is $V_0(y)$. The fault is assumed to be locked interseismically from the surface to depth L and creeping at constant shear stress below. After Li and Rice (1987).

Combining equations (6.3), (6.4), and (12.32) leads to the generalized Elsasser coupling equation:

$$\left(\frac{hH\mu}{\eta} + bH\frac{\partial}{\partial t} \right) \frac{\partial^2 u}{\partial y^2} = \frac{\partial u}{\partial t} - V_0(y). \tag{12.33}$$

At the plate boundary, the thickness-averaged stress is proportional to the net thickness averaged displacement $2u - \Delta u$, where $\Delta u(t)$ is the specified slip on the coseismic fault, assumed to be a periodic sequence of earthquakes as in figure 12.7. Thus, the appropriate boundary condition for the differential equation (12.33) is

$$\mu \frac{\partial u}{\partial y} = k(2u - \Delta u). \tag{12.34}$$

The stiffness k is estimated by analyzing a traction-free plate with a stress-free cut at $y = 0$ below depth L, loaded by stress σ. It can be shown that the thickness-averaged displacement satisfies $\partial u/\partial y = \sigma/\mu$ everywhere except at $y = 0$, where there is a jump in displacement of

$$[u] = \frac{2\sigma H}{\pi \mu} \ln \left[\frac{1}{\sin(\pi L/2H)} \right], \tag{12.35}$$

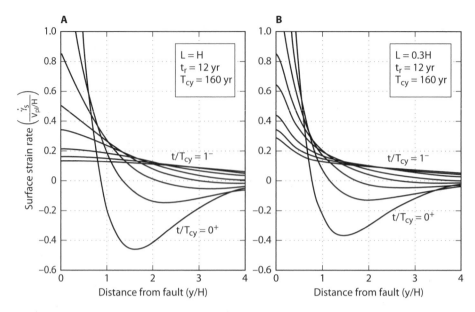

Figure 12.17. Surface strain rate as a function of normalized distance from the fault for the Li-Rice model at times 0.0, 0.1, 0.2, 0.3, 0.4, 0.6, 0.8, and 1.0 of the recurrence time. The particular calculations shown are for a repeat time of 160 years and a Maxwell relaxation time of 12 years. A: The fault is locked through the entire elastic lithosphere. B: The fault is locked only partway through the lithosphere, $L = 0.3H$. After Li and Rice (1987).

where the brackets indicate jump in displacement across the plate boundary at $y = 0$. Note that if the fault is locked throughout the elastic layer, $L = H$ and $[u] = 0$, but becomes unbounded as $L \to 0$. From equation (12.35), it is possible to infer the appropriate stiffness in (12.34) as

$$k = \frac{\pi \mu}{4H} \left\{ \ln \left[\frac{1}{\sin(\pi L / 2H)} \right] \right\}^{-1}. \tag{12.36}$$

The partially cracked plate analysis was also used to approximate the displacements at the earth's surface in terms of the thickness-averaged displacements. The surface strain rates are shown in figure 12.17 as a function of distance normal to the fault for two cases, one in which the coseismic ruptures extend through the entire lithosphere and one in which they break only the top 30% of the lithosphere. Note that in the latter case, the time-varying fault creep concentrates strain rate near the plate boundary, particularly late in the earthquake cycle.

Li and Rice (1987) compared the surface strain rates with the observed decay in strain rate following the 1906 and 1857 earthquakes in California (figure 12.18). Assuming a plate velocity of 35 mm/yr and an earthquake cycle time of 160 years, the data are consistent with lithospheric thickness in the range of 20 to 30 km and Maxwell relaxation times of 10 to 16 years. Recall that the simple layer over Maxwell half-space required a relaxation time on the order of 15 to 30 years when compared to the strain rate data following the 1906 earthquake. Thus, including the effects of stress-driven creep below the seismogenic fault does not appear to exert a major influence on the inferred Maxwell relaxation times.

Johnson and Segall (2004) examined an extension of the viscoelastic coupling model discussed in section 12.4.1, which includes stress-driven creep on a downdip extension of the coseismic fault. The approach is based on a boundary element method in which each element on the creeping portion of the fault slips in response to the elastic stress induced by coseismic

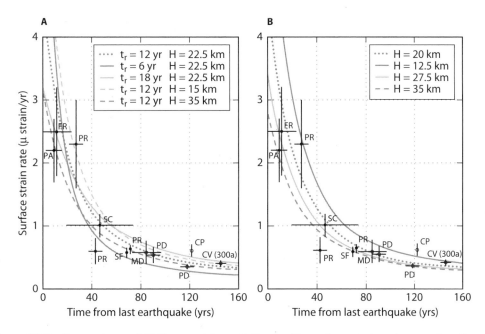

Figure 12.18. Comparison of Li-Rice model predictions with surface strain rate as a function of time since the last great earthquake. Data are from Thatcher (1983): closed circles are northern California, where time is post-1906 quake; open circles are southern California, where time is post-1857 earthquake. Curves with parameters indicated are compared to the data. Plate velocity is 35 mm/yr, and earthquake cycle time is 160 yr. A: $L = 9$ km. B: $L = 11$ km. $t_R = 14$ yr.

slip as well as slip on all other creeping elements, plus the viscoelastic stress resulting from both coseismic slip and creep on all elements for all past times. They compared models with kinematically imposed creep, as in section 12.4.1; creep at constant resistive stress, as in the Li-Rice model; as well as a linear-viscous rheology in which the slip rate on the creeping fault is proportional to the shear stress acting on the fault (see Figure 12.19).

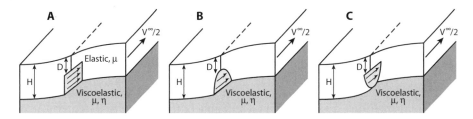

Figure 12.19. Three models of a vertical strike-slip fault with coseismic slip to depth D and creep below. A: Kinematically imposed creep at a constant rate. B: Stress-driven creep at constant shear stress. C: Viscous creep in which the slip rate across the downdip extension of the fault is proportional to the imposed stress. After Johnson and Segall (2004).

Of particular interest is the case in which the fault creep-rate downdip of the seismogenic zone is proportional to the imposed shear stress. This can be viewed as a linear approximation to steady-state friction. High stresses immediately after a large earthquake drive fast post-seismic slip (figure 12.20). Early in the cycle, the slip rate greatly exceeds the long-term

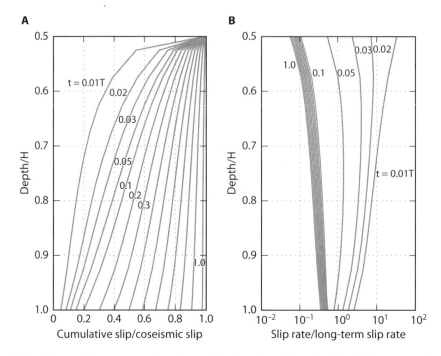

Figure 12.20. Slip and slip rate as a function of depth below the locked fault for a model in which the fault below the seismogenic zone exhibits a linear viscous rheology. The example shown is for the case of $D = 0.5H$—that is, the upper half of the elastic plate remains locked between earthquakes. $T/t_R = 10$. A: Cumulative slip normalized by the coseismic slip. B: Slip rate normalized by the long-term slip rate. Viscosity of the fault zone is 0.1% that of the Maxwell half-space, assuming a fault zone thickness of 1 cm. After Johnson and Segall (2004).

rate, whereas late in the cycle, the creep rate lags the long-term rate. Immediately prior to the next earthquake $t = T^-$, the entire creeping zone has slipped an amount equivalent to the coseismic slip.

The effect of the high slip rates on the deep fault is to increase the surface strain rates relative to a model in which the coseismic fault ruptures the entire elastic layer (figure 12.21) at all times through the earthquake cycle. The higher strain rate associated with downdip postseismic slip suggests that it should be possible to explain both the rapid strain rate decay following the 1906 earthquake and the broad-scale contemporary deformation field observed in the Carrizo Plain with a single mechanical model, and this indeed appears to be the case, as illustrated in figure 12.22. There are seven parameters in the inversion shown in figure 12.22, so perhaps it is not too surprising that a reasonable fit to the data is obtained. The parameter estimates are all geologically reasonable, however, so one can conclude that a combination of stress-driven postseismic slip combined with distributed viscoelastic flow is consistent with what is currently known about deformation along the San Andreas fault.

12.4.4 Viscoelastic Cycle Models for Dipping Faults

Savage (1983) presented a method for extending the back-slip concept for subduction zone settings to models including an elastic plate overlying a viscoelastic asthenosphere. The underlying assumption is that the process of subduction yields no long-term deformation of the hanging-wall plate. The premise is that plate tectonics involves only horizontal rigid body motion, with no vertical displacement.

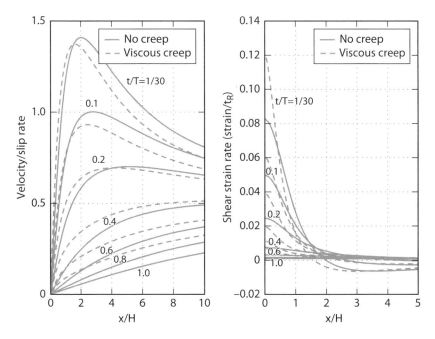

Figure 12.21. Surface velocity and strain rate as a function of normalized distance from the fault trace, comparing the case of a linear-viscous fault below the seismogenic zone (figure 12.20), to one in which the coseismic slip breaks the entire elastic lithosphere. After Johnson and Segall (2004).

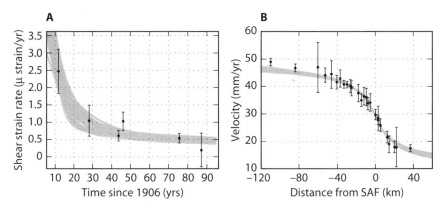

Figure 12.22. Comparison of deformation data from the San Andreas fault and a model including a linear viscous fault below the seismogenic zone. A: Fit to the post-1906 strain rate data. B: Fit to the contemporary GPS-derived velocity field in the Carrizo Plain region. The gray curves show the predicted data from a number of models generated by a Markov Chain Monte Carlo estimation. These data constrain the Maxwell relaxation time only to the range 24–600 years, the average recurrence interval in northern California to 210–460 years, and the average recurrence interval along the Carrizo Plain segment to 315–400 years. The slip rates were not significantly improved over the a priori bounds of 19–27 mm/yr for the Bay Area and 31–42 mm/yr along the Carrizo segment. Details of the data and estimation procedure are given in Johnson and Segall (2004).

In a reference frame fixed to the overriding plate, the displacement is

$$\mathbf{u}(\mathbf{x}, t) = \mathbf{u}_{bs}(\mathbf{x}, t) + s \sum_{n=0}^{\infty} \hat{\mathbf{u}}(\mathbf{x}, t + nT). \tag{12.37}$$

Here, $\mathbf{u}_{bs}(\mathbf{x}, t)$ is the displacement at the earth's surface due to back slip, while $\hat{\mathbf{u}}(\mathbf{x}, t)$ represents the surface displacement due to a single earthquake of unit slip at time $t = 0$, including both the coseismic elastic response and the postseismic viscoelastic response. It is assumed that the rate of back slip is appropriate to cancel the plate motion, $\dot{s} = v^{plate}$, and that the coseismic slip recovers the accumulated slip deficit, $s = \dot{s}T$, where T is the recurrence time of large plate boundary earthquakes.

Steady back slip alone produces steady-state surface deformation, which can be written as

$$\mathbf{u}_{bs}(\mathbf{x}, t) = \mathbf{v}_{bs}(\mathbf{x})t + \mathbf{u}_0(\mathbf{x}), \tag{12.38}$$

where $\mathbf{v}_{bs}(\mathbf{x})$ is the velocity at the earth's surface due to back slip, and $\mathbf{u}_0(\mathbf{x})$ is independent of time. It is convenient to measure deformation relative to the time immediately prior to the most recent earthquake, such that $\mathbf{u}(\mathbf{x}, t = 0^-) = 0$. Thus,

$$\mathbf{u}_0(\mathbf{x}) = -s \sum_{n=1}^{\infty} \hat{\mathbf{u}}(\mathbf{x}, nT), \tag{12.39}$$

since by definition, $\hat{\mathbf{u}}(\mathbf{x}, t = 0^-) = 0$. Thus,

$$\mathbf{u}(\mathbf{x}, t) = \mathbf{v}_{bs}(\mathbf{x})t + s\hat{\mathbf{u}}(\mathbf{x}, t) + s \sum_{n=1}^{\infty} [\hat{\mathbf{u}}(\mathbf{x}, t + nT) - \hat{\mathbf{u}}(\mathbf{x}, nT)] \qquad 0 \le t < T. \tag{12.40}$$

If there is no long-term deformation, the cumulative displacement at the end of each earthquake cycle must vanish—that is, $\mathbf{u}(\mathbf{x}, T) = 0$. Substituting this constraint into equation (12.40), and noting that all terms in the sum but the last cancel, we find that

$$\mathbf{v}_{bs}(\mathbf{x}) = -s\hat{\mathbf{u}}(\mathbf{x}, \infty)/T. \tag{12.41}$$

With this result, equation (12.40) becomes

$$\mathbf{u}(\mathbf{x}, t) = -\dot{s}t\hat{\mathbf{u}}(\mathbf{x}, \infty) + \dot{s}T\hat{\mathbf{u}}(\mathbf{x}, t) + \dot{s}T \sum_{n=1}^{\infty} [\hat{\mathbf{u}}(\mathbf{x}, t + nT) - \hat{\mathbf{u}}(\mathbf{x}, nT)] \qquad 0 \le t < T,$$

$$\tag{12.42}$$

where we have made use of the fact that $s = \dot{s}T$. The first term represents the deformation due to back slip, proportional to the fully relaxed response to a single earthquake, and accumulates linearly in time. The second term represents the coseismic and viscoelastic postseismic response to the most recent earthquake, while the sum gives the deformation due to all previous earthquakes. As we shall see, this sum is easily computed, so the computation of an infinite sequence of earthquakes involves no more overhead than computing the response to a single event.

An alternative approach was presented by Matsu'ura and Sato (1989), who argue that uplifted marine terraces along many subduction margins (figure 12.23) contradict the

Figure 12.23. Marine terraces on the Mahia Peninsula on the east coast of the North Island of New Zealand record uplift adjacent to the Hikurangi subduction zone. Four terraces on the coastal platform record coseismic uplift during the last 3,500 years. Marine terraces above the coastal platform record high sea-level stands during the last 130,000 years. Photo by D. L. Homer, *EOS*, vol. 65, no. 1, 3 January 1984.

assumption of negligible long-term deformation. Instead, they compute the long-term deformation as resulting from steady sliding along the entire plate interface in a layered elastic–viscoelastic earth. Matsu'ura and Sato (1989) take the surface deformation to be the sum of (1) steady sliding over the entire plate interface, representing the long-term deformation, (2) steady back slip over the seismogenic portion of the fault, and (3) the complete viscoelastic response to an infinite sequence of prior coseismic slip events. They show that the deformation due to steady sliding is equivalent to the fully relaxed response to a single slip event on the same part of the fault. We have already seen that this is true for strike-slip faulting (equation [12.31]) and show later in this section that it is also true for dip-slip faulting. With this result, it is possible to write the cumulative displacement at the earth's surface as

$$\mathbf{u}(\mathbf{x}, t) = \dot{s}t\hat{\mathbf{u}}_s(\mathbf{x}, \infty) - \dot{s}t\hat{\mathbf{u}}(\mathbf{x}, \infty) + s \sum_{n=0}^{\infty} \hat{\mathbf{u}}(\mathbf{x}, t + nT). \qquad (12.43)$$

The first term denotes the steady forward slip, the second denotes back slip, and the sum represents the displacement due to all past earthquakes. The distinction between the first and second terms is that $\dot{s}t\hat{\mathbf{u}}_s(\mathbf{x}, \infty)$ represents forward slip over the entire plate interface, whereas $\dot{s}t\hat{\mathbf{u}}(\mathbf{x}, \infty)$ corresponds to back slip only on the portion of the plate interface that is locked between large earthquakes. Sato and Matsu'ura (1988) show that slip within a Maxwell asthenosphere, or any elastic layers beneath the asthenosphere, does not contribute to the fully relaxed deformation at the earth's surface. This is intuitive in that shear stresses vanish in a fully relaxed Maxwell medium, but also see problem 4. Again measuring deformation relative to the time of the last earthquake, $\mathbf{u}(\mathbf{x}, t = 0^-) = 0$, we have

$$
\mathbf{u}(\mathbf{x}, t) = \dot{s}t[\hat{\mathbf{u}}_s(\mathbf{x}, \infty) - \hat{\mathbf{u}}(\mathbf{x}, \infty)] + \dot{s}T\hat{\mathbf{u}}(\mathbf{x}, t) + \dot{s}T \sum_{n=1}^{\infty} [\hat{\mathbf{u}}(\mathbf{x}, t + nT) - \hat{\mathbf{u}}(\mathbf{x}, nT)]
$$

$$
0 \leq t < T. \tag{12.44}
$$

Comparing to equation (12.42), we see that the difference between the two models is the forward slip $\dot{s}t\hat{\mathbf{u}}_s(\mathbf{x}, \infty)$ in equation (12.44). It will come as no surprise then that the cumulative deformation at the end of the cycle is

$$
\mathbf{u}(\mathbf{x}, T) = \dot{s}T\hat{\mathbf{u}}_s(\mathbf{x}, \infty). \tag{12.45}
$$

The coseismic slip and backslip exactly cancel, so that only the steady-state motion over the entire plate interface contributes to the long-term deformation.

If earthquakes cut the entire lithosphere with uniform slip, then $\hat{\mathbf{u}}_s(\mathbf{x}, \infty) = \hat{\mathbf{u}}(\mathbf{x}, \infty)$, because, as discussed previously, steady slip below the lithosphere does not contribute to the fully relaxed deformation. In this case, equation (12.44) reduces to an infinite series of forward-slip events, without any back slip:

$$
\mathbf{u}(\mathbf{x}, t) = \dot{s}T\hat{\mathbf{u}}(\mathbf{x}, t) + \dot{s}T \sum_{n=1}^{\infty} [\hat{\mathbf{u}}(\mathbf{x}, t + nT) - \hat{\mathbf{u}}(\mathbf{x}, nT)] \qquad 0 \leq t < T. \tag{12.46}
$$

Furthermore, if points on the fault either slip in repeating earthquakes (by amount $\dot{s}T$) or creep at a constant rate $\dot{s}$, then $\hat{\mathbf{u}}_s(\mathbf{x}, \infty) - \hat{\mathbf{u}}(\mathbf{x}, \infty)$ vanishes everywhere the fault is locked and is equivalent to forward slip at a rate $\dot{s}$ everywhere the fault is creeping; there is no back slip if the fault is creeping at the long-term rate. Thus, equation (12.44) can equally well be interpreted as a sum of forward earthquake slip events over the seismogenic fault, with steady forward creep on the remaining portions of the lithospheric fault.

We now turn to developing explicit expressions for the various terms in equations (12.42) and (12.44). As in chapter 6, the displacements at the earth's surface are written in the Fourier domain. Equation (6.64) gives the displacement field due to a single earthquake with unit slip as

$$
u_i(k, t, \mathbf{x}_s) = u_i^e(k, \mathbf{x}_s)\mathcal{H}(t) + \sum_{j=1}^{N} \Upsilon_i^{(j)}(H, k, \mathbf{x}_s)[e^{-t/\tau_j(k)} - 1], \tag{12.47}
$$

where u_i^e are the elastic displacements due to a unit slip at $\mathbf{x}_s$, k is the horizontal wavenumber, H is the elastic layer thickness, N is the number of independent relaxation times $\tau_j(k)$, and $\mathcal{H}(t)$ is the Heavyside function. $\Upsilon_i^{(j)}(H, k, \mathbf{x}_s)$ are the complicated functions that depend on the component of displacement i, wavenumber k, and source location $\mathbf{x}_s$, which result from the matrix operations in equations (6.63). Equations (12.47) are written for unit slip localized at source coordinate $\mathbf{x}_s$. To obtain the full displacement, these must be integrated over the seismogenic portion of the fault

$$\hat{u}_i(k, t) = \mathcal{H}(t) \int_\Sigma u_i^e(k, \mathbf{x}_s) d\mathbf{x}_s + \int_\Sigma \sum_{j=1}^N \Upsilon_i^{(j)}(H, k, \mathbf{x}_s)[e^{-t/\tau_j(k)} - 1] d\mathbf{x}_s,$$

$$= u_i^e(k)\mathcal{H}(t) + \sum_{j=1}^N F_i^{(j)}(H, k)[e^{-t/\tau_j(k)} - 1], \tag{12.48}$$

where

$$F_i^{(n)}(H, k) = \int_\Sigma \Upsilon_i^{(n)}(H, k, \mathbf{x}_s) d\mathbf{x}_s,$$

$$u_i^e(k) = \int_\Sigma u_i^e(k, \mathbf{x}_s) d\mathbf{x}_s. \tag{12.49}$$

It is understood here that $u_i^e(k, \mathbf{x}_s)$ represents the elastic response due to a point source, while $u_i^e(k)$ represents the elastic response due to uniform slip over the lithospheric fault. Substituting equations (12.48) into (12.42) and making use of the sum (12.24) yields

$$u_i(k, t) = -\dot{s}t \left[u_i^e(k) - \sum_{j=1}^N F_i^{(j)}(H, k) \right] + \dot{s}T u_i^e(k)\mathcal{H}(t)$$
$$+ \dot{s}T \sum_{j=1}^N F_i^{(j)}(H, k) \left[e^{-t/\tau_j(k)} - 1 \right] \frac{e^{T/\tau_j(k)}}{e^{T/\tau_j(k)} - 1} \quad 0 \le t < T. \tag{12.50}$$

The term in brackets is the fully relaxed response due to unit slip on the fault. $\dot{s}T u_i^e(k)\mathcal{H}(t)$ is the elastic response due to the most recent earthquake, while the last term represents the viscoelastic response due to all previous earthquakes. It is left to the reader to verify that the displacements vanish for $t = 0^-$ and $t = T$. The corresponding expression in which the long-term deformation is given by the fully relaxed, faulted elastic lithosphere is easily obtained by modifying the term proportional to $\dot{s}t$ according to equation (12.44).

The surface velocity field is found by differentiating equation (12.50) with respect to time:

$$v_i(k, t) = -\dot{s} \left[u_i^e(k) - \sum_{n=1}^N F_i^{(n)}(H, k) \right] - \dot{s}T \sum_{n=1}^N \frac{F_i^{(n)}(H, k)}{\tau_n(k)} \left[\frac{e^{T/\tau_n(k)}}{e^{T/\tau_n(k)} - 1} \right] e^{-t/\tau_n(k)}. \tag{12.51}$$

The velocities (12.51) correspond to the case with no long-term deformation, which we may refer to as the *rigid plate model*, in that only rigid body displacements remain at the end of the earthquake cycle. The model in which the long-term deformation corresponds to the fully relaxed asthenosphere is obtained by modifying the steady-state terms in equation (12.51). We refer to this as the *deforming plate model*, in that it predicts finite long-term deformation. In this case, the average long-term surface velocity is found from equation (12.45), noting that the displacements accumulate over the cycle time T. Thus, $\mathbf{v}^{lt} = \dot{s}\hat{\mathbf{u}}_s(\mathbf{x}, \infty)$, where the superscript

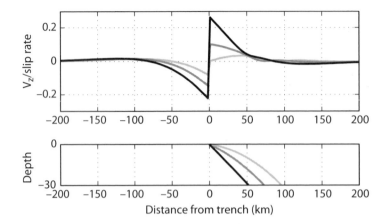

Figure 12.24. Long-term vertical uplift rate for the deforming plate model, equation (12.52). Results are compared for a planar fault with constant dip of 30 degrees and two curved faults. For the curved faults, the dip increases linearly with depth from 5 degrees and 15 degrees to 30 degrees. Calculation courtesy of K. M. Johnson.

refers to *long term*. From equation (12.48),

$$v_i^{lt}(k) = \dot{s} \left[u_i^e(k) - \sum_{n=1}^{N} F_i^{(n)}(H, k) \right], \tag{12.52}$$

where it is implicit that the integrations in equation (12.49) are for uniform slip over the entire lithosphere.

Figure 12.24 shows the predicted long-term uplift rate for a thrust fault with a constant dip of 30 degrees. Note that the uplift rate at the trench is roughly one-fourth of the plate velocity, corresponding to approximately 10 mm/yr for a convergence rate of 40 mm/yr. This is certainly far too rapid relative to observed long-term uplift rates. Sato and Matsu'ura (1992) taper the dip so that the fault intersects the earth's surface with lower dip, which minimizes the uplift rate at the fault (figure 12.24). They find uplift rates of ~1 mm/yr at a distance of 135 km from the trench, in reasonable agreement with long-term uplift rates. While the subducting plate must bend continuously as it descends into the mantle, the upper plate in these models is fully elastic, and thus fails to account for inelastic deformation in the bending plate. Inelastic deformation will also occur in the overriding plate. A significant amount of the permanent deformation in the upper plate is likely to occur by faults that splay off the main plate boundary thrust. This splay faulting complicates comparison of model predictions with observed terrace uplift rates.

The long-term velocity field within the earth for a fault, infinite along strike with constant dip, is illustrated in cross section in figure 12.25. The far-field velocity exhibits uniform convergence at the plate rate within the elastic lithosphere but decays with depth within the viscoelastic asthenosphere. A prominent downwelling occurs beneath the downdip extension of the fault within the asthenosphere. Note that this downwelling is driven by the kinematically imposed repeating earthquakes at the plate boundary, not by density differences between the subducting slab and the surrounding material.

As a final step, we construct the solution for steady creep on a dipping fault. As in the strike-slip case, the velocity field for a steadily creeping fault has an elastic component $\dot{s}u_i^e(k)$ and a viscous component that can be constructed from an infinite sequence of infinitesimal slip

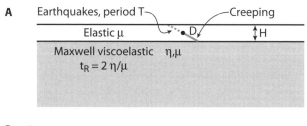

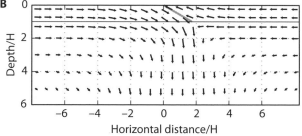

Figure 12.25. Long-term displacements in both the elastic plate and the underlying Maxwell half-space for the deforming plate model. The upper part of the fault $z < D$ slips episodically with repeat time T, whereas the lower part of the fault creeps steadily at rate $\dot{s}$. After Johnson (2004).

events with recurrence time $\delta\tau$ (figure 12.11). The latter can be derived from the time derivative of equation (12.47). The response to a single earthquake with slip Δu at time τ is

$$v_i(k, t, \mathbf{x}_s) = -\Delta u \sum_{j=1}^{N} \Upsilon_i^{(j)}(H, k, \mathbf{x}_s) \frac{e^{-(t-\tau)/\tau_j}}{\tau_j}. \tag{12.53}$$

Let $\Delta u = \dot{s}\delta\tau$ and integrate over all past slip events, again evaluating the velocity at $t = 0$,

$$v_i(k, \mathbf{x}_s) = -\dot{s} \sum_{j=1}^{N} \Upsilon_i^{(j)}(H, k, \mathbf{x}_s) \int_{-\infty}^{0} e^{\tau/\tau_j} \frac{d\tau}{\tau_j},$$

$$= -\dot{s} \sum_{j=1}^{N} \Upsilon_i^{(j)}(H, k, \mathbf{x}_s). \tag{12.54}$$

The velocity field at the earth's surface due to a steadily creeping fault in the elastic layer is thus given by integrating

$$v_i(k, \mathbf{x}_s) = \dot{s} \left[u_i^e(k, \mathbf{x}_s) - \sum_{j=1}^{N} \Upsilon_i^{(j)}(H, k, \mathbf{x}_s) \right] \tag{12.55}$$

over the creeping part of the fault. If the fault creep is uniform over the entire lithospheric fault, then integration of equation (12.55) is equivalent (12.52). In this limit, the long-term deformation is entirely due to forward fault creep, and there is no fault locking and earthquake slip, so the two results must agree. As in the strike-slip case, the constant creep-rate response is equivalent to the fully relaxed response to a single earthquake.

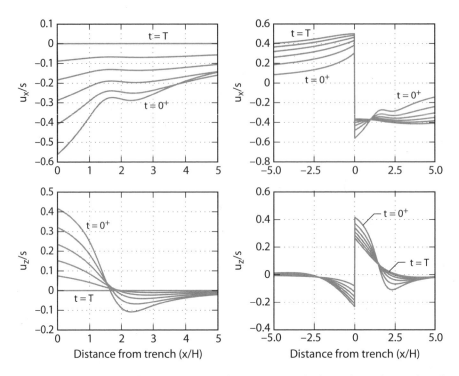

Figure 12.26. Accumulated displacements at regular time intervals throughout the earthquake cycle for a 30-degree dipping thrust fault in an elastic layer overlying a Maxwell half-space. Curves are shown for $t = 0+$, $0.2T$, $0.4T$, $0.6T$, $0.8T$, and T^-, where T is the earthquake cycle time. $\mu T / \eta = 2.5$. Left: The rigid plate model, equation (12.42), in which there is no long-term deformation. Right: The deformable plate model, equation (12.44), with finite long-term deformation. The plate thickness H is 30 km, and the depth of fault locking $D = H$. Fault dips to the right. Computations employ the surface gravity approximation.

Cumulative displacements throughout the earthquake cycle are shown in figure 12.26. Note that for the rigid plate model, the displacements are shown only on the hanging-wall side of the fault and are referenced to a distant point on that plate. In this model, the coseismic displacement is completely negated by interseismic deformation so that at $t = T$ there is no displacement of the hanging wall. At the end of the cycle, the foot-wall plate (not shown) has translated horizontally by $v^{plate}T$ toward the hanging wall. For the deforming plate model, the hanging wall near the fault subsides interseismically, but at a slower rate than in the rigid plate model, so that there is a net uplift of the hanging wall at the end of the cycle. In contrast, the coseismic subsidence on the hanging wall near $x/H \sim 2$ not only recovers, it accumulates net uplift by the end of the cycle. The relative far-field horizontal motion in the deformable plate model is $s \cos(\delta)$, which for a 30-degree dip is $\sim 0.87s$. Figure 12.27 illustrates the cumulative displacements for the same ratio T/t_R when only the upper half of the lithospheric fault is locked between earthquakes. The lower portion of the fault slips at a steady rate $\dot{s}$. The general patterns are similar to those in figure 12.26; however, the coseismic, and therefore interseismic deformation is concentrated closer to the fault.

Surface velocities for both the rigid plate and deforming plate models are shown in figures 12.28, 12.29, and 12.30. Figure 12.28 contrasts the two models for coseismic ruptures that break the entire elastic layer $D = H$, with $T/t_R = 2.5$. Although velocities decrease with time in both models, there are clear differences, particularly in the vertical rates. The rigid plate

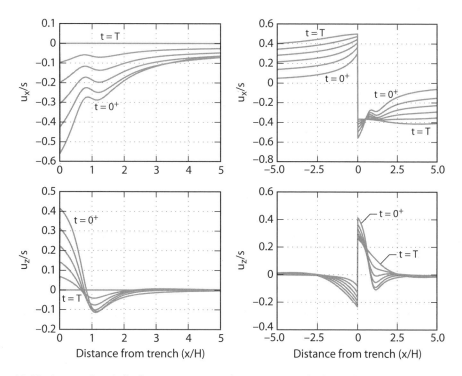

Figure 12.27. Accumulated displacements at regular time intervals throughout the earthquake as in figure 12.26, but with only the upper half of the lithospheric fault locked between earthquakes, $D = 0.5H$. The plate thickness H is 30 km, and the fault dips 30 degrees to the right through an elastic layer that overlies a Maxwell half-space. Curves are shown for $t = 0+$, $0.2T$, $0.4T$, $0.6T$, $0.8T$, and T^-, where T is the earthquake cycle time. $\mu T/\eta = 2.5$. Left: The rigid plate model, equation (12.42), in which there is no long-term deformation. Right: The deformable plate model, equation (12.44), with finite long-term deformation. Computations employ the surface gravity approximation.

model predicts that coseismic uplift near the trench fully recovers during the earthquake cycle, so the subsidence rate in this region exceeds that predicted by the deforming plate model at all times. Conversely, the deforming plate model predicts that the landward coseismic subsidence not only recovers, as it does in the rigid plate model, but ultimately leads to long-term uplift (figure 12.26). Thus, the uplift rate in this region, $\sim 1.5H < x \lesssim 2.5H$, is greater in the deforming plate model than it is in the rigid plate model. The horizontal velocities are more similar. Within one lithospheric thickness of the fault, the velocities are positive—that is, toward the overriding plate. At short times, transient postseismic deformation causes the more distal hanging wall to move toward the fault, $v_x < 0$. At later times in the cycle, interseismic strain accumulation causes the entire hanging wall to move toward the fixed overriding plate.

As is the case at strike-slip plate boundaries, with relatively long relaxation times, $T/\tau_R < 1$ relaxation effects are not significant, and the interseismic velocity does not vary significantly throughout the earthquake cycle. Conversely, transient postseismic effects are more pronounced with short relaxation times. Figure 12.29 illustrates the surface velocities for the same geometry as Figure 12.28, except that $T/t_R = 5$. In this case, transient effects are more pronounced, in both the vertical and horizontal velocities. Because of this, the velocities predicted by the deforming plate and rigid plate models tend to be more similar early in the earthquake cycle. Nevertheless, the subsidence of the hanging wall near the fault is more rapid in the rigid plate model than in the deforming plate counterpart.

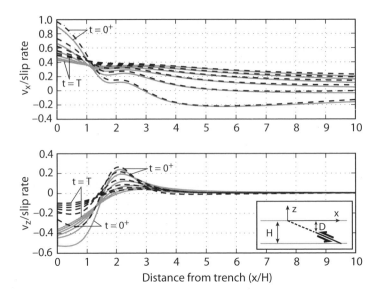

Figure 12.28. Variation in velocity on the hanging wall through the earthquake cycle for a thrust fault in an elastic layer overlying a Maxwell viscoelastic half-space. Top: horizontal velocity. Bottom: Vertical velocity. Both are normalized by fault slip rate $\dot{s}$. $T/\tau_R = 2.5$ and $D/H = 1.0$. Rigid plate model velocities are shown with solid lines, and deforming plate velocities are shown with dashed lines for $t = 0+, 0.2T, 0.4T, 0.6T, 0.8T$, and T^-. Computations employ surface gravity approximation. T is the earthquake repeat time, D is the depth of coseismic slip, and H is the elastic layer thickness.

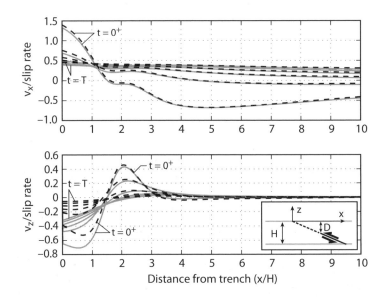

Figure 12.29. Variation in velocity on the hanging wall through the earthquake cycle for a thrust fault in an elastic layer overlying a Maxwell viscoelastic half-space. Top: Horizontal velocity. Bottom: Vertical velocity. Both are normalized by fault slip rate $\dot{s}$. $T/\tau_R = 5$ and $D/H = 1.0$. Rigid plate model velocities are shown with solid lines, and deforming plate velocities are shown with dashed lines for $t = 0+, 0.2T, 0.4T, 0.6T, 0.8T$, and T^-. Computations employ surface gravity approximation. T is the earthquake repeat time, D is the depth of coseismic slip, and H is the elastic layer thickness.

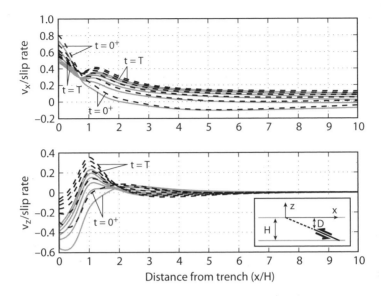

Figure 12.30. Variation in velocity on the hanging wall through the earthquake cycle for a thrust fault in an elastic layer overlying a Maxwell viscoelastic half-space. Top: Horizontal velocity. Bottom: Vertical velocity. Both are normalized by fault slip rate $\dot{s}$. $T/\tau_R = 2.5$ and $D/H = 0.5$. Rigid plate model velocities are shown with solid lines, and deforming plate velocities are shown with dashed lines for $t = 0+, 0.2T, 0.4T, 0.6T, 0.8T,$ and T^-. Computations employ surface gravity approximation. T is the earthquake repeat time, D is the depth of coseismic slip, and H is the elastic layer thickness.

Figure 12.30 illustrates the surface velocity fields when the coseismic slip and hence interseismic locking extend only halfway through the elastic plate. Otherwise, the parameters are the same as in figure 12.28, including the ratio of earthquake repeat time to the Maxwell relaxation time. The steady creep at depth causes the velocity distribution distant from the fault to be more nearly steady state in comparison to the model in which coseismic faulting cuts the entire elastic layer. The uplift pattern above the downdip end of the fault is also substantially different. The majority of the uplift occurs late in the cycle (figure 12.27), so the uplift rate is low early in the cycle but increases with time toward the end of the cycle. This appears to be a diagnostic feature of downdip creep: uplift rates increase with time, whereas if the fault cuts the entire elastic plate, the uplift rates decrease with time (figure 12.28). At any given time, the uplift rate is markedly higher in the deforming plate model than the rigid plate model. The subsidence rate near the fault is greater in the rigid plate model, as it is for all parameters considered here.

Thatcher and Rundle (1984) employed a viscoelastic back-slip model with no long-term deformation, similar to those presented here, and compared to repeated leveling observations from the Nankai Trough region of southwest Japan summarized by Thatcher (1984). They concluded that viscoelastic coupling models fit the inland data reasonably well, but the data near the trench "hardly at all." They argue that near-trench deformations immediately following the earthquake are better explained by postseismic slip downdip of the coseismic rupture zone. This is reminiscent of the difficulty in distinguishing between postseismic slip and distributed viscoelastic flow for strike-slip faults. For dip-slip faulting, however, these processes can be distinguished with measurements near the fault, as demonstrated by Hsu and others (2007) with postseismic data following the Chi-Chi earthquake in Taiwan.

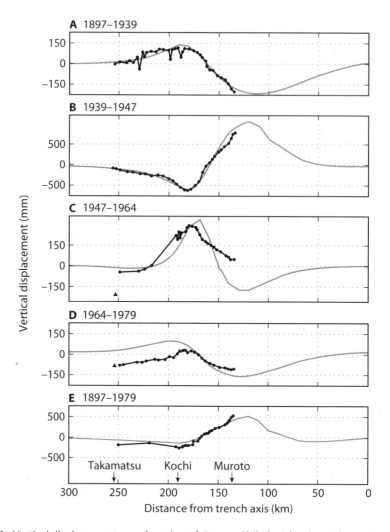

Figure 12.31. Vertical displacements as a function of time on Shikoku Island, southwest Japan, adjacent to the Nankai trough. A: Preseismic. B: Coseismic. C: Postseismic. D: Interseismic. E: Total change from 1897 to 1979. Data was assimilated by Thatcher (1984). Model predictions are from Sato and Matsu'ura (1992). After Sato and Matsu'ura (1992).

Sato and Matsu'ura (1992) fit the Thatcher (1984) data set for Shikoku reasonably well assuming finite long-term deformation (figure 12.31). They assume an elastic layer thickness of 35 km and an asthenospheric viscosity of 5×10^{18} Pa-s. In order to fit the observed leveling data, they require that coseismic slip does not extend through the entire elastic layer.

Whether the rigid or deforming plate approximations are more appropriate to modeling interseismic deformation in subduction environments is still a question of debate. There is, however, interest in applying models of this type to areas of continental collision with significant seismic hazard and rapid rates of geologic deformation, such as the Himalayas or Taiwan. While it seems reasonable that the deforming plate model should be more appropriate in these geologic settings, it should be emphasized the neither model discussed here accounts for inelastic deformation of the lithosphere on long timescales. Elastic strain cannot accumulate indefinitely, and in the shallow, brittle crust, any inelastic deformation is likely to occur as faulting. Over geologic time, the effects of erosion and sedimentation, as well as the consequent loading and unloading effects (King et al. 1988), must be accounted for. In subduction zones, there is also the potential of accretion of material to the hanging-wall plate. These issues are beyond the present discussion, but see Sato and Matsu'ura (1992) for discussion.

12.5 Rate-State Friction Earthquake Cycle Models

Viscoelastic earthquake cycle models are kinematic in the sense that plate boundary slip is imposed at selected times. In order for earthquakes to occur naturally, the model must include a friction law capable of exhibiting unstable slip. As discussed in chapter 11, rate- and state-dependent friction laws coupled to elastic continua naturally lead to unstable sliding under appropriate conditions. Furthermore, the frictional resistance to sliding recovers naturally during extended interseismic periods when sliding rates are low. Earthquake cycle models incorporating rate and state friction are not easily amenable to analytical treatment, so that numerical analysis is required. On the other hand, the predictions of these models (e.g., Tse and Rice 1986) bear great similarity to observations in nature, so they warrant discussion.

To a great extent, the behavior is dependent on the temperature, and hence depth dependence of frictional properties, as shown for example in figure 11.12. The laboratory data for sliding of granite surfaces exhibit significant scatter but are generally steady-state velocity weakening at temperatures between 100°C and 300°C, and increasingly velocity strengthening with increasing temperature above 300°C. The temperature dependence is mapped to depth using geotherms appropriate to the tectonic environment of interest, as for example by Blanpied et al. (1991) for the San Andreas fault. A summary of the assumed distribution in frictional properties based loosely on Rice (1993) is shown in figure 12.32. The fault is steady-state velocity weakening at depths less than 14 km. Rice (1993) included a velocity-strengthening region at depths less than 2 km, but we take the fault to be velocity weakening for all $z \leq 14$ km, to minimize surface creep. In the calculations here, $a = 0.01$ for $z \leq 17$ km, but increases slowly below that.

As discussed in chapter 11, laboratory values of d_c typically range from a few tens of microns to one hundred microns. These values correspond to nucleation dimensions of a few meters to a few tens of meters. It is important that the spatial grid in the numerical scheme resolves features on the scale of the critical nucleation dimension L_c (equation [11.46]), which scales linearly with d_c (chapter 11). For this reason, many computations are done with values of d_c far larger than the laboratory range. This may be justified if d_c scales with fault roughness, since faults in nature are generally rougher than those in laboratory experiments. On the other hand, the occurrence of very small earthquakes $M < 0$ suggests that at least in some areas, d_c is comparable to laboratory estimates. It is important to keep in mind, however, that interesting aspects of the simulations may depend strongly on the values of d_c chosen. The effective stress is chosen here following Rice (1993), assuming hydrostatic pore pressure to depths of roughly 3 km, with a constant effective stress of 50 MPa at greater depth (figure 12.32).

For antiplane strain geometry appropriate for an infinitely long strike-slip fault, a symmetric slip distribution about $z = 0$ ensures that this plane is traction free (e.g., chapter 4). Because the fault is everywhere slipping at all times, albeit often at a vanishingly low rate, it is efficient to use Fourier transform methods described in chapter 4 to compute stress changes resulting from fault slip. In calculations here, the computational domain extends to a depth of 24 km; slip at a constant v^∞ is imposed over $24 \leq z \leq 96$ km. The calculations here employ the radiation damping approximation of elastodynamics (chapter 11). Similar

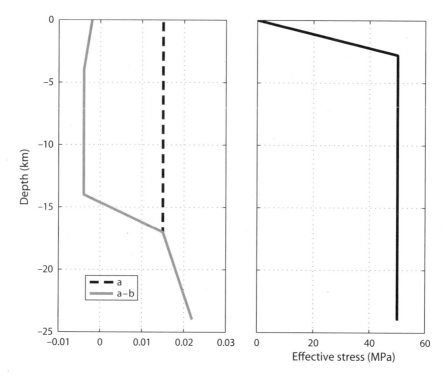

Figure 12.32. Variation in frictional properties (left) and effective normal stress (right) for a rate-state friction-based earthquake cycle model. For depths $z \geq 17$ km, $b = 0$, so for this region, $a - b = a$.

computations including full elastodynamics are given in Lapusta and others (2000). At the time of this writing, there are only limited laboratory data on frictional properties of rocks at high sliding velocities. This, when combined with the fact that other processes including thermal weakening are likely to become important during fast sliding, indicates that the models are less representative of nature during periods of fast slip.

The slip rate as a function of depth is shown in figure 12.33. Solid lines are drawn every 10 years during the interseismic period. At depths below 23 km, the slip rate is nearly constant, and slip accumulates at a steady rate. In the velocity-weakening region, the fault is essentially locked interseismically, causing stress to accumulate. Fast-slip nucleates in the depth range 8 to 12 km. Interestingly, in these calculations, every third earthquake is preceded by a creep event in the depth range of roughly 8 to 15 km. This delays the onset of the ultimate earthquake, causing it to nucleate at shallower depth, 8 to 10 km, and to have greater coseismic slip. The events preceded by a deep creep event have roughly 8 meters of slip and occur every 235 years, while the other events have roughly 6 meters of slip and occur every 169 years. Each of the model earthquakes is followed by substantial postseismic creep below the seismogenic zone.

The slip rate as a function of depth is shown in figure 12.34. Between earthquakes, the velocity-weakening portion of the fault slips at four or more orders of magnitude less than the plate velocity and is thus effectively locked. The deeper, velocity-strengthening fault slips at a more nearly steady rate. Leading up to the earthquake, the nucleation zone accelerates to slip speeds of order 1 m/s, at which point, the rupture expands vertically. Following the earthquake, the fault beneath the coseismic slip zone undergoes rapid transient slip, which gives rise to time-dependent deformation at the free surface. In this particular simulation, the

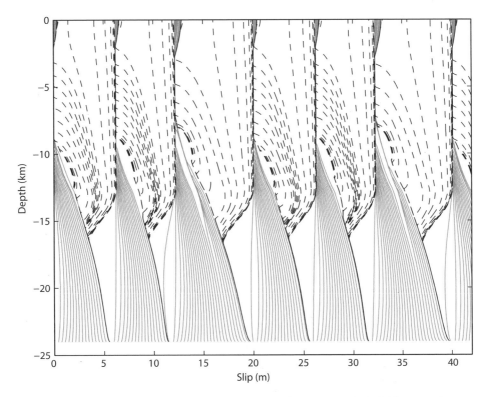

Figure 12.33. Slip profiles as a function of depth at specified time intervals. Solid lines are drawn every 10 years during the interseismic period, and dashed lines are every 2 seconds when the maximum slip speed exceeds 1 cm/s. $d_c = 15$ mm in these calculations, far greater than observed in laboratory friction experiments.

deep creep rates are two orders of magnitude greater than the average slip rate one week after the earthquake.

Displacement at the free surface is readily computed using methods described in chapter 2 and is shown in figure 12.35. One year after the earthquake, the surface velocities near the fault substantially exceed the long-term rates. The surface velocities decay rapidly with time but are still slightly elevated after five years. By 35 years post-earthquake, the velocity field has stabilized to a steady interseismic distribution. The velocity distribution is qualitatively similar to what has been observed in viscoelastic models with high strain rates near the fault immediately after the earthquake, but decreasing with time.

12.6 Summary and Perspective

The rates of interseismic crustal deformation directly constrain future seismic hazard. Proper interpretation of these data, however, requires mathematical models that accurately describe the physical processes operating on active plate boundaries. (Intraplate earthquakes remain far more enigmatic and are not discussed here.) Kinematic elastic models provide a useful starting point but ignore stress relaxation at depth and realistic fault constitutive laws. Back-slip models, whether elastic or viscoelastic, are useful in that they separate quasi-steady block-like motions, which change only on geologic timescales, from the alternating strain accumulation and release associated with the earthquake cycle. In subduction zones, elastic back-slip models

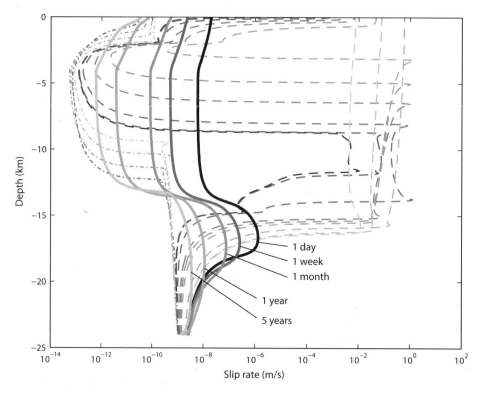

Figure 12.34. Slip-rate profiles as a function of depth for selected times. Solid lines are drawn at 1 day, 1 week, 1 month, 1 year, and 5 years following the model earthquake. The remaining dash-dot curves are drawn at $0.25T$, $0.5T$, $0.75T$, and $0.95T$, where T is the recurrence interval. Dashed lines are drawn every 5 seconds when the maximum slip speed exceeds 1 cm/s.

predict interseismic contraction of the overriding plate and subsidence near the trench, with uplift farther inland.

Crustal motions distant from plate boundaries are well described by rotation of spherical caps. Elastic block models satisfy these far-field kinematics and include localized deformation resulting from plate boundary locking. Such descriptions are straightforward at plate tectonic scales but can become more ambiguous at the scale of complex plate boundary zones, where inelastic deformation and time-dependent evolution of the fault geometry must be considered. Complications in fully elastic models arise due to irregular fault geometry, as well as areas where the predicted block motion is oblique to strike-slip faults, implying localized convergence or extension.

Viscoelastic models allow for time-dependent relaxation of stress at depth in the earth. Kinematic earthquake cycle models with an infinite sequence of plate boundary ruptures repeating at regular time intervals are derived from the solution for a single dislocation suddenly introduced into an elastic layer overlying a Maxwell viscoelastic half-space, as developed in chapter 6. The predicted behavior depends strongly on the ratio of the earthquake repeat time to the material relaxation time. When this ratio is close to one or less, the surface velocities are nearly steady throughout the earthquake cycle. In contrast, when the cycle time is long compared to the relaxation time, surface strain rates near the fault are higher than average early in the earthquake cycle but decrease to lower than average rates midway through the cycle. The approach can be extended in a straightforward manner to include multiple viscoelastic layers.

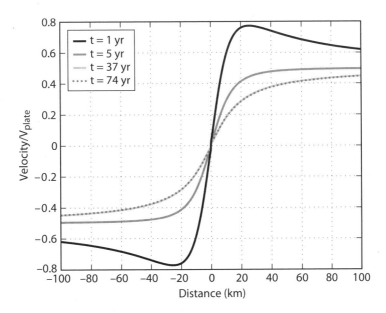

Figure 12.35. Fault parallel velocity (normalized by the plate rate) as a function of distance perpendicular to the fault resulting from the slip-rate distribution shown in figure 12.34, for different times after the model earthquake. Times are labeled in years. The velocity distributions at $t = 37$ yrs and $t = 74$ yrs are indistinguishable on this plot.

Plausible viscosity structures can lead to velocity distributions with apparent asymptotes that are well below the average fault slip rate. Failure to account for this effect can lead to biased estimates of the long-term fault slip rate.

Different models have been developed to include the effects of creeping faults at depth in viscoelastic cycle models. Creep at constant imposed rate is easily included from the result for repeating earthquakes, in the limit that the size and repeat time become vanishingly small. More physically realistic descriptions allow the creep rate to vary with the imposed stress. A linear viscous fault rheology can be considered as a crude approximation to steady-state velocity strengthening friction, with strength varying linearly, rather than logarithmically, with slip rate. Such models predict substantially higher strain-rates immediately after the earthquake, consistent with some observations, although other interpretations are permissible as discussed in chapter 6.

For dip-slip faults, the procedure is similar. We again start from the viscoelastic response to a single earthquake; however, in this case, we use the propagator-matrix-based solution in the Fourier transform domain. The summation for an infinite sequence of repeating earthquakes is simply accomplished for each Fourier mode. There is, however, more ambiguity about the long-term deformation field. We identify the rigid plate model in which the deformation integrated over a full earthquake cycle involves only horizontal, rigid body translation and a deforming plate model in which the long-term deformation is equivalent to steady slip on a fault overlying a relaxed asthenosphere. The vertical velocity fields predicted by the two models are significantly different within one elastic layer thickness of the fault.

Earthquake cycle models based on rate- and state-dependent friction and the radiation damping approximation to elastodynamics can be computed using standard numerical methods. Calculations based on observed temperature, and hence depth, dependence of frictional properties are consistent with many seismological and geodetic observations, including significant afterslip at depth below the seismogenic zone.

The utility of rate and state friction-based models is not that they provide a better fit to surface deformation data compared to viscoelastic or other models. Rather, it is that they integrate laboratory friction data into a mechanically consistent framework that reconciles a number of observations of faults in nature. Future work will undoubtedly integrate rate- and state-dependent frictional behavior with distributed viscoelastic flow at depth. Models that include inelastic deformation related to repeated faulting in the brittle crust, as well as at geometric complexities and fault terminations, should lead to improved forecasting of future seismic behavior.

12.7 Problems

1. Show that equation (12.4) can be written in a form in which the site velocities are linearly related to the angular velocity vector, as in equation (12.5).

2. Show that the temporal behavior for the $n = 2$ term in the Savage-Prescott model $T_2(t/t_R, T/t_R)$ is given by

$$T_2(t/t_R, T/t_R) = \frac{v^\infty T}{t_R} e^{-t/t_R} \left(\frac{e^{T/t_R}}{e^{T/t_R} - 1} \right) \left[\frac{t}{t_R} + \frac{T}{t_R} \left(\frac{1}{e^{T/t_R} - 1} \right) \right]. \tag{12.56}$$

Hint: For the sum

$$\sum_{k=0}^{N} \frac{kT}{t_R} e^{-kT/t_R}, \tag{12.57}$$

premultiply both sides by $(e^{T/t_R} - 1)^2$.

3. Show that the average slip rate in each equivalent depth interval in the Savage-Prescott model is v^∞ for all modes (i.e., for all n). The apparent slip rate is given by equation (12.21).

   Hint: A change of variables allows the average slip rate to be written in terms of the incomplete Gamma function

$$\gamma(n, x) = \int_0^x e^{-t} t^{n-1} dt. \tag{12.58}$$

Carrying out the summation and noting $\Gamma(n) = (n - 1)!$ for n integer leads to the desired result.

4. Show for an antiplane, strike-slip fault *within* a Maxwell viscoelastic half-space overlain by an elastic layer that the surface displacements vanish in the fully relaxed state, $t \to \infty$. Make use of the solution for a line dislocation in a layered elastic medium given in equation (5.77). Note that one can take advantage of the limiting properties of Laplace transforms (equation [A.17]), so it is not actually necessary to invert the resulting transform to verify the limiting behavior.

5. Show that the surface velocities for a dip-slip earthquake cycle model (12.51) can be derived by first differentiating equation (12.42) with respect to time and making use of equations (12.48).

6. In chapter 6, we found that the postseismic deformation due to a vertical strike-slip fault cutting fully through an elastic layer of thickness H overlying a Maxwell viscoelastic half-space could be written in the Fourier domain as

$$u_3(k, t) = -i \Delta u k \left(1 - e^{-at} e^{-H|k|}\right),$$ (12.59)

equation (6.49), where

$$a = \frac{1 - e^{-2H|k|}}{t_R}.$$ (12.60)

From this result, show that the surface velocity field due to an infinite sequence of earthquakes with repeat time T is

$$\frac{v_3(k, t)}{\dot{s}} = \frac{-iT}{k} a e^{-H|k|} \frac{e^{-a(t-T)}}{e^{aT} - 1}.$$ (12.61)

Compare this result to the solution based on the infinite series solution (equation [12.19]). Note that this procedure can be extended to included multiple viscoelastic layers with differing viscosities, as in Johnson et al. (2007).

12.8 References

Abromowitz, M., and I. A. Stegun. 1972. *Handbook of mathematical functions*. New York: Dover Publications.

Altamimi, Z., X. Collilieux, J. Legrand, B. Garayt, and C. Boucher. 2007. ITRF2005: a new release of the International Terrestrial Reference Frame based on time series of station positions and earth orientation parameters. *Journal of Geophysical Research* **112**, B09401, doi:10.1029/2007JB004949.

Blanpied, M., D. Lockner, and J. Byerlee. 1991. Fault stability inferred from granite sliding experiments at hydrothermal conditions. *Geophysical Research Letters* **18**, 609–612.

Hsu, Y.-J., P. Segall, S.-B. Yu, L.-C. Kuo, and C. A. Williams. 2007. Temporal and spatial variations of afterslip following the 1999 Chi-Chi, Taiwan, earthquake. *Geophysical Journal International* **169**, 367–379, doi: 10.1111/j.1365-246X.2006.03310.x.

Johnson, K. M. 2004. Mechanical models of interseismic deformation in California and Taiwan. PhD Thesis, Stanford University.

Johnson, K. M., G. E. Hilley, and R. Burgmann. 2007. Influence of lithosphere viscosity structure on estimates of fault slip rate in the Mojave region of the San Andreas fault system. *Journal of Geophysical Research* **112**, B07408, doi:10.1029/2006JB004842.

Johnson, K. M., and P. Segall. 2004. Viscoelastic earthquake cycle models with deep stress-driven creep along the San Andreas fault system. *Journal of Geophysical Research* **109**, B10403, doi:10.1029/2004JB003096.

Kenner, S. J., and P. Segall. 2000. Postseismic deformation following the 1906 San Francisco earthquake. *Journal of Geophysical Research* **105**, 13,195–13,209.

King, G.C.P., R. S. Stein, and J. B. Rundle. 1988. The growth of geological structures by repeated earthquakes: 1. Conceptual framework. *Journal of Geophysical Research* **93**(B11), 13,307–13,318.

Lapusta, N., J. R. Rice, Y. Benzion, and G. T. Zheng. 2000. Elastodynamic analysis for slow tectonic loading with spontaneous rupture episodes on faults with rate- and state-dependent friction. *Journal of Geophysical Research* **105**, 23,765–23,789.

Li, V., and J. R. Rice. 1987. Crustal deformation in great California earthquake cycles. *Journal of Geophysical Research* **92**(B11) 11,533–11,551.

Matsu'ura, M., and T. Sato. 1989. A dislocation model for the earthquake cycle at convergent plate boundaries. *Geophysical Journal International* **96**, 23–32.

McCaffrey, R. 2002. Crustal block rotations and plate coupling. In S. Stein and J. Freymueller (Eds.), Plate Boundary Zones. *American Geophysical Union Geodynamics Series* **30**, 101–122.

Meade, B. J., and B. H. Hager. 2005. Block models of crustal motion in southern California constrained by GPS measurements. *Journal of Geophysical Research* **110**(B3), B03403.

Murray, M. H., and P. Segall. 2001. Modeling broadscale deformation in northern California and Nevada from plate motions and elastic strain accumulation. *Geophysical Research Letters* **28**(22), 4315–4318.

Prawirodirdjo, L., Y. Bock, R. McCaffrey, J. Genrich, E. Calais, C. Stevens, S. S. O. Puntodewo, C. Subarya, J. Rais, P. Zwick, and P. Fauzi. 1997. Geodetic observations of interseismic strain segmentation at the Sumatra subduction zone. *Geophysical Research Letters* **24**, 2601–2604.

Rice, J. R. 1993. Spatio-temporal complexity of slip on a fault. *Journal of Geophysical Research* **98**, 9885–9907.

Sato, T., and M. Matsu'ura. 1988. A kinematic model for deformation of the lithosphere at subduction zones. *Journal of Geophysical Research* **93**, 6410–6418.

———. 1992. Cyclic crustal movements, steady uplift of marine terraces, and evolution of the island arc-trench system in southwest Japan. *Geophysics Journal International* **111**, 617–629.

Savage, J. C. 1983. A dislocation model of strain accumulation and release at a subduction zone. *Journal of Geophysical Research* **88**(NB6), 4984–4996.

Savage, J. C., and W. H. Prescott. 1978. Asthenosphere readjustment and the earthquatke cycle. *Journal of Geophysical Research* **83**, 3369–3376.

Schwartz, D. P., D. Pantosti, K. Okumura, T. J. Powers, and J. C. Hamilton. 1998. Paleoseismic investigations in the Santa Cruz Mountains, California: implications for recurrence of large-magnitude earthquakes on the San Andreas fault. *Journal of Geophysical Research* **103**, 17,985–18,001.

Sieh, K. E., and R. H. Jahns. 1984. Holocene activity of the San Andreas fault at Wallace Creek, California. *Geological Society of America Bulletin* **95**(8), 883–896.

Thatcher, W. 1983. Nonlinear strain buildup and the earthquake cycle on the San Andreas fault. *Journal of Geophysical Research* **88**, 5893–5902.

———. 1984. The earthquake deformation cycle on the Nankai trough, southwest Japan. *Journal of Geophysical Research* **89**, 5674–5680.

Thatcher, W., and J. B. Rundle. 1984. A viscoelastic coupling model for the cyclic deformation due to periodically repeated earthquakes at subduction zones. *Journal of Geophysical Research* **89**, 7631–7640.

Tse, S. T., and J. R. Rice. 1986. Crustal earthquake instability in relation to the depth variation of frictional slip properties. *Journal of Geophysical Research B* **91**(9), 9452–9472.

WGCEP (Working Group on California Earthquake Probabilities). 1995. Seismic hazards in southern California: probable earthquakes, 1994–2024. *Bulletin of the Seismological Society of America* **85**(2), 379–439.

———. 1999. Earthquake probabilities in the San Francisco Bay region: 2000–2030—a summary of findings, Open File Report 99–517. Washington, DC: U.S. Geological Survey.

APPENDIX A

Integral Transforms

A.1 Fourier Transforms

We define the Fourier transform as

$$\mathcal{F}[f(x)] = \bar{f}(k) = \int_{-\infty}^{+\infty} f(x)e^{-ikx}dx, \tag{A.1}$$

where k is the *wavenumber*, related to wavelength λ by $k = 2\pi/\lambda$. The corresponding inverse transform is

$$\mathcal{F}^{-1}[\bar{f}(k)] = f(x) = \frac{1}{2\pi} \int_{-\infty}^{+\infty} \bar{f}(k)e^{ikx}\,dk. \tag{A.2}$$

A proof that $\mathcal{F}^{-1}\{\mathcal{F}[f(x)]\} = f(x)$ is given, for example, by Sneddon (1951). A heuristic demonstration follows from the Fourier transform of the *Dirac delta function* $\delta(x - \xi)$. (See figure 3.5 for a description of the Dirac delta function.)

$$\int_{-\infty}^{+\infty} \delta(x - \xi)e^{-ikx}\,dx = e^{-ik\xi}. \tag{A.3}$$

If the inverse transform (A.2) is to yield the original function, we conclude that

$$\frac{1}{2\pi} \int_{-\infty}^{+\infty} e^{-ik(\xi-x)}\,dk = \delta(x - \xi), \tag{A.4}$$

which is a statement of the orthogonality of harmonic functions of differing frequencies. Substituting equation (A.1) into (A.2) yields

$$f(x) = \frac{1}{2\pi} \int_{-\infty}^{+\infty} \left[\int_{-\infty}^{+\infty} f(\xi)e^{-ik\xi}\,d\xi\right] e^{ikx}\,dk,$$

$$= \int_{-\infty}^{+\infty} f(\xi)\,d\xi \, \frac{1}{2\pi} \int_{-\infty}^{+\infty} e^{-ik(\xi-x)}\,dk,$$

$$= \int_{-\infty}^{+\infty} f(\xi)\delta(x - \xi)\,d\xi = f(x), \tag{A.5}$$

demonstrating that given equation (A.4), application of (A.2) to the output of (A.1) yields the original function.

The Fourier transform of a derivative is easily computed

$$\mathcal{F}[f'(x)] = \int_{-\infty}^{+\infty} \frac{d\,f(x)}{dx}e^{-ikx}\,dx. \tag{A.6}$$

Integrating by parts,

$$\mathcal{F}[f'(x)] = ik \int_{-\infty}^{+\infty} f(x)e^{-ikx}\,dx = ik\mathcal{F}[f(x)]. \tag{A.7}$$

Repeating this process, $\mathcal{F}[f''(x)] = -k^2\mathcal{F}[f(x)]$, and so on, each higher order derivative multiplying by an additional factor of ik.

The *shift theorem* states that

$$\mathcal{F}[f(x-a)] = e^{-ika}\mathcal{F}[f(x)], \qquad (A.8)$$

which can be seen from (A.1) by making the substitution $y = x - a$.

A *convolution* is defined as

$$h(x) = \int_{-\infty}^{+\infty} f(y)g(x-y)\,\mathrm{d}y, \qquad (A.9)$$

and is sometimes written as $h = f \star g$. The transform of $h(y)$ is, making use of the shift theorem (A.8),

$$\begin{aligned}
\bar{h}(k) &= \int_{-\infty}^{+\infty} \left[\int_{-\infty}^{+\infty} f(y)g(x-y)\,\mathrm{d}y \right] e^{-ikx}\,\mathrm{d}x, \\
&= \int_{-\infty}^{+\infty} f(y)\,\mathrm{d}y \int_{-\infty}^{+\infty} g(x-y)e^{-ikx}\,\mathrm{d}x, \\
&= \int_{-\infty}^{+\infty} f(y)e^{-iky}\,\mathrm{d}y \int_{-\infty}^{+\infty} g(x)e^{-ikx}\,\mathrm{d}x, \\
&= \bar{f}(k)\bar{g}(k).
\end{aligned} \qquad (A.10)$$

Demonstrating that convolution becomes multiplication in the Fourier domain.

We will also have need to use *two-dimensional Fourier transforms*, which can be defined as

$$\begin{aligned}
\mathcal{F}(k_1, k_2) &= \int_{-\infty}^{+\infty} \int_{-\infty}^{+\infty} f(x_1, x_2) \exp^{-i(k_1 x_1 + k_2 x_2)}\,\mathrm{d}x_1 \mathrm{d}x_2, \\
f(x_1, x_2) &= \frac{1}{(2\pi)^2} \int_{-\infty}^{+\infty} \int_{-\infty}^{+\infty} \mathcal{F}(k_1, k_2) \exp^{i(k_1 x_1 + k_2 x_2)}\,\mathrm{d}k_1 \mathrm{d}k_2,
\end{aligned} \qquad (A.11)$$

where k_1 and k_2 are the wavenumbers in the 1- and 2-direction, respectively.

Alternative conventions are used. In particular, setting $k = 2\pi k'$ in equations (A.1) and (A.2) leads to

$$\begin{aligned}
\mathcal{F}[f(x)] &= \bar{f}(k') = \int_{-\infty}^{+\infty} f(x)e^{-2\pi ik'x}\mathrm{d}x, \\
\mathcal{F}^{-1}[\bar{f}(k')] &= f(x) = \int_{-\infty}^{+\infty} \bar{f}(k')e^{2\pi ik'x}\mathrm{d}k',
\end{aligned} \qquad (A.12)$$

where now $k' = 1/\lambda$.

A.2 Laplace Transforms

Taking $s = ik$ in equation (A.1) and restricting consideration to functions that vanish for time $t < 0$ yields the Laplace transform:

$$\mathcal{L}[f(t)] \equiv \bar{f}(s) = \int_0^\infty f(t)e^{-st}\mathrm{d}t. \qquad (A.13)$$

The Heavyside step function $H(t)$ is defined as zero for $t < 0^-$ and unity for $t > 0^+$. From equation (A.13), the Laplace transform of the Heavyside function is

$$\mathcal{L}[H(t)] = \frac{1}{s}. \tag{A.14}$$

It may be directly shown, via integration by parts, that the transform of a derivative is given by

$$\mathcal{L}\left[\frac{\partial}{\partial t} f(t)\right] = s\,\bar{f}(s) - f(t = 0^+), \tag{A.15}$$

where $f(t = 0^+)$ is the value of f at $t = 0^+$.

Equation (A.15) can be used to derive two very useful limiting results. Take the limit of both sides of equation (A.15) as $s \to \infty$. Since the integration is independent of s, the limit may be brought inside the integral. The left-hand side thus vanishes, leading to the *initial value theorem*:

$$\lim_{s\to\infty} \int_0^\infty \dot{f}(t) e^{-st} \mathrm{d}t = \lim_{s\to\infty} s\,\bar{f}(s) - f(t = 0^+),$$

$$\int_0^\infty \lim_{s\to\infty} \dot{f}(t) e^{-st} \mathrm{d}t = \lim_{s\to\infty} s\,\bar{f}(s) - f(t = 0^+),$$

$$f(t = 0^+) = \lim_{s\to\infty} s\,\bar{f}(s). \tag{A.16}$$

Taking the limit of both sides of equation (A.15) as $s \to 0$ yields the *final value theorem*:

$$\lim_{s\to 0} \int_0^\infty \dot{f}(t) e^{-st} \, \mathrm{d}t = \lim_{s\to 0} s\,\bar{f}(s) - f(t = 0^+),$$

$$\int_0^\infty \dot{f}(t) \, \mathrm{d}t = \lim_{s\to 0} s\,\bar{f}(s) - f(t = 0^+),$$

$$f(t \to \infty) = \lim_{s\to 0} s\,\bar{f}(s). \tag{A.17}$$

The convolution theorem for Laplace transforms is

$$\mathcal{L}\left[\int_0^t f(\tau) g(t - \tau) \, \mathrm{d}\tau\right] = \mathcal{L}[f(t)]\mathcal{L}[g(t)]. \tag{A.18}$$

The inverse Laplace transform is given by the *Bromwich integral*:

$$f(t) = \mathcal{L}^{-1}[\bar{f}(s)] = \frac{1}{2\pi i} \int_{-i\infty+\varepsilon}^{i\infty+\varepsilon} e^{st} \bar{f}(s) \, \mathrm{d}s \tag{A.19}$$

(e.g., Latta 1974), where the integration path is parallel to the imaginary axis, but offset to the right of any poles at $s = 0$ by an amount ε.

The integral in equation (A.19) can often be conveniently evaluated using the *residue theorem*. For a function that is *analytic* within a closed contour C, except for isolated singularities, the residue theorem (e.g., Carrier et al. 1966) states that the integral around C is equal to $2\pi i$ times the sum of the residues inside C. We illustrate by example. Consider

$$\phi(s) = \frac{f(s)}{(s + a)(s + b)(s + c) \ldots}, \tag{A.20}$$

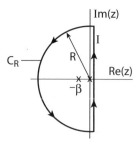

Figure A.1. Contour path for evaluating the inverse Laplace transform equation (A.24). Simple poles are located at $z = 0$ and $z = -\beta$. C_R represents the path closed in the left half-plane with radius R.

where $f(s)$ is analytic—that is, contains no singularities or branch cuts. In this case, $\phi(s)$ is said to contain simple poles at $s = -a$, $s = -b$, $s = -c$, and so on. The residues of $\phi(s)$ are

$$\frac{f(-a)}{(b-a)(c-a)\ldots}, \quad \frac{f(-b)}{(a-b)(c-b)\ldots}, \quad \frac{f(-c)}{(a-c)(b-c)\ldots}, \quad \ldots \qquad (A.21)$$

Next, consider the case where the singularity is raised to a power n:

$$\psi(s) = \frac{f(s)}{(s+a)^n}, \qquad (A.22)$$

where $f(s)$ is again analytic. $\psi(s)$ is said to have a pole of order n at $s = -a$. In this case, the residue is given by

$$\frac{f^{(n-1)}(-a)}{(n-1)!}, \qquad (A.23)$$

where the superscript $^{(n-1)}$ indicates differentiation with respect to s of order $n-1$.

The power of the residue theorem is illustrated by solving for the inverse transform of the function

$$\mathcal{L}^{-1}\left[\frac{1}{s}\left(\frac{\beta}{s+\beta}\right)\right] = \frac{1}{2\pi i}\int_{-i\infty+\varepsilon}^{i\infty+\varepsilon} \frac{1}{s}\left(\frac{\beta}{s+\beta}\right)e^{st}ds, \qquad (A.24)$$

where β is assumed to be real and positive. Closing the contour on the left side so that it encloses the simple poles at $s = 0$ and $s = -\beta$, we have

$$I + C_R = 2\pi i(residue|_{s=-\beta} + residue|_{s=0}), \qquad (A.25)$$

where I is $2\pi i$ times the integral in equation (A.24), and C_R is a semicircular contour (figure A.1). For s on C_R, $s = R\exp(i\theta)$ and $ds = iR\exp(i\theta)d\theta$. Thus, in the limit as $R \to \infty$, the integral becomes, for $t > 0$,

$$\frac{\lim}{R \to \infty}\int_{C_R} \frac{i\beta}{R\exp(i\theta)}e^{R\exp(i\theta)t}d\theta = 0, \qquad (A.26)$$

since C_R is restricted to the left half-plane, where the real part of s is negative $[\pi/2 < \theta < 3\pi/2,$ such that $\cos(\theta) < 1]$. Thus, the integral of interest is simply the sum of the residues,

$$residue|_{s=-\beta} + residue|_{s=0} = \frac{\beta}{-\beta}e^{-\beta t} + 1 = \left(1 - e^{-\beta t}\right). \qquad (A.27)$$

For $t < 0$, we close the contour on the right side. In this case, the integrand is analytic everywhere within the contour. Since the contribution from C_R vanishes, the Bromwich integral must be zero. The function in the time domain thus vanishes for $t < 0$. This is true of all functions in the Laplace domain with no poles in the positive half-plane.

As an example with a higher order pole, consider the inverse transform:

$$\mathcal{L}^{-1}\left(\frac{1}{s^n}\right) = \frac{1}{2\pi i}\int_{-i\infty+\varepsilon}^{i\infty+\varepsilon}\frac{1}{s^n}e^{st}ds. \tag{A.28}$$

There is a pole of order n at the origin $s = 0$. Completing the contour as in the preceding example, the contribution from C_R vanishes. Taking $n - 1$ derivatives of e^{st} and evaluating at $s = 0$ yields t^{n-1}, so that

$$\mathcal{L}^{-1}\left(\frac{1}{s^n}\right) = \frac{t^{n-1}}{(n-1)!}. \tag{A.29}$$

We now turn to evaluating the inverse transform (6.35), where β is real and positive:

$$f(t) = \mathcal{L}^{-1}\left[\frac{1}{s}\left(\frac{\beta}{s+\beta}\right)^n\right] = \frac{1}{2\pi i}\int_{-i\infty+\varepsilon}^{i\infty+\varepsilon}\frac{e^{st}}{s}\left(\frac{\beta}{s+\beta}\right)^n ds. \tag{A.30}$$

The integrand contains a simple pole at $s = 0$ with residue 1, and a pole of order n at $s = -\beta$. The latter has residue $\phi^{(n-1)}(s = -\beta)/(n-1)!$, where $\phi = \beta^n e^{st} s^{-1}$. Computing the derivative $\phi^{(n-1)}$:

$$\phi^{(n-1)} = \frac{\beta^n e^{st}}{s}\left[t^{n-1} - \frac{(n-1)t^{n-2}}{s} + \frac{(n-1)(n-2)t^{n-3}}{s^2}\cdots\frac{(-1)^{n-1}(n-1)!}{s^{n-1}}\right]. \tag{A.31}$$

Evaluating the residue at $s = -\beta$ leads to the inverse transform:

$$\mathcal{L}^{-1}\left[\frac{1}{s}\left(\frac{\beta}{s+\beta}\right)^n\right] = 1 - e^{-\beta t}\sum_{m=1}^{n}\frac{(\beta t)^{n-m}}{(n-m)!}, \tag{A.32}$$

where the initial term of unity results from the simple pole at $s = 0$.

A.3 References

Carrier, G. F., M. Krook, and C. E. Pearson. 1966. *Functions of a complex variable.* New York: McGraw-Hill.

Latta, G. E. 1974. Transform methods, in *Handbook of applied mathematics*, C. E. Pearson (Ed.). New York: Van Nostrand Reinhold, pp. 585–644.

Sneddon, I. N. 1951. *Fourier transforms.* New York: McGraw-Hill.

APPENDIX B

A Solution of the Diffusion Equation

We seek the solution of the diffusion equation

$$\kappa \frac{\partial^2 u}{\partial x^2} = \frac{\partial u}{\partial t},$$

(B.1)

subject to initial and boundary conditions

$$u(x > 0, t = 0) = 0,$$
$$u(x = 0, t) = \Delta u,$$
$$u(x = \infty, t) = 0.$$

(B.2)

There are a number of approaches, some involving integral transforms. We make use here of the Fourier transform pair defined in equations (A.1) and (A.2). This requires that we extend the problem over the full domain $-\infty < x < \infty$. To do so, we modify the initial condition to be

$$u(x, t = 0) = \Delta u[2 - 2H(x)],$$

(B.3)

where $H(x)$ is the Heavyside function. Note from figure B.1 that this guarantees that $u(0, t) = \Delta u$ for all t.

Fourier transforming the differential equation leads to

$$-\kappa k^2 \hat{u} = \frac{d\hat{u}}{dt},$$

(B.4)

where k is the wavenumber. The boundary condition transforms as follows. The Fourier transform of the constant $2\Delta u$ is $2\Delta u \delta(k)$, whereas the transform of the Heavyside function is $1/(ik)$, such that the initial conditions in the transform domain are

$$\hat{u}(t = 0) = 2\Delta u \left[\delta(k) - \frac{1}{ik} \right].$$

(B.5)

The solution to equation (B.4) is simply

$$\hat{u}(k, t) = \hat{u}(t = 0)e^{-\kappa k^2 t}.$$

(B.6)

Applying the initial condition (B.5) leads to

$$\hat{u}(k, t) = 2\Delta u \left[\delta(k) - \frac{1}{ik} \right] e^{-\kappa k^2 t}.$$

(B.7)

The inverse transform is thus

$$u(x, t) = \frac{\Delta u}{\pi} \int_{-\infty}^{\infty} \left[\delta(k) - \frac{1}{ik} \right] e^{-\kappa k^2 t} e^{ikx} dk,$$

$$= \frac{\Delta u}{\pi} \left(1 - \int_{-\infty}^{\infty} \frac{1}{ik} e^{-\kappa k^2 t + ikx} dk \right),$$

(B.8)

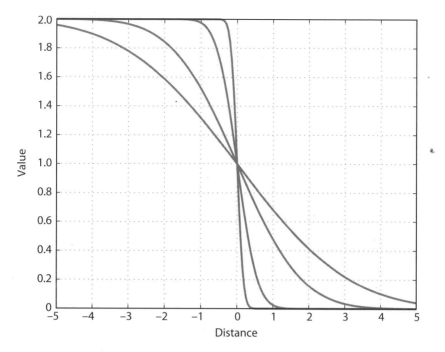

Figure B.1. Initial condition for a solution of the diffusion equation. Choosing an initial condition given by $2 - 2H(x)$ guarantees that the value of the function on the boundary is always unity.

since the integral over the delta function extracts the value of the integrand at $k = 0$, which is one.

The first trick to completing the inverse transform is to take the spatial derivative so that

$$\frac{\partial u(x, t)}{\partial x} = -\frac{\Delta u}{\pi} \int_{-\infty}^{\infty} e^{-\kappa k^2 t + ikx} dk. \tag{B.9}$$

Next consider the argument of the exponential. Complete the square as

$$-\kappa k^2 t + ikx = -\kappa t \left(k - \frac{ix}{2\kappa t} \right)^2 - \frac{x^2}{4\kappa t}, \tag{B.10}$$

such that the integral becomes

$$\frac{\partial u(x, t)}{\partial x} = -\frac{\Delta u}{\pi} e^{-x^2/4\kappa t} \int_{-\infty}^{\infty} \exp \left[-\kappa t \left(k - \frac{ix}{2\kappa t} \right)^2 \right] dk. \tag{B.11}$$

Now introduce a change of variables such that $k' = (\kappa t)^{1/2}[k - (ix/2\kappa t)]$, which leads to

$$\frac{\partial u(x, t)}{\partial x} = -\frac{\Delta u}{\sqrt{\pi \kappa t}} e^{-x^2/4\kappa t}, \tag{B.12}$$

where we have made use of the fact that

$$\int_{-\infty}^{\infty} e^{-k'^2} dk' = \sqrt{\pi}. \tag{B.13}$$

It is now a relatively straightforward matter to integrate to recover the displacement. Since the displacement at $x = 0$ is constrained to Δu, we can write

$$u(x, t) = \Delta u \left(1 + \int_0^x \frac{\partial u}{\partial x'} dx' \right). \tag{B.14}$$

Introducing equation (B.12) into the preceding leads to

$$u(x, t) = \Delta u \left(1 - \frac{1}{\sqrt{\pi \kappa t}} \int_0^x e^{-x'^2/4\kappa t} dx' \right). \tag{B.15}$$

Last, a change of variables such that

$$y = \frac{x'}{\sqrt{4\kappa t}} \tag{B.16}$$

leads to

$$u(x, t) = \Delta u \left(1 - \frac{2}{\sqrt{\pi}} \int_0^{x/\sqrt{4\kappa t}} e^{-y^2} dy \right) = \Delta u \operatorname{erfc} \left(\frac{x}{2\sqrt{\kappa t}} \right). \tag{B.17}$$

APPENDIX C

Displacements Due to Crack Model of Strike-Slip Fault by Contour Integration

This appendix treats the integral in equation (4.24):

$$u(x) = \frac{\Delta\tau x}{\mu\pi} \int_{-a}^{a} \frac{\sqrt{a^2 - \xi^2}}{x^2 + \xi^2} d\xi. \tag{C.1}$$

The integral can be evaluated analytically using contour integration methods. First, notice that the integrand is symmetric, with simple poles at $\pm i|x|$. We choose a branch cut for the term $(a^2 - z^2)^{1/2}$ between $-a < \Re(\xi) < a$ (see figure C.1). The branch cut causes the displacements to be discontinuous across the cut as required if it is to model a crack.

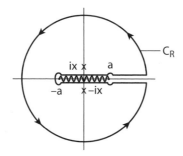

Figure C.1. Contour path for integral in equation (4.24).

The displacement on the crack surface, half the slip, is $u(z) = \Delta\tau/\mu(a^2 - z^2)^{1/2}$. Write $(a^2 - z^2)^{1/2} = -i[(z-a)(z+a)]^{1/2}$. Now, erect polar coordinates at each branch point $z = \pm a$ (figure C.2), where $-\pi < \theta_j < \pi$, $j = 1, 2$. Now, $(a^2 - z^2)^{1/2} = -i(r_1 r_2)^{1/2} \exp i(\theta_1 + \theta_2)/2$. For a

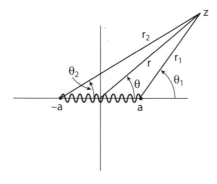

Figure C.2. Polar coordinate systems. r, θ mark the position with respect to the origin, r_1, θ_1 mark the position with respect to the right end of the crack, and r_2, θ_2 mark the position with respect to the left end.

point immediately above the center of the crack, $r_1 = r_2 = a$, $\theta_1 = \pi$, $\theta_2 = 0$, and $u(0^+) = -ia\Delta\tau/\mu \, \exp(i\pi/2) = a\Delta\tau/\mu$. For a point immediately below the center of the crack, $\theta_1 = -\pi$, $\theta_2 = 0$, so that $u(0^-) = -ia\Delta\tau/\mu \, \exp(-i\pi/2) = -a\Delta\tau/\mu$. The slip in the center $s = u(0^+) - u(0^-)$ is thus given by equation (4.18).

According to the residue theorem (appendix B), the value of the closed contour surrounding the two simple poles is $2\pi i$ times the sum of the residues:

$$\oint \frac{\sqrt{a^2 - z^2}}{x^2 + z^2} dz = 2\pi i \left[residue|_{z=i|x|} + residue|_{z=-i|x|} \right], \tag{C.2}$$

$$= 2\pi i \left[\frac{\sqrt{a^2 + x^2}}{2i|x|} - \frac{\sqrt{a^2 + x^2}}{-2i|x|} \right] = \frac{2\pi}{|x|} \sqrt{a^2 + x^2}. \tag{C.3}$$

The minus sign before the second term occurs because the pole at $-ix$ is on the opposite side of the branch cut. For points on the imaginary axis, $r_1 = r_2 = r$. Above the branch cut, $\theta_1 + \theta_2 = \pi$, so that $(a^2 - z^2)^{1/2} = -ir \, \exp(i\pi/2) = r = (a^2 + x^2)^{1/2}$. Below the branch cut, $\theta_1 + \theta_2 = -\pi$ so that $(a^2 - z^2)^{1/2} = -ir \, \exp(-i\pi/2) = -r = -(a^2 + x^2)^{1/2}$.

If we label the integral from $-a$ to a as I and the remote contour as I_R, then the complete contour is the sum of twice the integral of interest and the remote contour:

$$\oint \frac{\sqrt{a^2 - z^2}}{x^2 + z^2} dz = 2I + I_R. \tag{C.4}$$

The factor of two arises because the integrand changes sign across the branch cut and the direction of integration reverses. Thus, from equations (C.3) and (C.4):

$$I = \frac{1}{2} \left[\frac{2\pi}{|x|} \sqrt{a^2 + x^2} - I_R \right]. \tag{C.5}$$

Consider next the integral over the remote path I_R. In the limit that $|z|$ is much greater than either a or x, $r_1 \sim r_2 \sim r$ and $\theta_1 \sim \theta_2 \sim \theta$, where $z = re^{i\theta}$ and θ is measured from the origin (figure C.2). In this limit, we can write $(a^2 - z^2)^{1/2} \sim -ir \, \exp i\theta$. For fixed r along the path C_R, $dz = ire^{i\theta} d\theta$. Thus,

$$I_R = \lim_{R \to \infty} \oint_{C_R} \frac{\sqrt{a^2 - z^2}}{x^2 + z^2} dz = -i \int_0^{2\pi} \frac{re^{i\theta}}{r^2 e^{2i\theta}} \left(ire^{i\theta} d\theta \right),$$

$$= \int_0^{2\pi} d\theta = 2\pi. \tag{C.6}$$

Thus, finally,

$$I = \frac{\pi}{|x|} \sqrt{a^2 + x^2} - \pi, \tag{C.7}$$

and from equation (C.1)

$$u(x) = \frac{\Delta\tau}{\mu} \left[sgn(x) \sqrt{a^2 + x^2} - x \right]. \tag{C.8}$$

Author Index

General Index